Biological and Medical Physics, Biomedical Engineering

This series is intended to be comprehensive, covering a broad range of topics important to the study of the physical, chemical and biological sciences. Its goal is to provide scientists and engineers with textbooks, monographs, and reference works to address the growing need for information. The fields of biological and medical physics and biomedical engineering are broad, multidisciplinary and dynamic. They lie at the crossroads of frontier research in physics, biology, chemistry, and medicine.

Books in the series emphasize established and emergent areas of science including molecular, membrane, and mathematical biophysics; photosynthetic energy harvesting and conversion; information processing; physical principles of genetics; sensory communications; automata networks, neural networks, and cellular automata. Equally important is coverage of applied aspects of biological and medical physics and biomedical engineering such as molecular electronic components and devices, biosensors, medicine, imaging, physical principles of renewable energy production, advanced prostheses, and environmental control and engineering.

More information about this series at http://www.springer.com/series/3740

Stephen P. Cramer

X-Ray Spectroscopy with Synchrotron Radiation

Fundamentals and Applications

Springer

Stephen P. Cramer
Advanced Light Source Professor emeritus
University of California
Davis, CA, USA

Senior Research Scientist
SETI Institute
Mountain View, CA, USA

ISSN 1618-7210 ISSN 2197-5647 (electronic)
Biological and Medical Physics, Biomedical Engineering
ISBN 978-3-030-28549-4 ISBN 978-3-030-28551-7 (eBook)
https://doi.org/10.1007/978-3-030-28551-7

This Springer imprint is published by the registered company Springer Nature Switzerland AG
The registered company address is: Gewerbestrasse 11, 6330 Cham, Switzerland

To my father, who always asked about the "book" and
To my wife, Sybil, who put up with me writing it

Preface

When I first visited the Stanford Synchrotron Radiation Project as a naïve graduate student in 1974, I had no idea that I was joining a revolution that would continue for the next ~45 years—and that is still under way. At the time, I was a "parasite," grudgingly allowed to use the synchrotron radiation that was a nuisance by-product of storing relativistic charged particles in circular storage rings. Back then, the important science was high energy physics with colliding electron and positron beams. It was exciting to visit the SPEAR control room and watch the gradually and then sharply rising cross sections as they discovered new particles. The ring operators received "palm frond awards" for their efforts, while a Nobel prize was soon awarded to Stanford's Burt Richter and Brookhaven's Samuel Ting for discovery of the J/Ψ particle [1].

Over time, the parasites devoured the host, first taking over the SPEAR storage ring and eventually the entire linear accelerator, which is now the LCLS free electron laser. A similar scenario played out in Germany, where the PETRA ring originally built for positron-electron collisions evolved into the PETRA-III synchrotron radiation lab, and the technology that was to be used in the planned linear collider helped lead to the FLASH and European XFEL free electron lasers. Biology has a word for parasites that consume their hosts: "parasitoids."

Since then (and even earlier), the progress in synchrotron radiation technology and research has been rapid and unrelenting. A Web of Science search yields nearly 50,000 papers under the "synchrotron radiation" topic, and no doubt many more papers based on synchrotron work are buried under other headings. Several Nobel prizes have been based in part on data gathered at synchrotron sources, for structural studies of ATP synthase, the green fluorescent protein, G-protein-coupled receptors, and ribosomes. Ironically, the prizes have all been for structural biology—hardly a consideration when the first storage rings were being designed.

Synchrotron radiation has certainly gone mainstream, not just on the covers of *Physics Today* and *C&En News*. The storage ring SESAME in Jordan is one small step towards peace in the Middle East. The Advanced Light Source in Berkeley appeared in a bad movie—*The Hulk*, and synchrotron radiation even showed up as special advice in a good TV series—*Breaking Bad.*

This text has evolved out of a course given in the Applied Science Department at UC Davis. I have tried to achieve a balance between the synchrotron radiation production side and the spectroscopy that is enabled with these photons. For those who care about the production, transport, and detection of synchrotron radiation, start at the beginning. For those who only care about the spectroscopic applications, you can dive in half way at Chap. 6.

Some comments about style (and substance). In many cases, the derivation of equations from first principles to describe synchrotron radiation or its applications is complex and lengthy. Without apology, I usually present the equations and reference other sources where such derivations have already been done. For the equations themselves, I have adopted the approach used in the ALS Manual, where for practical equations the appropriate units are listed in brackets []. However, the powers of those units are considered obvious and are not listed in the brackets. In general, I have tried as much as possible to refrain from links to websites because it would just be a matter of time before all of the links would be obsolete. But, of course, there are a few exceptions. Another moving target has been facility specifications. Several examples of storage ring properties have changed because of upgrades, but rewriting the text to keep up with changes would be a fool's errand.

I hope you enjoy this first edition, and I hope that it occasionally stimulates what Emerson called "creative reading."

"There is then creative reading as well as creative writing. When the mind is braced by labor and invention, the page of whatever book we read becomes luminous with manifold allusion."

Los Altos, CA
October, 2020

Stephen P. Cramer

Acknowledgements

I have turned to many friends and colleagues for help in converting a rough draft into the final product. I especially thank Graham George, Frank deGroot, Ercan Alp, David Attwood, Kwang-Je Kim, and Ingolf Lindau for looking at the draft manuscript and making helpful comments and corrections. They are not to blame for the errors that persist. For the preparation of the book, I thankfuly acknowledge expert help with figures from both Birgit Deckers and Kung-Min (Leo) Lin. I am grateful to the Chemistry Department at Williams College (especially Amy Gehring, Dave Richardson, and Jay Thoman) for providing a wonderful venue (including the now departed Bronfman library) for writing a book. I also thank Jay Pasachoff for the use of his house as another place to work productively in Williamstown.

As far as learning about X-rays, I thank Keith Hodgson, Seb Doniach, Brian Kincaid, George Brown, Peter Eisenberger, and Herman Winick for advice during my early days with EXAFS. Later, Jerry Hastings and Peter Siddons showed me how to do proper X-ray fluorescence and inelastic scattering experiments, while my introduction to soft X-rays and XMCD came from C. T. Chen, George Sawatzky, Francesco Sette, and John Fuggle. When I embarked on nuclear adventures, Ercan Alp, Yoshitaka Yoda, and Uwe Bergman were there to help. I have also depended on Helmut Wiedemann for his expertise about machine physics and Orren Tench for many years of advice about X-ray detectors. And of course, none of this would have been possible without many years of support from DOE, NSF, and NIH and beam time from SSRP/SSRL, NSLS, ALS, APS, ESRF, PETRA-III, UVSOR, KEK-AR, and SPring-8. I thank all of the students and postdocs who spent long hours at the beamlines while I went home to bed. They were a joy to work with and helped make this journey possible. Finally, I thank my wife, Sybil, for editing numerous papers and proposals over the years and for proofing the galleys of this book with exquisite attention to detail.

Contents

Chapter 1
Introduction and Historical Background

1.1 X-rays and the Electromagnetic Spectrum

This is a book about X-rays—how they are created with synchrotron radiation and how they are used in X-ray spectroscopy. For many of us, our first encounter with X-rays is as a child in a dentist's chair, or with a broken bone in the doctor's office, or perhaps at an airport baggage inspection. Compared to other forms of electromagnetic radiation, such as visible light that we see with, infrared radiation that we feel as warmth on our skin, or microwaves that cook our food, X-rays have always seemed somewhat mysterious—perhaps vaguely associated with Superman in our unconscious mind. However, as illustrated in Fig. 1.1, X-rays are just another part of the continuous spectrum of electromagnetic radiation.

There are many different definitions of the X-ray region. From a chemist's point of view, X-ray radiation is involved in excitation of the core electrons of atoms. This ranges from the 55 eV required to dislodge a 1s electron in lithium to the 116 keV required to do the same for uranium. On the low-energy side, the extreme ultraviolet (EUV) community has laid claim to photons from 30 to 250 eV, while on the high-energy side there are γ-rays, often produced during nuclear reactions and typically involving hundreds of keV or more. Even with a conservative definition of 100 eV to 100 keV, X-rays vary in energy or wavelength by three orders of magnitude, compared to the ~two-fold variation for all of visible light (Fig. 1.1).

Another fuzzy boundary is between "soft X-rays" and "hard X-rays." X-rays are called "soft" when they do not penetrate deeply. For our purposes, soft X-rays are those with energies below ~2 keV, a range normally served by vacuum beamlines and grating monochromators, while hard X-rays are above ~2 keV and generally use crystal monochromators. Some have tried to define a middle region as "tender" X-rays; this seems more cute than useful.

S. P. Cramer, *X-Ray Spectroscopy with Synchrotron Radiation*, Biological and Medical Physics, Biomedical Engineering, https://doi.org/10.1007/978-3-030-28551-7_1

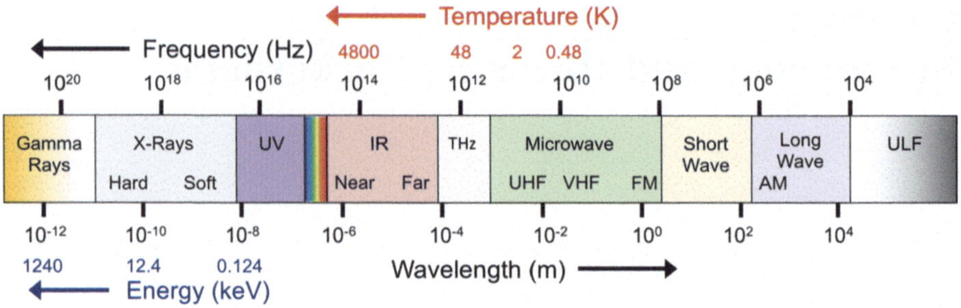

Fig. 1.1 The position of X-rays in the electromagnetic spectrum. The Kelvin scale refers to the temperature of a blackbody that would have maximum radiation at that wavelength

1.2 Sources of X-rays

From Maxwell's equations, we know that all electromagnetic radiation comes from accelerating charges. The most common X-ray sources, so-called X-ray tubes, involve a high-voltage electron current passing through a vacuum from the cathode to anode. In these sources, the broadband radiation results from a deceleration of electrons as they pass by atomic nuclei, hence the German name *bremsstrahlung* or "braking radiation" (Fig. 1.2). These tubes also produce narrow lines of X-ray fluorescence, the result of electronic transitions between different atomic core levels. Respectively, these are examples of *rapid (negative) acceleration* and *bound-state transitions* on a microscopic scale. For many applications, these sources have one major drawback—the radiation is emitted essentially in all directions and from a relatively large volume.

1.2.1 Synchrotron Radiation

As shown in Fig. 1.2, synchrotron radiation results from the *transverse acceleration* of a *relativistic charged particle*. In contrast with conventional X-ray tubes, which radiate from a large source in all directions, in a synchrotron radiation source, the X-rays are emitted from a small bunch of electrons in a narrow cone along the direction of the moving particles. As we will see later, the fact that the particles are relativistic—traveling very close to the speed of light—gives rise to these special radiation properties. Collimation and small source size make synchrotron radiation much more useful than the other sources in Fig. 1.2. To be more quantitative about these benefits, we need to define some figures of merit.

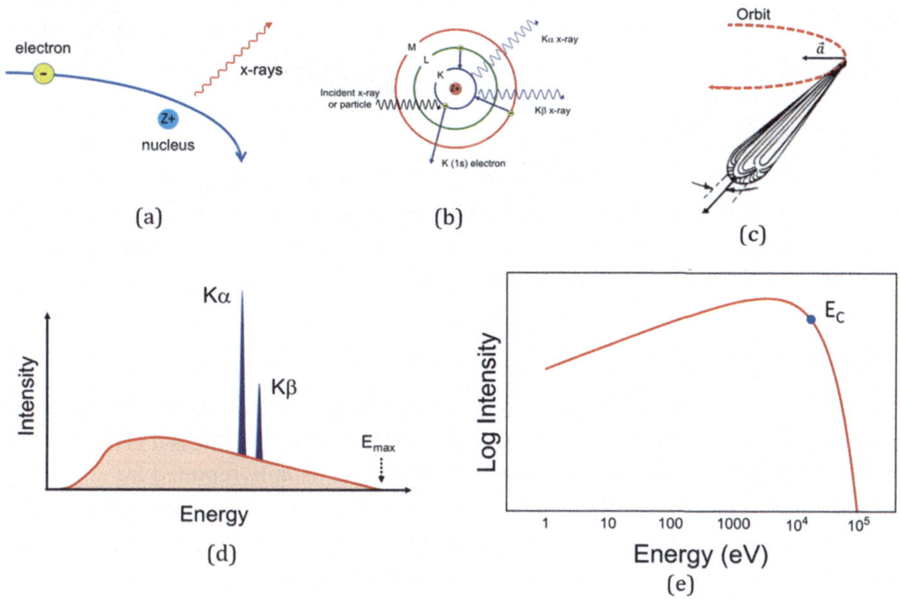

Fig. 1.2 Production of synchrotron radiation vs. other X-ray sources. (**a**) Bremsstrahlung; (**b**) an oversimplified view of Kα and Kβ X-ray atomic transitions; (**c**) synchrotron radiation; (**d**) a generic spectrum from an X-ray tube, exhibiting both bremsstrahlung and characteristic Kα and Kβ emission, the maximum energy $E_{\max}$ is that of the incident electron beam; (**e**) a generic synchrotron radiation source spectrum. Half of the source power is above or below the critical energy E_C

1.3 Brightness: A Figure of Merit

Just like people, some light sources are dim, and some are bright. We all have an intuitive sense of *brightness*. A computer projector emits more photons than an argon laser, yet while lasers require special eye protection and safety classes, we don't have to be certified to show powerpoint presentations (at least not yet). An electrical space heater emits more power than many beamlines, but a space heater cannot drill though metal as can happen with a focused synchrotron beam.

Brightness is associated with the effectiveness and usability of the radiation and it is a figure of merit commonly used to compare X-ray sources. In X-ray optics, it has a formal definition—the brightness $\mathscr{B}$ is the flux $\mathscr{F}$ in photons per unit time per unit solid angle per unit area of source (Fig. 1.3):

$$\mathscr{B} = \frac{d^4\mathscr{F}}{d\theta d\psi dx dy} \cong \frac{\mathscr{F}\left[\text{photons}\,\text{s}^{-1}\right]}{\Delta\Omega\left[\text{mrad}^2\right]\Delta A\left[\text{mm}^2\right]} \tag{1.1}$$

For most applications, one cares about the number of photons in a particular energy range, so an even more important quantity is the "spectral brightness" $\mathscr{B}_s$

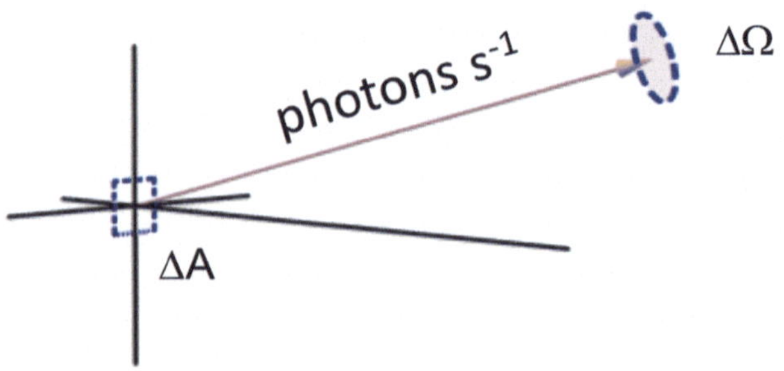

Fig. 1.3 Schematic of quantities involved in definition of X-ray brightness. The spectral brightness is the brightness in a 0.1% $\Delta E/E$ bandwidth

(Eq. 1.2). It is usually clear from the context whether the quantity referred to is brightness or spectral brightness, so we will drop the subscript unless it is needed for clarity. In practical units, X-ray spectral brightness is usually reported as:

$$\mathscr{B}_\mathrm{s} = \frac{\mathrm{photons\,s}^{-1}}{(\mathrm{mm}^2)(\mathrm{mrad}^2)(0.1\%\Delta E/E)} \tag{1.2}$$

1.3.1 Spectral Brightness of a Blackbody Source

Why not just use a very hot lamp as our X-ray source? To make such a comparison, we use a formula derived by Attwood for the brightness of a blackbody:

$$\mathscr{B}_\mathrm{s} = 3.146 \times 10^{11} \times \left(\frac{kT}{eV}\right)^3 \times \frac{(\hbar\omega/kT)^3}{\exp(\hbar\omega/kT) - 1} \times \frac{\mathrm{photons/s}}{\mathrm{mm}^2 \times \mathrm{mrad}^2 \times (0.1\%\Delta E/E)} \tag{1.3}$$

If we plug in the temperature for the surface of our sun, 5778 K, this corresponds to ~0.498 eV (see Appendix A for conversion factors), and we find that the maximum brightness is at 1.4 eV (or 8860 Å) with a value of ~4×10^{10} photons s^{-1} mrad^{-2} mm^{-2}/(0.1%$\Delta E/E$). In the X-ray range, a synchrotron will have a spectral brightness more than ten orders of magnitude higher! Despite the fact that the sun puts out a prodigious amount of energy, it is not a very "bright" source.

Brightness and spectral brightness are useful terms for comparing sources because they are not changed by ideal optical elements such as lenses and mirrors. (The inevitable losses in real optical elements can only diminish brightness.) Once the brightness of a source is set, there are no clever optical tricks for making a source brighter. In contrast, other quantities such as power density depend on your distance from the source or on how the beam is focused. Brightness and spectral brightness are thus intrinsic properties or *source invariants*, and we can use them to objectively compare different X-ray facilities.

1.3.2 Why Is Spectral Brightness Important?

For an X-ray experiment, photons are only of value if you can put them on the sample. If the photons are from a large source and are diverging in all directions, it is difficult to focus them back to a small point. Brighter beams are more useful because they are easier to (a) collect, (b) monochromate, and (c) focus. As we will see in Chap. 4 on X-ray optics, brightness is especially important for X-rays, because of severe limitations on X-ray mirrors, lenses, and monochromators. For X-ray spectroscopy, spectral brightness is even more important, because most X-ray spectroscopy is done using one photon energy at a time, so it helps when the source is tuned to the energy that is desired.

Finally, a couple of warnings about word usage. In some X-ray literature (particularly European sources), the word *brilliance* is used for what we have defined as brightness. In some UV-visible optics literature, what we call brightness is referred to as *radiance*. However, if Born and Wolf were happy to call it brightness [2], then so are we.

1.4 The Synchrotron Radiation Revolution

In the second half of the twentieth century, something very special happened to our ability to produce X-rays—synchrotron radiation sources became available. The existence of synchrotron radiation ("SR") had been predicted since the turn of the century. It was finally seen as visible radiation from a glass synchrotron chamber at General Electric research (GE) in 1947 (Fig. 1.4) [3]! Thanks to the technological development of synchrotron radiation sources, the brightness of available X-ray sources began to double on average approximately every year—a trend noted in the mid-1980s by Munro and Marr [4]. This exponential rate of increase allowed enormous improvements in the quality and quantity of X-ray experiments. The rate of improvement has been faster than Moore's law—the ~18-month doubling time for computer chip density and speed that held for ~40 years from 1975. If airplanes had made the same exponential progress over the last 30 years, we would now be flying faster than the speed of light! (Fig. 1.5).

Fig. 1.4 Left: the GE synchrotron design team, left to right: Robert Langmuir, Frank Elder, Anatole Gurewitsch, Ernest Charlton, and Herb Pollock. Right: Synchrotron light from the 70-MeV electron synchrotron at GE

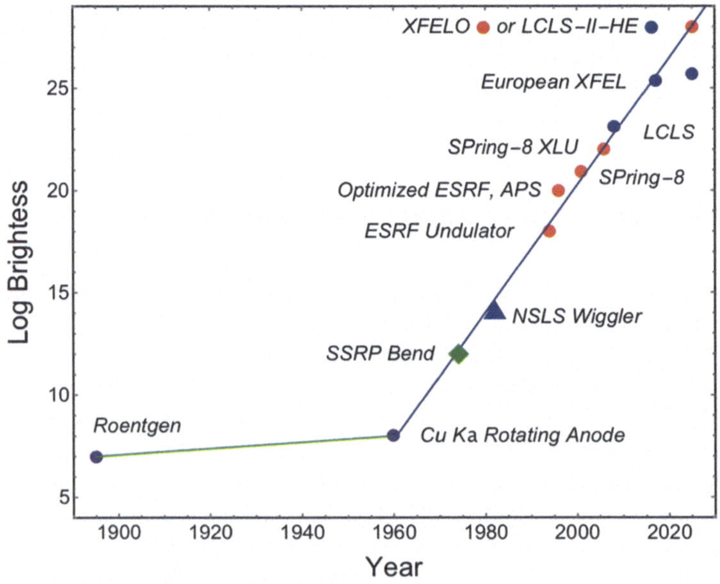

Fig. 1.5 Evolution of average X-ray spectral brightness over time. Symbol code: first generation (◆), second generation (▲), third generation (●), and fourth generation (●). Synchrotron source generations are explained in Chap. 2. Dates and brightnesses are approximate

A Selective X-ray and Synchrotron Radiation History
1895—Röntgen discovers X-rays [5]
1944—Ivanenko and Pomeranchuk predict energy loss in synchrotrons [6]
1947—Blewett observes visible SR at GE synchrotron in Schenectady [3]
1949—Schwinger publishes complete theory [7]
1956—Tomboulian and Hartman use SR for spectroscopy at Cornell [8]
1968—SRC –Tantalus—First fully dedicated SR facility
1976—INS-SOR: First storage ring built solely for SR
1981—SRS at Daresbury: First high-energy storage ring built for SR
1982—NSLS: First storage ring with high brightness (Chasman-Green) lattice
1994—ESRF: First third-generation SR source optimized for undulators
1996—APS: 7 GeV third-generation SR source
1997—SPring-8: 8 GeV third-generation SR source
2001—Tesla test facility (FLASH)—Soft X-ray free-electron laser
2009—LCLS hard X-ray free-electron laser
2017—European XFEL—High repetition rate hard X-ray free-electron laser
2018—MAX-IV MBA lattice approaches diffraction limit

The dramatic change in X-ray brightness from synchrotron radiation drew a new community of scientists into the X-ray field [9]. However, these sources are large and expensive (Fig. 1.6); they have inevitably been "user facilities" no longer under the control of individual scientists. Although working around the clock at a distant lab not under their complete control was a common experience for high-energy physicists, it was less familiar to many chemists and biologists. What has emerged is a new species of scientist—the synchrotron radiation user. There are now many thousands of SR users around the world, ranging from casual visitors to those whose entire careers are based at these facilities.

1.5 Synchrotron Radiation Laboratories

The worldwide inventory of synchrotron radiation sources is a moving target, with new projects constantly coming on line and with some older facilities eventually being retired. As of 2020, there were scores of active or proposed storage ring-based synchrotron facilities around the world (Appendix B). There were also more than a dozen free electron laser facilities based on linear accelerators (Chap. 12).

1.6 Synchrotron Radiation from Outer Space

Humans do not have a monopoly on synchrotron radiation. Nature bends the trajectories of relativistic charged particles in a variety of settings, and the resulting emission has the characteristic spectrum and polarization properties of synchrotron radiation. One example is the Crab Nebula (Fig. 1.7). This is thought to be the

Fig. 1.6 The four largest storage ring dedicated synchrotron radiation sources in the world. Top left: the Advanced Photon Source (APS) at Argonne National Lab near Chicago. Top right: European Synchrotron Radiation Facility (ESRF) in Grenoble, France. Lower left: PETRA-III in Hamburg, Germany. Lower right: SPring-8 near Osaka in Japan

remnants of a Type I supernova that exploded in 1054 CE. At that time it was so bright it could be seen in the daytime, and Chinese astronomers recorded it as a "guest star." Nowadays, at the heart of the nebula is a pulsar or "neutron star," which emits high-energy electrons into the surrounding magnetic field of ~10^{-4} Gauss. The electrons have extraordinarily high energies (up to ~1000 TeV), resulting in synchrotron X-rays up to ~100 keV.

Jupiter is also a source of synchrotron radiation, although at much lower energies (Fig. 1.7). Here, the magnetic field is stronger (~1 Gauss), but the electron energies are lower (~10 MeV) so that the bulk of the radiation is in the microwave region: 0.1–15 GHz. On a vastly different scale, extragalactic "accretion disks" produce synchrotron radiation when jets of relativistic particles spiral around magnetic field lines.

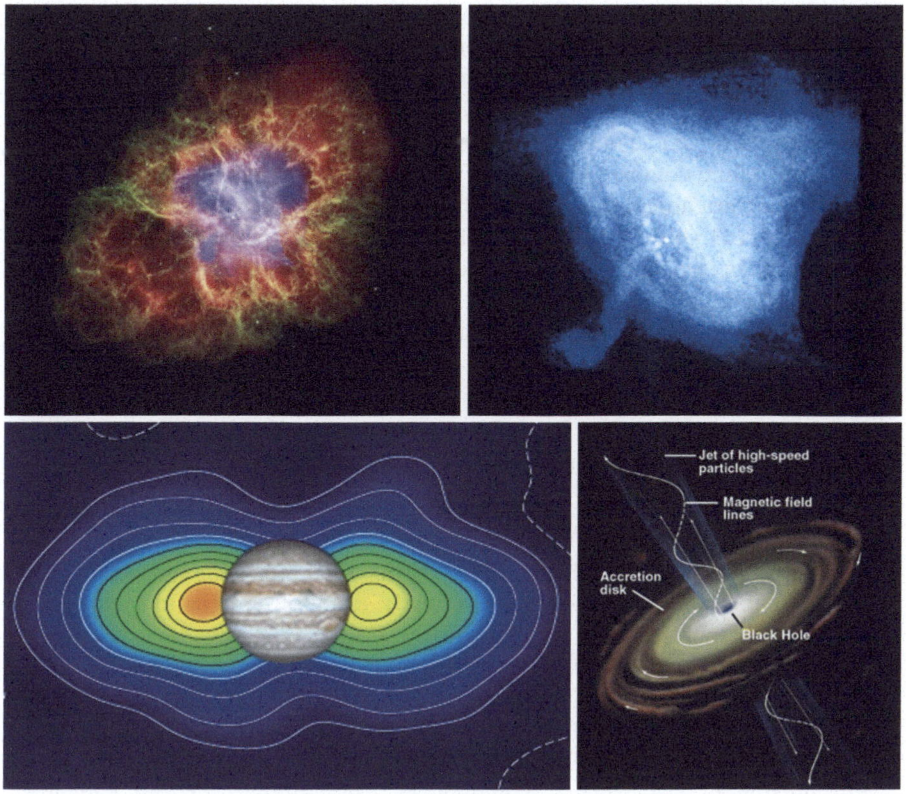

Fig. 1.7 Top left: Crab nebula in visible light. Top right: X-ray emission from the Crab nebula, imaged by the Chandra satellite. Bottom left: Microwave synchrotron radiation from Jupiter. Bottom right: synchrotron radiation from an accretion disk near a black hole

1.7 Suggested Exercises

1. Rigel A is a blue-white star in the constellation Orion with surface gases around ~12,000 K. The way astronomers use the word, Rigel is the 7th "brightest" star in the night sky. Stellar "apparent brightness" is based on the apparent visual magnitude as perceived by the human eye, so it does not separately take into account the distance or size of the star. Assuming a blackbody spectrum, calculate the peak spectral brightness of Rigel A.

   An additional metric used by astronomers is "luminosity." By some estimates, the luminosity of Rigel is >100,000 suns (https://en.wikipedia.org/wiki/Rigel). Compare the astronomical apparent brightness, peak spectral brightness, and luminosity of Rigel and our sun. Another star in Orion, Bellatrix, has a temperature of 21,500 K. Among stars in the sky, it ranks about 22nd on an astronomy apparent stellar brightness scale. How does it compare with Rigel on an intrinsic brightness scale? For more fun, consider Eta Carinae, which is ~180 times the radius of the sun, and has a surface temperature is 36,000–40,000 K.

2. A common value for the average spectral brightness of a synchrotron source is ~10^{20} photon s^{-1} mm^{-2} $mrad^{-2}$/0.1% bandwidth. What temperature is required for a blackbody source to have the same spectral brightness at 10 keV? How does that temperature compare with that of a hydrogen bomb?
3. Apart from brightness and spectral brightness, a third metric for synchrotron sources is peak spectral brightness. This is the spectral brightness over the time interval when the maximum flux occurs. If the average spectral brightness of a synchrotron source is 10^{20} photon s^{-1} mm^{-2} $mrad^{-2}$/0.1% bandwidth, what is the peak spectral brightness if the bunch length is 50 ps and the repetition rate is 1 MHz?

1.8 Reference Books and Review Articles

1. "Synchrotron Radiation—A Powerful Tool in Science" in *Handbook on Synchrotron Radiation*, E.-E. Koch, D. E. Eastman, and Y. Farge, North-Holland, Amsterdam, 1–63 (1983)—overview of early synchrotron radiation sources and references to more detailed accounts. ISBN 978-0444864253.
2. *Particle Accelerator Physics—Basic Principles and Linear Beam Dynamics*, H. Wiedemann, Springer-Verlag, New York, 4th edition (2015)—chapters 2 and 3 contain histories of linear and circular particle accelerators, with extensive references. ISBN 978-3-319-18316-9.
3. *Introduction to Synchrotron Radiation*, G. Margaritondo, Oxford University Press, New York (1988)—chapter 1 has a brief history of X-ray science. ISBN 0-19-504524-6.
4. *An Introduction to Synchrotron Radiation*, P. Willmott, John Wiley and Sons (2011)—chapter 1 has a brief history of X-ray science. ISBN 978-0-470-74579-3.
5. *Synchrotron Radiation News*, Vol. 28, 2015—a special issue devoted to personal reminiscences.
6. "Fifty years of synchrotron science: achievements and opportunities", *Phil. Trans. Royal Soc.* **2019**, *377, A*, S. S. Hasnain and C. R. Catlow, eds.—a special issue based on a 2018 conference
7. *A Skeleton in the Darkroom—Stories of Serendipity in Science*, G. Shapiro, Harper & Row, San Francisco (1986). Ch. 1 recreates Röntgen's discovery of X-rays. ISBN 0-06-250778-8.

Chapter 2
The Storage Ring Complex

Synchrotron radiation sources are among the largest and most expensive scientific instruments ever built. At first glance, the scale and complexity of these facilities, as shown in Figs. 1.6 and 2.1, can be daunting. However, when we look "under the hood," we find that they all have the same essential components, and furthermore, many of these components are repeated over and over again. The key pieces can be divided into two main categories: *radio frequency systems* to accelerate the particles and also to replenish their energy and *magnets* to bend and focus the charged particle beams. Of course, there needs to be a charged particle source, a massive vacuum system to contain the charged particles, and a safety infrastructure to contain and control the high-energy particles and X-radiation.

Why do we care? In our case of synchrotron spectroscopy:

- The storage ring determines the properties of the particle beam.
- The particle beam sets limits on the properties of the photon beam.
- The photon beam determines how well we can do our experiments.

Most synchrotron radiation sources around the world are based on electron storage rings. These facilities include not only the storage ring itself but the initial source of high-energy electrons, invariably a linear accelerator, and often an intermediate device to raise the particle energy, a "booster synchrotron." A diagram for a typical storage ring complex is shown in Fig. 2.1.

Although a view of the interior (Fig. 2.1) of a storage ring complex can at first be bewildering, it becomes understandable if you break it up into smaller components, many of which are repeated over and over again. We begin the description of a storage ring at the electron source and then follow the charged particles as they flow through the complex.

S. P. Cramer, *X-Ray Spectroscopy with Synchrotron Radiation*, Biological and Medical Physics, Biomedical Engineering, https://doi.org/10.1007/978-3-030-28551-7_2

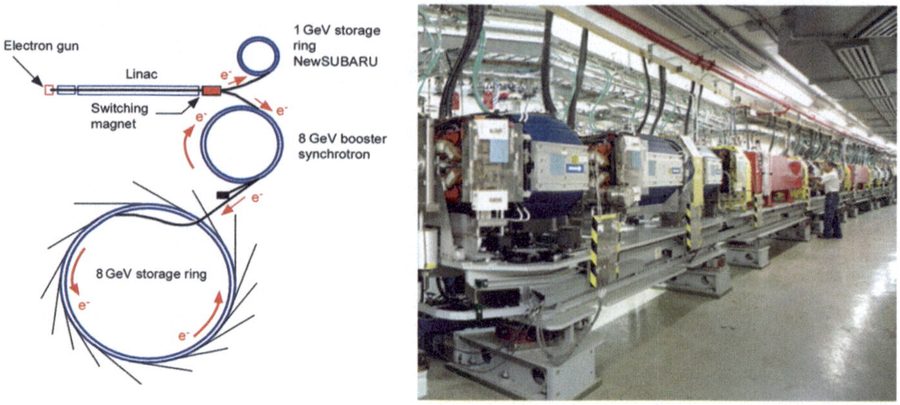

Fig. 2.1 Left: a diagram of the SPring-8 synchrotron facility in Japan. Right: inside the Advanced Photon Source (APS) storage ring

2.1 How to Accelerate and Steer Charged Particles: The Lorentz Force

In order to create our relativistic charged particle beam circulating around our storage ring, we need to apply forces to (1) provide and replenish the particles' energy and (2) bend them in a circle and also to keep them focused. The Lorentz force equation describes how charged particles in accelerators interact with electric ($\vec{E}$) or magnetic ($\vec{B}$) fields, and the net *Lorentz force* $\vec{F}$ on a particle with charge q and velocity $\vec{v}$ is given by (in SI units):

$$\vec{F} = q\left(\vec{E} + \vec{v} \times \vec{B}\right) \tag{2.1}$$

It is important to note that the force and acceleration from an electric field are along the direction of that field, while the force and acceleration from the applied magnetic field are perpendicular to both the field and the particle motion. Since the change in energy of the particle is given by the force times the change in position parallel to that force (Eq. 2.2), *only the electric field imparts energy to the particle*:

$$\Delta E_{\text{kin}} = \int \vec{F} \cdot d\vec{s} \tag{2.2}$$

$$\Delta E_{\text{kin}} = q \int \vec{E} \cdot d\vec{s} \tag{2.3}$$

In principle an electric field could also be used to steer the particles (as in a cathode ray tube). However, in practice, for relativistic charged particles, the force from a magnetic field of 1 Tesla (which is easy to produce) is about the same as from an electrical field of three million V/cm (which is impossible to produce). *The net result is that electric fields are used to provide energy to the particles in the beam, while magnetic fields are used to guide, bend, and focus the particles.*

2.2 The Linear Accelerator

The first requirement is to generate a stream of high-energy particles for filling the storage ring. In the early days of particle accelerators, all of the energy was added in a single step by application of a high dc voltage across a vacuum gap. However, acceleration by static voltages is limited by arcing and the breakdown of materials. Although the arcing made for good science fiction movies, machine physicists eventually turned to ac acceleration by radio frequency fields. The components of an ac linear accelerator involve a particle source, a power source, and an accelerating structure.

2.2.1 Particle Sources and Bunchers

Where do the electrons for a synchrotron source come from? In most cases, the source of electrons is an electron "gun" similar to what is found in an old-fashioned TV tube (Fig. 2.2). An electron gun consists of three major parts: a negative cathode, a positive anode, and a control grid. The negatively-charged cathode is made of a material (such as cesium or tungsten) where electrons are essentially "boiled off" the surface; this is so-called *thermionic emission*. The emitted electrons are accelerated toward the positive anode by an electrical potential applied across the gun and then make their way into the linear accelerator.

At the ALS, the applied voltage on the control grid is pulsed open for about 25 ns, during which about 1.5×10^{11} electrons pass through the grid. These electrons are then accelerated by the main anode to about 120,000 eV (120 keV), passing through a hole in the anode before they enter the bunchers and linac.

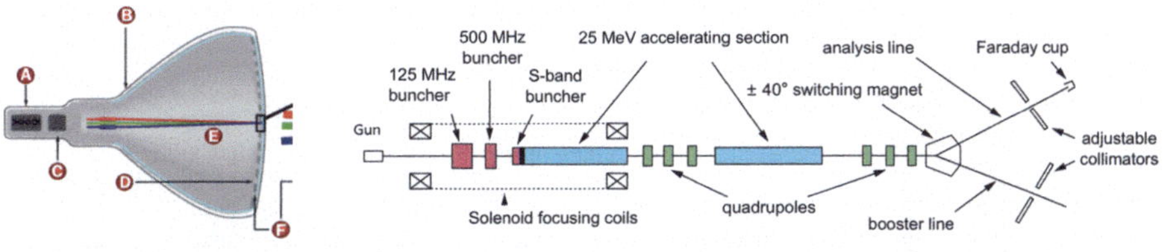

Fig. 2.2 Left: electron gun in a cathode ray tube (CRT) for color TV. *A*: cathode, *B*: conductive coating, *C*: cathode, *D*: phosphor-coated screen, *E*: electron beams anode, *F*: shadow mask. Right: the ALS linac components from gun to output [10]

Since electrons have to arrive at the linear accelerator or synchrotron within certain time intervals to be properly accelerated, the continuous or partially bunched electron stream from a gun is usually put into bunches by various *prebunchers*. For example, the ALS linac has three sections that sequentially divide the beam into 0.125, 0.5, and ~3 GHz bunches (Fig. 2.2).

Pulsed laser electron sources can be considerably brighter than conventional electron guns. In these devices, a high-intensity laser is focused onto a "photocathode." The light causes photoemission from a very small spot, resulting in a much lower emittance source, which is critical for free electron lasers (Chap. 12).

2.2.2 *RF Power and Waveguides*

In a conventional linear accelerator ("linac"), the energy that goes into accelerating the electrons comes from microwave rf fields produced by klystron sources. The klystron tube, invented at Stanford before World War II, takes energy from a dc electron beam and uses it to create a high-voltage microwave output beam. The wavelength of the microwaves produced depends on the dimensions of the metallic cavities that are used, and the most popular frequencies for microwave production range from 3 to 10 GHz. For example, at the ALS, the linear accelerator uses 2.9979 GHz. Microwaves from the klystron are transmitted to the accelerator by rectangular waveguides (Fig. 2.3).

In the linac, the waveguide component is a series of metallic electromagnetic cavities that are driven by the klystron microwave sources. (It is not that different from your home microwave oven, except a million times more powerful.)

2.2.3 *The Linac and Particle Acceleration*

The common way to describe a linac is to draw an analogy with a surfer on a water wave. Although this comparison is a good starting point, there are a couple things going on that are not immediately obvious. First of all, free electromagnetic (EM) waves have electric and magnetic fields perpendicular to the propagation

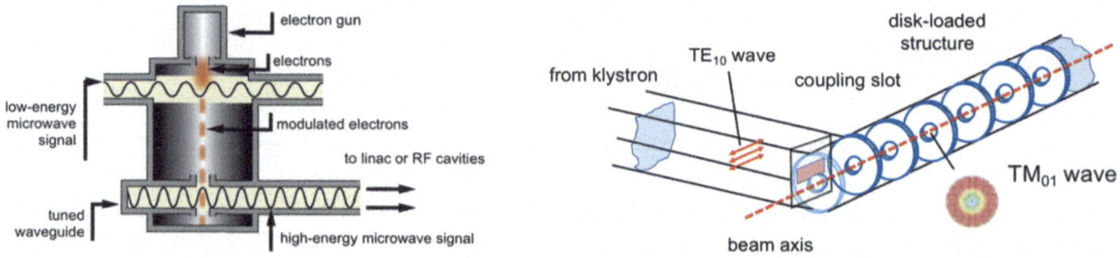

Fig. 2.3 Left: schematic of klystron operation. Right: a klystron feeding the accelerator cavity

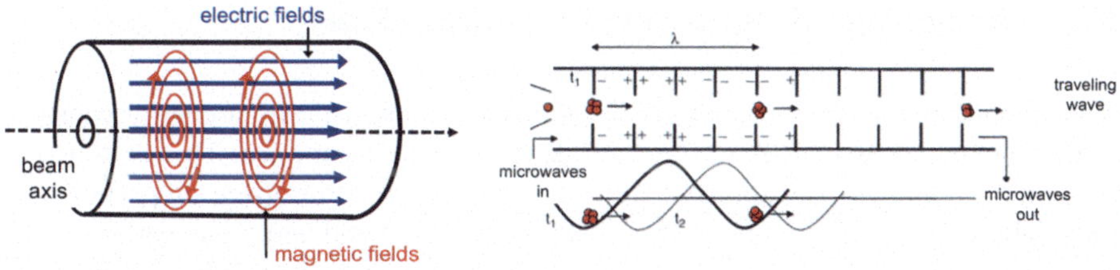

Fig. 2.4 Left: schematic of the electromagnetic fields inside a typical circular waveguide. Right: illustration of a disc-loaded waveguide sequence with electrons arriving at proper time for acceleration

direction—they are the so-called *transverse* waves. But when we discussed the Lorentz forces, we argued that to add energy to the particle, we need an electric field along the direction of motion; hence we need a longitudinal component to the electric field.

As illustrated in Fig. 2.4, the first trick that machine physicists employ in designing accelerator cavities is to use special, usually cylindrical, shapes so that electric field components are allowed only along the direction of propagation. The example shown is for a cylindrical "pillbox" cavity, for which the z-component (along the beam axis) E_z of the electric field of an appropriate travelling wave is given (in cylindrical coordinates) by:

$$E_z = E_0 J_0(k_c r) \cdot \exp[i(\omega t - kz)] \tag{2.4}$$

where ω is the angular frequency, k is the wave number, k_c is the cutoff wave number, r is the radial coordinate, and J_0 is a Bessel function (Appendix D).

The second issue has to do with timing. The particles are moving at close to the speed of light, but the phase velocity of the traveling electromagnetic wave is actually greater than the speed of light. To slow that velocity down so that the particles can successfully "surf" the cavity, machine physicists insert metal irises into the cavity at periodic intervals. The mathematical description of such a "disc-loaded" waveguide is complicated [11], but the bottom line is that the particle bunch arrival in each section can be timed to result in acceleration by the longitudinal electric field.

As the particle bunch propagates down the cavity, some early particles are ahead of the voltage maximum, while some slightly later particles experience greater accelerating voltages. As shown in Fig. 2.4, the net results are that (a) *particles become more tightly bunched* and (b) *the bunches gain energy* from the electromagnetic wave as they are propelled down the linac. In a typical modern instrument, they gain about ~10–20 MeV per meter. For example, each of the pair of 3.5 m final sections of the linac for the ALBA source provides an energy gain of 52 MeV or ~15 MeV/m [12].

2.3 The Storage Ring: Inside the Shield Walls

Having gone to the great expense of producing a relativistic high-energy charged particle beam, we would like to use the particles more than once. This is done by bending the particles in an approximate circle, hence the term *storage ring*. These are essentially large high vacuum vessels, with magnets to bend and focus particles and with one or more rf cavities to replace the energy lost to synchrotron radiation. Machine physicists usually build magnets that have primarily one multipole term, *dipole*, *quadrupole*, or *sextupole*, the fields for which are summarized in Appendix C. We will see that each serves a different purpose.

2.3.1 Dipole Magnets

The most basic magnet component in a storage ring, the one that bends particles in a circular orbit, is a dipole magnet often referred to as a *bend magnet.* An ideal bend magnet for a horizontal orbit employs a constant vertical magnetic field with no horizontal component. Most storage ring dipoles are electromagnets, employing steel pole pieces and cooled copper coils. For example, a typical bend magnet employs a ~1 kA current to achieve a 1.58 Tesla field, and weighs more than 2 tons (Fig. 2.5).

A horizontally moving relativistic electron ($\beta = v/c \sim 1$) in such a field follows a circular trajectory with a bend radius ρ in practical units:

$$\rho[\mathrm{m}] = \frac{3.3E[\mathrm{GeV}]}{B[\mathrm{Tesla}]} \tag{2.5}$$

For the ALS, with a ring energy of 1.9 GeV, this yields a bend radius of 4 m. Since the physical circumference of the ring is ~200 m (a radius of 32 m), there is clearly a lot of extra real estate between bend magnets. We will now see what else is involved in that space.

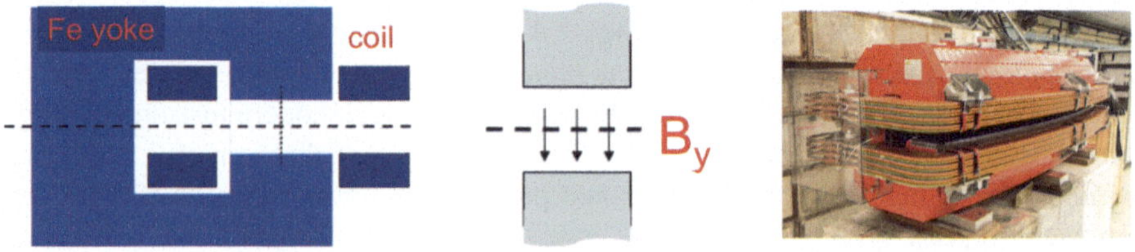

Fig. 2.5 Left: schematic of the windings, yoke, and pole pattern of a dipole ("bend") magnet. Middle: the resulting magnetic field. Right: practical dipole magnet at SESAME

2.3.2 Quadrupole Magnets

The particles traveling in bunches around the storage ring all have slightly different positions and trajectories, and without additional restoring forces, the beam would become more and more diffuse. Without focusing, particles would rapidly leave their stable orbits and be lost from the ring. The key components for focusing charged particles are *quadrupole magnets*.

Quadrupole fields are achieved by juxtaposing two electromagnets, each with a pair of hyperbolic pole pieces, as illustrated by Fig. 2.6. They are usually arranged to give either "upright" or "skew" (rotated) fields (Appendix C). The focusing action of quadrupoles stems from the fact that they possess a field gradient, so that the angular deflection of a particle is proportional to its distance from the beam center. Obviously, a larger field gradient results in a stronger "lens." Quantitatively, the *focusing strength* k of a quadrupole magnet is defined by $k = (e/\beta E)g$, where g is the field gradient, or in practical units:

$$k\left[\mathrm{m}^{-2}\right] = 0.2998\frac{g[\mathrm{T/m}]}{\beta E[\mathrm{GeV}]} \tag{2.6}$$

The focal length f of a quadrupole magnet with path length l is then given by:

$$f[\mathrm{m}] = \frac{1}{kl} = \frac{3.3E[\mathrm{GeV}]}{g[\mathrm{T/m}]l[\mathrm{m}]} \tag{2.7}$$

Since practical field gradients and path lengths are on the order of 10 T/m and 0.5 m, respectively, a typical focal length for 2 GeV electrons is ~1.3 m.

While optical lenses focus visible light in both horizontal and vertical planes, a minor hitch for focusing charged particles with quadrupole magnets is that they can only focus in one plane at a time—they are actually defocusing in the perpendicular plane. To achieve net focusing in both horizontal and vertical planes, pairs or triplets of quadrupoles are arranged together. Two magnetic lenses can be treated just like glass lenses in geometrical optics in the thin lens approximation, where the focal

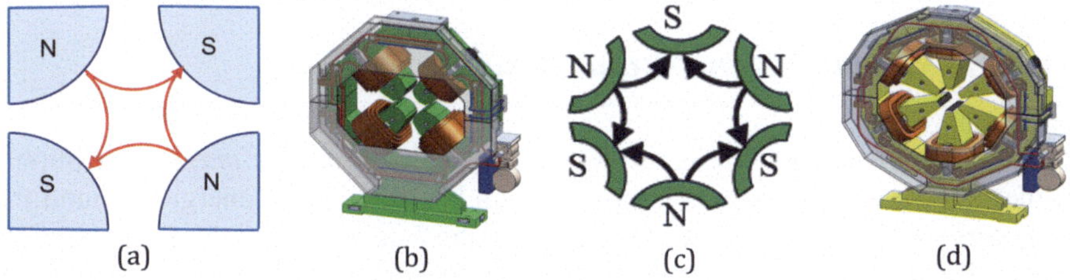

Fig. 2.6 Left to right: (**a**) quadrupole magnetic fields, (**b**) schematic of a quadrupole magnet for SESAME, (**c**) magnetic field in a vertical sextupole, (**d**) schematic of a sextupole magnet for SESAME

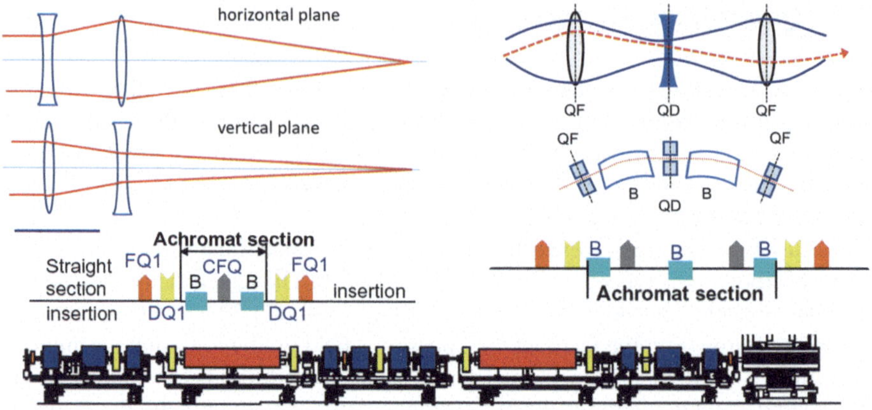

Fig. 2.7 Top left: combined focusing and defocusing forces for quadrupole magnets can achieve net focusing in both directions. Top right: two representations of a FODO cell. Middle: basic components for double bend and triple bend achromats. FQ and DQ are focusing and defocusing quadrupoles, respectively, B is bend magnet. Bottom: strings of quadrupole magnets (blue) in triplets and pairs of doublets surround the dipole magnets (red) in the APS lattice. The yellow magnets are sextupoles

length of a lens pair with individual focal lengths f_1 and f_2 separated by distance d is given by:

$$\frac{1}{f^*} = \frac{1}{f_1} + \frac{1}{f_2} - \frac{d}{f_1 f_2} \tag{2.8}$$

If we arrange a pair of quadrupoles with opposite magnetic fields, so that in both vertical and horizontal planes $f_1 = -f_2 = f$, then the first two terms in the above equation will cancel, and the total focal length will be $f* = f^2/d$. In practice, the above result is only good in the thin lens approximation, and since quadrupole magnets are not really "thin," the horizontal and vertical focal lengths will be different. This is sometimes cured by using three quadrupoles, to make a *quadrupole triplet*, with a central quadrupole of length l bounded by two smaller quadrupoles of length $l/2$ (Fig. 2.7).

2.3.3 Sextupole Magnets (and Beyond)

The last type of magnet commonly employed is the *sextupole magnet*. Sextupole magnets are needed because the particle beam has a spread of energies around the ideal energy E. If we look at the expression for the focusing strength of a quadrupole magnet (Eq. 2.6), we see that it depends on the beam energy—the focal length will be shorter for the lower energy particles. Such a lens is said to have *chromatic aberrations*, in the same way that a simple glass lens has a different focal length for red and blue light.

Sextupole magnets can be considered as "quadrupoles with varying focal strength across the horizontal aperture" [13]. Thanks to aberrations induced by the quadrupoles, higher-energy particles tend to be to the outside of the ideal orbit, while lower-energy particles tend to be inside that orbit. Sextupoles can reverse chromatic aberrations due to the quadrupoles by providing more focusing power on the outside of the orbit and less on the inside. (Going beyond sextupoles, one of the world's most recent storage rings, Max-IV, uses a lattice that even includes *octupole magnets* [14–16].)

2.3.4 Storage Ring Lattices

The array of magnets that guides a charged particle beam along the desired trajectory in the storage ring is called the *lattice*. To simplify the design of storage rings, lattices are divided into sectors that contain a particular sequence of magnets. These *superperiods* are repeated a number of times to form most of the complete ring. There are often smaller groups of magnets, called *cells*, which also repeat within a superperiod. One superperiod of the APS lattice is shown in Fig. 2.7. The superperiod structure determines the fundamental properties of the storage ring, such as the size and divergence of the beam as well as its stability.

The complexity of magnet lattices continues to increase as machine physicists seek to reduce the *emittance* (see below) of their storage rings. A state-of-the-art storage ring has nearly 1000 magnets in the lattice! For example, NSLS-II uses 60 dipoles, 240 quadrupoles, 260 sextupoles, and ~270 "large aperture" and "corrector magnets" [17]. The MAX-IV storage ring has 20-fold symmetry, and each superperiod employs 7 dipole, 16 quadrupole, and 18 sextupole magnets [14–16]. Not to be outdone, the new lattice for the upgraded ALS, "ALS-U," will have nine dipole magnets in each superperiod.

2.3.5 Beam Perturbations and Loss Mechanisms

In practice, none of the electrons in a storage ring follow the ideal orbit for long. In part, this is because electrons lose energy by synchrotron radiation as they travel around the ring. The electrons are restored to the correct orbit by the focusing effect of the quadrupole magnets and by differential acceleration in the rf cavity. Thanks to these restoring forces, the particles oscillate around the ideal orbit. The longitudinal motions around the ideal position are called *synchrotron oscillations*, while the transverse motions are called *betatron oscillations*. The number of horizontal or vertical betatron oscillations that occur during one revolution around the ring is called the *tune*.

Machine physicists work hard to avoid integer values for the tune. Why? Suppose there is some perturbation in the ring that causes an unwanted motion of the particle beam. If the tune is an integer, then this perturbation will add in-phase during each particle orbit, ultimately leading to loss of the beam. This is circumvented by using tune values that avoid integers, as well as integer ratios of x and y tunes, as illustrated in the "tune diagram" of Fig. 2.8.

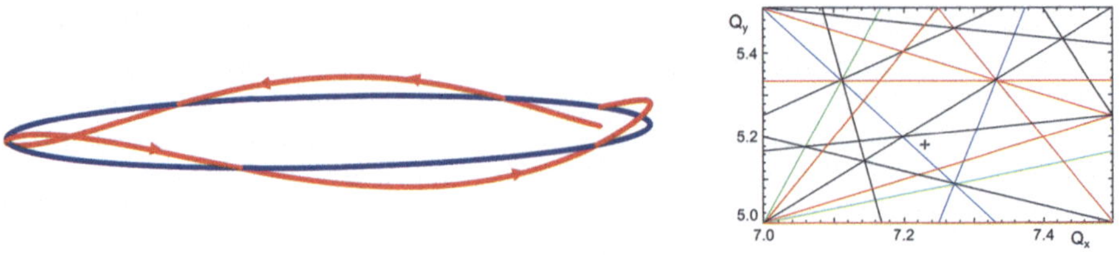

Fig. 2.8 Left: betatron oscillations (red) perpendicular to the ideal particle orbit (blue). Right: a typical tune diagram, in this case for the SESAME ring. The operating point is marked by a +

A Selective History of Storage Ring Design

The earliest storage ring lattices were developed for high-energy physics experiments, and they were simple repeats of focusing (F) and defocusing (D) magnets, with drift sections (O) containing bend magnets in between, hence the name "FODO" lattice. Over time, machine physicists developed increasingly sophisticated designs to improve the brightness of synchrotron radiation sources. From 1975 to 1977, Renate Chasman and G. Kenneth Green were the first to design a ring to minimize the emittance of the stored particle beam [18]. With contributions by Sam Krinsky, they invented the so-called Chasman-Green lattice that was implemented in the first facility built for brightness—NSLS. Part of the secret was to introduce sextupole magnets to reduce the chromatic aberrations of conventional designs that had only dipoles and quadrupoles, and such a design is often called a double-bend achromat or DBA. Sadly, they both died in 1977 before construction of NSLS began [19].

Renate Chasman and Kenneth Green

Another option, the "triple-bend achromat" or TBA, added an extra bend magnet to the superperiod and was used for the first time at the ALS. This design,

(continued)

which gave more flexibility for the beam parameters in the insertion device region, was also used by the Swiss light source (SLS). However, subsequent designs such as NSLS-II, Diamond, and SOLEIL stayed with a DBA lattice. DBA and TBA designs remained the rule for more than three decades.

Since the emittance of a storage ring synchrotron light scales with the third power of the bend magnet deflection angles, it was often thought that for very low emittance, a ring had to be very large, with many small deflection magnets. However, the ring designers in Sweden realized that the same principle could be applied to a modest size ring, and they developed a "multi-bend achromat" (MBA) design with 7 bend magnets in each of 20 achromat cells [20]. This new lattice design was implemented in 2016 with the commissioning of the Swedish light source MAX-IV [14], while the ALS-U upgrade envisions a 9-bend MBA lattice [21]. Plans for MBA upgrades now exist for many of the third-generation sources [22], all in a quest for the ultimate "diffraction-limited storage ring" or DLSR [23].

Apart from small orbit deviations caused by synchrotron radiation, particles can be scattered so far from the ideal orbit that they are lost from the beam entirely. This scattering can be caused by (a) collisions with residual gas molecules in the chamber—bunch-gas scattering, (b) scattering between electrons within the same bunch—Touschek scattering, and (c) quantum excitations due to synchrotron radiation. Whether or not these events result in loss from the beam depends on the "transverse acceptance" of the ring—how large a deviation from the ideal orbit can occur with the electron still in a stable orbit.

Each of these loss mechanisms is associated with a lifetime. The vacuum lifetime associated with bunch-gas scattering can be divided between τ_{cs} for Coulomb or elastic scattering and τ_{bs} for bremsstrahlung or inelastic scattering, and both of these are inversely related to the residual gas pressure in the ring. The Touschek lifetime, τ_{T}, is one of the factors that determines how dense a charge can be placed in a given bucket. Finally, the quantum lifetime, τ_{q}, is proportional to the aperature of the ring. The net effect of these loss mechanisms is to give the particle beam an effective lifetime τ described by the combination of these terms [24]:

$$\tau^{-1} = \tau_{\mathrm{q}}^{-1} + \tau_{\mathrm{T}}^{-1} + \tau_{\mathrm{cs}}^{-1} + \tau_{\mathrm{bs}}^{-1} \tag{2.9}$$

2.3.6 *Describing the Stored Beam: Phase Space and Emittance*

A harmonic oscillator such as a pendulum undergoes sinusoidal oscillations in time according to: $x = a\sin(kt + \delta)$. To describe such motion in *phase space*, one maps the simultaneous position and momentum of a particle throughout its trajectory.

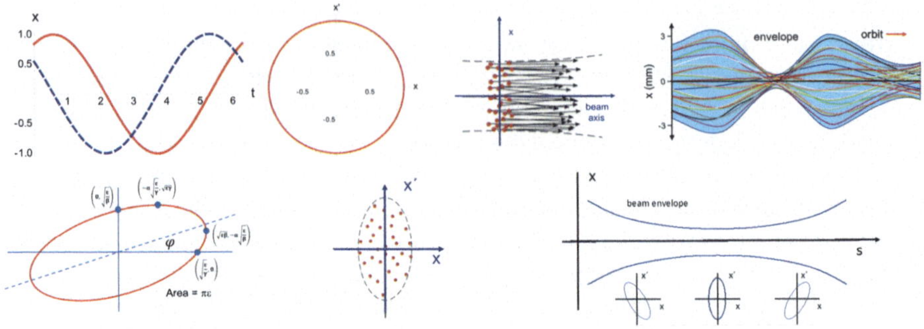

Fig. 2.9 Top left: position (red line) and speed (blue dashed line) vs. time description compared with phase space (x' vs. x) representation of a harmonic oscillator. Top middle: trajectories in one particle bunch at tightest focus. Top right: sketch of different trajectories for individual particles in a bunch as it travels around the ring. Bottom left: the phase space ellipse and Twiss parameters for a particle envelope at one position around the ring. Bottom middle: distribution of particles in the phase space ellipse at focus. Bottom right: beam size through a bend magnet or drift space, along with associated phase space ellipses

Assuming constant mass, one can also map position and velocity, in this case: $dx/dt = ak\cos(kt + \delta)$. As shown in Fig. 2.9, the pendulum can then be viewed as moving around on an orbit in phase space. Accelerator physicists often transform the description of particles in a ring to equations similar to a harmonic oscillator. Analogous to the harmonic oscillator, the particle in a storage ring can then be described as moving in a volume in phase space.

Because of synchrotron radiation and other perturbations, there is an approximately Gaussian distribution of particle positions (x, y) and trajectories (x', y') around the ideal orbit. To describe this collection of particles that constitute the stored beam, for each pair of position and momentum coordinates, one surrounds a fraction of the particles (usually 1 standard deviation or ~68%) by an ellipse called the phase ellipse. The α, β, and γ parameters that describe the phase ellipse are often called *Twiss parameters*.

$$\gamma_x x^2 + 2\alpha_x x x' + \beta_x x'^2 = \varepsilon_x \text{ and } \gamma_y y^2 + 2\alpha_y y y' + \beta_y y'^2 = \varepsilon_y \tag{2.10}$$

At various positions around the ring, the orientation and eccentricity of this phase space ellipse will change, depending on whether the beam is being focused or defocused. Thus, the Twiss parameters are actually a function of position: $\alpha(s)$, $\beta(s)$, and $\gamma(s)$ (Fig. 2.10). At a particular point in the lattice, the beam size σ_x or σ_y and divergence $\sigma_{x'}$ or $\sigma_{y'}$ can be calculated from these parameters via:

$$\sigma_x = \sqrt{\varepsilon_x \beta_x} \text{ and } \sigma_y = \sqrt{\varepsilon_y \beta_y} \tag{2.11}$$

$$\sigma_{x'} = \sqrt{\varepsilon_x \gamma_x} \text{ and } \sigma_{y'} = \sqrt{\varepsilon_y \gamma_y} \tag{2.12}$$

A key point is that although the phase ellipse can be rotated and skewed, the area of the ellipse does not change. This result can be derived from "Liouville's theorem,"

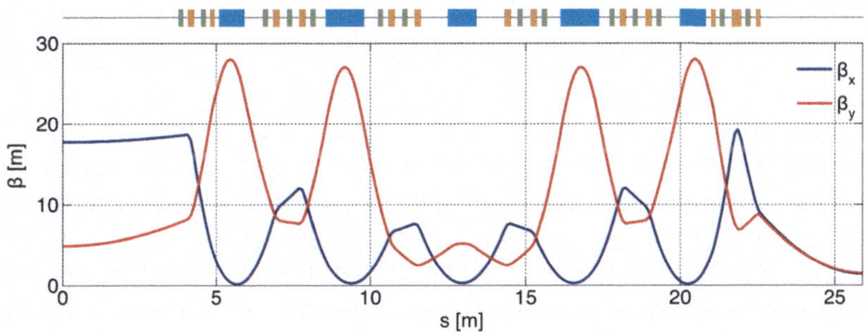

Fig. 2.10 The Twiss β_x and β_y parameters vs. orbit positions for the Sirius ring [25]

but we will just take it as a given. This conserved area (within a factor of π) is called the *emittance*, and there are generally different values for the horizontal and vertical emittances—ε_x and ε_y. At the point of tightest focus of the beam, the emittance can also be described as the product of the physical size of the beam in the horizontal (σ_x) or vertical (σ_y) direction, and its angular divergence ($\sigma_{x'}$ or $\sigma_{y'}$) in that dimension. Thus, at the beam waist:

$$\varepsilon_x = \sigma_x \sigma_{x'} \tag{2.13}$$

$$\varepsilon_y = \sigma_y \sigma_{y'} \tag{2.14}$$

The horizontal emittance ε_x is increased by the synchrotron radiation caused by the bend magnets, and it is often ~10–100 times larger than the vertical emittance ε_y. The ratio $\varepsilon_x/\varepsilon_y$ is called the *coupling ratio.*

2.3.7 The Diffraction Limit

From the Heisenberg uncertainty principle, a light source has a natural limit to the product of the source size and divergence, in both x and y directions, known as the *diffraction limit.* If we represent the source sizes and divergences respectively in either the horizontal or vertical planes using $\sigma_{x,y} = \sigma_r/\sqrt{2}$ and $\sigma_{x',y'} = \sigma_{r'}/\sqrt{2}$, where the subscript r refers to the radial source parameters, as defined in Wiedemann, then we can write the following equation for the photon beam emittance:

$$\varepsilon_{photon,r} = \sigma_r \sigma_{r'} \cong \frac{\lambda}{2\pi} \tag{2.15}$$

For an undulator source, for each dimension, this reflects to the following diffraction limited photon beam divergence and diffraction limited source size:

$$\sigma_{r'} = \sqrt{\frac{\lambda}{L}} \text{ and } \sigma_r = \sqrt{\lambda L}/(2\pi) \tag{2.16}$$

To summarize, even with infinitesimal electron beam size or divergence, the photon beam will have a natural divergence and apparent size that depend on the length of the source and the wavelength of the photons being used. Almost from the first use of storage rings for synchrotron radiation, the dream has been to build a "diffraction limited storage ring" or "DLSR" that would achieve the ultimate possible source brightness. A final word of warning: there are many conflicting definitions of the limiting photon properties and emittance, all of which are approximate because the true photon distribution is not Gaussian.

2.3.8 Putting Energy Back in: Storage Ring RF Cavities

A storage ring also needs one or more rf cavities to pump microwave energy into the electron beam. The goal is primarily to restore the energy lost by synchrotron radiation (although in some cases, the particle energy is also raised after injection). Thus, a storage ring can be viewed as a massive microwave $\rightarrow$ X-ray transducer.

In contrast with linac rf waveguides, which often involve traveling electromagnetic waves, the storage ring rf cavity contains a *standing* electromagnetic wave. In the simplest case of a short ($L < 2.03R$) "pillbox" cavity with length L and radius R, the electric and magnetic fields of the simplest TM_{010} mode in cylindrical coodinates, respectively, E_z, E_θ, E_r, H_z, H_θ, and H_r, are expressed in terms of Bessel functions J_0 and J_1 (Appendix D):

$$E_z(r,t) = E_0 J_0(kr) \cdot \mathrm{e}^{i\omega t} \text{ and} \tag{2.17}$$

$$H_\theta(r,t) \sim -iE_0 J_1(kr) \cdot \mathrm{e}^{i\omega t} \tag{2.18}$$

where the remaining field components are zero, and the resonant frequency ω_0 is determined by the boundary condition that the electric field is zero at the the cavity wall (Fig. 2.11):

$$\omega_c = kc = \frac{2.405c}{R} \tag{2.19}$$

In order for one bunch of particles to remain in a stable orbit, it must arrive at the same phase of the rf cavity cycle each time. Thus, the rf cavity frequency f_{rf} must be a multiple h of the particle orbit frequency $f_{particle}$, so that the cavity goes through an integral number of cycles during particle orbits, and the particles always arrive at the right time to be accelerated. For practical purposes, the bunch velocity is the speed of light, so the period T for one orbit is the ring circumference L (actually the orbit length L) divided by c. An acceptable rf frequency is thus:

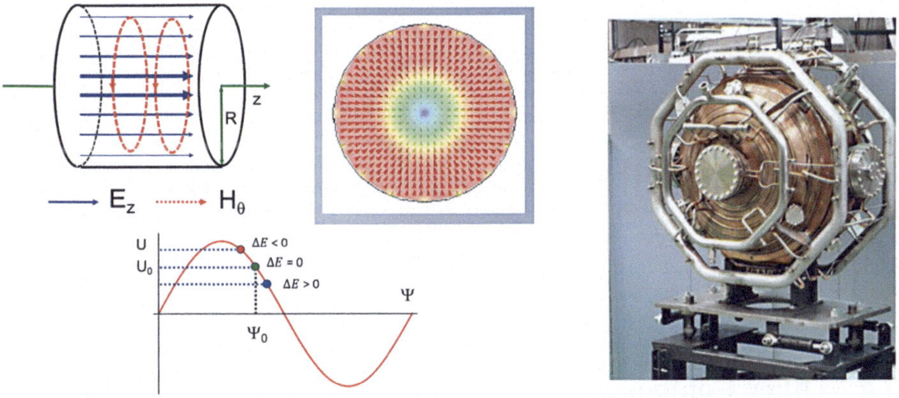

Fig. 2.11 Top left and middle: side view of *E* and *H* fields within an rf cavity; end on view of *E* field intensity for TM_{010} mode—purple is most intense, and there is a node at the boundary. Lower left: particle timing and potentials they experience [26]. It is counterintuitive at first, but lower-energy electrons arrive earlier, because their orbit is slightly smaller. Right: ALS rf cavity before placement in the ring

$$f_{\mathrm{rf}} = h \cdot f_{\mathrm{particle}} = \frac{h}{T_{\mathrm{orbit}}} = \frac{h \cdot c}{L} \tag{2.20}$$

The above requirement is referred to as the *synchronicity condition*, and h is called the *harmonic number*. If $h > 1$ (and it is usually several hundred), then h bunches of particles can be placed around the ring, with a minimum spacing such that the cavity goes through just one cycle. At the ALS, L is 196.8 m and f_{rf} is nominally 500 MHz, so $h = 328$. In principle, one could fill the ring with 328 bunches, with a separation given by the time for one rf cycle—2 ns or ~0.6 m between bunches. The locations where particles can be stored are referred to as *buckets*, and in practice, not all buckets are filled.

Electrons traveling the ideal orbit should arrive at a phase such that they will receive W_0, the average amount of energy lost per cycle due to synchrotron radiation. The remainder of the bunch also needs to arrive in the downward sloping portion of that cycle (Fig. 2.11). This is so that electrons that have less energy than the ideal value will receive slightly more energy in the cavity, while electrons with more energy than the ideal value receive slightly less of a boost.

The configuration of buckets that contain electrons is called the *fill pattern*, and it is a beam property that the user might be able to dictate or perhaps vote on. Some experiments just need raw flux, in which case the highest current is usually achieved by filling nearly all of the buckets (Fig. 2.12). In other cases, a timing experiment might need 100 ns or even 500 ns between pulses, in which case a string of empty buckets is put in place.

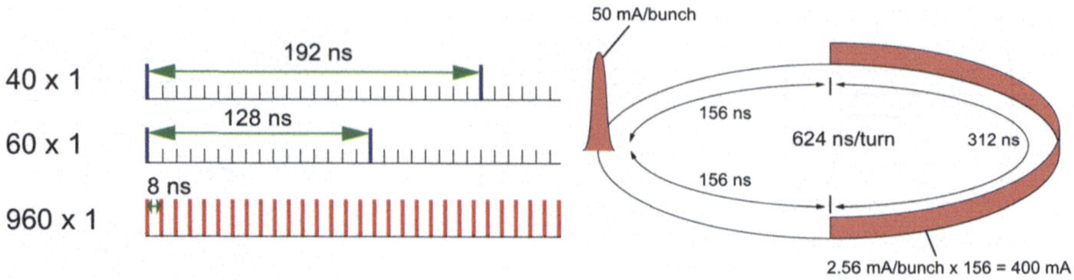

Fig. 2.12 Typical fill patterns. Left: some schemes at PETRA-III. Thin lines are empty buckets. Right: fill pattern at SPring-8 that is a compromise for both high flux and timing experiments

2.4 Other Components

2.4.1 Booster Synchrotrons

A linear accelerator with the same final particle energy as the storage ring is usually unjustifiably large and expensive. A common and cheaper approach is to use a *booster synchrotron* as an intermediate step in raising the particle energy. A synchrotron is simply a rapidly tunable storage ring. Electrons are injected into the synchrotron at relatively low energies, and as the particles acquire energy from the rf cavity, the fields in the bend magnets are ramped to maintain a constant bend radius. For example, at the ALS, the booster synchrotron dipoles are ramped from near zero to 1.2 Tesla in about 0.5 s. The magnets are then ramped down, the synchrotron refilled, and the process starts over again.

2.4.2 The Brief Flirtation with Positrons

If you read the early literature about third-generation high-energy sources, you will come across references to positron storage that might be confusing. The stored particle beam tends to make positive ions from collisions with residual gas molecules in the vacuum chamber. In theory, these ions form a cloud that is attracted to the negative charge of an electron beam, and collisions with these ions reduce the stored beam lifetime [27]. To avoid this "ion trapping," many thought it better to store positive particles than negative particles. So, in another layer of complexity, some of the brightest high-energy rings (APS and PETRA-III) planned to store positrons instead of electrons.

In these schemes, positrons are created by directing high-energy electrons onto a tungsten target, yielding positrons via bremsstrahlung pair production. The collected positrons are accelerated in their own short linac and stored in a small "positron accumulator ring" ("PAR") before injection into the booster synchrotron. Although positron storage had a certain elegance, over time, other solutions were found to the ion trapping problem. Both APS and PETRA-III switched back to electrons.

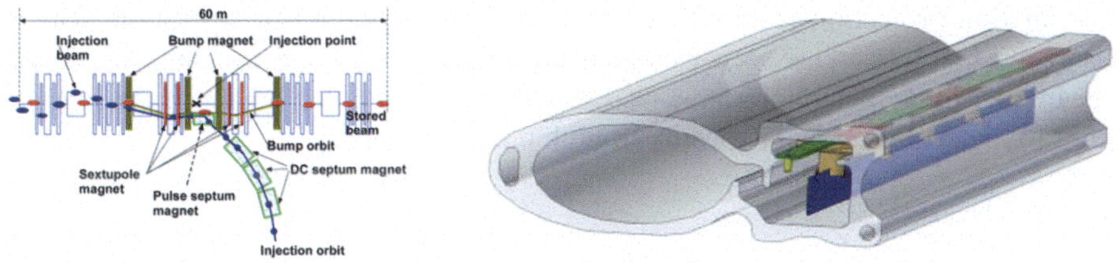

Fig. 2.13 Left: A SPring-8 diagram showing the injection system. Right: cross section of a storage ring vacuum vessel

2.4.3 *Injection Components*

Having accomplished the miracle of accelerating a particle beam to almost the velocity of light, how does one feed this beam into the circular orbit of a synchrotron or storage ring? A series of dc magnets bends the beam coming from the transport line into a trajectory almost parallel to the ring orbit. Then, a special, short pulse *septum magnet* is used to bend the beam the final degree or two, as close as possible to the stored beam orbit. As the electron beam circulates around the ring, a *kicker magnet* is used to give its orbit an outward deflection so that it can merge with the injected pulses. Then, the kicker magnet and septum magnet are rapidly turned off so orbits of subsequent bunches are not affected (Fig. 2.13).

2.4.4 *Vacuum, Thermal, and Radiation Protection Technology*

A massive infrastructure of more conventional technology is required to support the operation of a storage ring. All of the accelerator technology discussed so far requires ultra-high vacuum in the chambers containing the beams. Chemistry lab vacuum lines have pressures ranging from (10^{-2} to 10^{-4} torr) for routine Schlenk lines or better (10^{-4} to 10^{-7} torr) for high vacuum lines with diffusion pumps or turbo pumps. Storage rings typically operate at least several orders of magnitude better still, on the order of 10^{-10} torr [28].

To achieve and maintain this vacuum, the chambers are initially connected with turbomolecular pumps backed by dry mechanical pumps during the bakeout. During operations, the chamber vacuum is maintained with a combination of sputter ion pumps, non-evaporable getter (NEG) pumps, and titanium sublimation pumps (TSP).

Managing the thermal load from synchrotron radiation requires careful engineering, because without sufficient cooling the beam could melt most metals. The synchrotron radiation that does not go down the beamline is often absorbed by a massive water-cooled copper block known as a "crotch," with ion and titanium sublimation pumps placed close by.

Radiation protection is another critical factor for storage ring operation, which has to consider worst-case scenarios such as complete loss of electron bunches down a beamline. Many beamlines have lead shielding placed at the height of the storage ring to protect from direct exposure to high energy electrons.

2.5 Insertion Device Hardware

We saw in Chap. 1 (Fig. 1.2) that synchrotron radiation is produced when the trajectory of a relativistic electron is bent by a magnetic field. It is not a great leap to conclude that if bending by one magnet is good, then using 10 or even 100 magnets would be better. These arrays of magnets are called *insertion devices* (IDs), because they are inserted into "straight sections" of the lattice and ideally do not affect the overall orbit of the beam. Here we focus on the hardware for producing the requisite alternating N–S magnetic fields. In Chap. 3, we will illustrate the special properties of the radiation that is produced.

2.5.1 Superbends

The simplest insertion device for boosting synchrotron radiation is a *superbend* magnet—a very high field magnet that replaces one of the normal dipole bend magnets in a storage ring. Superbend magnets increase the amount of radiation, but more importantly, they increase the average energy of the photons produced. At the ALS, superconducting superbend magnets with a field of 5 Tesla have been used to replace the normal 1.3 Tesla bend magnets [29], and the modified lattice is shown in Fig. 2.14. The Nb-Ti superconductor magnet coils are maintained at close to 4 K by a Gifford-McMahon cryocooler. As backup in case of cryocooler failure, the magnet is immersed in an 85 liter liquid He vessel, which will support operation for about 18 h without refilling. (Technically, superbends are not really insertion devices—if they fail or are turned off, the beam no longer follows its proper orbit.)

2.5.2 Wavelength Shifters

To allow for an even greater magnetic field, machine physicists use a device called a *wavelength shifter*—which is essentially a very strong one-pole magnet preceded and followed by weaker dipoles that restore the beam to its original trajectory. At SPring-8, the orbit change produced by a superconducting 10 Tesla dipole magnet is balanced by two 1.9 Tesla magnets, resulting in no net deflection of the beam trajectory (Fig. 2.15).

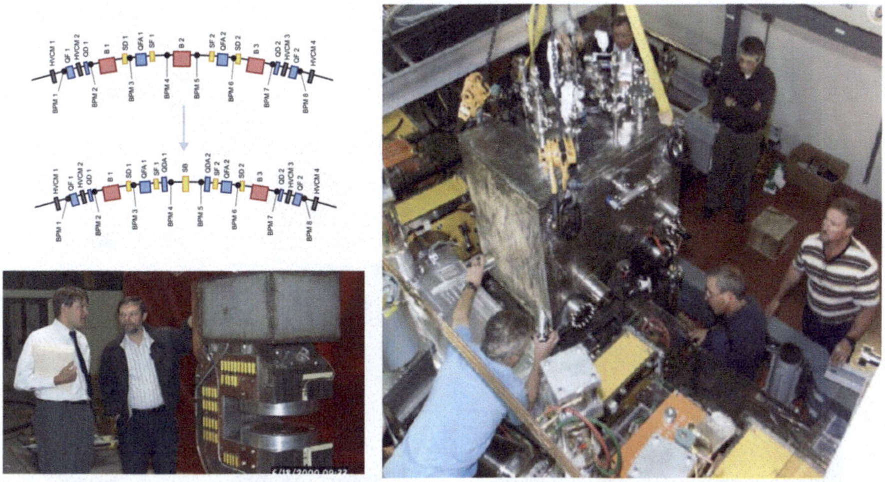

Fig. 2.14 Top left: ALS superbend modification of lattice. A pair of quadrupole magnets is also added, since the bend magnet being replaced had some focusing properties. Bottom left: the device before installation. Right: lowering the superbend into the ALS ring

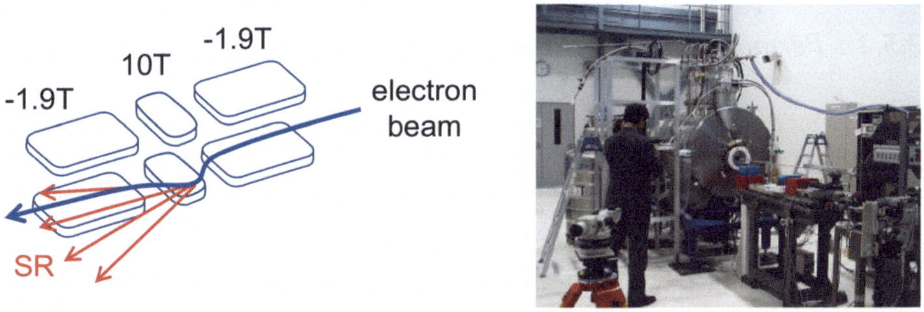

Fig. 2.15 Left: particle trajectory in the SPring-8 Wavelength Shifter. Right: the SPring-8 device before installation in the ring

2.5.3 Electromagnet Wigglers

Wigglers generally consist of a series of ~10–20 powerful dipole bend magnets arranged at a constant spacing. Many wigglers use electromagnets based on coils and pole pieces (Fig. 2.16). Weaker magnets are placed at each end, so that the particle beam emerges from the device in the same direction that it entered.

It turns out that electromagnets are great for producing high fields with long periods. However, when a shorter period is desired, the current density required for the magnet coils becomes unreasonably large, and a different technology is required [30]. For this reason, a new kind of insertion device based on permanent magnets was invented.

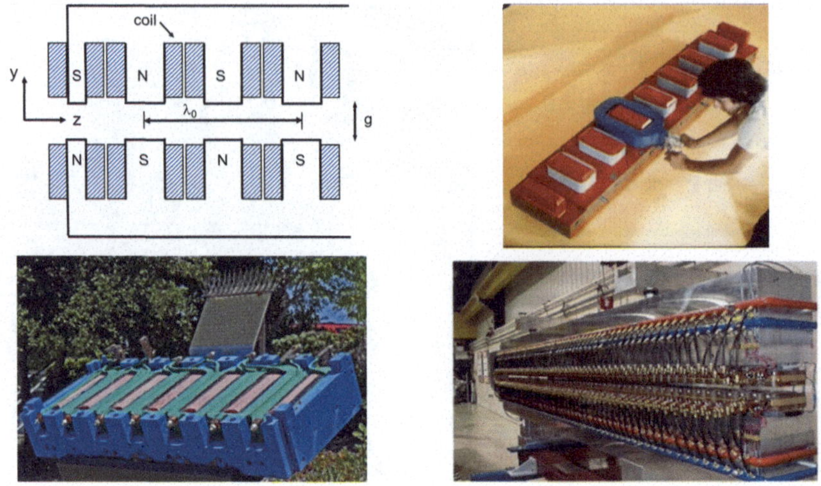

Fig. 2.16 Top left: schematic of components for electromagnetic insertion devices. Top right: Assembly of the 7 pole 45 cm period 1.9 T SSRL wiggler. Note the two additional "half-wave poles" at each end. Bottom left: electromagnetic wiggler now relegated to an educational display at SLAC. Bottom right: Wisconsin "6 EM" electromagnetic undulator

2.5.4 *Permanent Magnet Insertion Devices*

To allow for production of shorter period insertion devices, the community turned to insertion devices based on rare earth permanent magnets, using some of the same concepts employed in the kitchen magnets on your refrigerator. The use of permanent magnets was first proposed by Klaus Halbach in 1980 [31] and implemented soon after in the Exxon/SSRL/LBNL 54-pole wiggler in 1983 [32,33].

These "hard" (high "coercivity") permanent magnets have the nice property that magnetic fields for an assembly can be approximated as the linear sum of fields from the individual magnets [30]. Although they can be arranged many different ways, for "pure permanent magnet" (PPM) insertion devices the most common pattern is a so-called Halbach array (Fig. 2.17), which involves a cycle of four magnets with fields rotated by 90° from one block to the next [31].

The field in a PPM ID can be derived analytically from the "superposition principle," which means you can just add the fields from different blocks together. The detailed equations depend on the type and shape of the magnetic material and can be found elsewhere [34]. In the case of four blocks with square cross sections per period, block height h equal to one quarter the undulator period λ_u, then for a gap g between upper and lower jaws of the device, the peak field B_0 is given by Eq. 2.21, assuming a typical $SmCo_5$ remanent field of 0.9 T [34]: .

$$B_0(T) = 1.28 \exp\left(-\pi g/\lambda_u\right) \tag{2.21}$$

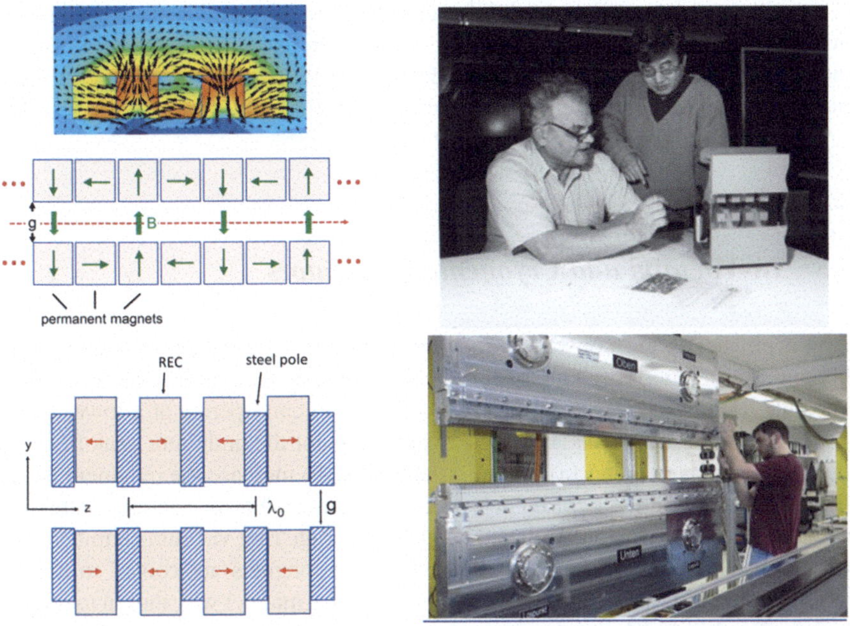

Fig. 2.17 Top left: flux lines in a "Halbach array" and associated magnet arrangement in a PPM ID. Lower left: magnet layout and pole pieces in a hybrid ID. Top right: Halbach and Kim discussing model of a permanent magnet undulator in 1986. Bottom right: adjusting a permanent magnet undulator in one of the world's newest facilities, PETRA-III

From the above, it is clear that by changing the gap g between the undulator magnets, one can tune the magnetic field that is applied to the particle beam. In the above example, for an undulator period of 7 cm, a maximum field of ~1.2 T is achieved at a gap of 0.6 cm.

An alternative to PPM IDs is the so-called hybrid design. As shown in Fig. 2.17, these devices use steel poles in conjunction with rare earth magnets, and depending on the period, they can produce about 10–50% larger field amplitudes. Another advantage of the hybrid design is that the peak field is less sensitive to variations in the angle of magnetization of the rare earth magnets. One can also incorporate tuning studs to trim the field under the individual poles and so produce a very uniform field quality. For hybrid insertion devices, there is no analytical solution for the field, but approximate parameterized equations have been extracted from the numerical modeling, and the field dependence on the gap is still very close to a simple exponential (Eq. 2.22). For the samarium-cobalt device discussed below:

$$B_0[T] = 3.33 \exp\left[-g/\lambda_u(5.47 - 1.8g/\lambda_u)\right] \tag{2.22}$$

The first hybrid device, the Exxon/LBL/SSRL 54-pole wiggler, still in operation at SSRL, has a magnetic period of 70 mm and a minimum gap of 8 mm [33]. It uses

$SmCo_5$ with a remanent field of 0.9 T and vanadium-permendur alloy pole pieces, yielding a maximum field of 1.63 T. A more modern example of a hybrid insertion device is the 2 m long U-29 standard undulator for PETRA-III (Fig. 2.17). With an undulator period of 28 mm, at the minimum gap of 9.5 mm, it provides a peak field B_0 of 0.81 Tesla and radiates about 3 kW of power [35].

2.5.5 *In-Vacuum and Cryogenic Undulators*

Machine physicists are constantly striving to build insertion devices with higher fields and shorter periods. Part of the driving force is economics, because higher fields and shorter periods mean that the storage ring can operate at a lower energy and produce the same X-rays. From the expressions for the undulator peak fields, we find that one needs to either (a) decrease the gap or (b) increase the remanent field or (c) both. To decrease the gap, ID designers have resorted to in-vacuum undulators, where the magnets are actually placed inside the storage ring vacuum chamber. They have also used *cryogenic undulators* in which the magnets are cooled to increase their remanent field [36]. For example, a NdFeB alloy magnet was shown to increase B_r to 1.58 T at 148 K. The cooling can provide almost 75% more photon flux from an ID. As illustrated below, Danfysik employed both approaches for their 226-pole 17.7 mm period cryogenic in-vacuum undulator at the Diamond Light Source in England [37] (Fig. 2.18).

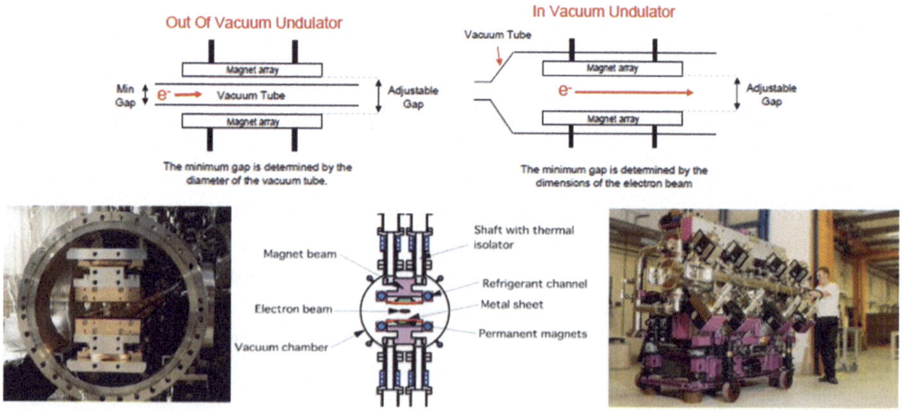

Fig. 2.18 Top: comparison of out of vacuum and in-vacuum undulators. Lower, left to right: an undulator inside the storage ring vacuum; design for a combined in-vacuum and cryogenic undulator; the Danfysik in vacuum undulator [37]

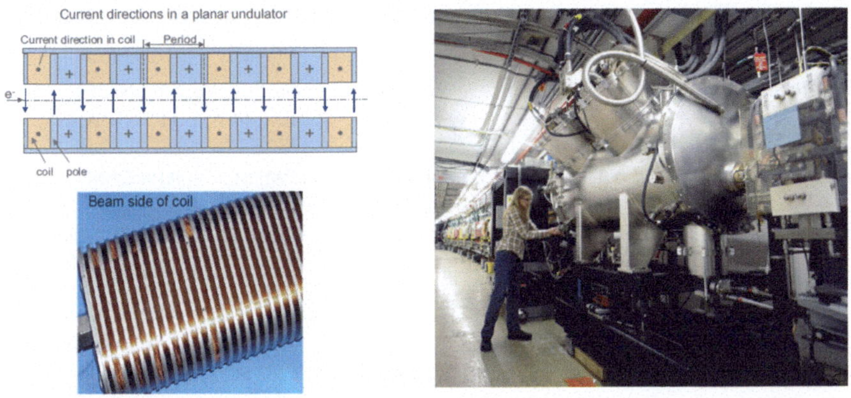

Fig. 2.19 Top left: general layout of a superconducting undulator [38]. Lower left: physical windings for a "superconducting undulator" ("SCU") [39]. Right: the SCU1 at the APS

2.5.6 Superconducting Undulators

The use of superconducting magnets is the latest step in the quest for shorter periods and higher fields. In principle, superconducting coils can provide higher field strength for the same gap and period length, which translates to more flux from the device. As can be seen in Fig. 2.19, the technology is quite different from other devices mentioned so far.

As one example, a superconducting undulator "SCU1" is in operation at the APS . It is wound with round NbTi superconducting wire, which allows a period of 18 mm and a maximum field of 0.976 T at 450 A of current. The 1 m long magnet has 69.5 periods and is suspended by Kevlar strings within a liquid helium cryostat (Fig. 2.19).

2.5.7 Insertion Devices for Circular Polarization

So far, we have emphasized insertion devices that produce sinusoidal vertical magnetic fields, which results in sinusoidal horizontal motion of the particle beam. As we will see in the next chapter, this horizontal acceleration results in a linearly polarized X-ray source.

Why do we want special insertion devices for circular polarization?

- Circularly polarized X-rays allow us to study the magnetic properties of materials.
- In the hard X-ray region, there are quarter wave plates analogous to those used in UV-vis spectroscopy, but in the soft X-ray region, it is hard to transform linear polarization into circular polarization, so it is better to start with circular polarization at the source.

Table 2.1 Summary of the more popular types of insertion devices (there are many more)

Name	Deflection parameters	Polarization	Magnets[a]	Example
Wavelength shifter	$K \gg 1$	Linear	SC	SWLS [56]
Planar wiggler	$K \gg 1$	Linear	EM, PM, SC	TPS MPW [57]
Planar undulator	$K_x \approx 1\ K_y = 0$	Linear	PM, SC	APS CPMUs and SCUs [58]
Elliptical undulator (EPU)	$K_x \neq K_y \approx 1$	Any	PM	APPLE-2 [59] APPLE-III [60] Delta [61] APPLE-X [62] APPLE-KNOT [63]
Helical undulator	$K_x = K_y \approx 1$	Circular	PM, SC	BL23SU [64]
Crossed undulators	$K_{x1} = K_{y2} \approx 1$ $K_{x2} = K_{y1} = 0$	Any	PM	HELIOS [65]

[a]*EM* electromagnet, *SC* superconducting magnet, *PM* permanent magnet

For some experiments (such as XMCD that we discuss in Chap. 7), one needs a circularly polarized X-ray beam. As you might expect, one way to obtain circular polarization is to have the particle beam rotate as it travels, yielding a helical trajectory. To accomplish this, ID designers use the same basic tools as with horizontal insertion devices: electromagnets, permanent magnets, or superconducting magnets. The plethora of designs is summarized in Table 2.1, and it has been described in review articles [40–42] and the many books on insertion devices (see references at end of chapter).

To achieve a helical trajectory, the magnetic field also needs to vary in a helical manner. Initial approaches using electromagnets included "asymmetric wigglers" [44] and "elliptical wigglers"[45], but these have been superseded by permanent magnet devises. For example, a permanent magnet *helical undulator* [46] uses sets of magnets that have both vertical and horizontal field components. The magnet arrays can consist of pairs of horizontal and vertical Halbach arrays (Fig. 2.20) or horizontal and vertical hybrid magnet assemblies. The goal of a helical field can also be achieved with continuous superconducting wires running in opposite directions in a so-called *bifilar helical magnet* [47].

A less obvious way to produce circular polarization is the so-called *crossed undulator* [48,49]. This device employs two conventional undulators at 90° to each other, along with an intervening "phase delay" (Fig. 2.20). It turns out that a monochromator stretches the radiation pulses in the axial direction, causing the radiation from two undulators to overlap in time. Depending on the relative phase of the two devices, the resulting polarization can range from left to right circular as well as linear (although at ±45° to the horizontal and vertical planes.) An example is the device deployed at BESSY [50].

One problem with the use of twin pairs of magnet arrays is the asymmetric nature of the particle beam distribution. The ellipsoidal beam is much larger in the horizontal direction, which limits the gap that can be used for the arrays that produce the

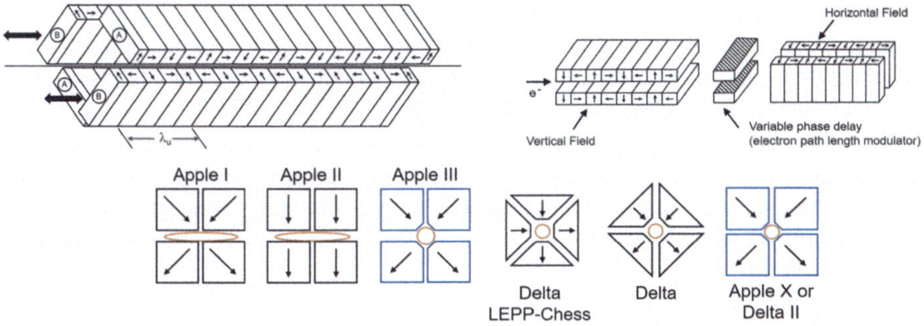

Fig. 2.20 Top left: helical undulator based on Halbach arrays [43]. Top right: crossed undulator concept. Bottom: different arrangements for magnets in elliptically polarizing undulators

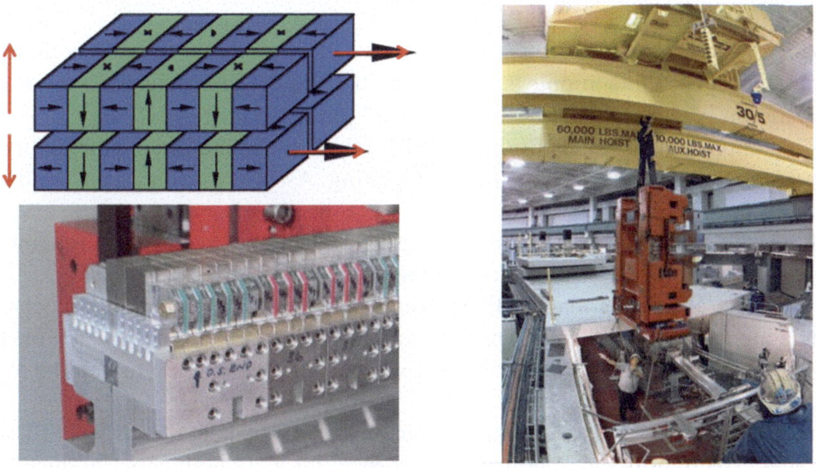

Fig. 2.21 Top left: use of "phased" split Halbach arrays to produce a helical magnetic field. Lower left: bottom half of EPU, showing two rows of Halbach arrays. Right: EPU installation at the ALS

horizontal magnet field. The crossed undulator approach, although elegant, takes up twice as much precious real estate in the storage ring straight section. However, for free-electron laser facilities (Chap. 12) where space is more plentiful and the bunch is smaller and more symmetrical, the crossed undulator approach is making a comeback [51,52].

For storage ring sources, the most popular approach has been the so-called APPLE-2 (Advanced Planar Polarizer Emitter-2) invented by Sasaki [53]. In an APPLE-2 *elliptically polarizing undulator* or "EPU," there are four banks of magnets in Halbach arrays—two on top and two below (Fig. 2.21). The peak energy of the undulator output is changed by varying the vertical separation between the magnet assemblies, a so-called "gap scan", while the polarization is varied by changing the relative positions (phases) of adjacent rows of magnets—a "row scan." In the case of the EPU on ALS beamline 4.0.2, the polarization can be changed from left to right circular polarization in a few seconds, and the peak energy can be varied as quickly as the monochromator scans [54,55].

A final note—why are the frames for these IDs so huge? It turns out that when the gaps are closed down, the magnetic forces between the jaws are enormous—several thousand Newtons per meter of length. For a 3 meter insertion device, this translates to over 10,000 Newtons or over a ton of force. No wonder the overall assemblies can weigh ~9000 kg!

2.6 Suggested Exercises

1. What is the bend radius for a dipole magnet with a field of 2 Tesla for an electron beam with an energy of 2 GeV?
2. What is the approximate focal length of a quadrupole doublet where the field gradient for each quadrupole is 10 Tesla/m, the length of the magnets is 0.3 m, the separation between the magnets is 3 m, and the beam energy is 2 GeV?
3. Using approximate values from Fig. 2.10, calculate the minimum and maximum horizontal beam sizes for the Sirius storage ring. Assume the horizontal emittance is 0.28 nm rad.
4. The MAX-IV ring has a circumference of 528 m and runs with a microwave frequency of 99.931 MHz. Calculate the harmonic number h.
5. Consider a $Nd_2Fe_{14}B$ magnet PPM insertion device with rectangular blocks, an undulator period of 68.3 mm, and a minimum gap of 23.82 mm. If we assume that the magnet remanent field was 1.26 T, what would the maximum field be at the device center?
6. Compare a $SmCo_5$ hybrid magnet undulator with a remanent field $B_r = 0.9$ T with a $Nd_2Fe_{14}B$ hybrid with a remanent field $B_r = 1.3$ T. In both cases, assume high permeability (vanadium-permendur) alloy pole pieces. Calculate the fields at a 1.2 cm gap if the undulator magnetic period is 7.00 cm. (You will have to research the proper formula for the latter device.)
7. One sometimes reads that high-energy electrons are at the front of the bunch, because they have a higher velocity. Calculate the difference in transit times for electrons of energy 2.9 and 3.1 GeV, if they were to travel along a perfect circle with a circumference of 500 m.

2.7 Reference Books and Review Articles

1. *Particle Accelerator Physics—Basic Principles and Linear Beam Dynamics*; H. Wiedemann, Springer-Verlag: Heidelberg, fourth Edition, **2015**, ISBN 978-3-662-02903-9—in-depth treatment of accelerator and storage rings.

2. *Synchrotron Radiation Sources—A Primer*, H. Winick, ed. World Scientific, Singapore, **1994**, ISBN 978-9-810-218-560—more tutorial, with more practical details about operations.
3. *Synchrotron Radiation*; H. Wiedemann, Springer-Verlag: New York, **2003**, ISBN 978-3-540-433-927.
4. S. Krinsky, M. L. Perlman, and R. E. Watson, "Characteristics of Synchrotron Radiation and Its Sources", *Handbook of Synchrotron Radiation*, Vol. 1, Chapter 2, pp. 65-171, North-Holland, New York, **1983**, 0-444-86-709-0—prescient, and still useful.
5. *Undulators, Wigglers and Their Applications*, Hideo Onuki and Pascal Elleaume, eds., Taylor & Francis, New York, **2003**, ISBN 978-0415280402—good chapters on undulator radiation and technology.
6. *Synchrotron Radiation*, Philip Duke, Oxford University Press, New York, **2000**, ISBN 978-0198517580—good chapters on both synchrotron radiation and storage ring physics.
7. *Synchrotron Radiation News*, Vol. 31, No. 3, **2018**—an entire issue summarizing the state of the art for undulator technology.
8. *Synchrotron Light Sources and Free-Electron Lasers,* J. Jaeschke, S. Khan, J. R. Schneider, and J. B. Hastings, eds., Springer: Switzerland, 1st Edition, **2016**, ISBN 978-3-319-143-958—50 chapters! Excellent chapter by Bartolini about modern storage ring design.
9. *Synchrotron Radiation News*, Vol. 27, No. 6, **2014**—"*Ultra-Low-Emittance Sources*", an entire issue with news about developments about the next generation of storage rings .
10. *Synchrotron Radiation News*, Vol. 31, No. 3, **2018**—"*Undulators*", an entire issue with news about developments in undulator technology around the world.
11. *Synchrotron Radiation News*, Vol. 32, No. 3, **2019**—"*Facility Upgrades*", upgrade plans for PETRA-IV, FLASH2020+, and an MBA lattice for the ALS.

2.8 Machine Physics Software

A variety of software tools are available from the US Particle Accelerator School, including a program "BeamOptics" that will let you design your own storage ring! http://uspas.fnal.gov/materials/materials-downloads.shtml

Borland, M. (2000). Elegant: a Flexible SDDS-Compliant Code for Accelerator Simulation. Technical Report LS-287. Advanced Photon Source, Argonne, IL, USA.

Chapter 3
Synchrotron Radiation Fundamentals

3.1 Introduction

Synchrotron radiation has many useful properties, including *high-average power*, *narrow angular collimation*, and a *spectral range* that can include peak output across the X-ray region. These properties all derive from the fact that the particles are *relativistic*, traveling very close to the speed of light. For most users, a qualitative understanding of the terms and concepts in the next section will suffice.

Although there are no sharp divides among synchrotron sources, for practical purposes, we can distinguish between broadband sources, such as bend magnets and wigglers and more narrow band sources such as undulators and free-electron lasers (Fig. 3.1). In this chapter, we present a qualitative treatment of the broadband sources, followed (without derivation) by rigorous equations and graphs for the source properties. The same approach is then used for undulator radiation. We save free electron lasers for Chap. 12.

What do we really need to know? To understand synchrotron radiation, the most important fact is that the accelerated electrons are traveling close to the speed of light—they are "relativisitic." In fact, the particle speed is so close to the speed of light that the ratio of $v/c = \beta = 0.99999\ldots$ (typically 7 or 8 9's). Thus, a more useful measure of the relativistic nature of an electron is the Lorentz factor γ, which can be expressed in terms of the particle velocity and the speed of light, or as the ratio of its total energy E to its rest mass energy m_0c^2.

$$\gamma = \sqrt{1 - \frac{v^2}{c^2}} = \sqrt{1 - \beta^2} = \frac{E}{m_0c^2} = \frac{E[\text{GeV}]}{0.511[\text{MeV}]} = 1957E[\text{GeV}] \tag{3.1}$$

S. P. Cramer, *X-Ray Spectroscopy with Synchrotron Radiation*, Biological and Medical Physics, Biomedical Engineering, https://doi.org/10.1007/978-3-030-28551-7_3

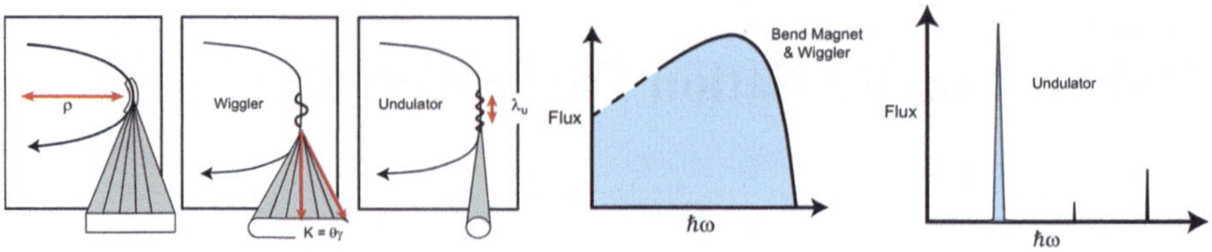

Fig. 3.1 Left: the three main types of storage ring synchrotron sources: bend magnets, wigglers, and undulators. Middle: typical bend magnet or wiggler spectrum. Right: typical undulator spectrum

3.2 Bend Magnet Radiation: A Qualitative Description

The bend magnet is the simplest synchrotron radiation source in a storage ring. For a couple of decades, they were the only synchrotron sources available, and even now, they are excellent sources for a variety of experiments. What makes bend magnets such good X-ray sources? As explained by Kim, it has to do with relativity [66, 67].

A well-known result from electromagnetic theory is that acceleration of charged particles results in electromagnetic radiation, with the electric field, $\vec{E}$, proportional to the observed acceleration, $\vec{a}$, thus $\vec{E} \propto \vec{a}$. In addition, the radiated power P per unit area is proportional to the so-called Poynting vector $\vec{S}$, which is proportional to the square of the electric field: E^2, and thus:

$$P \propto \vec{S} \propto E^2 \propto a^2 \tag{3.2}$$

For a simple dipole antenna, the observed acceleration is proportional to cos θ, where θ is the angle with respect to the plane perpendicular to the dipole axis (Fig. 3.2). Hence the radiated power varies as $\cos^2\theta$. Now consider a relativistic electron traveling through a magnet with velocity $v = \beta c$. An observer traveling with the same speed in a reference frame parallel to the electron sees a horizontal acceleration in which the electron travels back and forth through a distance of length ρ/γ^2 in a time interval $\Delta t' = 2\rho/\gamma c$ (Fig. 3.3). In this reference frame, the same donut-shaped pattern is expected as for the dipole antenna.

Now consider the reference frame of the synchrotron radiation user (Figs. 3.2 and 3.3). Special relativity tells us that time passes at different rates in these two frames. Because the particle is traveling with velocity βc, the apparent time interval Δt in the laboratory frame, the "observer time," is related to a time interval $\Delta t'$ in the particle frame, the "emitter time", by the equation:

$$\Delta t = \kappa \Delta t' = (1 - \beta \cos \theta)\Delta t' \tag{3.3}$$

Without too much math (see Exercise 3.1), one can show that:

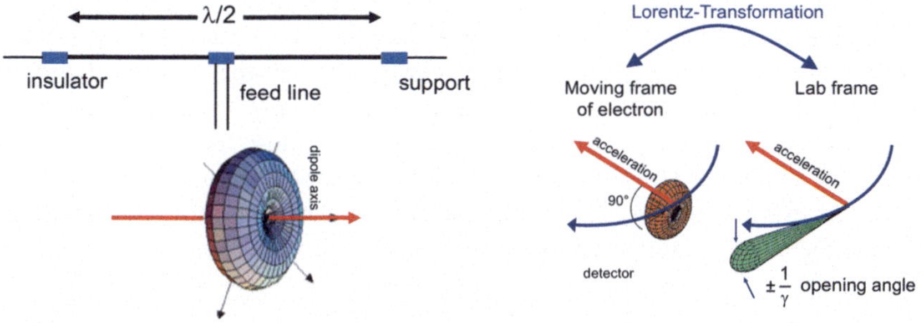

Fig. 3.2 Left: the radiation pattern from nonrelativistic horizontal acceleration of electrons, such as in a dipole antenna. Right: the effect of a Lorentz transformation on the radiation pattern

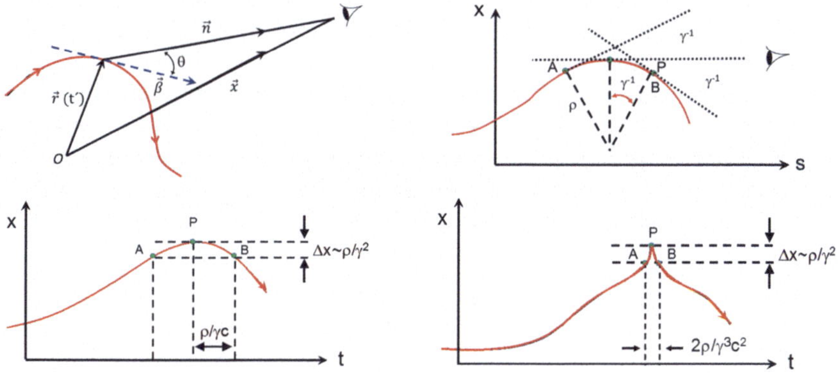

Fig. 3.3 Top left: coordinate system describing electron trajectory and observer. Top right: the arc over which time squeezing is significant. Lower left: apparent motion in emitter time. Lower right: apparent motion in observer time (redrawn from [67])

$$\kappa \cong \frac{1}{2}\left(\frac{1}{\gamma^2} + \theta^2\right); \ \theta \ll 1 \tag{3.4}$$

Thus, a dramatic contraction in Δt in the observer frame compared to $\Delta t'$ in the particle frame can occur when γ is large and θ is small, in other words at high particle energies and in the forward direction. For example, at a typical 3 GeV ring, $\gamma \cong 6000$ and for θ close to 0, $\kappa \cong 3.6 \times 10^6$! Furthermore, the apparent acceleration will depend on the square of the "time-squeezing factor" κ, via:

$$\frac{d^2x}{dt^2} \approx \frac{\Delta x}{(\Delta t^2)} \approx \frac{\Delta x}{(\kappa \Delta t')^2} \approx \frac{\Delta x}{\kappa^2 (\Delta t'^2)} \approx \gamma^4 \frac{c^2}{\rho} \tag{3.5}$$

The apparent acceleration and electric field in the laboratory frame are increased by γ^4, which is more than 10^{12} even for a modest 0.5 GeV storage ring! To estimate the amount of power radiated, we remember (Eq. 3.2) that the power radiated per unit area is proportional to the square of the observed acceleration (and hence γ^8!). However, for the average power over time and space, we lose factors of $1/\gamma^2$ for the reduced solid angle and $1/\gamma^2$ for the reduced time interval. Overall, for a fixed bend radius ρ, the average power of the emitted radiation is still enhanced by a very respectable factor of γ^4 compared to our non-relativistic case.

$$P \propto \gamma^4 \tag{3.6}$$

We can also use qualitative arguments to estimate the angular extent over which SR is observed and the average energy of the photons emitted. The angular opening is on the order $\theta \leq 1/\gamma$, because it is only for small angles that the time-squeezing factor κ is small. We thus get most of the radiation as the electron passes through an arc of $1/\gamma$, which takes an apparent time on the order of $\Delta t \sim 2\rho/\gamma^3 c$ (Fig. 3.3).

From Fourier transform theory, the average frequency in a pulse of duration Δt is $\omega_{\text{typ}} = 1/\Delta t$. If we use $\gamma = 6000$ for our generic 3 GeV storage ring and a bend radius of 10 m, this yields a typical angular frequency of $\omega_{\text{typ}} = 3.2 \times 10^{18}\,\text{s}^{-1}$, for a photon energy of 13 keV. This turns out to be a surprisingly good estimate for the peak in the synchrotron radiation spectrum. The key result is that the median photon energy scales as E^3 and inversely with the bend radius.

$$E_{\text{typ}} \propto \frac{\gamma^3 c}{\rho} \tag{3.7}$$

For starters, this is all you really need to know about source properties, but if you want to go deeper, read on.

3.3 Bend Magnet Radiation: The Details

As mentioned before, detailed calculation of the radiation from a bend magnet is not for the faint of heart. Here the key equations are presented for reference, without rigorous step-by-step derivation. For those who want more, the details are available in excellent articles [67,68], in classic works such as the paper by Schwinger [7] or the text by Jackson [69], as well as in more recent books [70].

Our coordinate system describes the observation position by horizontal angle ϕ and vertical angle ψ, as shown in Fig. 3.4. We also need to consider the polarization, which can have horizontal $\hat{\varepsilon}_\sigma$ and vertical $\hat{\varepsilon}_\pi$ components (Fig. 3.4).

We now compare some exact results with those derived qualitatively.

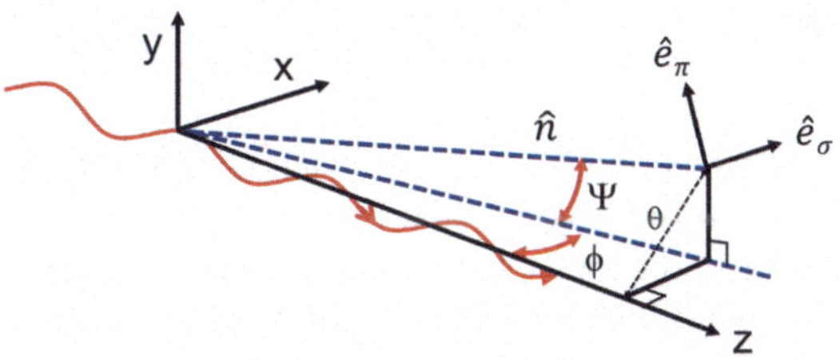

Fig. 3.4 A coordinate system for angles of observation of synchrotron radiation and electric field polarization components. For bend magnets, there is only a single arc (redrawn from [67])

3.3.1 Power Density and Power

In our qualitative derivation, we argued that the power radiated should scale with γ^4, which in terms of beam energy means E_e^4. This turns out to be correct. In practical units, the *angular power density* is given by:

$$\frac{d^2P}{d\phi^2}\left[\mathrm{W\,mrad}^{-2}\right] = 5.42E_e^4[\mathrm{GeV}]B[\mathrm{T}]I[\mathrm{A}] \tag{3.8}$$

As a reminder, the beam energy E_e and the strength of the magnetic field B are related to the bend radius ρ, via:

$$\rho[\mathrm{m}] = 3.336\frac{E_e[\mathrm{GeV}]}{B[\mathrm{T}]} \tag{3.9}$$

If we integrate over horizontal and vertical angles, we finally obtain, in practical units, a formula for the power emitted over an arc of length l:

$$P_{\mathrm{bm}}[\mathrm{kW}] = 1.27E_e^2[\mathrm{GeV}]B^2[\mathrm{T}]I[\mathrm{A}]l[\mathrm{m}] \tag{3.10}$$

3.3.2 Energy Loss Per Revolution

From the above equations, one can derive an interesting formula provided by Wiedemann for the energy loss per revolution for one electron [13]:

$$\Delta E[\mathrm{GeV}] = C_\gamma\frac{E_e^4[\mathrm{GeV}]}{\rho[\mathrm{m}]} \tag{3.11}$$

where:

$$C_\gamma = 8.8575 \times 10^{-5} \frac{\mathrm{m}}{\mathrm{GeV}^3} \tag{3.12}$$

If we apply this to a ring such as SPEAR3 with a bend radius of 7.86 m and an energy of 3 GeV, we find that the energy loss per turn is about 0.91 MeV or about 0.03% of the total electron energy. Since electrons are traveling around the ring more than a million times per second, they would quickly be lost from a stable orbit without energy replenishment in the rf cavities.

3.3.3 The Critical Energy: E_c

The energy distribution of power in the synchrotron radiation spectrum is defined by the critical energy E_c, which is the photon energy at which half of the power is below and half is above. Our qualitative derivation predicted that the typical photon energy would vary as γ^3/ρ. More precisely, the critical frequency is $\omega_c = 3\gamma^3 c/2\rho$, and the critical energy is given in Eq. 3.13. The position of E_c in a synchrotron spectrum was shown in Fig. 1.2.

$$E_c[\mathrm{keV}] = 0.665 E_e^2\left[\mathrm{GeV}^2\right] B[\mathrm{T}] = 2.22 \frac{E_e^3\left[\mathrm{GeV}^3\right]}{\rho[\mathrm{m}]} \tag{3.13}$$

3.3.4 The Bend Magnet Spectrum

The polarization and energy spectrum for a bend magnet depend on the observation angle. If we use the variable $X = \gamma\psi$ to capture the beam energy and observation angle (Fig. 3.4), then the overall result is that for a storage ring with a current I and bend radius ρ, the angle-dependent σ and π components of the photon flux $\mathcal{F}$ are given by (as derived in Kim[67]):

$$\begin{pmatrix} \dfrac{d^2\mathcal{F}_\sigma}{d^2\Omega} \\ \dfrac{d^2\mathcal{F}_\pi}{d^2\Omega} \end{pmatrix} = \frac{3\alpha}{4\pi^2}\gamma^2 \frac{\Delta\omega}{\omega}\frac{I}{e}\left(\frac{\omega}{\omega_c}\right)^2 \left(1+X^2\right)^2 \begin{pmatrix} K_{2/3}^2(\eta) \\ \dfrac{X^2}{1+X^2} K_{1/3}^2(\eta) \end{pmatrix} \tag{3.14}$$

where:

$$\eta = \left(\frac{\omega}{2\omega_c}\right)\left(1+X^2\right)^{3/2} \tag{3.15}$$

In the above, $K_{2/3}$ and $K_{1/3}$ are special functions known as *modified Bessel functions of the second kind*. These functions are illustrated in Appendix D, and their origin and properties are nicely explained by Duke [70].

If we add the two polarizations together, then the angular density of total photon flux at angular frequency ω and at observation angles θ and ψ, $d^2\mathcal{F}_{bm}/d\theta d\psi$, is given by:

$$\frac{d^2\mathcal{F}_{\mathrm{bm}}}{d\theta d\psi} = \frac{3\alpha}{4\pi^2}\gamma^2\frac{\Delta\omega}{\omega}\frac{I}{e}y^2\left(1+X^2\right)^2 \times \left(K_{2/3}^2(\xi) + \frac{X^2}{1+X^2}K_{1/3}^2(\xi)\right) \tag{3.16}$$

where the additional symbols involved are α = fine structure constant, e = electron charge = 1.602×10^{-19} Coulomb, ω_c (critical frequency) = $3\gamma^3/2\rho$, $y = \omega/\omega_c$, and $\xi = y(1 + X^2)^{3/2}/2$. In the forward direction, $\psi = 0$, hence there is only a σ component to the flux, and a formula in practical units [photons s^{-1} mr^{-2} (0.1%ΔE/E)$^{-1}$] is:

$$\left.\frac{d^2\mathcal{F}_{\mathrm{bm}}}{d^2\Omega}\right|_{\psi=0} = 1.33 \times 10^{13} E_{\mathrm{e}}^2[\mathrm{GeV}] I[\mathrm{A}] H_2(\omega/\omega_c) \tag{3.17}$$

where, as illustrated in Fig. 3.5:

$$H_2(y) = y^2 K_{2/3}^2(y/2) \tag{3.18}$$

In practical units, the critical photon energy $E_c = h\omega_c/2\pi$, is given by:

$$E_c[\mathrm{keV}] = 0.665 E_{\mathrm{e}}^2[\mathrm{GeV}]B[\mathrm{T}] = 2.218\frac{E_{\mathrm{e}}^3[\mathrm{GeV}]}{\rho[\mathrm{m}]} \tag{3.19}$$

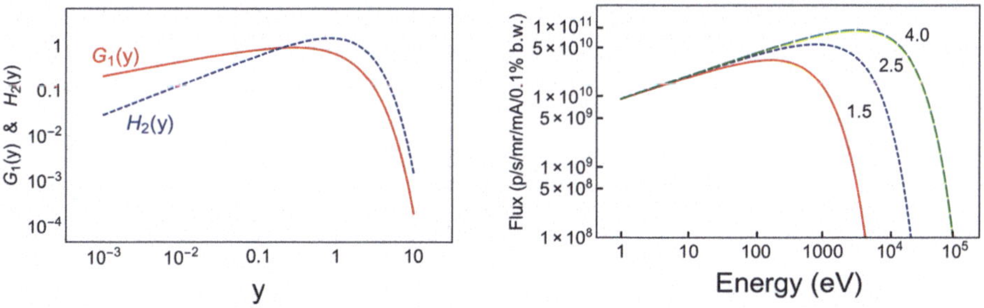

Fig. 3.5 Left: the functions $H_2(y)$ and $G_1(y)$, where $y = \omega/\omega_c = E/E_c$. Right: spectrum of the vertical angle integrated flux for a bend magnet source with a bend radius of 12.7 m and different electron beam energies in GeV

3.3.5 Bend Magnet Angular Density of Spectral Flux

Many experiments seek to use all the available photons, in which case the quantity of interest is the photon flux integrated over the vertical angle ψ. The final result is:

$$\frac{d\mathcal{F}_{\mathrm{bm}}}{d\theta} = \frac{\sqrt{3}}{2\pi}\alpha\gamma\frac{\Delta\omega}{\omega}\frac{I}{e}\frac{\omega}{\omega_c}\int_{\omega/\omega_c}^{\infty} K_{5/3}(y')dy' \tag{3.20}$$

or, in practical units (Fig. 3.5):

$$\frac{d\mathcal{F}_{\mathrm{bm}}}{d\theta} = 2.46\times10^{13}E_e[\mathrm{GeV}]I[\mathrm{A}]G_1(\omega/\omega_c) \tag{3.21}$$

where $G_1(\omega/\omega_c)$ (Fig. 3.5) is referred to as the *synchrotron radiation universal function*:

$$G_1\left(\frac{\omega}{\omega_c}\right) = \left(\frac{\omega}{\omega_c}\right)\int_{\omega/\omega_c}^{\infty} K_{5/3}(y')dy' \tag{3.22}$$

3.3.6 Angular Divergence

In our qualitative derivation, we saw that the apparent acceleration seen by an observer in the laboratory frame depends critically on the observation angle. The flux and spectral distribution will thus change with observation angle—higher energy photons will have a narrower opening angle than those at lower energy. The vertical angular divergence can be approximated by a Gaussian function, for which the rms deviation σ evaluated at $\Psi = 0$ is given by:

$$\sigma(\psi) = \sqrt{\frac{2\pi}{3}\frac{1}{\gamma}}\left(\frac{\omega_c}{\omega}\right)^{-1}\frac{\int_{\omega/\omega_c}^{\infty} K_{5/3}(y')dy'}{K_{2/3}^2\left(\frac{\omega}{2\omega_c}\right)} \tag{3.23}$$

The opening angle varies by about one order of magnitude for a 100-fold variation in photon energy (Fig. 3.6).

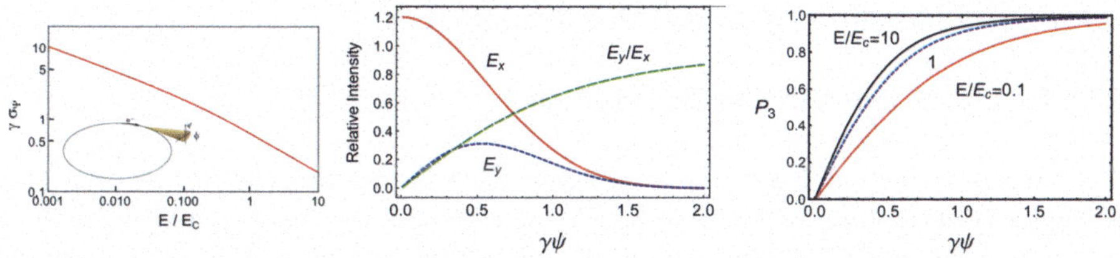

Fig. 3.6 Left: vertical divergence (σ) of bend magnet radiation (in units of $\gamma\sigma$) as a function of energy E with respect to the critical energy E_c. Middle: relative strengths of horizontal and vertical electric field components as a function of the observation angle from a bend magnet source. Right: degree of circular polarization of bend magnet radiation, P_3, as a function of the vertical angle $\gamma\psi$, for photon energies below ($E/E_c = 0.1$), equal to ($E/E_c = 1$), or greater than ($E/E_c = 10$) the critical energy

3.3.7 *Brightness of a Bend Magnet Source*

In order to evaluate the brightness of a bend magnet source, some assumptions need to be made about the electron beam source size and divergence (the emittance). In most calculations, people assume that the contribution of the curvature of the electron trajectory to the source size is negligible and that the angular spread in electron trajectories is small compared to the natural opening angle of synchrotron radiation, σ_ψ. One is thus left with three terms: the betatron oscillations, the dispersion for electrons with different energies, and finally σ_r, the contribution of diffraction to the apparent source size. The first two terms have been discussed in Chap. 2, and the last term is given by:

$$\sigma_r = \frac{\lambda}{4\pi\sigma_\psi} \tag{3.24}$$

The effective x and y source dimensions are then, as detailed by Hulbert and Weber, or Kim[75]:

$$\sum\nolimits_x = \left(\underbrace{\varepsilon_x\beta_x}_{\text{betatron}} + \underbrace{\eta_x^2\sigma_E^2}_{\text{dispersion}} + \underbrace{\sigma_r^2}_{\text{diffraction}}\right)^{1/2}$$

$$\sum\nolimits_y = \left(\varepsilon_y\beta_y + \sigma_r^2 + \frac{\varepsilon_y^2 + \varepsilon_y\gamma_y\sigma_r^2}{\sigma_\psi^2}\right)^{1/2} \tag{3.25}$$

Thus, if we use the angular density of flux from Eq. 3.16, then the in-plane brightness of a bend magnet source is given by:

$$B_{\text{bm}} = \frac{\frac{d^2F_{\text{bm}}}{d\theta d\psi} \mid \psi = 0}{2\pi\sum_x\sum_y} \tag{3.26}$$

3.3.8 *Polarization of Bend Magnet Sources*

3.3.8.1 Linear and Circular Polarization

A circularly polarized X-ray has oscillating electric and magnetic fields that are 90 degrees out of phase with each other. In the convention of Born and Wolf [2], the instantaneous electric field E_{rcp} for a right circularly polarized photon propagating in the z direction resembles a right-handed screw:

$$E_{\mathrm{rcp}} \equiv E_0 \left\{ \sin\left[\omega t - kz + \phi_0\right]\widehat{i} + \cos\left[\omega t - kz + \phi_0\right]\widehat{j} \right\} \qquad (3.27)$$

In this equation, $\omega = 2\pi\nu$ is the angular frequency, $k = 2\pi/\lambda$ is the wave number, where λ is the wavelength, ϕ_0 is an arbitrary phase shift, and $\widehat{i}$ and $\widehat{j}$ are unit vectors along the x and y axes, respectively. With the above definition, it turns out that left circularly polarized photons carry $+\hbar$ angular momentum. Although the Born and Wolf convention is standard for optics and chemistry literature, most physics literature uses the opposite definition, and one should check how the polarization is defined if the sign of a dichroism effect is to be meaningful. The papers of deGroot and Brouder generally use the Born and Wolf convention, while those of Thole, van der Laan, and Carra use the physics or "Feynman" definition [71]. The pitfalls of describing circular polarization are cogently described in Kliger et al. [72].

For an observer in the plane of the bend magnet trajectory, the observed particle acceleration is purely horizontal and perpendicular to the particle velocity. Therefore, the resulting electric field is also purely horizontal, and the radiation is linearly polarized along the x-axis. As the observer moves above or below the orbit plane, the observed acceleration acquires a vertical component, and the resulting radiation becomes elliptically polarized. The polarization asymptotically approaches pure right circular polarization at large angles above the storage ring (assuming a clockwise orbit when viewed from above) and left circular polarization at large angles below the ring.

To derive a quantitative expression for the polarization, we need to go back to the x- and y-components of the electric field. As described by Kim [67,73], from a bend magnet, these relative field amplitudes are of the form:

$$\begin{pmatrix} E_x \\ E_y \end{pmatrix} \propto \begin{pmatrix} K_{2/3}(\eta) \\ \dfrac{i\gamma\psi}{\sqrt{1 + (\gamma\psi)^2}} K_{1/3}(\eta) \end{pmatrix} \qquad (3.28)$$

where, as a reminder, $X = \gamma\psi$ and η was defined in Eq. 3.15.

The imaginary character of the E_y (vertical) field component implies that it is 90° out of phase with the E_x (horizontal) field component. This further implies that when

$\psi \neq 0$, the radiation is elliptically polarized, with the ratio of minor to major ellipse axes given by $r = E_y/iE_x$. Finally, in terms of r, the degrees of polarization are:

$$P_1 = \frac{1 - r^2}{1 + r^2}, \quad P_2 = 0, \quad P_3 = \frac{2r}{1 + r^2} \tag{3.29}$$

These results are summarized graphically in Fig. 3.6.

3.3.9 Superbends and Wavelength Shifters

These devices are primarily used to generate more synchrotron radiation at higher energies. There is no mystery here. The spectra are the same as bend magnet spectra, except that the critical energies are shifted to higher values because of the larger magnet field. In the case of wavelength shifters, there are two critical energies, one for the strong bend and another for the weaker bends that compensate for changes in beam trajectory.

3.4 Insertion Device Comparisons

The brightest synchrotron X-ray sources in the world are based on *insertion devices. Insertion devices* are magnetic structures "inserted" into the storage ring lattice for the production of synchrotron radiation. Again, why do we care? For the spectroscopist, compared to bend magnets, insertion devices can result in:

- *Higher-energy* X-rays.
- *Better collimated* X-rays.
- *Variable polarization* X-rays.

In short, insertion devices provide more of the X-rays needed for your experiment. With more X-rays, you can do better experiments (and sometimes even go home sooner). The hardware for various types of insertion device has already been discussed in Chap. 2. We saw that wigglers and undulators are both periodic arrays of magnets that create sources of synchrotron radiation. We now compare the source characteristics of wigglers, which produce a wide angular swath of radiation and a broadband spectrum, with undulators, which yield a more tightly focused beam and a spectrum with sharp peaks. These two basic types of insertion devices are compared (again) in Fig. 3.7.

In fact, there is no sharp dividing line between the two categories, and a given device can sometimes act as a wiggler and sometimes as an undulator. Both devices bend the electron trajectory back and forth many times, so that more radiation is produced, while the net deflection remains zero. To quantify how insertion devices make better sources, we consider a formula for the total synchrotron radiation power

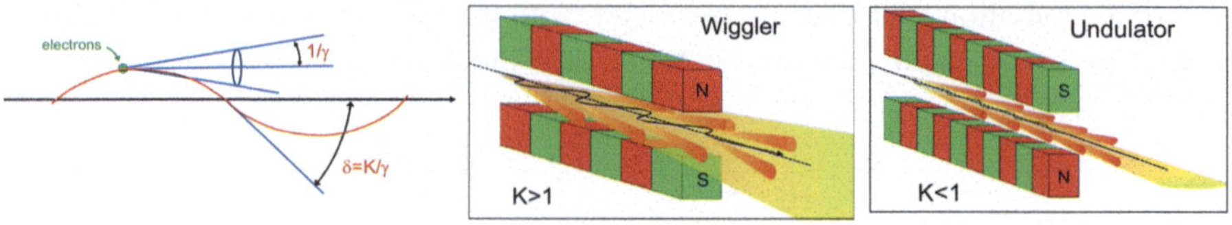

Fig. 3.7 Left: illustration of the deflection parameter *K*. Middle: a wiggler producing a relatively wide fan. For the wiggler, the angular deviations of the electron trajectory are greater than the natural opening angle of the synchrotron radiation. Right: superposition of cones from an undulator

radiated by a magnetic device of length L and field $B(s)$ in Tesla for a ring current of I amperes of electrons of energy E in GeV [74]:

$$P[\mathrm{kW}] = 1.265E_e^2[\mathrm{GeV}]I[\mathrm{A}]\int_0^L B^2(s)[\mathrm{T}]\,ds \tag{3.30}$$

The same equation (Eq. 3.30) applies to any magnetic structure—bend magnet, wiggler, or undulator. Since the ring energy is pretty much fixed, to generate more radiated power, there are only two options: (a) *increase the magnetic field* and (b) *increase the length over which the magnetic field operates*. If total power radiated were the best metric for the quality of a synchrotron source, then this formula would argue for long and very high field wigglers. However, for many spectroscopy experiments, spectral brightness is more important, and a third option (c) increasing the number of poles becomes more important. We will develop this further in the section on undulators.

3.4.1 K: The Deflection Parameter

The properties of undulators and wigglers lie on a continuum, characterized by the value of K, the "deflection parameter." This is the ratio of the angular excursion of the particle beam (δ) to the natural opening angle of the synchrotron radiation ($1/\gamma$). The maximum angular deflection δ depends on the beam energy E_e, the maximum field B, and the magnetic period λ_0, and in practical units, for $\rho = 3.335E(\mathrm{GeV})/B(T)$:

$$\delta = \lambda_0/(2\pi\,\rho_0) = \frac{\lambda_0 B[\mathrm{T}]}{20.95E_e[\mathrm{GeV}]} \tag{3.31}$$

$$K = \frac{\delta}{\gamma} = \frac{eB_0\lambda_0}{2\pi mc} = 0.934\lambda_0[cm]\,B_0[T] \tag{3.32}$$

If the deflection parameter $K \gg 1$, then the angular deflection δ is large compared to the natural opening angle of the synchrotron radiation ($1/\gamma$), and there are no significant interference effects between the different magnetic poles. A wiggler thus behaves like a series of bend magnet sources, except that the sinusoidal field means that the bends have variable radii. At the opposite extreme, in the small deflection or low K limit, there is interference between radiation from different poles, and the properties of undulators are very different from bend magnets and wigglers.

3.5 Wiggler Radiation

Wiggler magnets have been used to increase the flux at synchrotron radiation sources for more than 40 years. Because they deflect the electron beam first in one direction and then back in the opposite direction, they can employ stronger magnets than used for the bend dipoles, and they can produce this flux from dozens of deflections. The resulting source spectra resemble bend magnet dipole spectra with one significant difference: the apparent magnet strength and bend radius depend on the angle of observation.

3.5.1 *Wiggler Power*

The power generated by a wiggler can be obtained from Eq. 3.30 by replacing the integral over B^2 with $B_0{}^2/2$ and the arc l with the wiggler length L. The factor of ½ enters from averaging the square of the field over a sinusoidal period.

$$P\,[\mathrm{kW}] = 0.63\,E_{\mathrm{e}}^{2}[\mathrm{GeV}]\;I[\mathrm{A}]\;B_{0}^{2}[\mathrm{T}]\,L[\mathrm{m}] \tag{3.33}$$

3.5.2 *Wiggler Spectrum and Spectral Brightness*

The spectrum from a wiggler insertion device can be described as the incoherent sum of N bend magnet spectra. Because of the sinusoidal variation in magnetic field, the critical energy varies with observation angle (actually, ϕ in Fig. 3.4), as given by Eq. 3.34) (using $E_{c\text{-}max}$[keV] = 0.665 E_e^2 [GeV] B_0 *[T]*) (Fig. 3.8):

$$E_c(\phi)\,[keV] = E_{c-max}[keV]\sqrt{1 - (\phi/\delta)^2} \tag{3.34}$$

The polarization of radiation from a wiggler is less straightforward [75]. In the plane of the storage ring, wigglers produce horizontally polarized radiation. Out of the plane, the polarization of wiggler radiation remains linear, because of the

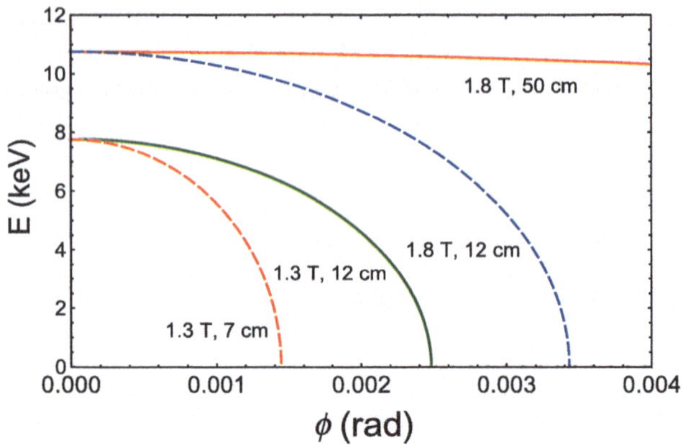

Fig. 3.8 The critical energy for a wiggler source on a 3 GeV ring, as a function of observation angle ϕ, for different values of magnetic field B in Tesla and magnet period λ_0 in cm

interference between radiation from pairs of poles with particles traveling in opposite directions. However, the direction of this polarization varies rapidly with angle, so that when averaging over finite angles, the degree of polarization, becomes, as given by Kim:

$$P_{\text{pol}} = \frac{(1+X^2)K_{2/3}^2(\eta) - X^2K_{1/3}^2(\eta)}{(1+X^2)K_{2/3}^2(\eta) + X^2K_{1/3}^2(\eta)} \tag{3.35}$$

where $X = \gamma\psi$ and $\eta = \frac{1}{2}\frac{\omega}{\omega_C}\left(1+X^2\right)^{3/2}$ [75]. In practice, the polarized nature of off-axis wiggler radiation is never really used.

3.6 Undulator Radiation: A Qualitative Approach

When is an insertion device an undulator instead of a wiggler? We saw that if the field is high and the period is long, the angular deflection of the beam is relatively large, K is »1, and the device is called a *wiggler*. When $K \leq 1$, δ is relatively small, and the device is called an *undulator*. Although a wiggler might radiate more power, that radiation comes from a large source and is spread out over a range of angles and energies. In undulator mode, an insertion device puts most of its radiation in a small range of angles and energies.

For many applications, brightness or spectral brightness is a better metric for the quality of a synchrotron source. For a given amount of power, we can improve the brightness by (1) decreasing the source size and (2) decreasing the beam divergence. Furthermore, we can improve the spectral brightness by (3) putting more of the source spectrum into a narrower bandpass. We will see below how an undulator accomplishes all three of these goals.

We start with a qualitative analysis of planar undulator radiation. We then provide the exact equations without detailed derivations, which are well documented in the articles and books cited at the end of this chapter. After planar undulators, the *elliptical undulator* is treated as a flexible device that can produce variable linear or circular polarization. Finally, we briefly mention the *helical undulator* as a special case for circular polarization.

In an undulator, the output exhibits interference effects between the radiation from different pole regions. This is because the natural opening angle of synchrotron radiation is greater than the angular deflection of the electron trajectory, and an observer cannot distinguish which pole is the source of the radiation. As with a diffraction grating, in order to calculate the final intensity, one has to *add the fields* from different sources to compute the total field and then square to get the intensity.

The basic spectrum of a planar undulator is actually easier to understand than that of a bend magnet. Suppose one is traveling in a reference frame along with the relativistic electron as it encounters the undulator structure. In the limit of very small electron horizontal excursions, the observer sees an electron oscillating back and forth with the spatial period of the undulator, λ_u, but shortened by the relativistic length contraction to $\lambda_1 = \lambda_u/\gamma$. In this reference frame, the observer sees a conventional dipole pattern (Fig. 3.9).

For an observer in the storage ring rest frame, this radiation has an additional wavelength contraction from the relativistic Doppler shift: $\gamma(1 - \beta\cos\theta)$. An approximate formula for the observed wavelength in the forward direction is thus (using $1-\beta \cong 1/2\gamma^2$):

$$\lambda = \gamma(1 - \beta\cos\theta)\lambda_1 = \gamma(1 - \beta\cos\theta)\frac{\lambda_u}{\gamma} \cong \frac{\lambda_u}{2\gamma^2} \tag{3.36}$$

where $\lambda = \lambda_u/2\gamma^2$ is the wavelength observed in the forward direction in the limit of very small excursions. Since γ is already a large number, on the order of 10^3 to 10^4, a wavelength reduction by $2\gamma^2$ has a profound effect. As an example, at the ALS, the

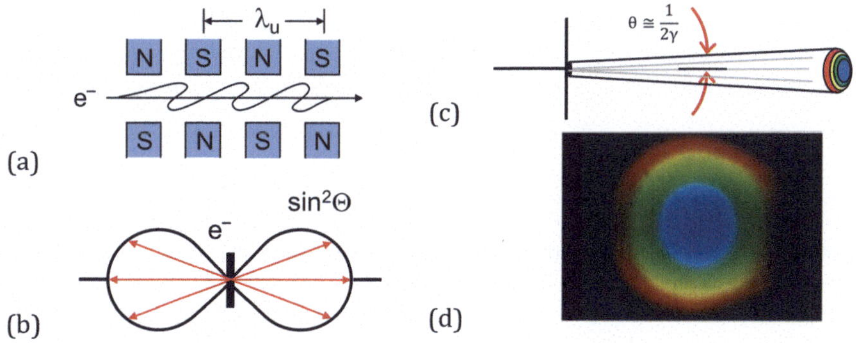

Fig. 3.9 Left: (**a**) the approximately sinusoidal magnetic field and the resulting sinusoidal electron trajectory and (**b**) the dipole radiation pattern observed in the traveling reference frame. Right: (**c**) the pencil-like radiation pattern observed in the laboratory frame and (**d**) the visible radiation pattern from an undulator on a low-energy ring. Notice how the photon energy decreases as the observation angle increases [76]

beamline 4 undulator period is 5 cm, and with the ring operating at 1.9 GeV, $\gamma = 3718$. Hence the output wavelength is as short as 18 Å or an energy of 689 eV!

3.7 Planar Undulator Radiation: More Exact Formulae

3.7.1 The Undulator Fundamental Wavelength

We already have Eq. 3.36 as an approximate expression for the wavelength of undulator radiation. Without too much extra work, we can get a more accurate formula. Remember that the Lorentz force from the magnetic field is always perpendicular to the electron velocity. Initially, the electron motion is completely in the forward direction, along the z-axis; hence the Lorentz force is horizontal, along the x-axis. This horizontal component to the electron motion in turn results in a front-to-back Lorentz force along the z-axis. Note that the energy of an electron is not changed by a magnetic device—γ remains constant. The speed of the electron remains βc, but the β_x and β_z components vary, so that β_z oscillates around an average velocity $(\beta_z c)^2 = (\beta c)^2 - (\beta_x c)^2$ where:

$$\frac{dx}{dt} = v_x = \beta_x c \cong \beta c \frac{K}{\lambda} \cos\left(\frac{2\pi z}{\lambda_0}\right). \tag{3.37}$$

After some tedious algebra, the effective average β for the z-direction, $\beta_{z,\mathrm{av}}$, is given by:

$$\beta_{z,\mathrm{av}} = \beta\left(1 - \frac{K^2}{4\gamma^2}\right)C \tag{3.38}$$

If we use this expression for the effective $\beta_{z,\mathrm{av}}$ in the previous derivation for the undulator wavelength (Eq. 3.36), then expanding and throwing away any small terms, we have:

$$\lambda \approx \frac{\lambda_u}{2\gamma^2}\left(1 + \frac{K^2}{2} + \gamma^2\theta^2\right) \tag{3.39}$$

The above equation is known as the *undulator equation*. Notice that K can be controlled (up to a point) by varying the magnetic field, since K varies linearly with B_0 (Eq. 3.32). An undulator is therefore a tunable source of X-rays—a spectroscopist's dream come true!

Although one might expect a stronger magnetic field to produce a higher-energy peak in the spectrum, in the undulator regime, the result is just the opposite—the fundamental energy decreases with K. From the previous discussion, the reason should be clear—as K increases, the effective average β for the z-direction decreases.

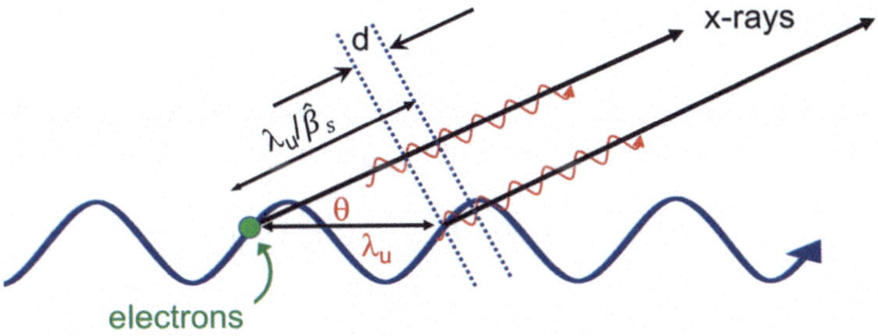

Fig. 3.10 Conditions for constructive interference from successive poles in an undulator. The angle θ is exaggerated for clarity; in undulator mode, the cones from different magnets overlap

In terms of energy or wavelength in practical units, the first harmonic from an undulator is given by:

$$E[\mathrm{keV}] = 0.950 \frac{E_{\mathrm{e}}^2[\mathrm{GeV}]}{(1 + K^2/2 + \gamma^2\theta^2)\lambda_{\mathrm{u}}[\mathrm{cm}]} \tag{3.40}$$

$$\lambda[\mathring{A}] = 13.06 \frac{\lambda_u[\mathrm{cm}](1 + K^2/2 + \gamma^2\theta^2)}{E_{\mathrm{e}}^2[\mathrm{GeV}]} \tag{3.41}$$

Another way to derive the radiation from an undulator is to consider the conditions for constructive interference from successive poles. As seen in Fig. 3.10, the time for an electron to travel through one period of the undulator is $\lambda_{\mathrm{u}}/c\beta_{z,\mathrm{av}}$, while during this time, the radiation wavefront will move by $\lambda_{\mathrm{u}}/\beta_{z,\mathrm{av}}$. Finally, the separation between wavefronts is:

$$d = \frac{\lambda_{\mathrm{u}}}{\beta_{z,\mathrm{av}}} - \lambda_{\mathrm{u}} \cos\theta \tag{3.42}$$

and using

$$\beta_{z,\mathrm{av}} = \beta* = \beta\left(1 - \frac{K^2}{4\gamma^2}\right) = 1 - \frac{1}{2\gamma^2}\left(1 + \frac{K^2}{2}\right) \tag{3.43}$$

one finally gets an expression for the radiation wavelength

$$\lambda = \frac{\lambda_u\theta^2}{2} + \frac{\lambda_u}{2\gamma^2} + \frac{\lambda_u K^2}{2\gamma^2} = \frac{\lambda_u}{2\gamma^2}\left(1 + \frac{K^2}{2} + \theta^2\gamma^2\right) \tag{3.44}$$

which is the same undulator equation (Eq. 3.39).

3.7.2 Integrated Power

Our general expression for the power from an insertion device was Eq. 3.30. In a planar undulator, integrating the square of the sinusoidal field leads to:

$$P[\mathrm{kW}] = 0.63E_{\mathrm{e}}^{2}[\mathrm{GeV}]I[\mathrm{A}]B_{0}^{2}[\mathrm{T}]L[\mathrm{m}] \tag{3.45}$$

3.7.3 Harmonics

For very small K-values and very close to on-axis observation, only the fundamental energy is observed. As K increases, higher harmonics begin to appear. If observed strictly on-axis, symmetry dictates that only odd harmonics have significant intensity. (For even harmonics, the radiation from one pole is out of phase with that from adjacent poles.) The intensities of the odd harmonics as a function of K are proportional to what Walker calls the "on-axis angular flux density function" $F_n(K)$, as illustrated in Fig. 3.11:

$$F_n(K) = \frac{n^2K^2}{\left(1+K^2/2\right)^2}\left[J_{(n-1)/2}\left(\frac{nK^2}{4\left(1+K^2/2\right)}\right) - J_{(n+1)/2}\left(\frac{nK^2}{4\left(1+K^2/2\right)}\right)\right]^2 \tag{3.46}$$

for n odd, and 0 for n even. As a reminder, the expressions J_x refer to "Bessel functions of the first kind," and these are illustrated in Appendix D.

For a real-world experiment, an important metric is the total flux for the nth harmonic, $\mathscr{F}^n$, available in the central cone of undulator radiation. The "undulator flux function," "$Q_n(K)$," takes this into account and provides a better estimate of usable flux (Fig. 3.11). The functions $F_n(K)$ and $Q_n(K)$ are connected with practical values for density of flux in Eqs. 3.54 and 3.55.

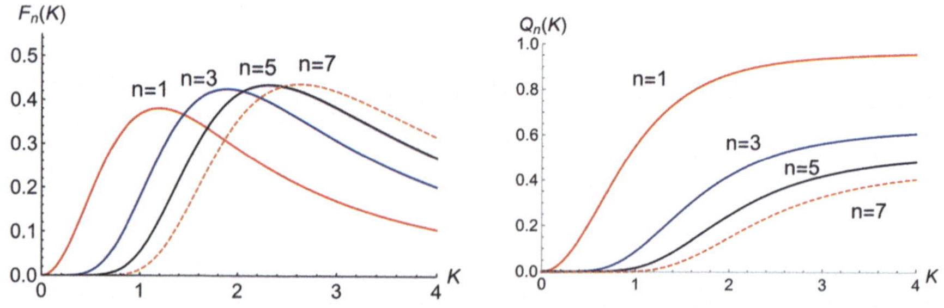

Fig. 3.11 Left: "on-axis angular flux density function" $F_n(K)$. Right: "undulator flux function" $Q_n(K)$

$$\mathscr{F}^n \sim Q_n(K) \qquad Q_n(K) = \left(1 + K^2/2\right)F_n(K)/n \tag{3.47}$$

The same logic that led to a reduced average β in the forward direction implies that there will be some z-motion at $2x$ the frequency of horizontal motion. The factor of 2 comes from the fact that a complete front-to-back oscillation occurs for each center-to-side horizontal excursion. Of course, the variable z-motion will generate an additional variable Lorentz force in the x-direction, which in turn leads to modified x-motion, a new z-motion *ad infinitum*. The situation is properly described by a pair of coupled differential equations that can only be solved numerically:

$$\begin{aligned} \frac{d^2x}{dt^2} &= \frac{dz}{dt}\frac{e}{m\gamma}B_y(z) \\ \frac{d^2z}{dt^2} &= \frac{dx}{dt}\frac{e}{m\gamma}B_y(z) \end{aligned} \tag{3.48}$$

If one imagines traveling along with the electron beam at average β, then the particle bunch oscillates around the mean position in a figure-8-like trajectory (Fig. 3.12). Note that in the laboratory frame, the z-motion is only directly observable at off-axis angles.

From the argument above, it seems that the second harmonic resulting from the front-to-back oscillation of the electron beam will only be observable out of the plane of the electron orbit. However, this is only true for a hypothetical particle beam with no divergence. In practice, the particles that are traveling slightly off-axis will contribute to second harmonic radiation along the undulator axis.

How much actual motion occurs in a typical undulator? For a typical undulator with a 5 cm period and at $K = 2$ on a 2 GeV ring, the x oscillation amplitude turns out to be 4 μ. Despite these small excursions, such a device with 50 poles will produce kW of X-ray power!

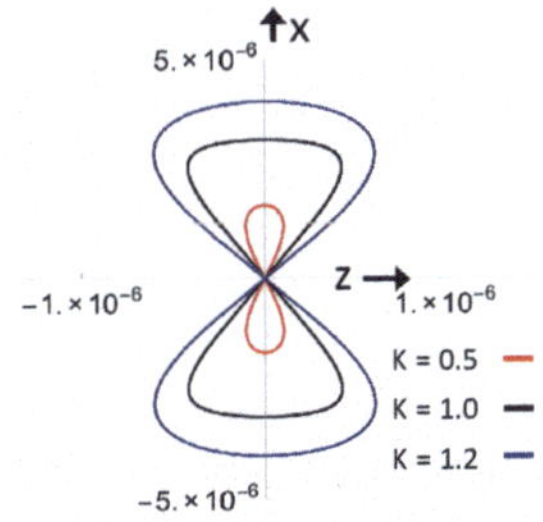

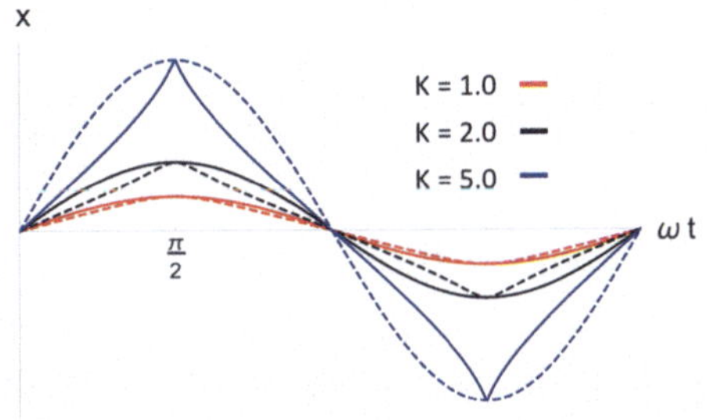

Fig. 3.12 Left: motion of a particle relative to coordinate system moving with the average velocity, illustrated for $K = 0.5$, 1.0, and 1.2. Notice that z-amplitude scale is different to enhance visibility in this direction. Right: observed motion in moving time frame (---) vs. observer time frame (—)

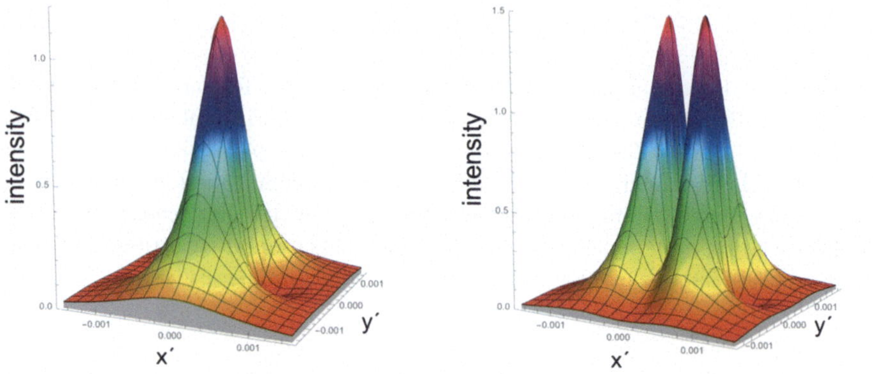

Fig. 3.13 Left: distribution of the first harmonic flux in the $x - y$ plane. Right: distribution of the second harmonic flux in the $x' - y'$ plane. There is a node along $x = 0$ [77]

Walker has provided a nice expression for the relative amplitude motions in the x' and z' directions of the frame traveling with the average electron speed βc:

$$\frac{z' \text{amplitude}}{x' \text{amplitude}} = \frac{K}{8\left(1 + K^2/2\right)^{1/2}} \tag{3.49}$$

3.7.4 Angular Properties of Undulator Radiation

The angular excursions of the electron beam in an undulator are small. In addition, the observed relativistic effects and interference effects depend critically on the observation angles in the horizontal and vertical planes. Together, these result in radiation patterns that are narrow cones, as opposed to the fan-like patterns from bend magnets and wigglers (Fig. 3.13).

3.7.5 Spectral Bandwidth

Our expressions for fundamental wavelength and harmonics are correct, but the actual spectra are broadened by the fact that the undulator has a finite number of magnetic periods N, usually on the order of -100–200. Thus, the sinusoidal electric field output of the device is truncated in time, and it can be viewed as the product of a sinusoid with a "box" function. There is an important theorem in Fourier transform mathematics that says that the Fourier transform of a *product* is the *convolution* of the Fourier transforms of the functions being multiplied. The result is that the shape

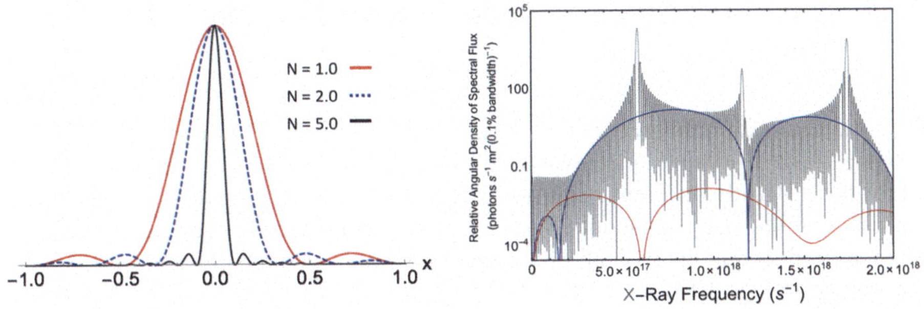

Fig. 3.14 Left: the lineshape function $S_N(x)$, scaled to constant peak height. Right: a theoretical undulator spectrum at a single observation angle. The undulator parameters were $\lambda_u = 3.65$ cm, $N = 50$, and $K = 1$. Electron energy was $E_e = 1.5$ GeV and $I = 0.4$ A. Observation angles were $\phi = 2.1 \times 10^{-5}$ and $\theta = 4.0 \times 10^{-5}$ [67]. Such a spectrum is never observed because of angular averaging due to slit size and finite electron emittance.

of the spectrum is not a δ function, but instead the square of a *sin* function. This is referred to as the "lineshape function" S_N (Fig. 3.14).

$$S_N\left(\frac{\Delta\omega}{\omega_1}\right) = \left(\frac{\sin N\pi\,\Delta\omega/\omega_1}{N\pi\,\Delta\omega/\omega_1}\right)^2 \tag{3.50}$$

The spectral bandwidth $\Delta\lambda_n/\lambda$ thus depends on the number of periods N and the order of the harmonic n, via:

$$\Delta\lambda/\lambda_n \cong 1/nN \quad \Delta\omega/\omega_n \cong 1/nN \tag{3.51}$$

3.7.6 Spectrum at an Arbitrary Angle

The intensity from an undulator at frequency ω ultimately depends on the observation angle, defined by φ and ψ in Fig. 3.4. In theory, at one particular angle, the spectrum of radiation from a planar undulator is actually quite complex, as shown in Fig. 3.14. Kim notes that the functions are "easily evaluated numerically" [67], and here we present the results without derivation. The observation angle has a large effect on the frequency and polarization of the observed radiation. If one approximates the distribution in angle by a Gaussian with standard deviation σ_Θ, then for the nth harmonic:

$$\sigma_\theta \cong \frac{1}{2\gamma}\sqrt{\frac{1+\frac{K^2}{2}}{nN}} = \sqrt{\frac{\lambda_n}{2L}} \tag{3.52}$$

The opening angle of radiation from an undulator is reduced from that of a bend magnet by a factor $\approx 1/\sqrt{nN}$. Since the number of periods can be greater than 100, this is a big deal!

3.7.7 Combined Effects of Finite N, Emittance, and Solid Angle

In practice, one never observes undulator spectra like Fig. 3.14. The sharp features become blurred because one invariably integrates over a cone of angles, each of which has a slightly different spectrum. The features are also broadened by the finite emittance of the electron beam, which effectively contributes to angular averaging. The net result is that the typically observed spectrum from an undulator appears like that shown in Fig. 3.15.

For many spectroscopy experiments, one needs a variety of different photon energies. As we mentioned earlier, by tuning the magnetic field, one changes K and the resulting maximum in the output spectrum. For a fixed ring energy, each undulator will have a range of energies that can be reached, but the flux will fall off steeply at the highest energy as $K \rightarrow 0$.

At the nth harmonic, the on-axis peak angular density of flux is given by:

$$\left.\frac{d^2\mathscr{F}^n}{d^2\Omega}\right|_{\phi,\psi=0} = \alpha N^2\gamma^2\frac{\Delta\omega}{\omega} \times \frac{I}{e} \times F_n(K) \tag{3.53}$$

In practical units of photons s^{-1} mrad^{-2} (0.1% bandwidth)$^{-1}$, this yields:

$$\left.\frac{d^2\mathscr{F}^n_{\mathrm{und}}}{d^2\Omega}\right|_{\phi,\psi=0} = 1.74 \times 10^{14} N^2 E_{\mathrm{e}}^2[\mathrm{GeV}] I[\mathrm{A}] F_n(\mathrm{K}) \tag{3.54}$$

A set of envelopes for the on-axis angular density of flux is shown in Fig. 3.15.

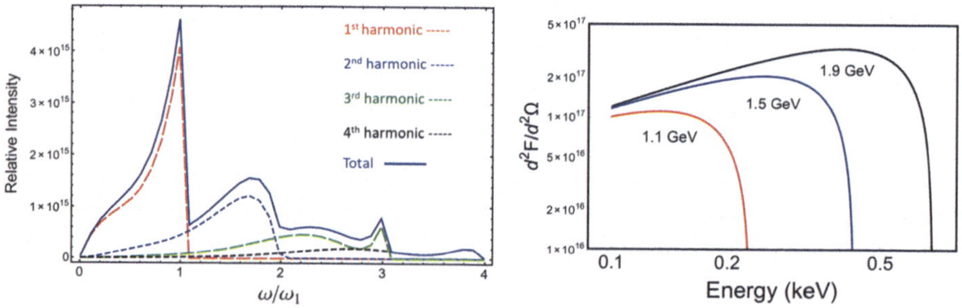

Fig. 3.15 Left: calculated angle-integrated ALS undulator spectrum for $K = 1$. Parameters: beam energy 1.5 GeV, number of periods 134, undulator period 3.65 cm, current 400 mA. Right: on-axis angular density of flux envelopes for an ALS 5 cm undulator at different electron beam energies

3.7.8 Integrated Flux and Spectral Brightness Envelopes

Of course, in a real-world experiment, one has to integrate the undulator radiation over a finite solid angle to obtain the maximum practical flux F of photons. To obtain this more realistic estimate of the flux, we use an approximate formula for the odd harmonic flux in the central cone. In practical units of photons s^{-1} (0.1% bandwidth)$^{-1}$, the integrated flux is at the nth harmonic:

$$\mathscr{F}^{n}_{\text{und}}\Big|_{\phi,\psi=0} = ½1.431 \times 10^{14} N\, I[\text{A}] Q_n(\text{K}) \qquad (3.55)$$

where $Q_n(K)$ was defined in Eq. 3.47.

Now that we have the undulator flux, we can obtain the practical brightness by dividing by the source area and divergence. We also need to consider contributions from the electron beam properties and possibly from diffraction effects. In deriving the bend magnet brightness, we ignored the angular divergence of the electron beam, because of the significant contribution from the natural opening angle $1/\gamma$. Since the natural divergence of undulator radiation is potentially very small, we now have to consider both the effective source sizes in the x and y dimensions, Σ_x and Σ_y, and the effective angular divergences, Σ'_x and Σ'_y. The various contributions yield the following expression for the on-axis brightness:

$$\mathscr{B}_{\text{und}} = \frac{\mathscr{F}_{\text{und}}}{(2\pi)^2 \sum_x \sum_y \sum'_x \sum'_y} \qquad (3.56)$$

A selection of spectral brightness envelopes is shown in Fig. 3.16. For the past several decades, the brightness of synchrotron sources has been steadily improved by reducing terms in the denominator of Eq. 3.56, i.e., by decreasing the electron beam emittance. However, source designers are close to achieving emittances so small that the wave nature of light itself comes into play, yielding the long sought diffraction-limited storage rings.

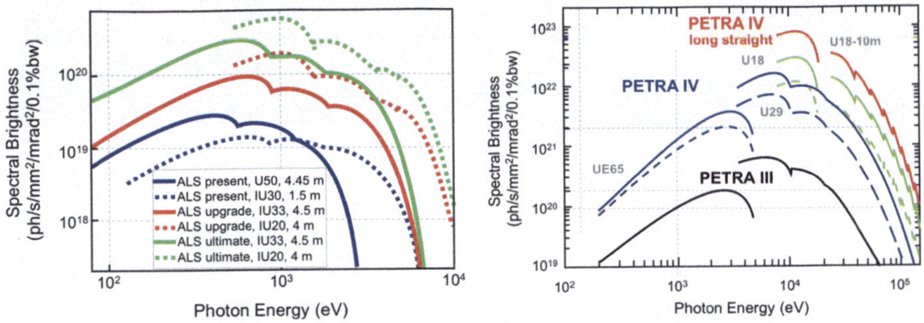

Fig. 3.16 Spectral brightness envelopes for storage rings now and after pending source upgrades. Left: for a small low-energy ring, such as ALS at 2 GeV. Right: for a large high-energy ring, PETRA-III/IV at 6 GeV

3.7.9 Polarization of a Planar Undulator

The polarization of radiation from a planar undulator is always linear, and it is obviously horizontal for the odd harmonics in the plane of the device. As with planar wigglers, the angle of polarization changes in a complicated fashion with the azimuthal angle. Since the even harmonics arise from z-components of the acceleration, it should not be too surprising that these components are vertically polarized on the y-axis. The complex interplay of undulator harmonic, observation angle, and polarization is illustrated nicely (Fig. 3.17) in the calculations reported by Elleaume and the photographs taken by Sasaki and coworkers [78].

3.8 Elliptical Undulator Properties

There are quite a few spectroscopy measurements that require variable polarization of the X-ray beam, not just the horizontal and vertical polarization seen in Fig. 3.17 but also left and right circular polarization. Scientists have invented a variety of undulators to manipulate the energy and polarization of the resulting synchrotron

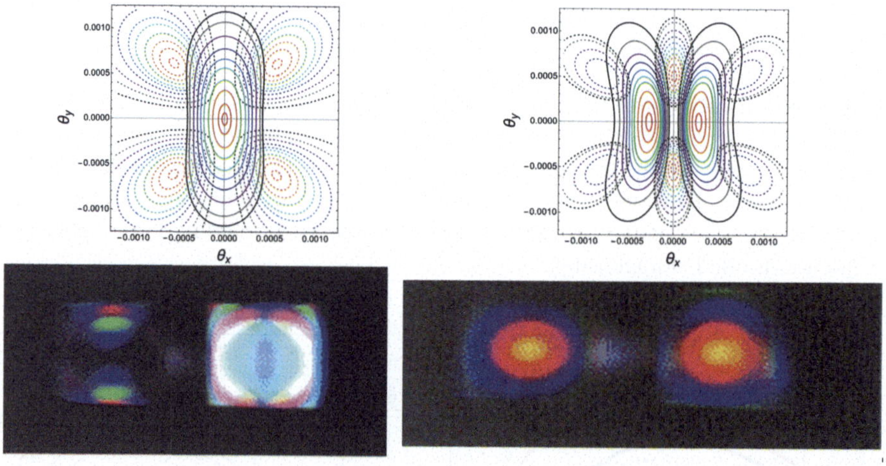

Fig. 3.17 Top: the distribution of horizontal (solid line) and vertical (dashed line) polarization from a planar undulator as a function of observation angles ($\gamma\psi_x$ and $\gamma\psi_y$) as calculated by Elleaume for first (left) and second harmonic (right) [79]. Contours are drawn (approximately) for every 10% reduction in intensity. Bottom: variable polarization from an elliptical undulator reported by Sasaki and coworkers [78], using the low-energy JSR ring operating at 138 MeV, which puts the radiation into the visible region of the spectrum. A Wollaston prism was used to select vertical and horizontal polarization components (left and right, respectively). Left: the undulator was run in horizontal polarization mode; right: undulator was run in elliptical polarization mode

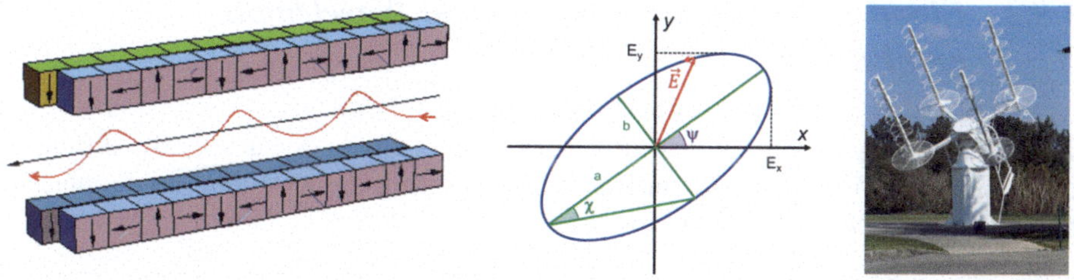

Fig. 3.18 Left: the electron trajectory as it passes through an elliptical undulator. Middle: a polarization ellipse. For a fixed point in space, the direction and amplitude of the electric field follow the ellipse over time. For linear polarization, the ellipse collapses to a straight line, while for circular polarization, the ellipse becomes a circle. For details refer to Appendix E.2. Right: an array of helical antennas for producing circular polarization at radio frequencies

radiation. The mechanical implementation of these schemes, including elliptical undulators, helical undulators, crossed undulators, and more exotic solutions, was described in Chap. 2 [80].

One of the most flexible insertion devices for providing variable polarization is the *elliptical undulator*, also called an elliptically polarizing undulator ('EPU'). It has a pair of perpendicular sinusoidal magnetic fields, which have the same periodicity but arbitrary relative phases and amplitudes, leading to a net on-axis magnetic field $\widehat{B}(s)$ (Eq. 3.57):

$$\vec{B}(s) = \widehat{B}_{x0} \sin\left(\frac{2\pi s}{\lambda_u} - \phi\right) + \widehat{B}_{y0} \sin\left(\frac{2\pi s}{\lambda_u}\right) \tag{3.57}$$

Since the elliptical undulator has two field components, it will have two K-values, with the horizontal K_x controlled by the strength of the vertical field and the vertical K_y controlled by the strength of the horizontal field (Fig. 3.18):

$$K_x = 0.934\widehat{B}_{y0}[\mathrm{T}]\lambda_u[\mathrm{cm}] \quad K_y = 0.934\widehat{B}_{x0}[\mathrm{T}]\lambda_u[\mathrm{cm}] \tag{3.58}$$

3.8.1 Elliptical Undulator Power

The equation for total power is a slight modification of that for a planar undulator, taking into account the two different fields.

$$P_{\text{e-und}}[\mathrm{kW}] = 0.633E_e^2[\mathrm{GeV}]\left(\widehat{B}_{x0}^2 + \widehat{B}_{y0}^2\right)[\mathrm{T}]I[\mathrm{A}]L(\mathrm{m}) \tag{3.59}$$

3.8.2 Elliptical Undulator Fundamental Wavelength and Energy

The on-axis wavelength of the first harmonic is given by a slightly different expression than for a planar undulator (Eq. 3.41), with the $K^2/2$ term now replaced by individual $K_x^2/2$ and $K_y^2/2$ terms [81]:

$$\lambda[\AA] = 13.06\,\frac{\lambda_u[\mathrm{cm}]}{E_e^2[\mathrm{GeV}]}\left(1+\frac{K_x^2}{2}+\frac{K_y^2}{2}\right) \tag{3.60}$$

Including the angular dependence, in practical units, the wavelength is given by:

$$\lambda[\AA] = 13.06\,\frac{\lambda_u[\mathrm{cm}]\left(1+\frac{K_x^2}{2}+\frac{K_y^2}{2}+\gamma^2\theta^2\right)}{E_e^2[\mathrm{GeV}]} \tag{3.61}$$

and in practical units, the energy is given by:

$$E_1[\mathrm{keV}] = 0.950\,\frac{E_e^2[\mathrm{GeV}]}{\lambda_u[\mathrm{cm}]\left(1+K_x^2/2+K_y^2/2+\gamma^2\theta^2\right)} \tag{3.62}$$

Notice that a planar undulator is a special case of an elliptical undulator, with $K_y = 0$.

3.8.3 Elliptical Undulator Polarization

The polarization of radiation from an elliptical undulator depends on the relative strengths and phases of the radiation horizontal and vertical electric fields $\widehat{E}_x$ and $\widehat{E}_y$, which in turn depend on the relative strengths and phases of the undulator vertical and horizontal magetic fields $\widehat{B}_x$ and $\widehat{B}_y$ as an electron spirals through the device. A common quantitative way to describe the polarization of a light source is the use of '*Stokes parameters*' (see Appendix A.4). Briefly, the polarization rates P_1 and P_2 together define the linear erect and linear skew polarizations, while P_3 refers to the amount of circular polarization. These polarization rates are given by [77]:

$$\begin{aligned} P_1 &= \frac{\widehat{E}_{x0}^2-\widehat{E}_{y0}^2}{\widehat{E}_{x0}^2+\widehat{E}_{y0}^2} = \frac{\widehat{B}_{y0}^2-\widehat{B}_{x0}^2}{\widehat{B}_{y0}^2+\widehat{B}_{x0}^2} \quad P_2 = \frac{2\widehat{E}_{x0}\widehat{E}_{y0}\cos\phi}{\widehat{E}_{x0}^2+\widehat{E}_{y0}^2} = \frac{2\widehat{B}_{x0}\widehat{B}_{y0}\cos\phi}{\widehat{B}_{y0}^2+\widehat{B}_{x0}^2} \\ P_3 &= \frac{2\widehat{E}_{x0}\widehat{E}_{y0}\sin\phi}{\widehat{E}_{x0}^2+\widehat{E}_{y0}^2} = \frac{2\widehat{B}_{x0}\widehat{B}_{y0}\sin\phi}{\widehat{B}_{y0}^2+\widehat{B}_{x0}^2} \end{aligned} \tag{3.63}$$

where ϕ is the difference in phase between y- and x-electric field components. From the above, we see that pure circular polarization (degree of circular polarization $P_3 = 1$) occurs when $\widehat{E}_{x0} = \widehat{E}_{y0}$: (hence $\widehat{B}_{x0} = \widehat{B}_{x0}$) and and $\phi = \pi/2$. In the convention of Born and Wolf, this is called right circular polarization. A nice example of the variable polarization from an elliptical undulator was reported by Sasaki and coworkers [78], using the low-energy JSR ring operating at 138 MeV which put the radiation in the visible region of the spectrum (Fig. 3.17).

3.9 Helical Undulators

A helical undulator is simply the special case of an elliptical undulator with equal $\widehat{B}_{x0}$ and $\widehat{B}_{y0}$ that are 90° out of phase. The net amplitude of the field stays constant, while the direction rotates around a circle along the axis of the device. Consequently, the electrons spiral in a circle as they travel down the device, and the resulting synchrotron radiation has pure circular polarization. Several special properties are worth pointing out.

3.9.1 Fundamental Energy

The on-axis fundamental energy for a helical undulator can be obtained from Eq. 3.62 by setting $K_x = K_y$ and the observation angle $\theta = 0$:

$$E_1[keV] = 0.950 \frac{E_e^2[GeV]}{(1 + K^2)\lambda_u[cm]} \tag{3.64}$$

3.9.2 Harmonics

The helical undulator possesses the interesting and useful property that there are no higher harmonics produced on-axis in the forward direction.

3.9.3 Power

For a given magnetic field B_0, the helical undulator produces twice the power radiated by a planar undulator. This is easy to understand by referring back to Eq. 3.30 and noting that the magnitude of the magnetic field does not change in this device, only the direction.

$$P[\text{kW}] = 1.266 E_e^2[\text{GeV}] I[\text{A}] B_0^2[\text{T}] L[\text{m}] \tag{3.65}$$

Finally, (from Kim) the power density radiated in the forward direction is:

$$\frac{d^2P}{d^2\phi} = \frac{Ne^2}{\pi\varepsilon_0 c}\gamma^4 \frac{I}{e}\omega_u \frac{K^2}{(1+K^2)^3} \tag{3.66}$$

or (from Clarke) in practical units:

$$\frac{d^2P}{d\Omega}\left[\text{W mrad}^{-2}\right] = 4626\,\frac{E_e^4[\text{GeV}]\, I[\text{A}]\, B_0^2[\text{T}]\, L[\text{m}]}{(1+K^2)^3} \tag{3.67}$$

Notice that the on-axis flux goes towards zero at high K-values, because most of the power goes into higher harmonics, which are forbidden in the forward direction.

3.10 Other Insertion Devices

As mentioned in Chap. 2, a bevy of other insertion devices have been designed and implemented, but none approach the popularity of the three main designs: wiggler, planar undulator, and EPU. For more information about the radiation properties of so-called *exotic insertion devices*, consult the excellent chapters by Onuki [82] and Sasaki [80].

3.11 Suggested Exercises

1. Do the math on $\Delta t = \kappa \Delta t' = (1 - \beta\cos\theta)\Delta t'$ to show that $\kappa \cong \frac{1}{2}\left(\frac{1}{\gamma^2} + \theta^2\right); \theta \ll 1$.
2. Do the math to derive an improvement over undulator equation $\lambda = \lambda_0(1 - \beta\cos\theta)$, by replacing β with $\beta*$, starting from:

$$\beta_{z,\text{av}} = \beta* = \beta\left(1 - \frac{K^2}{4\gamma^2}\right) = 1 - \frac{1}{2\gamma^2}\left(1 + \frac{K^2}{2}\right)$$

3. Consider a dipole magnet with a field of 2 Tesla on an electron storage ring with an electron energy of 2 GeV.
   (a) What is the critical energy for the synchrotron radiation from this magnet?
   (b) What is the degree of circular polarization at a vertical angle of 0.3 mrad for 100 eV photons?

4. SPring-8 operates at 8 GeV and about 100 mA. They have built a wavelength shifter to maximize output of high-energy photons.

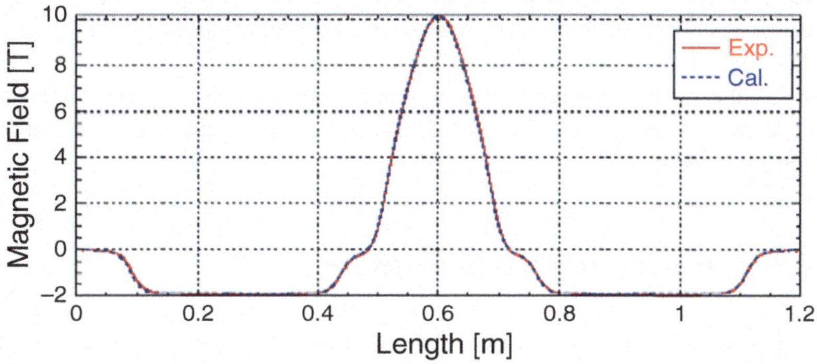

   (a) From the above field diagram, estimate (approximately) how much power is radiated by their wavelength shifter.
   (b) Estimate the critical energy on-axis and for a few angles off-axis.

5. SPring-8 also has an undulator with 781 periods over 37 m.

   (a) How much power is radiated by this device at its maximum field of 0.59 T?
   (b) What is the first-order energy radiated on-axis when $K = 1.5$?
   (c) What is the first-order energy radiated at 1 microradian when $K = 1.5$?

6. For a typical undulator with a 5 cm period and at $K = 2$ on a 2 GeV ring, calculate the power radiated by such a device with 50 poles.
7. Consider the initial properties of the Large Hadron Collider: 8.3 T bend magnets with a bend radius of 2813 m and protons at an energy of 7 TeV.

   (a) What is the critical energy and wavelength of the bend magnet synchrotron radiation?
   (b) How much power is radiated by this device at its maximum field of 0.59 T?
   (c) How much power would be radiated if the LHC stored electrons at 7 TeV?
   (d) What wavelength do you expect from a 2×280 mm period 5 Tesla undulator?

3.12 Reference Books and Review Articles

1. *The Science and Technology of Undulators and Wigglers*, James A. Clarke, Oxford Series on Synchrotron Radiation, Oxford University Press, New York, 2004, ISBN 0-19-850855-7.
2. *Undulators, Wigglers, and Their Applications*, H. Onuki and P. Elleaume, eds., Taylor and Francis, London, 2003, ISBN 0-415-28040-0.
3. *Synchrotron Radiation—Production and Properties*, Philip John Duke, Oxford University Press, New York, 2000, ISBN 0-19-851758-0.

4. *Synchrotron Radiation*, Helmut Wiedemann, Springer-Verlag, Heidelberg, 2003, ISBN 3-540-43392-9.
5. "Characteristics of Synchrotron Radiation", Kwang-Je Kim, *AIP Conference Proceedings*, **184**, 565 (1989)—clear derivations of bend magnet and insertion device properties, also includes practical formulae.
6. "Optical and power characteristics of synchrotron radiation sources", Kwang-Je Kim, *Optical Engineering*, **34**, 342–352 (1995)—more concise presentations of useful formulae.

3.13 Synchrotron Radiation Software

There is a veritable cottage industry in software that calculates synchrotron radiation properties of bend magnets and insertion devices. The fundamental programs include.

URGENT [83]. This was one of the earliest routines for calculating undulator spectra. "URGENT—A Computer-Program for Calculating Undulator Radiation Spectral, Angular, Polarization, and Power-Density Properties," R. P. Walker and B. Diviacco, *Rev. Sci. Inst.* **63**, 392 (1992).

SPECTRA [84]. A program for radiation from both bending magnets and insertion devices, written in C with its own GUI. **"**SPECTRA: a synchrotron radiation calculation code," Takashi Tanaka and Hideo Kitamura, *J. Syn. Rad.* **8**, 1221–1228 (2001).

The more recent trend in software is to include source calculations as part of larger packages that also model beamline optics, and some examples are:

WAVE [85]. "WAVE—A Computer Code for the Tracking of Electrons through Magnetic Fields and the Calculation of Spontaneous Synchrotron Radiation," M. Scheer, *Proc. ICAP2012.* This is a BESSY contribution to synchrotron radiation calculations. It is designed (a) calculate synchrotron source properties and to work with other programs to (b) track the influence of insertion devices on the storage ring and (c) follow the synchrotron radiation through subsequent optics (programs RAY and PHASE).

https://www.helmholtz-berlin.de/forschung/oe/fg/undulatoren/arbeitsgebiete/wave_en.html

XOP [86]. A toolkit developed at the APS, XOP (X-ray OPtics utilities) is a graphical user interface for executing programs of interest, such as XURGENT source calculations and ShadowVUI raytracing.

SRW [87] and *Sirepo* [88]. SRW calculates source spectra for relativistic electrons in arbitrary magnetic fields, and it can then follow wavefronts through optical systems of beamlines such as mirrors, refractive lenses, and zone plates. Initially an ESRF contribution, now available from Brookhaven via github: https://github.com/ochubar/SRW. *Sirepo* is a browser-based GUI for using SRW.

Chapter 4
X-ray Optics and Synchrotron Beamlines

4.1 Introduction

Storage rings and their associated insertion devices are wonderful sources of synchrotron radiation, but the X-rays are not useful unless they are brought to bear on a sample in an experiment. This generally involves transporting the X-rays out of the storage ring, selecting a part of the spectrum, and perhaps focusing it on the sample. All of this is the job of a beamline using X-ray optics. At first glance, X-ray beamlines can appear dauntingly complex and very different from UV-visible instruments (Fig. 4.1). However, on closer inspection, one finds mostly the same components as on a desktop grating spectrometer—slits, mirrors, and gratings (or crystals). In this chapter, we will discuss what makes X-ray optics look so different from more familiar components used in UV-visible and IR spectroscopies.

Why should we care about X-ray optics? As a fundamental part of your synchrotron spectroscopy experiment, some of the optics will be directly under your control. For example, you might be required to:

- Choose optimum crystals or gratings for your energy range.
- Move and bend mirrors to position and focus the beam.
- Move crystals or gratings to change photon energy.
- Adjust slits for appropriate energy resolution.

Although computer control might allow you to do all of these things without too much thinking, you will get better results if your understanding of X-ray optics gets to a deeper level than mere "knob knowledge." Understanding the optics will allow you to get more flux and a higher-quality beam—leading to a faster and better measurement.

There is no new physics in X-ray optics—the same phenomena of *refraction*, *reflection*, and *diffraction* occur at all frequencies, and the same general equations apply from the infrared regime to γ-rays. However, the parameters that govern the

S. P. Cramer, *X-Ray Spectroscopy with Synchrotron Radiation*, Biological and Medical Physics, Biomedical Engineering, https://doi.org/10.1007/978-3-030-28551-7_4

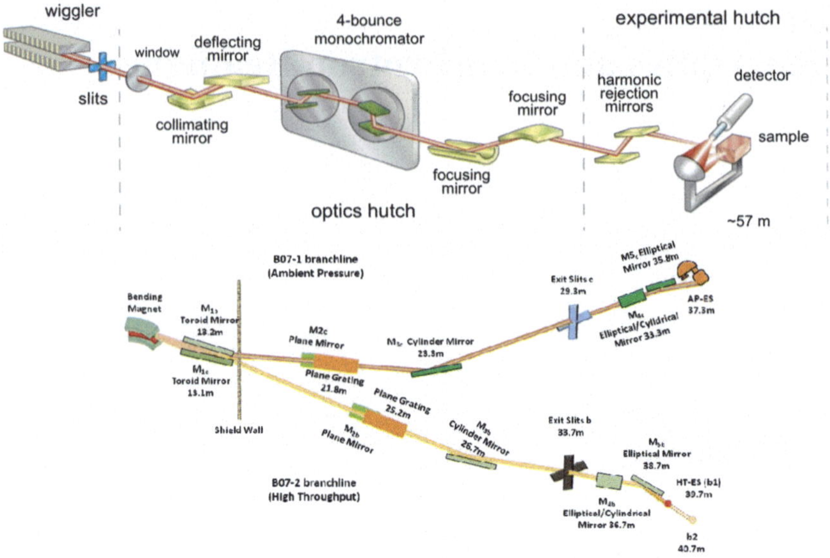

Fig. 4.1 Top: a hard X-ray beamline at Diamond Light Source. Bottom: a pair of soft X-ray beamlines at Diamond

response of materials in the X-ray region are very different from those for UV-visible and infrared light. This makes lenses impractical in most cases and forces one to use mirrors and gratings at glancing angles or nearly perfect crystals. Furthermore, the short wavelengths of X-rays and the high brightness of synchrotron sources necessitate exquisitely fine tolerances in the figure and finish of X-ray optics. As we shall see, the resulting beamlines for X-ray spectroscopy can stretch half a football field and cost millions of dollars.

4.2 X-ray Optical Constants and Equations

From high school physics, you probably remember that for *reflection* by mirrors, we only need to know that the angle of reflection is equal to the angle of incidence, $\theta_{\mathrm{i}} = \theta_{\mathrm{r}}$ (Fig. 4.2). To predict the direction of the refracted rays at an interface, in addition we require for both materials the *absolute index of refraction*, *n*, defined as the ratio of the speed of light in a vacuum, *c*, to that in the given material, *v*:

$$n = c/v \tag{4.1}$$

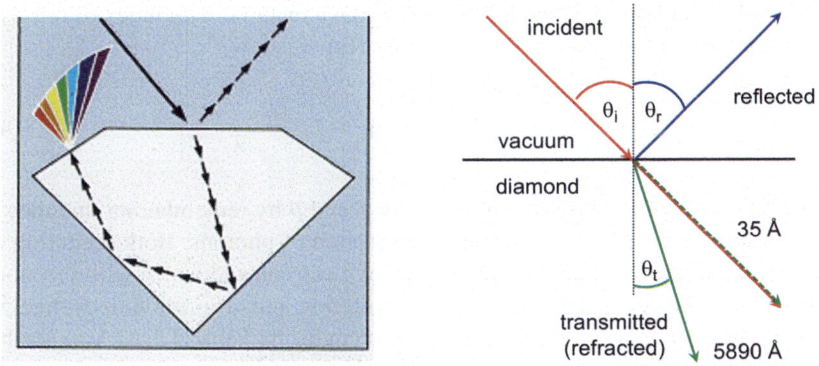

Fig. 4.2 Left: refraction and reflection for diamond with visible light. Right: comparison of refraction in diamond for yellow light and soft X-rays. With this scale, the deviation of the refracted X-ray from a straight line is not visible

As defined in Fig. 4.2, the angle of refraction θ_t or Θ_2 is given by Snell's law:

$$n_1 \sin \theta_1 = n_2 \sin \theta_2 \; or \; n_1 \sin \theta_i = n_2 \sin \theta_t \tag{4.2}$$

where n_1 and n_2 are the indices of refraction for both materials. In the X-ray region, n is very close to but slightly less than 1. Thus, it is simpler to use the *refractive index decrement* δ, which is the deviation of n from 1:

$$\delta = 1 - n \tag{4.3}$$

In the X-ray region, the angular deviations caused by refraction are miniscule (Fig. 4.2), and this means conventional lenses with large apertures are out of the question. Away from grazing incidence, the reflectivity for X-rays is also negligible, and this means that conventional mirrors with large deflection angles are also not feasible. Below we will see how scientists have overcome these issues to make practical X-ray optics.

4.2.1 The Complex Index of Refraction

The refractive index that we have used so far does not take into account an additional phenomenon—absorption. By combining n with the absorption coefficient β, one can define a *complex index of refraction* $\tilde{n}$:

$$\tilde{n} = n - i\beta = 1 - \delta - i\beta \tag{4.4}$$

For an electromagnetic wave traveling through a material with index of refraction ñ, the amplitude of the wave is then given by (Eq. 4.5):

$$A = A_0 \exp\left(\frac{-2\pi\beta x}{\lambda}\right) \exp\left[-2\pi i \frac{(nx - ct)}{\lambda}\right] \tag{4.5}$$

We can make some sense of the trends for δ and β by remembering that they are related, respectively, to the scattering or absorption of photons. Both scattering and absorption tend to increase with the number of electrons and hence with the atomic number Z, and the general trend is also for scattering and absorption to decline with increasing photon energy or decreasing wavelength. In Table 4.1 we can see how different X-ray parameters are from those for the visible region.

4.2.2 *The Fresnel Equations*

Simple equations like the mirror equation and Snell's law tell us about the angles for reflection and refraction, but they don't tell us how much power goes into the reflected and refracted beams. The fractions of the beam that are reflected and refracted can be calculated using the Fresnel equations. In these expressions, one defines electric field in-the-plane "p" or "π" components, as well as out-of-plane "s" or "σ" components, as illustrated in Fig. 4.3. This leads to four equations in four unknowns. Using four continuity constraints related to the two components of the

Table 4.1 Representative optical constants in the visible and X-ray regions[a]

Material	n (5890 Å)	n (35 Å)	δ (35 Å)	δ (1 Å)	β (35 Å)	β (1 Å)
Diamond	2.42	0.997	2.92×10^{-3}	2.98×10^{-6}	1.94×10^{-3}	2.04×10^{-9}
Silicon	3.97	0.9964	3.50×10^{-3}	3.17×10^{-6}	1.5×10^{-3}	3.17×10^{-8}
Gold	0.266	0.9937	6.2×10^{-3}	1.88×10^{-5}	8.3×10^{-3}	2.54×10^{-6}

[a]X-ray values from Henke Tables [89] via CXRO website

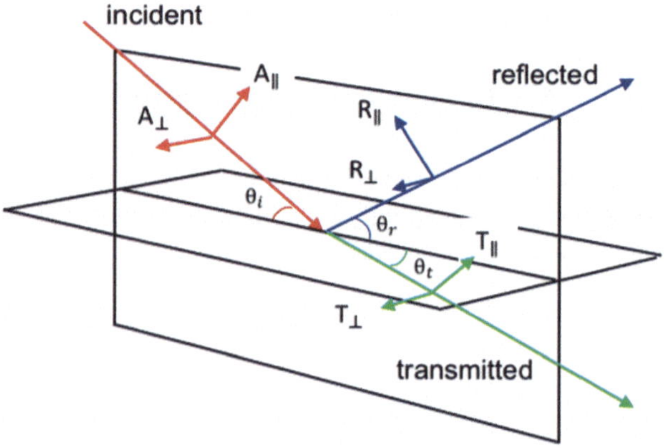

Fig. 4.3 Illustration of different electric field components involved in the Fresnel equations. Others may use "p" or π and "s" or σ subscripts to refer to field components, respectively, in and perpendicular to the scattering plane

electric and magnetic fields, one can solve for the σ and π components of the transmitted and reflected electric fields (Appendix E).

For synchrotron X-ray experiments, the most relevant form of these equations is for the case with radiation traveling from the vacuum with index of refraction unity to an absorbing medium with a complex index of refraction $\tilde{n}$. The details are summarized in Appendix E, so here we will concentrate on the results.

Since we are concerned here about mirrors, we first require definition of the reflectivity **R**. Here we mean the "intensity reflectivity," where if I_0 is the incident intensity, I is the reflected intensity, R is the reflected electric field amplitude (which can be complex), and $R*$ is its complex conjugate, then:

$$\mathbf{R} = I/I_0 \tag{4.6}$$

The reflectivity is often different for parallel and perpendicular components of the fields, so we need to define both p- and s-intensity reflectivities, where the fields are defined in Fig. 4.3:

$$\mathbf{R}_\mathrm{p} = \left(R_{\parallel}/A_{\parallel}\right)\left(R_{\parallel}/A_{\parallel}\right)* \tag{4.7}$$

and

$$\mathbf{R}_\mathrm{s} = (R_{\perp}/A_{\perp})(R_{\perp}/A_{\perp})* \tag{4.8}$$

Under normal incidence conditions, the Fresnel equations simplify so that a good approximation to the intensity reflectivity $\mathbf{R}_\mathrm{N}$ is:

$$\boldsymbol{R_N} = [(1-\tilde{n})/(1+\tilde{n})]^2 = \left(\delta^2+\beta^2\right)/\left[(2-\delta)^2+\beta^2\right] \tag{4.9}$$

Even putting in optimistic values, say those for gold at 35 Å (Table 4.1), gives $\mathbf{R}_\mathrm{N} \cong 4.6 \times 10^{-5}$. Clearly anything like normal incidence is not going to work! Let's now see which angles might still be viable for X-ray mirrors. More general forms of the Fresnel equations for amplitude reflectivities that work for all angles (from Michette), while still assuming that the external medium is a vacuum, are given in Eqs. 4.10:

$$R_\sigma = \frac{R_\perp}{A_\perp} = \frac{\sin\theta_i - \left(\tilde{n}^2 - \cos^2\theta_i\right)^{1/2}}{\sin\theta_i + \left(\tilde{n}^2 - \cos^2\theta_i\right)^{1/2}} \tag{4.10}$$

and

$$R_\pi = \frac{R_\parallel}{A_\parallel} = \frac{\tilde{n}^2\sin\theta_i - \left(\tilde{n}^2 - \cos^2\theta_i\right)^{1/2}}{\tilde{n}^2\sin\theta_i + \left(\tilde{n}^2 - \cos^2\theta_i\right)^{1/2}} \tag{4.11}$$

These more general equations can now be used to illustrate the properties of real-world X-ray mirrors.

4.3 Reflection: X-ray Mirrors

Snell's law (Eq. 4.2) can be rearranged to yield:

$$\theta_2 = \sin^{-1}\left(\frac{n_1}{n_2}\sin\theta_1\right) \tag{4.12}$$

When $n_1 > n_2$, there will be a value for θ_1 where $\theta_2 = 90$ degrees—the transmitted beam is tangent to the interface. This value for θ_1 is labeled θ_c and is referred to as the *critical angle*. One can easily derive:

$$\theta_c = \theta_1 = \sin^{-1}\left(\frac{n_2}{n_1}\right) \tag{4.13}$$

Thus, for rays that are more glancing than θ_c, there is no transmitted beam and all of the energy goes into the reflected beam. With visible light, this phenomenon is called *total internal reflection*. As we saw earlier in Fig. 4.2, diamond strongly refracts visible light. With the proper geometry, light entering the "table" of a diamond can undergo two total internal reflections before emerging from the "crown." Diamond cutters exploit this combination of reflection and refraction to give diamonds their sparkle, and otherwise sane people spend large amounts of money to observe this. Total internal reflection (combined with dispersion) also gives rise to rainbows, at considerably less expense to the viewer.

4.3.1 Total External Reflection

Since the index of refraction for X-rays is less than 1, somewhat paradoxically, the vacuum is considered the more dense medium, and there will be a glancing angle below which all X-rays are reflected. The phenomenon called total internal reflection when referring to diamonds and raindrops is now called *total external reflection*, but of course, it is the same physics. Because the difference in refractive indices is minute, the glancing angles for total external X-ray reflection are also very small.

Notice that Snell's law defines $\theta_1 = \theta_i$ and $\theta_2 = \theta_t$ with respect to the interface normal. Since X-ray mirrors operate very close to grazing incidence, it is often simpler to use Snell's law in terms of glancing angles:

$$n_1 \cos\theta_{1g} = n_2 \cos\theta_{2g} \text{ or } n_1/n_2 = \cos\theta_{2g}/\cos\theta_{1g} \tag{4.14}$$

Total reflection occurs when the refracted ray is predicted to be parallel to the surface:

$$\cos\theta_{\rm cg} = n_2/n_1$$

For transmission from vacuum, $n_1 = 1$, and by employing the approximation for small angle that $\cos\theta \cong 1 - \theta^2/2$, a useful formula is then:

$$\theta_{\rm cg} \cong \sqrt{2\delta} \tag{4.15}$$

If we plug in some typical values (Table 4.1), we find that for gold at 35 Å, the critical angle is about 6°, dropping to less than 0.5° (5–7 mrad) at 1 Å.

4.3.2 *Less than Total External Reflection*

So far, we have ignored the detailed shape of the reflectivity curve for angles above the critical angle. This will depend on the complex index of refraction and the surface roughness of the mirror. In Fig. 4.4 we illustrate (using Eqs. 4.10 and 4.11) the effects of different values for δ, β, and σ on the reflectivity as a function of glancing angle.

In the absence of absorption, stronger refraction leads to a higher angle for total external reflection. At most energies, higher *Z* elements will have larger values for δ. For this reason, SiO_2 X-ray mirrors are often coated with a metal such Ni or Au. On the other hand, increased absorption (larger β) diminishes the reflectivity. Both effects are shown in Fig. 4.4.

The third factor influencing mirror performance is surface roughness, which can be characterized by the rms surface displacement—σ. An approximate formula for the loss of reflectivity due to roughness is:

$$R = R_0 \exp\left[-(4\pi\sigma\theta_i/\lambda)^2\right] \tag{4.16}$$

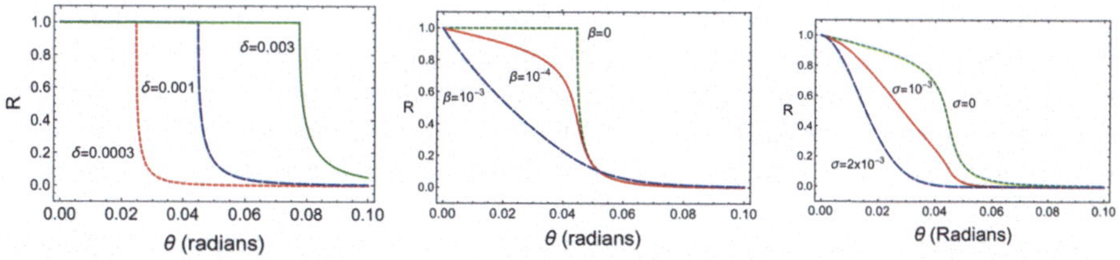

Fig. 4.4 Left: mirror p-reflectivity vs. glancing angle for different refractive index coefficients δ. Middle: p-reflectivity vs. glancing angle for different absorption coefficients β. Right: reflectivity vs. glancing angle for different amounts of surface roughness σ

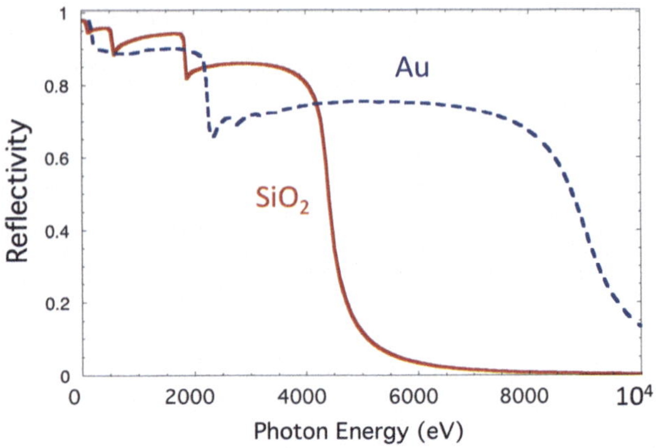

Fig. 4.5 Reflectivity vs. energy for an SiO_2 (glass) mirror at 0.4° and a gold-coated mirror at 0.5°. Notice the structure resulting from the Si K-edge and Au M-edges, as well as how the Au coating extends the useful range to nearly 10 keV, calculated using CXRO website

From this we can see that for a 1 degree beamline at 1 Å, the mirror roughness σ should be on the order of a few Å—about 1 atomic layer. The tolerances are obviously more forgiving for longer wavelengths or shallower glancing angles. The effects of increasing σ on mirror performance can also be seen in Fig. 4.4.

Since β increases sharply above an X-ray absorption edge (Chap. 6), the reflectivity of a mirror will drop as the X-ray energy passes through that edge. Despite these losses, the larger δ helps enhance the reflectivity at higher energies. These effects are both illustrated in Fig. 4.5.

4.3.3 Mirror Shapes and Aberrations

Grazing incidence X-ray mirrors are often shaped to collimate or focus the X-ray beam. Just as for visible region optics, spherical shapes are easier to produce than aspheres such as ellipsoids and paraboloids. However, because of the grazing incidence geometries, the "astigmatism" aberrations for spherical mirrors are severe, as shown in Fig. 4.6.

Astigmatism means that the *sagittal rays* in the horizontal plane come to a different focus than do the *tangential rays* in the vertical plane. It can be reduced by (a) using two different mirrors with different radii for horizontal and vertical focusing, (b) using one mirror with two different radii of curvature, or (c) employing two different elliptical mirrors to achieve focusing in the horizontal and vertical planes (Fig. 4.6).

The last of these combinations is called a Kirkpatrick-Baez pair (Fig. 4.7) [90]. It has the wonderful property that near-perfect focusing can be achieved in both horizontal and vertical planes. Furthermore, it is "achromatic"—its focal length does not change with the X-ray wavelength. Elliptical shapes were originally

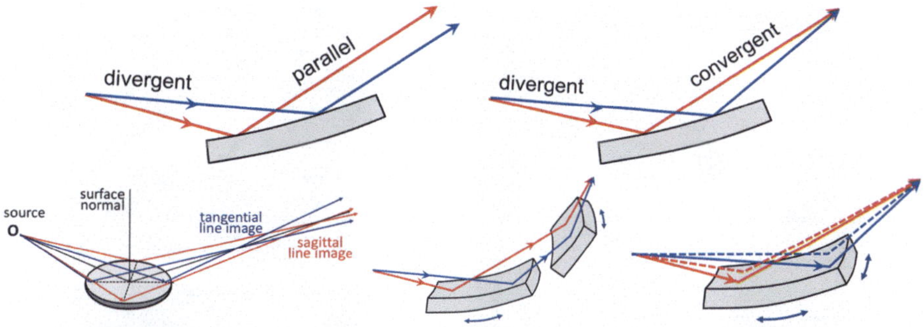

Fig. 4.6 Shapes used in X-ray mirrors. Top: a parabolic mirror will convert a divergent beam into a collimated parallel beam (or vice versa), while an elliptical mirror focuses from a point to a point. Bottom left: astigmatism of a spherical mirror at glancing incidence. Bottom right: two approaches to solving astigmatism—discrete vertical and horizontal focusing vs. bent cylinder

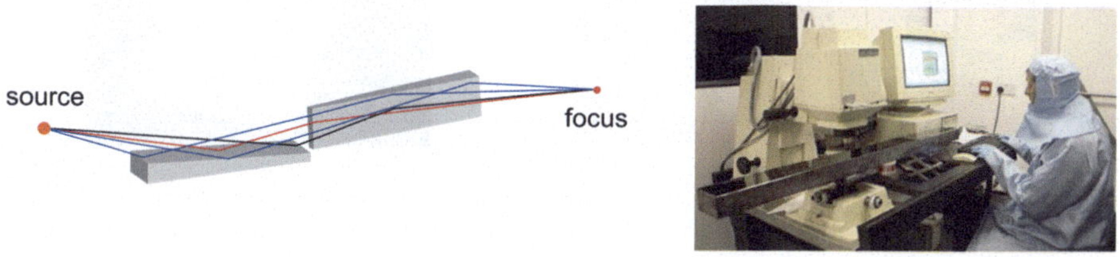

Fig. 4.7 Left: sequential focusing by two elliptical mirrors in KB (Kirkpatrick-Baez) optics. Right: a technician measures the finish quality (surface roughness) of a mirror

achieved by clever bending schemes [91], referred to as "dynamic figuring." Nowadays the best elliptical mirrors use "static figuring" and are machined with liquid jet or other technologies directly to an ellipsoidal shape with sub-nanometer figure accuracy and ~1 Å rms roughness [92, 93].

4.3.4 Practical Mirror Fabrication

A common expression for optics builders is that "you can only make what you can measure." The two important quantities to measure are referred to as "figure" and "finish." Figure refers to the long range quality of the surface—how close it is to the design shape, while surface finish or surface roughness is a measure of local irregularities on a surface. Mirror makers go to extraordinary lengths to check the quality of their mirrors (Fig. 4.7).

To optimize the reflectivity of mirrors in the X-ray region, designers typically begin with a smooth Si, glass (Zerodur®), or Ultra Low Expansion (ULE) fused Si

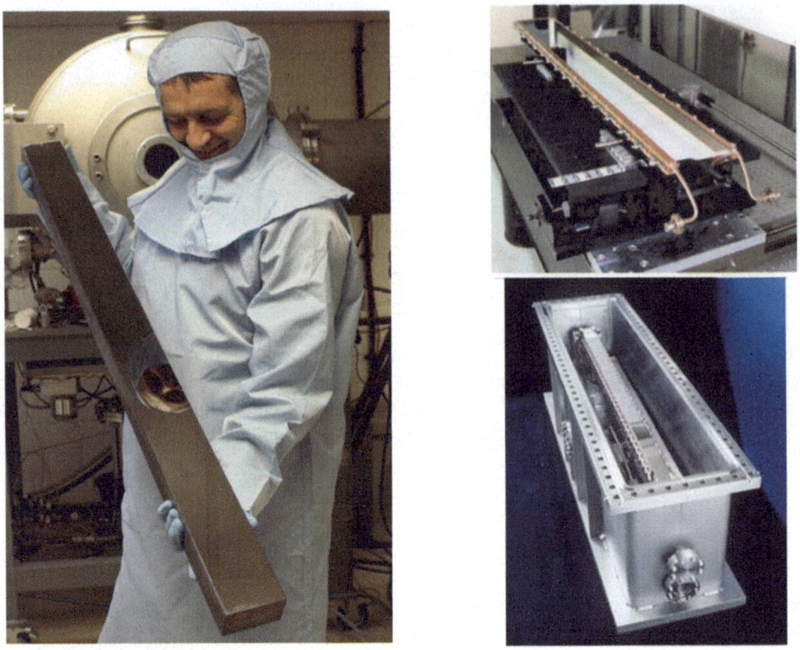

Fig. 4.8 Glancing incidence mirrors. Left: plane mirror before coating. Top right: a mirror in a copper channel for flowing cooling water. Bottom right: a mirror installed in its vacuum chamber

template and then coat the original surface with a metal such as Ni or Au. The metal coating serves to increase the electron density at the surface (and hence δ and the critical angle) and also to reduce the surface roughness. The resulting objects are a thing of beauty (Fig. 4.8). But don't drop them—they cost as much as a Lamborghini and sometimes take 6 months for delivery.

4.3.5 Capillary Optics

For those who cannot afford a Lamborghini, we mention "capillary optics." These devices are really just multi-bounce X-ray mirrors, using a drawn capillary that gradually changes diameter. A single capillary optic consists of a slightly conical tube with a very smooth inner surface. The maximum cone angle depends on the maximum angle of total external reflection, which of course depends on the material and photon energy. A clever variation is the so-called Kumakhov lens, which is not really a lens but rather a collection of capillaries arranged to capture a larger solid angle (Fig. 4.9).

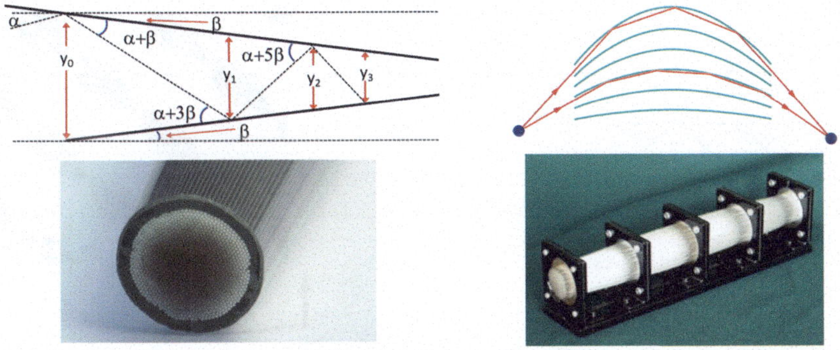

Fig. 4.9 Top left: angles involved in tapered capillary optics, redrawn from [94]. At some point the angles are too steep for reflection. A tapered capillary can mitigate this problem. Top right: variable tapers in a bundle of capillaries. Bottom left: entrance to capillary bundle. Bottom right: commercial Kumakhov lens

4.4 Refraction: X-ray Lenses

For X-rays coming from a vacuum into a solid material, $n_1 < n_2$ and hence X-rays are bent away from the normal. Thus, a focusing X-ray lens will be concave instead of convex! Such X-ray lenses can be made by drilling holes in low *Z* materials. The focal length of a single "hole-based" X-ray lens with radius of curvature *R* is given by:

$$1/f = 2\delta/R \tag{4.17}$$

This often yields an optic with an impractically long focal length. However, by combining a number "*N*" of closely spaced holes together, a useful device can be produced—a *compound refractive lens* or CRL [95]. In the thin lens approximation, this results in a shorter focal length, given by:

$$1/f = N2\delta/R \tag{4.18}$$

These devices are used to reduce the divergence of undulator beams at third-generation sources. The undulator source is placed at the focal length of the lens, and the beam that emerges is collimated (focused at infinity). Refractive optics are becoming a popular and efficient method for collimating undulator beams, yielding reduced divergence without much loss in intensity. Refractive optics are in place on half of the ESRF's beamlines and are in wide use at a half dozen other synchrotrons (Fig. 4.10).

Nowadays, CRLs are made by reactive ion etching (RIE) of holes into Si or even diamond. It turns out that parabolic surfaces produce a better focus when the incident beam in collimated. It is an active area of research.

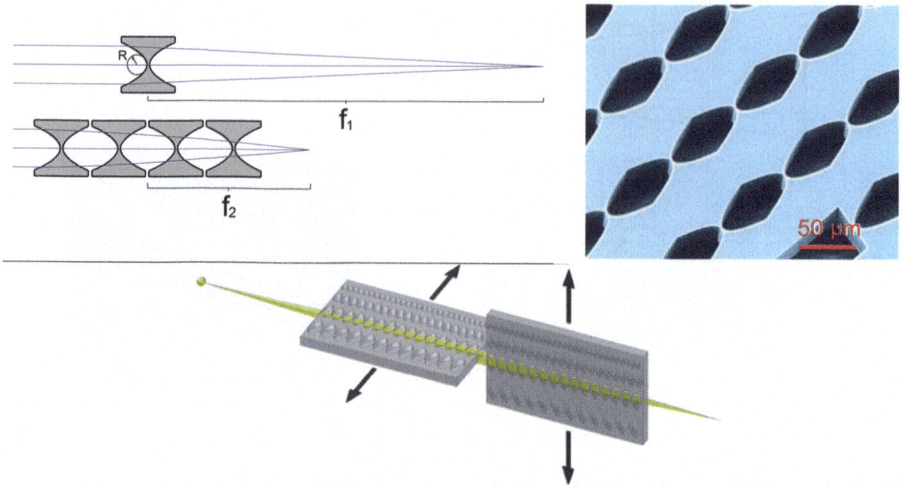

Fig. 4.10 Top left: single and compound X-ray lenses. Stacking lenses can yield a more manageable focal length. Top right: compound X-ray lenses in two dimensions. Bottom: two lens plates mounted on perpendicular stages to form a point-focus CRL

4.5 Diffraction: Gratings and Zone Plates

Diffraction normally refers to scattering of waves into well-defined directions by an ordered array of scatterers. Macroscopic arrays of lines or circles are used to make gratings and Fresnel lenses, while microscopic arrays of planes of atoms are used in X-ray diffraction from crystals and multilayers. The physics and mathematical derivations for diffraction are well documented in the texts listed at the end of this chapter. In this section, we mostly state those results and how diffraction is used for X-ray optics.

4.5.1 Gratings

A diffraction grating is a periodic arrangement of holes (transmission grating) or reflecting surfaces (reflection grating), as shown in Fig. 4.11. In synchrotron X-ray spectroscopy, gratings are typically used for the energy range from 50 to 2000 eV, where the wavelengths are too large for practical crystal monochromators. For describing diffraction from a grating, the convention is to, respectively, define angles α and β for the incident and diffracted beams with respect to the surface normal (note that β is considered negative if it is to the right of the normal). By requiring that the optical path length difference λ for rays from different grating lines spaced d apart is an integral number m of wavelengths λ, one arrives at the *grating equation*:

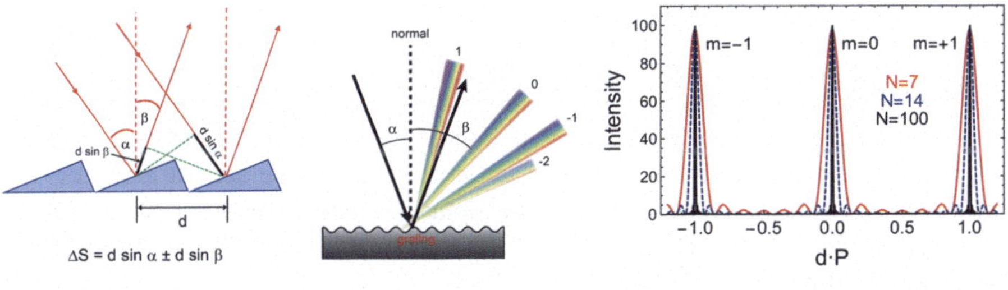

Fig. 4.11 Left: definition of angles in the grating equation. Angles to the right of the normal are negative. Middle: a reflection grating diffracting visible light. The numbers refer to values of m in the grating equation. Right: line shapes for diffraction from a grating with 7, 14, or 100 lines. d·P is the path length difference between neighboring wavefronts. Intensities are normalized to same height for visibility

$$m\lambda = d(\sin\alpha + \sin\beta) \tag{4.19}$$

For a grating with N lines illuminated, addition of the N reflections gives rise to interference where a particular diffracted beam maximizes with a profile given by:

$$I \propto \frac{\sin^2(N\pi\Lambda/\lambda)}{\sin^2(\pi\Lambda/\lambda)} \tag{4.20}$$

Since the integrated intensity is proportional to N, and the spectral line width goes as $1/N$, the peak intensity goes as N^2. (Notice the similarity to the properties of an undulator.) From Eq. 4.20, the intrinsic resolution R_m for a grating of width W can be derived from the number of lines illuminated and the diffraction order m:

$$R_m = \lambda/\Delta\lambda = E/\Delta E \cong Nm = Wm/d \tag{4.21}$$

The angles α and β are arbitrary, and by imposing different conditions on them, one arrives at different modes for scanning an X-ray monochromator. One of the most commonly employed designs imposes a *constant included angle* $2\theta = \alpha - \beta$, which allows for fixed in and out directions, and leads to a grating equation:

$$m\lambda = 2d\cos\theta\sin(\theta + \beta) \tag{4.22}$$

For practical monochromators, there are also geometric contributions to the resolution, because of the finite size of entrance and exit slits and the resulting divergence of the entrance and exit beams. For an entrance or exit slit of respective height S_1 or S_2 at a distance r_1 or r_2 from the grating:

$$\Delta\lambda_{S1} = S_1 d\cos\alpha/mr_1 \text{ and } \Delta\lambda_{S2} = S_2 d\cos\beta/mr_2 \tag{4.23}$$

The total resolution is the vector sum of the individual slit contributions, along with contributions from optical aberrations and slope errors in the grating itself and, of course, the intrinsic resolution from diffraction.

4.5.2 Practical Grating Issues

The gratings used for X-ray beamline monochromators and analyzers usually consist of a series of indentations or grooves in a glass or metal substrate, and they are used in grazing incidence geometry, where each reflecting surface acts as a source. To maximize monochromator efficiency, the optics designer can work with (1) the coating material, (2) the geometry of the groove profile, and (3) the shape of the substrate and the groove spacing.

4.5.2.1 Coatings

As we saw with mirrors, a high *Z* coating on a grating yields a higher refractive index decrement δ and better reflectivity. Common coatings include gold, platinum, and rhodium. It is important to remember that (just as with mirrors) the coating material will introduce altered reflectivity in the region of its absorption edges. Even higher efficiencies are now being achieved by adding multilayer coatings (described later in this chapter) to the grating substrate [96–99]. The shape of the substrate will depend on whether one uses the grating solely for dispersing the X-ray energies or whether one also wants the grating to focus the X-rays.

4.5.2.2 Groove Profiles

Although the grating equation tells us where the diffraction peaks occurs, it does not tell us about the relative intensity of different diffraction orders. When doing a spectroscopy measurement in say first order, all of the photons going into the second and higher orders (as well as zeroth order) are wasted. However (as suggested by Rayleigh in 1874 [100]), by manipulating the shape of the individual grating lines, one can alter the distribution of order intensity. Below we illustrate three types of gratings: (a) "amplitude gratings," (b) "blazed gratings," and (c) "laminar gratings" (or "lamellar gratings") (Fig. 4.12).

The amplitude grating is the simplest type of grating, where only the lands between the grooves contribute to diffraction (Fig. 4.12). If the land/groove ratio is 1:1, even with 100% reflectivity, the theoretical *m*th-order efficiency ε_m is [102]:

$$\varepsilon_0 = 25\% \quad \varepsilon_m = 100/(m\pi)^2\% \quad m = \pm 1, \pm 3, \pm 5, \ldots \tag{4.24}$$

For such a grating, most of the radiation goes into the zero-order reflection, where the grating acts like a plane mirror, with no separation of different wavelengths.

A blazed grating has the diffracting lands cut at an angle, the "blaze angle," with respect to the mean grating surface. In principle, such a grating can approach 100%

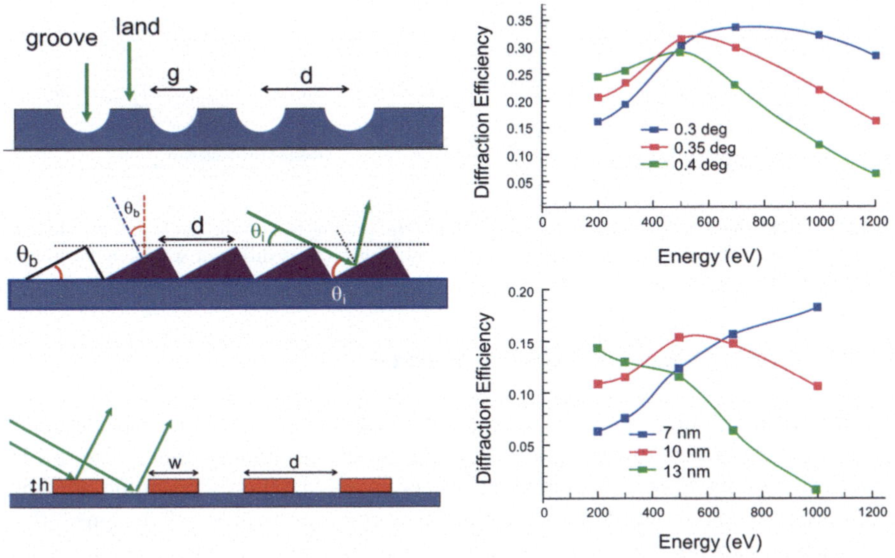

Fig. 4.12 Left (top to bottom): the shape of an amplitude grating and relevant parameters; the shape of a blazed grating and relevant parameters; shape of a laminar grating and relevant parameters. Top right: diffraction efficiency vs. energy for blazed gratings as a function of blaze angle. Bottom right: diffraction efficiency vs. energy for laminar gratings with a variety of groove depths. Redrawn from [101]

efficiency for a given order when the direction of the diffracted beam is the same as that for specular reflection from the land—the *blaze condition.* However, at the glancing angles required for X-ray reflectivity, one land shadows the next so that only a fraction of each land is illuminated. The effect of this shadowing is that the fraction f of each land that is illuminated is given by (where θ_1 is the angle of incidence on the land):

$$f = 1 - \frac{\tan\theta_b}{\tan\theta_1} \tag{4.25}$$

The blaze angle should be as small as possible in order to illuminate as much land as possible.

In a third approach, a *laminar* or *lamellar* or *phase grating*, the lands and grooves both contribute to the diffracted intensity. By appropriate choice of groove depth h and incidence angle, contributions from lands and grooves will interfere destructively, thus eliminating the wasteful zeroth-order radiation. The maximum efficiencies in a given order are in principle four times greater than the corresponding amplitude grating. However, just as with blazed gratings, at glancing incidence shadowing is significant. For laminar gratings, illumination of a groove is shadowed by the preceding land, and the radiation from the groove is shadowed by the subsequent land.

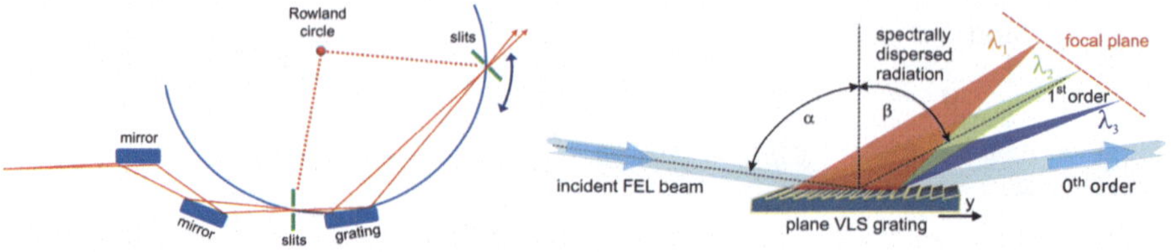

Fig. 4.13 Left: the Rowland circle geometry, where the grating lies on the circle and the slits are moved to collect different energies. Right: a VLS grating focuses different wavelengths from a diverging source [108]

4.5.2.3 Grating Shapes and Groove Spacings

The idea of combining a grating with a spherical surface was known in the nineteenth century. Even today, spherical gratings are commonly used (Fig. 4.13). Spherical gratings are a component of popular "Dragon" beamline designs [103–105] which continue to be built, as well as constant focal length designs [106].

Focusing can also be achieved with a variable line spacing (VLS) grating (Fig. 4.13). In these designs, the groove density along the grating surface follows a polynomial formula, $\sigma(y) = \sigma_0 + \sigma_1 y + \sigma_2 y^2 + \sigma_3 y^3$, where the polynomial coefficients are adjusted to minimize optical aberrations at a given wavelength [107, 108].

It seems that there is still no perfect solution, as beamlines continue to be built with blazed or laminar gratings, flat and spherical and VLS gratings. Combinations of curved gratings with variable line spacing are also being explored [109]. For a spectroscopist, the most important points are to know the (a) resolution and (b) spectral purity of the radiation being delivered.

4.5.3 Zone Plates

Zone plates (another invention in Lord Rayleigh's notebooks) are essentially circular, variable line space diffraction gratings (Fig. 4.14). They consist of a set of circular, concentric rings that become narrower and closer together as they go from the middle out to the edge. Their focusing properties are like a lens and are approximately described by the thin lens formula $1/p + 1/q = 1/f$, where p and q are the object and image distances and f is the focal length. Unlike a lens, a zone plate has different diffraction orders and therefore multiple focal spots.

As shown in Fig. 4.14, the equation describing the required zone plate spacing is:

$$r_n{}^2 = nf\lambda + n^2\lambda^2/4 \tag{4.26}$$

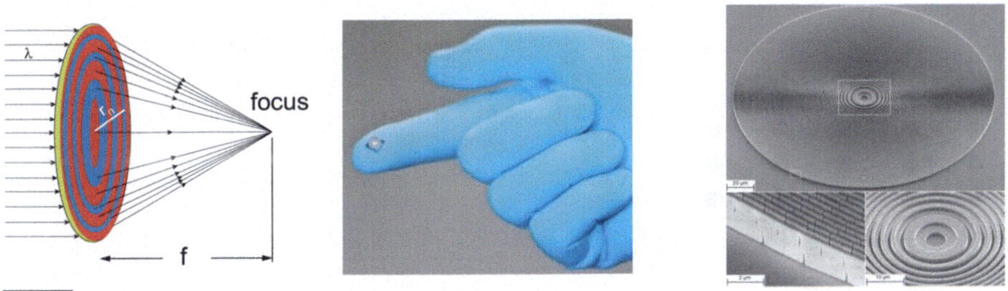

Fig. 4.14 Left: diffraction by zone plate with sequential zones of radius r_n. Middle: a zone plate for the LCLS [110]. Right: a zone plate with close-ups of central and outer rings [111, 112]

Practical zone plates are commonly made of gold rings, resting on top of a silicon nitride substrate. More recently, for high-power applications on free electron lasers, designers have switched to iridium rings (with a higher melting point) combined with a more conductive diamond surface [113]. The theoretical efficiency of an absorption-based zone plate in the first order is about 10%, and higher efficiency is obtained when, instead of an opaque zone, a transparent but phase-shifting zone is employed [114].

Zone plates are often used for micro-XES experiments (Chap. 8), where they compete with KB mirrors. Their upside is a relatively simple implementation and lower cost, but a downside is that they are "chromatic"—the focal length changes with X-ray wavelength and energy.

4.6 Diffraction: Crystals and Multilayers

As you might remember from freshman chemistry, the fundamental equation describing diffraction from crystals and multilayers is the Bragg equation:

$$\sin \theta = n\lambda/2d \tag{4.27}$$

where λ is the X-ray wavelength, d is the spacing between equivalent planes of atoms, and θ is the angle of the X-ray beam with respect to those planes (Fig. 4.15). (Note that if the planes are parallel to the crystal surface, then θ is a glancing angle as defined for mirrors.) Like with the grating equation, the Bragg equation can be derived by requiring an integral number of wavelengths between wavefronts reflected from one plane of atoms to the next.

In X-ray crystallography for structural analysis, the *kinematical approximation* assumes ideally imperfect crystals that are mosaics of small blocks (Fig. 4.15). In this approximation, it is assumed that diffraction is weak, so that the intensity remains constant as the X-rays pass through a crystal and also that multiple scattering within a crystal can be ignored. The result of scattering from a set of slightly

Fig. 4.15 Left: definition of terms in Bragg diffraction. Middle: an ideally imperfect crystal as a mosaic assembly of crystal blocks. Right: the resulting broad diffraction profile peak

misaligned crystals is a reflection with an approximately Gaussian profile and an integrated intensity proportional to the square of the *structure factor*: $I \propto |F|^2$ (Fig. 4.15).

4.6.1 Perfect Crystal Diffraction

The reflections observed from the "perfect" single crystals used as monochromators and analyzers are quite different—they exhibit nearly flat-topped profiles of almost 100% reflectivity with widths proportional to the structure factor: $I \propto |F|$ (Fig. 4.16). To explain this behavior, *dynamical diffraction theory* assumes one single crystal and adds two new considerations: (a) the loss in amplitude of the incident wave as it propagates through the crystal (due to reflection into the exit beam) and (b) the possibility that the reflected beam is rescattered into the direction of the incident beam (multiple scattering). Detailed descriptions and derivations of this theory can be found in the references at the end of this chapter; here as usual we concentrate on the results from those derivations.

As illustrated in Fig. 4.16, in dynamical theory with perfect crystals, it is possible to achieve 100% reflectivity over a range of angles, as opposed to the Gaussian profile of limited reflectivity that results from ordinary imperfect crystals. The reason for such high reflectivity is simple—the beam keeps penetrating until it finally scatters and leaves the crystal. The rationale for a width proportional to the structure factor is also straightforward—stronger scattering means a smaller number of atomic planes are required to scatter the beam (just as the width of a diffraction peak from a grating vary inversely with the number of lines illuminated.) When absorption is taken into account, the reflectivity drops slightly, and the *rocking curve* (intensity vs. angle) becomes somewhat more asymmetric, but reflectivities of >90% over a range of angles are not uncommon.

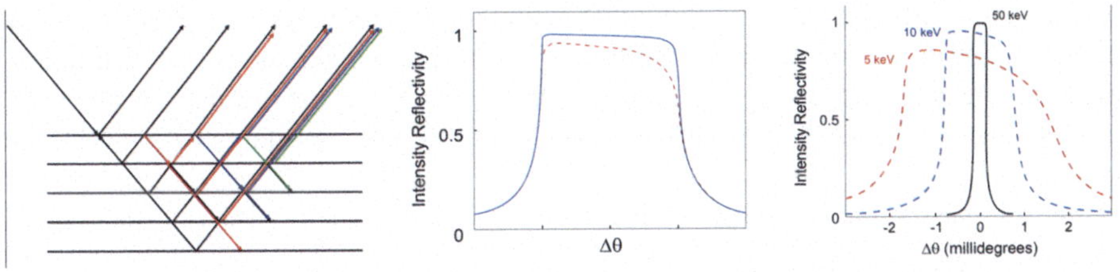

Fig. 4.16 Left: multiple scattering in perfect crystal diffraction. A proper calculation needs to sum over all possible scattering paths. Middle: diffraction profiles without absorption (solid line) and with absorption (dashed line). Right: typical diffraction profiles for different energies

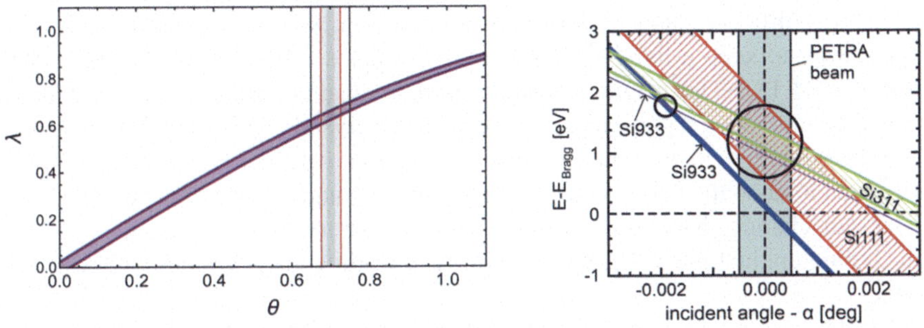

Fig. 4.17 Left: generalized $\lambda - \theta$ Dumond diagram superimposed on source and slit properties. The black vertical lines represent the source divergence and the red lines define the slit acceptance. Right: $\Delta E - \Delta\theta$ Dumond diagram for beamline P08 at PETRA-III [120]

For a perfectly parallel incident beam, the angular range $\Delta\theta_{\rm D}$ over which diffraction occurs is called the *angular Darwin width* ($\omega_{\rm D}$). Furthermore, from the Bragg equation, the resolution of a single crystal X-ray monochromator can be expressed as:

$$\Delta E/E = \Delta\lambda/\lambda = \cot\theta\Delta\theta \tag{4.28}$$

The final energy resolution will depend on a $\Delta\theta$ that is the geometric sum of the beam divergence on the crystal and the angular Darwin width $\omega_{\rm D}$.

A *Dumond diagram* (Fig. 4.17) is often used to visualize the effects of different angular factors on the energy (or wavelength) resolution of a diffracting crystal. If one plots the normalized wavelength for diffraction ($\lambda/2d$) vs. the diffraction angle θ, then from the Bragg equation, one obtains a sine wave plot which is approximately linear over small ranges. In the Dumond diagram, the sine curve is broadened to account for the fact that diffraction occurs over the angular Darwin width $\omega_{\rm D}$. The effects of the beam divergence or the inclusion of slits can then be represented by vertical lines corresponding to the range of allowed beam angles. The net energy resolution is seen as the region of overlap between these curves.

Since it is only the overlap region that matters, it is common to plot just the range of angles around the diffraction condition ($\theta - \alpha$), and to make the energy resolution explicit, one can use the range of energies ($\Delta E = E - E_0$) as the ordinate. Such a plot is illustrated in Fig. 4.17 for a variety of monochromator crystals at a PETRA-III beamline.

4.6.2 *Crystal Monochromators*

Although a single crystal is sufficient for creating a monochromatic beam, it is obvious that as the angle is changed for different energies, the exiting X-ray beam will change direction, and downstream its position will move significantly. The simplest way to keep a fixed-exit direction is to use a 2-bounce crystal monochromator (Fig. 4.18). The original designs used a "channel cut crystal" in which a channel was cut along a particular plane in a single crystal of Si or Ge. Using a single crystal presumably ensured that the planes would be parallel. However, because of the thermal load on the first crystal, its *d*-spacing and angle with respect to the second crystal can change. Nowadays, most monochromators have independent crystals with cooling on the first crystal and the ability to slightly vary the angle of the second crystal.

The channel cut geometry keeps the beam output angle constant, but the height of the beam will change as the pair of crystals is rotated. Some beamlines account for this by moving the entire experiment up and down to track the beam motion. A

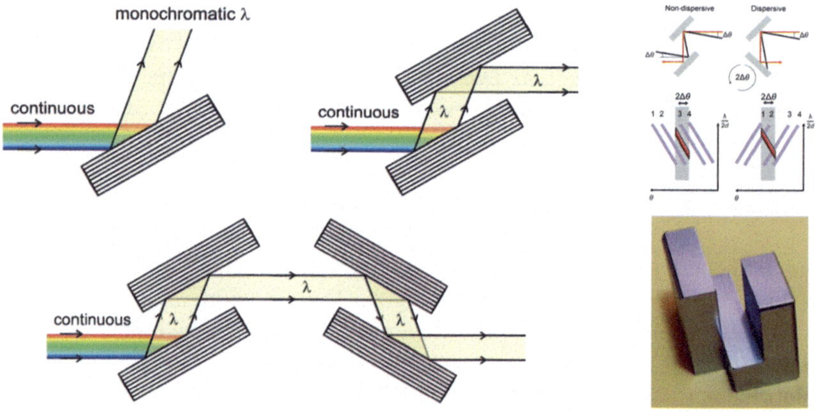

Fig. 4.18 Medium-resolution crystal monochromators. Top left: with a single reflection, the diffracted beam moves as the crystal rotates. Top middle: a double crystal monochromator, in non-dispersive geometry. Note that the beam will still change in vertical position as the monochromator rotates. Top right: comparison of Dumond diagrams for two crystals in non-dispersive and dispersive geometries. If the first crystal is rotated by θ, the second crystal must be rotated by 2θ. Lower left: a 4-bounce design. Lower right: channel cut crystals—the simplest 2-bounce monochromator

clever design for a "fixed-exit" beam monochromator was developed by Golevchenko and coworkers at Bell Labs [121]. By incorporating a rigid right angle rotating midway between the two crystals, with appropriate linkages, the horizontal position of the second crystal is adjusted to maintain a constant beam height. Instead of mechanical linkages, most beamline monochromators now accomplish a fixed-exit beam with computerized motion control.

For a 2-bounce monochromator in the symmetric or "non-dispersive" geometry, the resolution is not improved by the second crystal. This can be seen by looking at the overlap of slices in a Dumond diagram—when the two crystals are brought into alignment, their transmission functions are parallel and perfectly overlap (Fig. 4.18). Typical resolutions are 1.4×10^{-4} with Si(1 1 1) crystals and 3×10^{-5} with Si (3 1 1). The resolution can be improved by changing to a dispersive geometry, yielding a smaller overlap of transmission functions. However, with only two crystals, the angular deviation problem would return. All is solved by using two pairs of crystals, for a "4-bounce" monochromator, with the second and third crystals in the dispersive geometry, as shown in Fig. 4.18. In this geometry the resolution with Si(3 1 1) crystals can approach 10^{-5}.

Although 4-bounce designs with conventional crystals can achieve an impressive 0.1 eV resolution, some experiments such as IXS (Chap. 8) and NRVS (Chap. 10) require at least two orders of magnitude better resolution. This is normally achieved either with asymmetrically cut crystals or by operating in an extreme backscattering geometry.

4.6.2.1 Extreme Backscattering

The extreme backscattering approach exploits the fact that the angular Darwin width acceptance ω_D becomes quite large when the Bragg angle is close to 90°. It is given by:

$$\omega_D(\theta \sim 90^\circ) \cong 2\sqrt{|\chi|} \tag{4.29}$$

where the susceptibility $|\chi|$ is on the order of 10^{-6}. The acceptance can thus become on the order of milliradians (Table 4.2) [122]. By choosing a high-order reflection to yield such a Bragg angle, a resolution of ~1 meV can be obtained. For example, with the Si(11 11 11) reflection, an intrinsic resolution of 0.8 meV can be obtained at 21.75 keV—for a $\Delta E/E$ of 3.6×10^{-8}!

One complication is that close to backscattering, the diffracted wavelength $\lambda \cong 2d$, and the energy changes very little with angle. So, in extreme backscattering, the only practical way to scan such a monochromator is to change the d-spacing. This is done by changing the crystal temperature, using:

$$\Delta\lambda/\lambda = \Delta E/E = \Delta d/d = \alpha \Delta T \tag{4.30}$$

where α is the coefficient of thermal expansion. At room temperature, the lattice spacing for Si changes by 2.6×10^{-6} per degree Kelvin, so for the above reflection,

Table 4.2 Properties of some commonly used near-perfect crystals [115–117]

Material	F_H	d (Å)	t_{ext} (μ)	t_{abs} (μ)	ω_D (mrad)	α (10^{-6}/K)	k (W/cm/K)	Ref.
C (1 1 1) (diamond—300 K)	17.8	2.059	2.20	2900	0.01488	0.8	15–20	[115]
C (1 1 1) (diamond—100 K)	17.8	2.059	2.20	2900	0.01488	0.05	~50	[115]
Si (1 1 1) 300 K	60.6	3.135	1.50	252	0.02162	2.6	1.5	[115]
Si (1 1 1) 100 K	60.6	3.135	1.50	252	0.02162	−0.4	8.8	[115]
Si (4 0 0)	58.8	1.357	3.70	252	0.00921	2.6	1.5	[115]
Ge (1 1 1)	155.5	3.266	0.64	12.6	0.04897	5.9	0.58	[115]
Al_2O_3 (sapphire)		[c]				6.7[b]	23.1[b]	[118]
SiO_2 (quartz)		[c]				7.1[b]	10.7[b]	[119]
$LiNbO_3$		[c]				14.1[a]	4	[116]

[a]a-axis [b]|| c-axis [c]a wide range of reflections are used for these crystals in backscattering

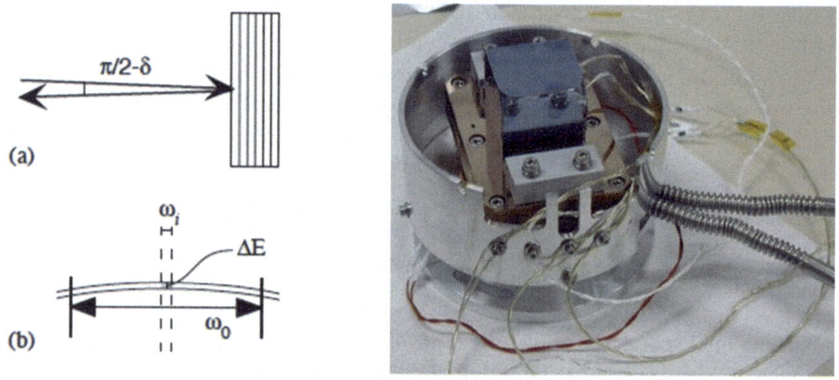

Fig. 4.19 Left: the geometry for extreme backscattering and the resultant Dumond diagram. Right: a monochromator for extreme backscattering with temperature control accessories

ΔT of 1 K corresponds to ΔE of ~57 meV, and a 100 meV vibrational spectrum can be scanned in less than 2 K. This reflection is used on the SPring-8 Beamline 35 for IXS experiments [123].

Sapphire crystals have been used in backscattering for NRVS experiments from 20 to 40 keV with ~1 meV resolution [124]. At ~37.13 keV, a 1 meV shift results from a ~0.15 K temperature change, so a 40 meV wide spectrum requires only a ~6 K temperature change. With further improvements in the quality of material available, sapphire looks promising for even higher-energy (>40 keV) experiments. Quartz (high-quality α-SiO_2) [125] and lithium niobate ($LiNbO_3$) have also been proposed as good alternatives for backscattering analyzers, and Gog et al. have published an extensive compilation of near backscattering reflections and relevant properties for all of these materials [116].

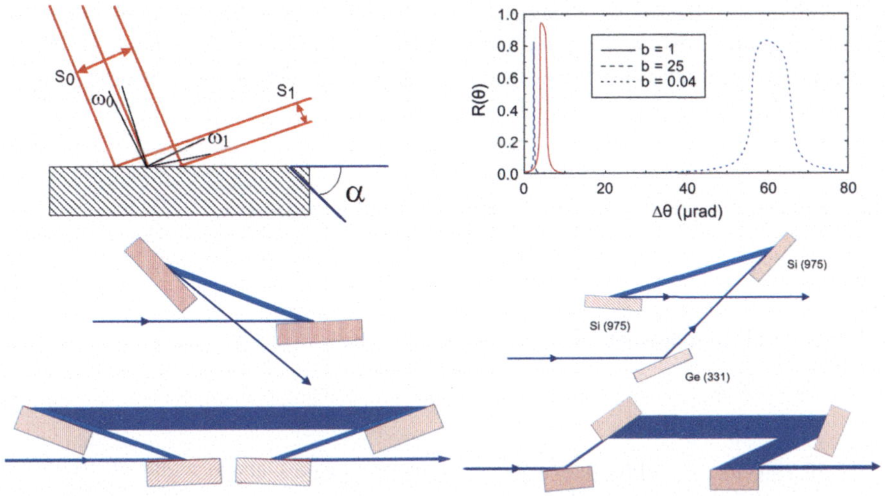

Fig. 4.20 Top left: the quantities involved in asymmetric diffraction. Top right: angular dependence of Si(8 4 0) reflectivity at 14.413 keV for three different asymmetry factors. Middle and bottom: a variety of ultra-high-resolution crystal monochromators [122, 126]

4.6.2.2 Asymmetric Diffraction

The other alternative for very-high-energy resolution involves asymmetric diffraction. In these designs, the crystal surface is not cut in the usual fashion parallel to the surface, but instead at an angle α called the "asymmetry angle" (Fig. 4.20). A special property of asymmetric diffraction is that it allows for trade-offs between beam size and angular divergence. This in turn allows monochromator designers to use very-high-order reflections with very-high-energy resolution but extremely narrow Darwin widths.

If one defines α as the angle between the crystal surface and the diffracting planes, then the incident and exiting glancing angles, θ_i and θ_e, are given by $\theta_\mathrm{i} = \theta + \alpha$ and $\theta_\mathrm{e} = \theta - \alpha$. A warning, some literature reverses the definition of the sign of α, yielding equations with opposite signs. One can then define an *asymmetry factor*, *b*, as:

$$b = \frac{\sin\theta_i}{\sin\theta_e} = \frac{\sin\,(\theta + \alpha)}{\sin\,(\theta - \alpha)} \tag{4.31}$$

The change in beam size, from the incident width H_i to the exit width H_e, is given by:

$$H_\mathrm{e} = H_\mathrm{i}/b \tag{4.32}$$

The change in beam size is accompanied by a change in the angular ranges for input and output. If ω_S is the intrinsic Darwin width of the reflection in the conventional symmetric geometry, then the angular acceptance ω_i is modified by:

$$\omega_i = \omega_S / \sqrt{b} \tag{4.33}$$

and the divergence of the reflected beam is also changed, by:

$$\omega_e = \omega_S \cdot \sqrt{b} \tag{4.34}$$

and the overall transformation of angles is given by:

$$\omega_e = b\omega_i \tag{4.35}$$

As an extreme example, the Si(9 7 5) reflection has a Darwin width of ~0.35 arc sec = 1.7 μrad at 14.4 keV. By using an asymmetry factor $b \sim 0.01$, the angular acceptance would improve approximately tenfold to ~17 μrad, while the exit beam divergence would be correspondingly reduced to ~0.017 μrad (Fig. 4.20).

Use of asymmetric diffraction gives a significant boost to the resolution. For example, in the popular 4-bounce geometry of Fig. 4.20, the energy resolution is predicted to be [127]:

$$\frac{\Delta E_{HRM}}{E} \cong \sqrt{b_1 b_2} \omega_S \cot \theta_B \tag{4.36}$$

Here you can see the multiple factors that allow for the exquisite resolution of asymmetric diffraction monochromators. With Bragg angles beyond 80°, cot θ_B is small and the Darwin widths are quite narrow. Finally, an extra boost is provided by the asymmetry factors. In one classic implementation, Ishikawa and coworkers used four Si(11 5 3) reflections at a Bragg angle θ_B of 80.4° and $b_1 = b_2 = 1/10.4$. With a Darwin width of 1.8×10^{-6} radians, they had a theoretical resolution $\Delta E/E$ of 8.6×10^{-9} at 14.4 keV or ~0.1 meV! Their experimental value was close to this: ~0.12 meV with a final photon flux of 10^7 photons/s [127].

4.6.2.3 Thermal Issues for Monochromators

The monochromator is one of the most important parts of the experiment that are under your control at the beamline. As the power and brightness of synchrotron sources has continually increased, heat load effects on the crystals have become an issue. Thermal distortion of the monochromator crystal can lower the reflectivity and distort the beam profile, so most third-generation synchrotron sources use a "high heat load monochromator" either alone or in conjunction with secondary monochromators (Fig. 4.21). The optimum crystal and cooling approach involves consideration of (a) how rapidly the heat can be removed and (b) the residual effects from

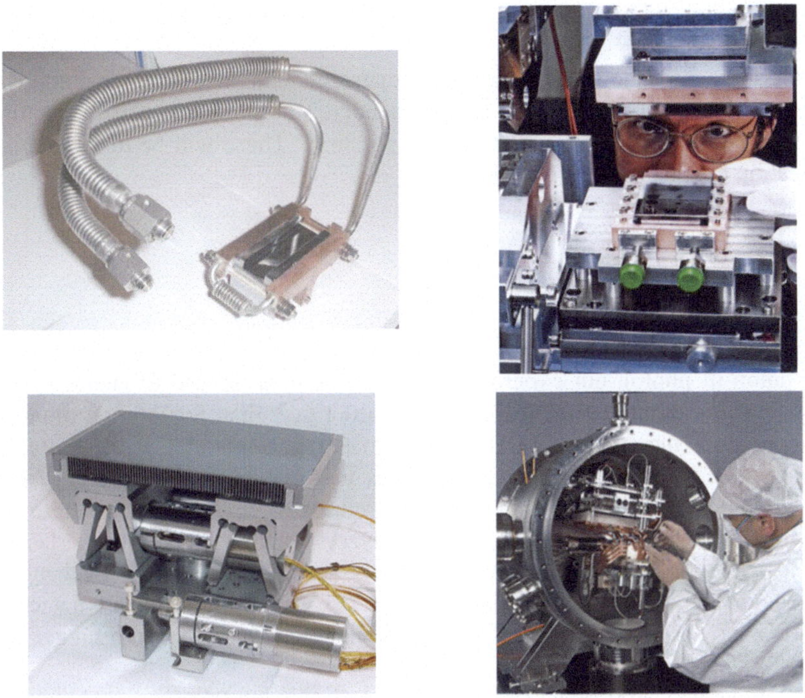

Fig. 4.21 Practical crystal monochromator components. Top left: a flat crystal mounted in a cooling assembly. Top right: looking in on a crystal. Lower left: a sagittal focusing crystal bender. Lower right: a technician adjusting monochromator before closing up

temperature variation that remains. Designers thus use the ratio of coefficients k for thermal conductivity to thermal expansion coefficient α as a figure of merit (Table 4.2).

One approach is diamond, which has the best thermal conductivity of candidate monochromator crystals. In many cases water cooling is sufficient. The simplicity and low cost of water cooling have made diamond crystals one of the popular options, and they have been used at a variety of sources [128].

An alternate and clever approach involves liquid nitrogen cooling of silicon crystals to ~80 K. The thermal conductivities of Si and Ge increase at low temperatures, and the coefficients of expansion for Si and Ge go to zero at 125 K and 50 K, respectively. Very stable high heat load LN_2-cooled monochromators are in use at ESRF [129], and a recent comparative study came out in favor of silicon over diamond [130] (Fig. 4.21).

The monochromators for NRVS are an example of extreme thermal sensitivity. Referring back to Eq. 4.30, a temperature change of 0.03 K will change the diffracted photon energy by more than 1 meV—the resolution of the experiment! For this reason, NRVS beamlines usually feature a separate hutch for the high-resolution monochromator, which is kept closed during the experiment, since re-equilibration can take hours.

Other issues for monochromator performance include the harmonic composition of the transmitted beam, backlash and encoders during scans, and the notorious monochromator artefacts cleverly called "glitches." These will be discussed in the application chapters where relevant.

4.6.3 Crystal Analyzers

For photon-in photon-out experiments, an *analyzer* is needed to characterize the energies of the emitted photons. This is a harder task than for the first monochromator, because the synchrotron source beam is nearly parallel, while the emitted photons come out in all directions. This forces analyzers to resolve different energies with a single reflection, so they all work relatively close to backscattering. To capture a reasonable solid angle, most analyzers arrange the crystal surface in a spherical (Rowland circle) geometry. This is accomplished either by bending a single crystal to the desired surface (Fig. 4.22), by arranging small flat crystals close to such a surface, or possibly by grouping such analyzers into bigger arrays (Figs. 8.23 and 8.26).

For moderate (~100 meV) resolution, the radius of the Rowland circle is usually on the order of 1–2 m. However, in some cases, say to see vibrational features, ~1 meV resolution is needed. In these cases, rather heroic 10–20 m Rowland circles are used. To cover a reasonable solid angle at that distance, some resort to multiple spherical analyzers. Even with a large bending radius, residual strain in the analyzer crystal would diminish the energy resolution. To avoid that strain, ultra-high-resolution analyzers are typically "diced," with a single crystal diced into thousands of pillars (Fig. 4.22). With typical dimensions of $0.9 \times 0.9 \times 5$ mm, these pillars mitigate the strain that would ordinarily result from bending a flat surface into a spherical one.

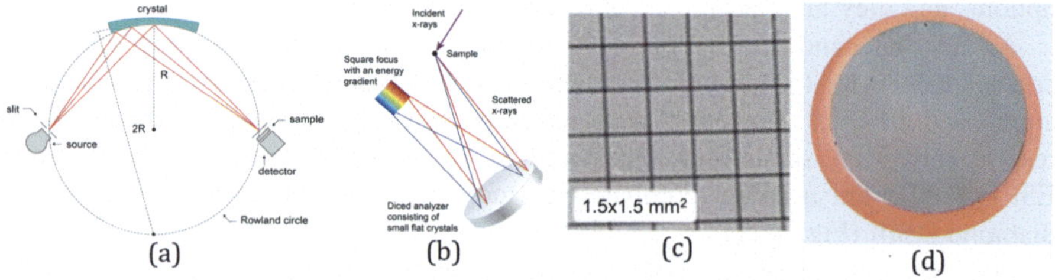

Fig. 4.22 (**a**) The Rowland circle geometry; (**b**) resolution is improved by combining a moderately diced analyzer with a position sensitive detector [131]; (**c**) diced optics with sub-mm lateral dimensions for high-resolution experiments; (**d**) a large crystal diced into thousands of individual pillars [132]

4.6.4 Crystal Polarizers

Crystals can also be used as quarter-wave plates to produce circularly polarized X-rays. As a reminder, a linearly polarized beam traveling in the z-direction can always be reconsidered as the sum of two perpendicular components, say:

$$\vec{E}(z,t) = \hat{i}E_\sigma \cos[\omega(t - n_\sigma z/c)] + \hat{j}E_\pi \cos[\omega(t - n_\pi z/c)] \tag{4.37}$$

where $\hat{i}$ and $\hat{j}$ are unit vectors in orthogonal directions, E_σ and E_π are the amplitudes in those directions, and n_σ and n_π are the refractive indices in those directions. For linear polarization, the relative amplitudes of E_σ and E_π determine the polarization orientation, and for ordinary materials, n_σ and n_π are equal. But, in some cases, materials can have different indices of refraction for different electric field directions—they are "birefringent." This "birefringence" causes a phase difference $\Delta\varphi$ to accumulate between the two components as the beam propagates through distance l:

$$\Delta\varphi = \frac{2\pi}{\lambda_0} l(n_\sigma - n_\pi) \tag{4.38}$$

If an optic introduces different phase shifts for these two directions, then it will change the polarization of the transmitted beam. (With visible light, the most familiar example is Iceland spar.) In particular, if the difference in phase shifts is $\pi/2$, then the optic is called a "quarter-wave plate," and the transmitted beam will have circular polarization.

It turns out that perfect or near-perfect crystals are birefringent in the vicinity of Bragg reflections [134, 135]. As illustrated in Fig. 4.23, suppose one has a purely linearly polarized incident beam, where E_π is the amplitude of the electric field in the

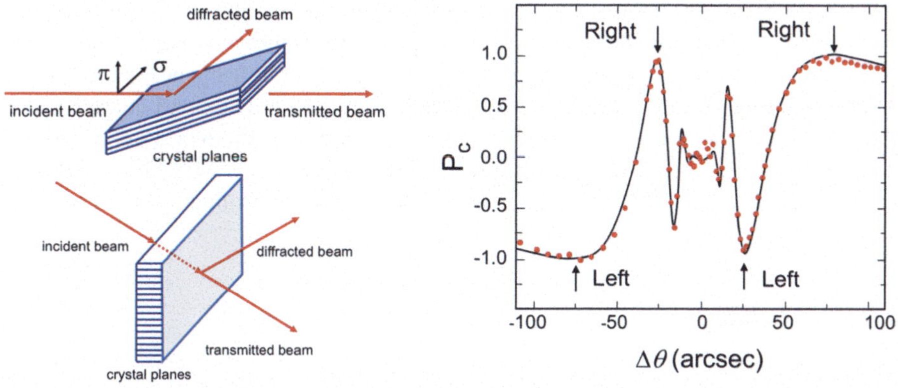

Fig. 4.23 Left: Bragg (top) and Laue (bottom) geometries used for crystal polarizers. Right: the degree of circular polarization (measured using the XMCD effect at Gd L-edge at 7.243 keV) vs. angular deviation from the Bragg angle for a 740 μ (1 1 1) diamond crystal, redrawn from [133]

diffraction plane of the phase plate and E_σ is the component perpendicular to the plane. If the diffraction planes of a crystal are rotated by an angle Ψ from with respect to the electric field of the incoming X-rays, and if the phase shift difference for the two components E_σ and E_π is Φ, then the degree of circular polarization is given by [133,135]:

$$P_3 = P_C = \left(\frac{I_R - I_L}{I_R + I_L}\right) = \sin 2\Psi \sin \Phi \tag{4.39}$$

Obviously, the maximum circular polarization is obtained when the diffracting planes are at $\pi/4$ or 45° with respect to the electric field. It can be shown that somewhat far from the exact Bragg condition, the phase shift difference Φ depends on the beam pathlength in the crystal t and the offset $\Delta\theta$, via [136]:

$$\Phi = \left(\frac{2\pi}{\lambda}\right)(n_\sigma - n_\pi) = \frac{r_e^2 \operatorname{Re}\left[F_h F_{\bar{h}}\right] \lambda^3 \sin(2\theta_B) t}{2\pi V^2 \Delta\theta} \tag{4.40}$$

where r_e is the classical electron radius, F_h is the structure factor of the *hkl* reflection, and V is the volume of the unit cell.

When the Ψ angle is $\pi/4$, as the crystal is rotated through θ, the degree of circular polarization P_C will be a function of that rotation (Fig. 4.23). The above equation has been formatted to emphasize that the phase shift depends on the beam pathlength in the crystal t and the angular offset. If $\Delta\theta_{\pi/2}$ is the angular deviation from the Bragg condition that produces a $\pi/2$ phase shift, then (assuming pure linear polarization to start with):

$$P_3(\Delta\theta) = P_C(\Delta\theta) = \sin 2\Psi \sin\left[(\pi/2)\left(\Delta\theta_{\pi/2}/\Delta\theta\right)\right] \tag{4.41}$$

Although birefringence is strongest close to the Bragg peak, then most of the beam is diffracted, and there are also artefacts from the angular dependence of the birefringence. Therefore, these crystals are thus usually employed in the wings of the reflection, where most of the beam is transmitted, the birefringence varies more slowly, and there is still a useful refractive index mismatch. Diamond quarter-wave plates are used to produce circularly polarized hard X-rays at the world's four largest storage ring synchrotron radiation sources, ESRF [133, 137], APS [138], SPring-8 [139], and PETRA-III [140], and at many other facilities around the world.

4.6.5 Multilayers

For the X-ray region, multilayers are assemblies of dozens of alternating layers of high- and low-Z materials, deposited on Si or another suitable substrate (Fig. 4.24). The individual layers have thicknesses of the order of nanometers. Multilayers have applications across the X-ray spectrum. At the lower end of the soft X-ray region, they can have high reflectivity even close to normal incidence. At the other extreme,

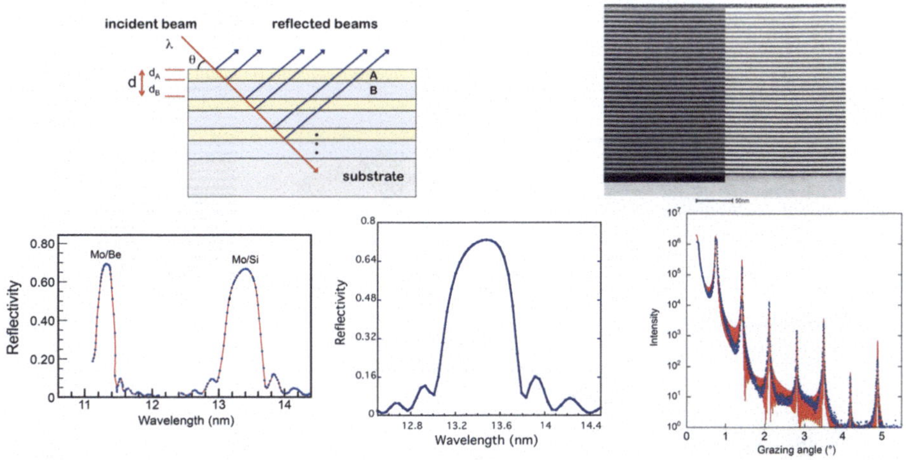

Fig. 4.24 Top left: the basic structure of a multilayer optic. Top right: electron microscope images of a Cr/B_4C multilayers—Cr is the dark layer [141]. Lower left: normal incidence reflectivity of CXRO multilayers consisting of 50 × 11.34 nm Mo/Be (left) or 40 × 13.42 Mo/Si bilayers (right) [142]. Lower middle: calculated performance (using CXRO website) of a Mo/Si multilayer. Lower right: high-energy reflectivity (at 0.154 nm) vs. angle for a Mo/Si multilayer [143]

they can improve the reflectivity of hard X-ray mirrors from 2000 to >50 keV. How can multilayer X-ray mirrors achieve such performance?

Multilayers are treated theoretically as stacks of partially reflecting mirrors, arranged so that weak individual reflections add in phase to produce good overall reflectivity. We saw earlier in this chapter that at large glancing angles, the intensity reflectivity R^2 might be ~10^{-4}. However, this also means that the amplitude reflectivity R might be 10^{-2}. Thus, if one can add the reflections from 100 layers in phase, the total reflectivity could approach 100%.

Apart from a small correction for refraction, if d is the overall period length, and $\bar{\delta}$ is the bilayer weighted real part of the refractive index, then the multilayer equation looks just like the Bragg equation:

$$m\lambda = 2d\sin\theta\sqrt{1 - \frac{2\bar{\delta}}{\sin^2\theta}} = 2d\sin\theta\sqrt{1 - \frac{4\bar{\delta}d^2}{m^2\lambda^2}} \qquad (4.42)$$

To produce a high efficiency multilayer reflector, one needs to stack materials having alternately high and low refractive indices, with thicknesses such that the path length difference for reflections from successive layer pairs is equal to one wavelength. These multilayers are thus quite similar to the "quarter-wave stacks" used for visible light coating technology. In the X-ray/EUV region, they can be used to improve the reflectivity of mirrors and/or to improve the performance of diffraction gratings (Fig. 4.25).

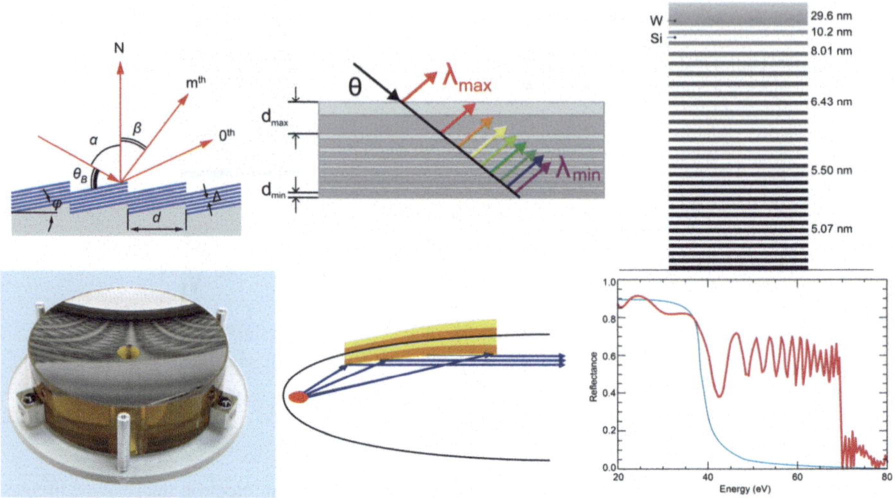

Fig. 4.25 Top left: multilayers applied to a diffraction grating to improve high-order performance. Top middle: depth-graded optics. Top right: schematic of variable spacing applied to an X-ray mirror. Lower left: a multilayer coated mirror made at the Center for X-ray Optics. Lower middle: graded multilayers applied to parabolic optics. The variation in bilayer spacing helps match the diffraction condition at different angles for the same wavelength. Lower right: improved reflectivity range from the multilayer coating shown above (red line), compared to a pure Pt coating (green line)

4.6.5.1 Graded Multilayers

Since the composition and thickness of the individual layers can be controlled down to atomic dimensions, it is possible to fabricate multilayer optics with a gradation of *d*-spacings. In one application, these "graded multilayers" are applied to ellipsoidal or paraboloidal surfaces and used to nearly continuously satisfy the diffraction equation for different angles of incidence (Fig. 4.25).

4.7 Putting It All Together: Typical Beamlines

So now we're ready to put all these components together to build a beamline. Well, maybe just understand a beamline. Using the reflective, refractive, and diffractive elements discussed so far, a beamline transfers photons from the synchrotron source to the sample—controlling the energy band pass and physical size and shape of the final beam. The required beam size and energy bandwidth of course depend on the experiment. But for spectroscopy and imaging, it is almost universally true that "the more photons, the better."

To efficiently use the bend magnet or insertion device radiation, it is necessary to match the "emittance" of the source with the "acceptance" of the optics. As an example, suppose the beamline uses a Si(4 0 0) crystal monochromator on a bend magnet source of a 5 GeV ring. We know from Chap. 3 that the vertical divergence is on the order of $1/\gamma = 10^{-4}$ or 100 μ-radians. But we also know from Table 4.2 that the Darwin width for high-resolution crystal reflection is ~1–10 μ-radians. To keep the optimal resolution of the crystal, one could insert slits to reduce the angular divergence of the beam, but such an approach would throw away most of the radiation. Instead, to match the source and the optic, one can insert a collimating mirror upstream of the monochromator that will reduce the vertical divergence of the source at the expense of a larger vertical beam size.

We can use a phase space approach to better understand beamline optics, in the same manner in which it was used to describe the properties of the electron beam. The emittance of the photon beam can be plotted vs. the acceptance of an optic, and the overlap represents the fraction of the beam transmitted. The output of the optic will then have its own phase space representation, which one can then map sequentially down the beamline.

As illustrated in Fig. 4.26, a diverging source will have a tilted ellipse in $y - y'$ space, where y is the vertical beam height and y' is the vertical angular deviation. In this representation, a slit is just a pair of vertical lines defining the range of heights accepted. In contrast, for a given wavelength, a flat crystal monochromator is a pair of horizontal lines whose separation is the Darwin width for that reflection. As for mirrors, a focusing optic will rotate the ellipse counterclockwise, while a collimating optic will produce an ellipse larger in y and smaller in y' (Fig. 4.26).

Referring to our original example, a slit would narrow the angular range of the source and preserve the intrinsic resolution of the Si(4 0 0) crystal, but at the expense of losing most of the available photons. Instead, the collimating mirror reduced the divergence by making the physical beam size larger (Fig. 4.26). Since crystals can be quite large, this is not a problem for the crystal acceptance. The phase space description of this approach is that collimation of the source achieves the best overlap with the crystal acceptance by rotation of the phase space ellipse.

In the following examples, for hard X-ray beamlines with crystal monochromators, the most popular approach is thus (1) collimate, (2) monochromate, and (3) refocus. Soft X-ray beamlines that use grating monochromators are sometimes a different story.

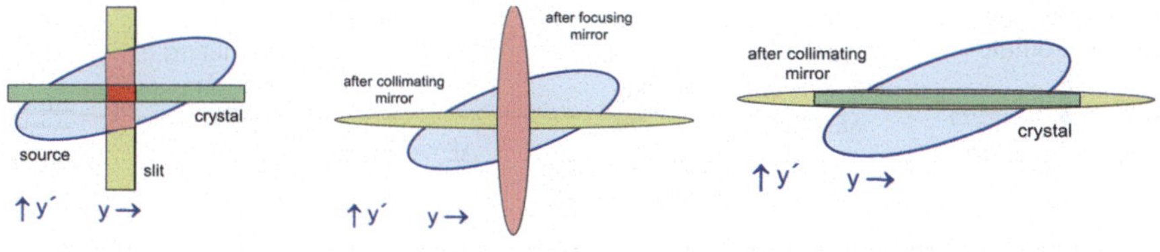

Fig. 4.26 Phase space approach to beamline optics. Left: phase space representations of the photon source emittance vs. slit and crystal acceptance. Middle: effects of focusing or collimating on the photon phase space. The area is conserved. Right: the collimating mirror successfully matches the beam divergence to the Darwin width of the crystal

4.7.1 Hard X-ray Beamline Examples

Hard X-ray beamlines for spectroscopy operate from ~2 to 60 keV (or higher), with resolutions on the order of 1 eV for garden variety EXAFS to less than 1 meV for IXS and NRVS, and even ~neV for Mossbauer spectroscopy (Chap. 9)! Silicon is by far the most popular monochromator material, but for energies below the Si(1 1 1) cutoff (~2400 eV), Ge, InSb, and YB_{66} are used. On the high-energy high-resolution side, sapphire and quartz are popular, and diamond is used in some high heat load monochromators. Three different approaches to hard X-ray beamlines are illustrated in Fig. 4.27.

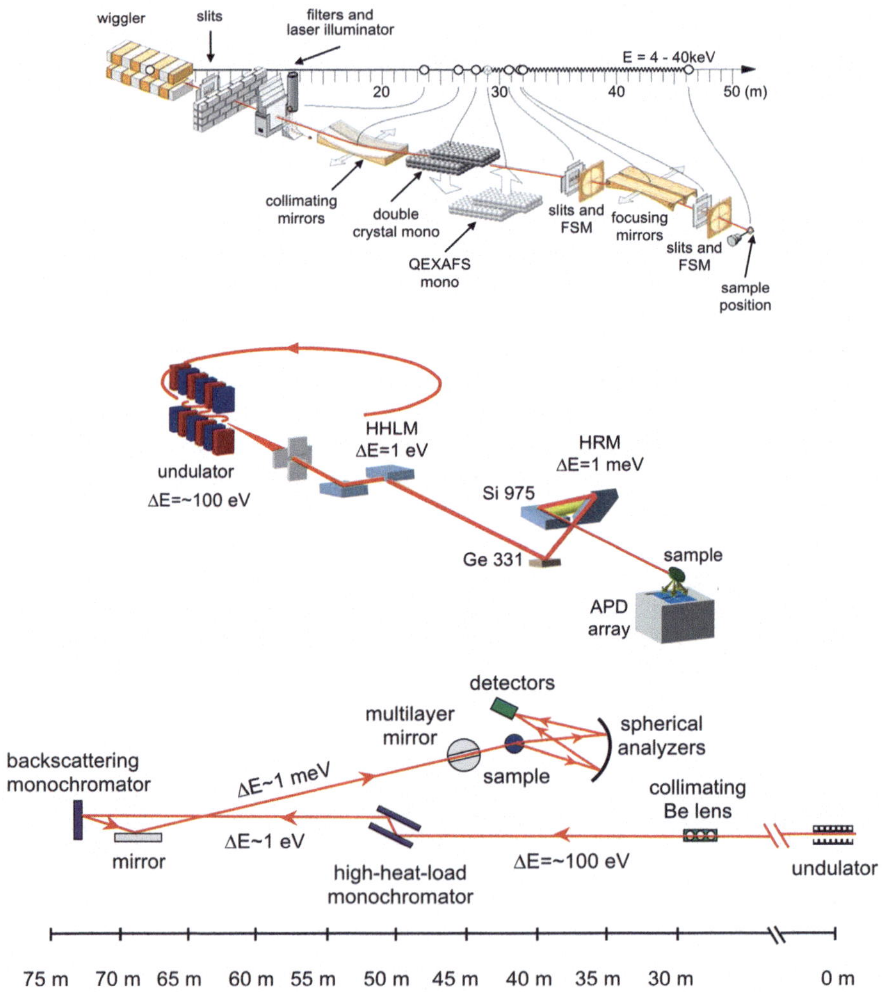

Fig. 4.27 Hard X-ray beamlines. Top: an X-ray absorption beamline at MAX-IV [144]. Middle: a beamline for nuclear resonance experiments at SPring-8 [147]. Bottom: a high-resolution inelastic X-ray scattering beamline at ESRF [146]

The "Balder beamline" at MAX-IV represents a popular approach for X-ray absorption spectroscopy beamlines [144]. The beam from a 38-pole 2.1 T wiggler source is collimated using a water-cooled mirror with either a Si or Ir surface (for different cutoff energies). This is followed by a liquid nitrogen-cooled monochromator with Si(1 1 1) or Si(3 1 1) crystals. The beam is refocused into a 100 × 100 μ^2 spot in the hutch with a toroidal mirror, again with either a Si or Ir surface. A huge amount of the ~1.2 kW power from the source is dissipated in the optics: ~700 W in the collimating mirror and up to ~500 W in the first crystal. The net result is a monochromatic beam with ~10^{12}–10^{13} photons s^{-1}.

Beamline BL09XU at SPring-8 is an example of a very productive beamline for nuclear spectroscopy experiments [145]. Since the very-high-resolution monochromators are heat sensitive, the source flux is first reduced with a liquid nitrogen-cooled high heat load Si (1 1 1) monochromator. The initial asymmetric Ge(3 3 1) crystal ($b = 1/2$) reduces the vertical divergence from 10 to 5 μrad while expanding the beam height from 0.5 to 1 mm. (This crystal also serves to set the output beam close to horizontal.) More collimation comes from reflection by the asymmetrically cut Si(9 7 5) crystal ($b = 1/10.8$), which narrows the angular divergence from 5 to 0.09 μrad, while expanding the vertical size of the beam from 1 to 11 mm. The second asymmetric Si(9 7 5) crystal demagnifies the vertical beam size back to ~0.5 mm ($b = 18$). The high-resolution monochromator is contained in its own separate hutch which keeps the temperature stable to ~0.02 K. This beamline now yields a 0.8 meV bandwidth beam at 14.4 keV with a flux of ~2.5 × 10^9 photons s^{-1}.

Beamline ID28 at ESRF uses a different approach for inelastic X-ray scattering experiments [146]. In this case the collimating element is chosen from a set of Be compound refractive lenses. The device with 24 holes reduces the vertical divergence to ~1 μrad. As with SPring-8, a Si (1 1 1) high heat load monochromator disposes of most of the unwanted flux. Then, a flat Si backscattering monochromator operating at 89.98° is used for high resolution: the Si (13 13 13) reflection can yield ~1.5 × 10^9 photons s^{-1} with ~0.5 meV resolution. The monochromatic beam is then refocused onto the sample with a 120 period Ru/B_4C multilayer mirror. The scattered radiation is analyzed with a set of spherically curved analyzers that each contain ~12,000 silicon (13,13,13) pixels that provide <1 meV resolution at 25,704 eV!

4.7.2 *Soft X-ray Beamlines*

Soft X-ray beamlines generally use grating monochromators over the range from the Li K-edge (~54 eV) to ~2000 eV. Although crystals such as YB_{66} and beryl can cover the 1–2 keV range, for example, at beamline BL2A at UVSOR [148], they are not "perfect," and the resulting linewidths are broadened by their mosaicity. The Si K-edge (~1839 eV) is in an overlap region which can also be covered well by InSb (1 1 1) crystal monochromators. Above this energy, crystal monochromators using

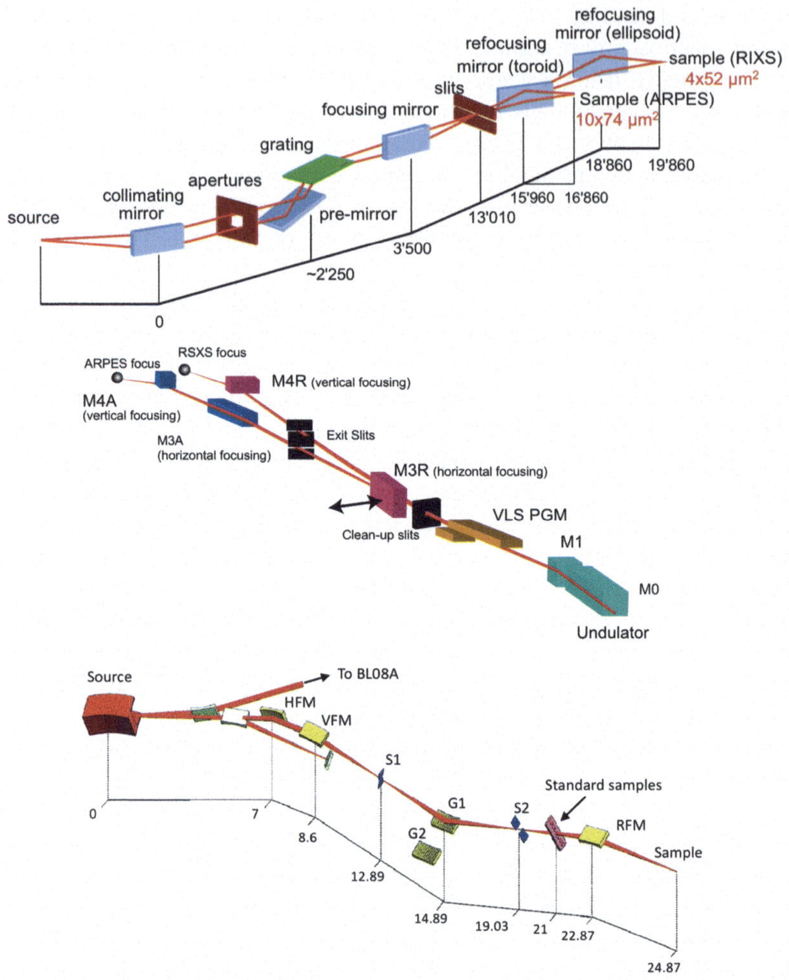

Fig. 4.28 Soft X-ray beamlines. Top: collimated PGM beamline ADDRESS at SLS [146]. Middle: VLS-PGM beamline at APS [155]. Bottom: a "Dragon" style SGM beamline such as at the Taiwan Photon Source [105]

Ge(1 1 1) crystals can take over, while above the P K-edge (2145 eV), Si(1 1 1) is preferred.

There is a long history of plane grating monochromator (PGM) designs (sometimes known as the SX-700) [149], for soft X-ray beamlines [150–152]. For example, the ADDRESS beamline at the SLS uses a collimated PGM design that is easy to compare with the previous hard X-ray beamlines [146]. The beam from the 75-period undulator source is collimated vertically (and focused horizontally) by a toroidal mirror (Fig. 4.28). A plane grating chosen from a set with different groove densities diffracts this parallel beam, and a cylindrical mirror focuses the radiation onto the exit slit. The monochromatic beam is then refocused onto the sample. As a representative performance, with the intermediate resolution (2000 l mm^{-1}) grating, the beamline delivers ~10^{12} photons s^{-1} at 750 eV with a $E/\Delta E$ resolution of ~15,000 or 50 meV. The ADDRESS analyzer for RIXS experiments [153] will be discussed in the Photon-In Photon-Out section (Chap. 8).

A variation on the PGM design is to use variable line spacing (VLS) gratings [154]. This allows for beamlines to operate without slits and to achieve very high resolution. The line spacing is varied so that the grating itself focuses a particular wavelength at a given distance. Such a beamline at the APS (Fig. 4.28) provides ~10^{10} s^{-1} over the 250–1700 eV range with an $E/\Delta E$ of 50,000 [155].

A popular alternative approach is based on a spherical grating monochromator (SGM) with a constant included angle in a variable Rowland circle geometry (Fig. 4.28). This was first implemented by Chen and Sette at NSLS as the "Dragon" beamline [103, 156]. In contrast with other beamlines discussed so far, it begins by focusing instead of collimating in the vertical and in the horizontal (Fig. 4.28). After passing through a slit, the beam is projected onto a spherical grating. The grating works not only to monochromate the beam but also to focus it in the vertical onto the exit slit. However, the focus position changes with the energy selected, so the exit slit has to move as the energy is scanned. The design has been commercialized by Horiba and is used at many light sources including the SRRC [157] and the Canadian Light Source [158] (Fig. 4.28).

4.8 Suggested Exercises

1. Using the small angle approximation for cos θ, derive the expression for the glancing angle for total external reflection: $\cos \theta_{cg} = n_2/n_1$.
2. Using tabulated indices of refraction, calculate the focal length and transmission at 35 Å of a bi-concave carbon lens with a radius of curvature of 81 μm.
3. The EPU beamline at the ALS has a pre-mirror at a glancing angle of 1.9 °.

   (a) Estimate the reflectivity at 35 Å for a Au mirror at this angle, using $\delta = 0.01$ and $\beta = 0.009$.
   (b) Using tabulated values, estimate the reflectivity of the same mirror at 2.6 KeV.
   (c) Would it be worthwhile to add a crystal mono for S K-edge work (see Chap. 7)?

4. Suppose you are trying to build a multilayer mirror for capturing emission at the nitrogen Kα fluorescence line (Appendix I). A Si/Ru multilayer can give a reflectivity of better than 8% at 20°. Find a combination of materials and spacings that will give (in principle) a better reflectivity. Why is the reflectivity better?
5. The Pohang Light Source has an undulator beamline with a 700 l/mm plane grating monochromator. Assuming a glancing angle of 1.7 °:

(a) At what grazing angle would the diffracted beam for the Cu L_3 edge occur?
(b) What is the best possible resolution for a 10 μ exit slit size at S2?
Suppose you would use a crystal monochromator instead.
(c) What energy beam would be diffracted by a Si(1 1 1) crystal at the same angle?
(d) What would be the resolution at that energy for a 1 mm S2 slit with such a mono?
(e) What would be the limiting resolution with a very small slit?

6. Suppose you want to use a zone plate for capturing nitrogen Kα fluorescence.

(a) You decide to purchase a zone plate with 1000 zones (500 periods). If your supplier can achieve a minimum zone width of 40 nm, what will be the approximate diameter D?
(b) What will the minimum focal length be at 400 eV?

7. Ishikawa and coworkers built a 4-bounce asymmetric diffraction monochromator using Si(11 5 3) reflections with $\omega_S = 1.8$ mrad and an asymmetry angle of $-78.4°$ for the first reflection. Calculate the asymmetry factor b. Then, calculate the angular acceptance ω_i and the exit beam divergence ω_e for the first reflection.

4.9 Reference Books and Review Articles

1. *Optical Systems for Soft X-rays*, A. G. Michette, Plenum, New York 1986, ISBN 0-306-42320-0 (hardcover) 1461293049 (paperback)—all you want to know about soft X-ray optics
2. *X-ray Optics for Third Generation Hard X-ray Synchrotron Radiation Sources*, D. M. Mills, Chapter 2, pp. 41-100, in *Third Generation Hard X-ray Synchrotron Radiation Sources,* edited by D. Mills, John Wiley & Sons, New York, 2002, ISBN 0-471-31433-1.
3. *Perfect Crystal Theory*, Chapter 14 in *X-ray Diffraction*, B. E. Warren, pp. 315-357, Dover Publications, New York, 1990, ISBN 0-486-66317-5.—a reprint from 1969, Warren understood perfect crystal diffraction long ago and presents it clearly.
4. *Diffraction by Perfect Crystals*, J. Als-Nielsen and D. McMorrow, Chapter 5 in *Elements of Modern X-ray Physics*, John Wiley & Sons, New York, 2002, ISBN 0-471-49859-0.
5. *Grating Monochromators and Optics for the VUV and Soft X-ray Region*, R. L. Johnson in *Handbook on Synchrotron Radiation*, Vol. 1a, Chapter 3, pp. 173 – 260, Elsevier, New York, 1983, ISBN 0-444-86709-0.
6. *Optical Engineering*, J. B. West and H. A. Padmore, in *Handbook on Synchrotron Radiation*, Vol. 2, Chapter 2, pp. 21 – 120, Elsevier, New York, 1987, ISBN 0-444-87046-6.
7. *X-ray Optics—High-Energy-Resolution Applications*, Y. Shvyd'ko, Springer, Berlin 2004, ISBN 3-540-21484-4.—especially strong for very high resolution crystal diffraction.

8. B. L. Henke in ALS Handbook, Lawrence Berkeley Lab PUB-643 Rev. 2, Berkeley (1989).
9. D. W. Lynch, "Optical Constants" in Handbook on Synchrotron Radiation, G. V. Marr, ed. Vol. 2, Chapter 11, pp. 723-783, North-Holland, Amsterdam (1987)
10. Other sources are:
11. E. Hecht, Optics, 4th Edition, Addison-Wesley, San Francisco (2002).—all the basics.
12. Born and Wolf, Principles of Optics, Pergamon Press, Oxford (1975). Good work never goes out of style.

4.10 X-ray Optics Software

Common interface packages that often combine source calculations with optics were described in Chap. 3, and these include XOP [86], SRW [87], SIREPO [88], SPECTRA [84], and OASYS [159].

In addition, there is the classic *Shadow* software, dating back to a package written by Cerrina in the early 1980s [160] and updated through SHADOW3 in 2011 [161] and 2014 [162]. One of the latest improvements is a new visual interface known as ShadowOui [163].

Chapter 5
X-ray Detectors and Electronics

5.1 Introduction

X-ray detectors are transducers—they convert X-rays into something more easily observed and/or measured: visible photons, a click on a loudspeaker, a grain of metallic silver, or, most commonly, an electrical voltage fed to a computer. Röntgen's eyes were part of the earliest X-ray detection system—he observed visible luminescence from a barium platinum cyanide screen that was being struck by X-rays from his cathode ray tube. Shortly after, he used film to record images (Fig. 5.1) [5, 164].

Detectors are a vital link in the chain of any successful synchrotron experiment. Although they have not yet reached the scale of detectors for high-energy physics installations, there is a growing appreciation that sometimes the most economical way to improve an experiment might be investing in X-ray detection as opposed to X-ray production.

An X-ray detector has to provide an accurate measure of X-ray intensity and, in some cases, X-ray energy, position, or even polarization. Detectors can be broadly divided into two classes according to how they are used: *integrating* detectors that provide an output corresponding to the integrated X-ray flux over time and *photon-counting* detectors that separately record the arrival of each X-ray. In fact, many detectors can be run in both modes, depending on the electronics used for readout.

5.2 Detector Properties

The characteristics of any X-ray detector depend on how it is used, including the energy and flux of the X-rays being detected, as well as the electronics that are employed. Among the most important parameters defining these characteristics are

S. P. Cramer, *X-Ray Spectroscopy with Synchrotron Radiation*, Biological and Medical Physics, Biomedical Engineering, https://doi.org/10.1007/978-3-030-28551-7_5

Fig. 5.1 Left: Roentgen's first X-ray image, supposedly of his wife's hand. Right: the ATLAS detector for the Large Hadron Collider (LHC). For scale, notice the man standing on platform

Table 5.1 Summary of most commonly used X-ray detectors for X-ray spectroscopy

Detector	$\Delta E/E$ (%) (5.9 keV)	Typical dead time	Max count rate (ph s^{-1} per element)	Typical application
Gas ionization (i.e., ion chamber)	n.a.		$\sim 10^{13}$	Beam line intensity monitor/ transmission XAS
Gas proportional	15	200 ns	$\sim 10^{5}$	Mössbauer spectroscopy
Scintillation/NaI (Tl)	40	250 ns	$\sim 2 \times 10^{5}$	Fluorescence-detected XAS
Ge or Si semiconductor	~3	~10 μs	$\sim 10^{5}$	Fluorescence-detected XAS/fluorescence elemental analysis
Si drift detector	~3	~1 μs	$\sim 10^{6}$	Fluorescence-detected XAS/fluorescence elemental analysis
Si strip detector	~3		$\sim 10^{6}$	Position-sensitive RIXS analysis
Avalanche photo-diode (APD)	20	~1 ns	$\sim 10^{8}$	Nuclear spectroscopies
Superconducting tunnel junction (STJ)	0.2 2[a]	~20 μs	$\sim 5 \times 10^{4}$	Fluorescence-detected XAS/soft X-ray fluorescence
Transition edge sensor (TES)	<0.03 0.2[a]	~1 ms	$\sim 10^{3}$	Fluorescence-detected XAS/soft X-ray fluorescence

[a]At ~500 eV

(a) *quantum efficiency* or *intrinsic efficiency*, (b) *energy resolution*, (c) *dead time* and *maximum count rate*, (d) *dark noise*, (e) *dynamic range*, and for certain applications (f) *spatial resolution*. Some representative properties for different devices are summarized in Table 5.1.

The *quantum efficiency* or *intrinsic efficiency* ε of an X-ray detector is the ratio of detected photons to the actual number incident on a device. This efficiency depends critically on the materials used in the active region for detection, the physical length of that active region, the nature of any physical or effective windows on the device, and finally the energy of the X-rays being detected. For example, an air-filled 20 cm ion chamber could have a 99% quantum efficiency around 3 keV but less than 1% efficiency at 30 keV.

The *energy resolution* R is the breadth of the measured energies vs. the actual energy: $R = \mathrm{FWHM}/E_0$, where FWHM is the full width at half maximum for a peak in the intensity vs. energy curve and E_0 is the mean value of the same peak. The energy resolution can also be defined in absolute energy terms simply using the FWHM. The *dead time* is the amount of time required to process one photon. If another photon arrives during the dead time, it will not be correctly counted and processed. As explained later in this chapter, the detector response during the dead time depends on whether it is a *paralyzable* or *non-paralyzable* device. The mathematics are illustrated Appendix F.

The *dark noise* is the amount of output readout in the absence of any incoming signal. With integrating detectors, for example, any electrical readout system will have voltage fluctuations known as thermal noise or "Johnson noise", $\overline{V_2} = 4_B kTR_L$ from the load resistor R_L. If the detector is a photodiode, there is also a *dark current* which is the reverse saturation current from random carrier generation in the diode. In pulse-counting systems, the dark noise is the number of counts observed in the absence of ionizing X-rays. The source of the dark noise might be purely electrical, but it might also include events produced by cosmic rays. The dark noise of a detector system sets a lower limit on the strength of an X-ray signal that can reasonably be measured in an experiment.

The *dynamic range* is the ratio between the largest and smallest signal intensities that a detector system can successfully process. The lower limit is set by the dark noise, while the upper limit is set by the detector properties and signal processing electronics.

We first discuss devices most commonly used as integrating detectors—which provide a signal related to the number of X-rays over a certain time interval. Examples include film, ion chambers, photodiodes, and charge-coupled devices (CCDs). This is followed by treatment of devices most commonly used as photon-counting detectors—which record the arrival of individual X-ray photons. In general, photon-counting is used for the best S/N ratio until the count rate exceeds the maximum linear response, at which point integrating devices such as ion chambers are employed.

5.3 Film and Image Plates

Photographic plates or films detect X-rays in the same manner in which they detect visible light. The energy from the X-ray causes the photoreduction of the silver in the silver halide, and the latent image is then chemically enhanced. Since the efficiency

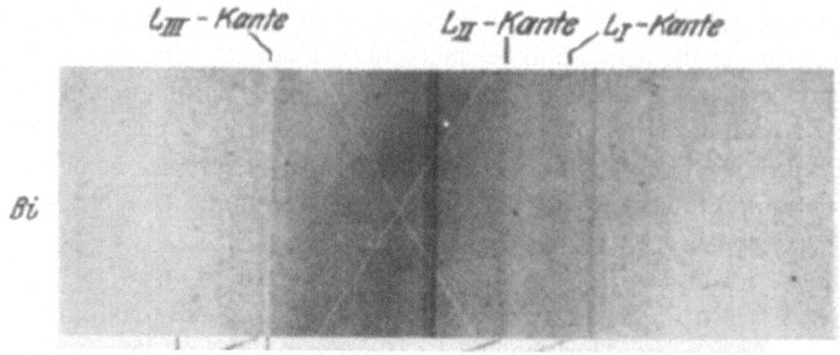

Fig. 5.2 The X-ray L absorption spectrum for Bi metal, recorded in an X-ray negative by Sandström in 1930 [166]. Notice the clear "white line" at the L_3-edge (L_{III}-Kante)

for stopping X-rays in film is often low, film was often combined with intensifying screens that efficiently stopped X-rays and converted them to visible light. Although film has a long history as the medium for X-ray detection (Fig. 5.2), dating all the way back to Roentgen's wife's hand, it has pretty much been displaced by electronic devices. Polaroid film and so-called *burn paper* are still used to monitor the X-ray beam position and shape when setting up experiments at the beamline [165].

"Image plates" are a modern electronic equivalent to film. Invented at Kodak but first commercialized by Fuji, they consist of small crystals of a *photo-stimulable* phosphor such as $BaFX{:}Eu^{2+}$ (X = Cl,Br), generally coated on a plastic support. Just as film stores a latent image after exposure to light, the image plate phosphor traps some of the X-ray energy as "color centers." These are created when X-ray excited electrons are trapped in regions with anion vacancies [167].

Instead of the chemical developer used for film, an image plate is read out by scanning a laser across the phosphor and recording the emitted light. When the color centers are illuminated by the laser beam, which could, for example, be a red HeNe laser, the trapped electron is released, and upon relaxation to the ground state, it emits a green or blue photon. The amount of luminescence is proportional to the X-ray dose, and high spatial resolution can be achieved by using a small laser spot.

5.4 Gas Ionization Chambers

A *gas ionization chamber* or *ion chamber* contains an X-ray-absorbing gas between a pair of electrodes. The electrodes are mounted on insulators and contained within a metal housing, with thin plastic windows to allow the entry and exit of X-rays (Fig. 5.3). Kapton® (polyimide), metalized Mylar® (polyester), polycarbonate, and polyethylene are popular window materials.

Ion chambers work by collecting the charges created by incident X-rays. Normally the gas is an insulator, but it becomes a conductor when absorption of X-rays leads to ionization of the gas molecules and production of electron-ion pairs. The

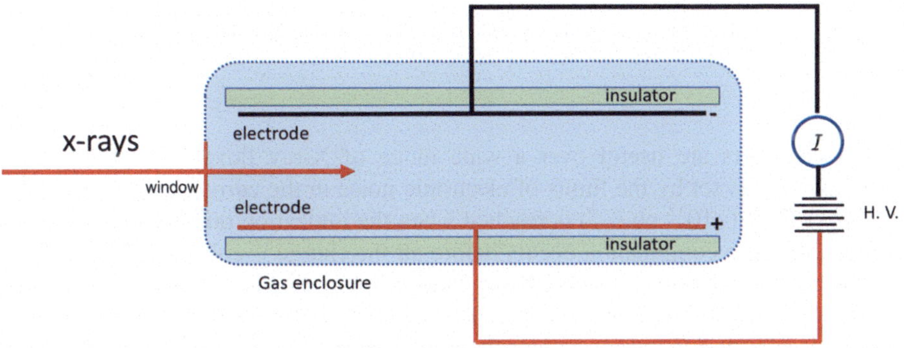

Fig. 5.3 Essential components of a gas ionization chamber

Table 5.2 X-ray-related properties of different gases [168, 169]

Gas	First IP (eV)	*W*-value (eV)	Fano factor
Helium	24.5	41	0.24
Nitrogen (N_2)	15.6	36	~0.3
Neon	21.6	36.3	0.13
Argon	15.7	26	0.2
Krypton	14.0	24	0.17
CH_4	14.5	~28	~0.3

voltage across the electrodes causes the electrons to be quickly collected at the anode, while the positive molecular ions are neutralized at the cathode.

The photoelectron (and Auger electrons) produced by an initial ionization event can proceed to ionize other gas molecules, creating a cascade that stops when all particles have energies below the gas ionization potential (IP). Thus, ideally, the number of ion pairs produced depends linearly on both the number and energy of the incident X-rays. Obviously, the X-ray energy, E, has to be higher than the first ionization potential of the gas molecules. However, because of the statistics of ion pair production, the average number of pairs produced is less than E/IP; rather it is E/W, where the "W-value" is the average energy dissipation per ion pair. Typical values for W for different gases are given in Table 5.2.

Thanks to their simplicity, economy, and wide range of linearity, ion chambers are useful for obtaining absolute fluxes at beamlines. If the voltage across the plates is such that all the electrons are collected, then the current can be used to calculate the absolute photon flux. Assuming I_0 is the incident intensity (photons s^{-1}), L is the length of the chamber (cm), ρ is the density of the gas (cm^{-3}), σ is the atomic or molecular X-ray cross section (cm^2), and $E_{X\text{-ray}}$ is the X-ray energy, then (neglecting losses due to recombination) the ion chamber current I_{ic} (A) can be obtained from the number of electron-ion pairs, and I_0 in photons s^{-1} can be determined by working backward from this equation:

$$I_{ic}[\mathrm{A}] = I_0\left[\mathrm{ph}\ s^{-1}\right]\ [1 - \exp\left(-\rho\sigma L\right)]\left(\frac{E_{x-ray}}{W}\right)/\left(6.242 \times 10^{18}\mathrm{electrons}\ C^{-1}\right) \tag{5.1}$$

Ion chambers are useful over a wide range of X-ray fluxes. The lower end (~10^4 ph s^{-1}) is set by the limits of electronic noise in the current amplifier, while the upper limit (~10^{13} ph s^{-1}) is reached when the ionization rate becomes so high that significant recombination occurs before all the charge is collected.

The time required to collect the X-ray-induced charges in an ion chamber depends on the distance to the plate and the net velocity of the charge. That motion is determined by two factors—random thermal motion associated with diffusion and the field-induced motion toward the oppositely charge plate, defined as the drift velocity v. A good estimate of the latter is given by:

$$v = \frac{\mu \varepsilon}{P} \tag{5.2}$$

where $\mathcal{E}$ is the electric field strength, P is the gas pressure, and μ is the *mobility*. For molecular ions, a typical value for μ is on the order of ~$1 \times 10^{-4}\ \mathrm{m}^2$ atm/Vs. Thus, for an ion chamber with a 1 cm distance between the beam and the plates, operating at 1 atm and 10^4 V/m, the collection time will be ~10 ms. The free electrons, because of their lower mass, have approximately 1000-fold higher mobilities; hence their collection times are typically in the microsecond range:

$$\frac{dn^+}{dt} = \frac{dn^-}{dt} = -\alpha n^+ n^- \tag{5.3}$$

where n^+ and n^- are the number densities for positive and negative charge carriers and α is the recombination coefficient. From this expression, one sees that recombination losses will go as the square of the incident flux, and the net current will not be a linear function of the beam intensity. Thus, in order to achieve reasonably good linearity between photon flux and electron current, the collection time has to be short enough so that recombination losses are negligible.

As illustrated in Fig. 5.4, with increasing voltage, the current increases because the recombination losses are being driven down. At some point, recombination becomes negligible and the ion chamber current plateaus. The experimenter should always check to make sure the ion chamber is operating in this flat region. Of course, there can be too much of a good thing, and increasing the voltage will eventually lead to a new operating regime involving secondary ionizations caused by the accelerated electrons. This will be discussed in the section on proportional counters.

The main electronic components required for ion chambers are an *electrometer* circuit, which produces an output voltage in proportion to the input current, as well as a stable high-voltage source. The electrometer directs the rather small input current from the ion chamber through a very large feedback resistor in order to produce a reasonable voltage at the output. The feedback resistor must be quite large, however. If the input current is 1 pA and the feedback resistor is 100 megohms, the output voltage will only be 100 μV.

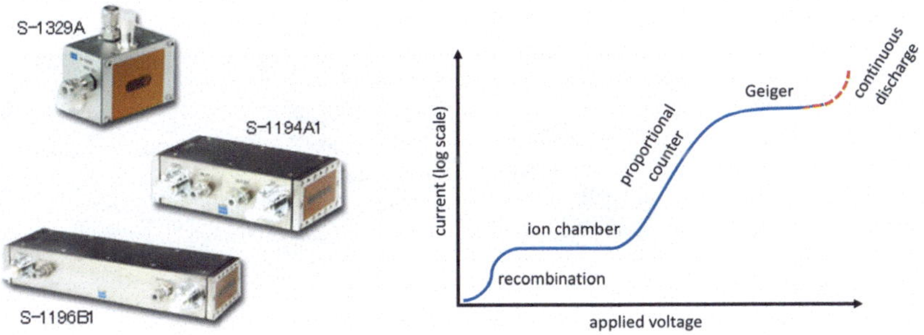

Fig. 5.4 Left: typical commercial ionization chamber devices (Ohyo Koken Kogyo Co., Ltd.). Right: different response regions for gas detectors. Typical ion chamber voltages are 300–500 V. Typical proportional counter voltages are ~1000–2000 V. Geiger-Müller counters usually operate at reduced pressure to move their operating voltages into that same range

5.5 Photodiodes and Diode Arrays

Photodiodes operating in integrated current mode are the solid-state equivalent of ion chambers. They normally consist of a p-type material at one surface and an n-type material at the other, with a depleted or "intrinsic" region in between, resulting in a "PIN" structure. A reverse bias is applied across the diode so that, as with ion chambers, the gap between electrodes normally functions as an insulator. (Some charge carriers are generated by thermal excitations, and the devices work better when cooled.) When X-rays are absorbed within the depletion region, charge carriers (holes and electrons) are freed and are swept to their respective collecting electrode by the electric field (Fig. 5.5).

The average number of electron-hole pairs generated by an X-ray is related to the X-ray energy, the band gap of the semiconductor, and the amount of heat produced (Table 5.3). As with gases, the average electron-hole pair energy is referred to as "*W*," and it is much smaller for semiconductors than for typical gases. The distribution of the number of electrons is reduced from Poisson statistics by the "Fano factor" "*F*" [170]. With a sufficient bias, nearly all the charge carriers can be collected, and if an external circuit is connected between the P- and N-layers, the average current will be a measure of the photon flux. The charge involved in discrete current pulses can be collected to reveal the X-ray energy, as discussed in Sect. 5.10.

Photodiodes are ideal for absolute measurements of the X-ray flux in the soft X-ray region under UHV conditions. They can also yield spatial resolution. For example, a series of diodes can be fabricated on a single piece of silicon, such as the quadrant detector in Fig. 5.5.

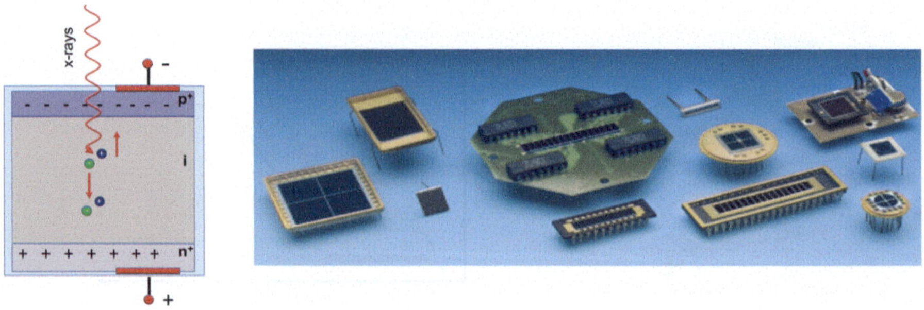

Fig. 5.5 Left: simplified photodiode structure. Multiple mobile electrons are produced depending on X-ray energy and material properties (see below). Right: a variety of commercial photodiodes from International Radiation Detectors, SXUV Series IRD

Table 5.3 Properties of different semiconductors

Material	Band gap (eV)	Energy/e-h pair (eV)—*W*	Fano factor—*F*
Silicon	1.12	3.61	0.115
Germanium	0.74	2.98	0.13
Diamond	5.48	~13	0.08
GaAs	1.43	5.2	0.12
CdTe	1.47	4.43	0.15
HgI_2	2.13	6.5	0.19 [171]

5.6 Charge-Coupled Devices (CCDs)

Charge-coupled devices or "CCDs" are in every home with a digital camera. They are lithographically produced by two-dimensional arrays of photodiodes connected to form the electronic equivalent of a "bucket brigade" (Fig. 5.6), and the "charge-coupling" refers to this motion of charges between adjacent locations. Boyle and Smith shared part of the 2009 Nobel Prize in Physics *for the invention of an imaging semiconductor circuit—the CCD sensor.*

In some cases, the diode material itself is used to stop the incident X-rays. Since the front of the CCD has an inert passivation layer as well as polysilicon structures and metal electrodes, in the soft X-ray region, CCDs are often "back-illuminated," with a thinned passivation layer for even better sensitivity.

Instead of window losses, hard X-ray detection often has the opposite problem—obtaining sufficient stopping power to collect most of the X-rays. Space scientists have developed 300-micron-thick Si CCDs that have better than 90% efficiency between 0.5 and 10 keV, but their budgets are of course "astronomical."

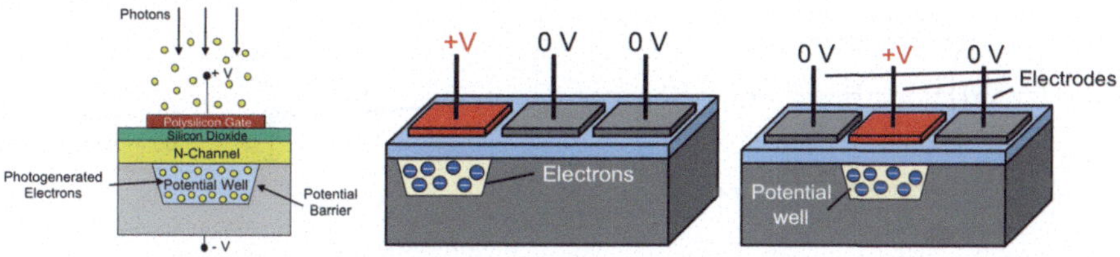

Fig. 5.6 Left: schematic of CCD structure. Electrons are generated by photon absorption in the Si-layer (blue), and they are trapped by the local charge. Right: electrons are moved between wells by stepping the voltage from one well to another

When used directly as an X-ray detector, the spatial resolution of a CCD used for depends on the pixel size, which is on the order of 10 μm × 10 μm. The maximum pixel count rate in integrating mode depends on the highest amount of charge that can be accommodated and the frequency with which the device can be read out.

5.7 Geiger Counters and Gas Proportional Detectors

The *Geiger counter* was the first device capable of detecting individual X-ray photons. It was developed in 1908 by Geiger and Rutherford at the Cavendish lab for recording α-particles [172], and with modifications by Geiger and Müller, it became sensitive to gamma-rays and X-rays [173]. A "Geiger-Müller tube" is similar to an ion chamber, but instead of flat plates, it commonly employs a cylindrical chamber containing a noble gas and a thin wire central anode (Fig. 5.7). As opposed to the uniform electric field in a flat plate ionization chamber, in a cylindrical geometry with cathode inner radius b and anode wire radius a, the magnitude of the electric field at radius r is given by:

$$\varepsilon(r) = \frac{V}{r \ln\left(\frac{b}{a}\right)} \tag{5.4}$$

Absorption of a photon in the gas yields an electron-ion pair. However, the combination of a high voltage (~500 V), high electric field strength in the vicinity of the wire anode, and reduced gas pressure (~0.1 atm) gives the free electron an especially large acceleration and sufficient energy to ionize additional gas molecules by collision. These additional electrons can in turn ionize additional gas molecules, yielding a so-called Townsend avalanche of current in the vicinity of the anode wire. Even more avalanches are produced by production and subsequent absorption of UV photons, so that one initial X-ray can produce on the order of 10^9–10^{10} electron-ion pairs. The process eventually stops when a space charge builds up from the relatively motionless positive ions. Additional time, ~100 μs, is needed for these ions to reach

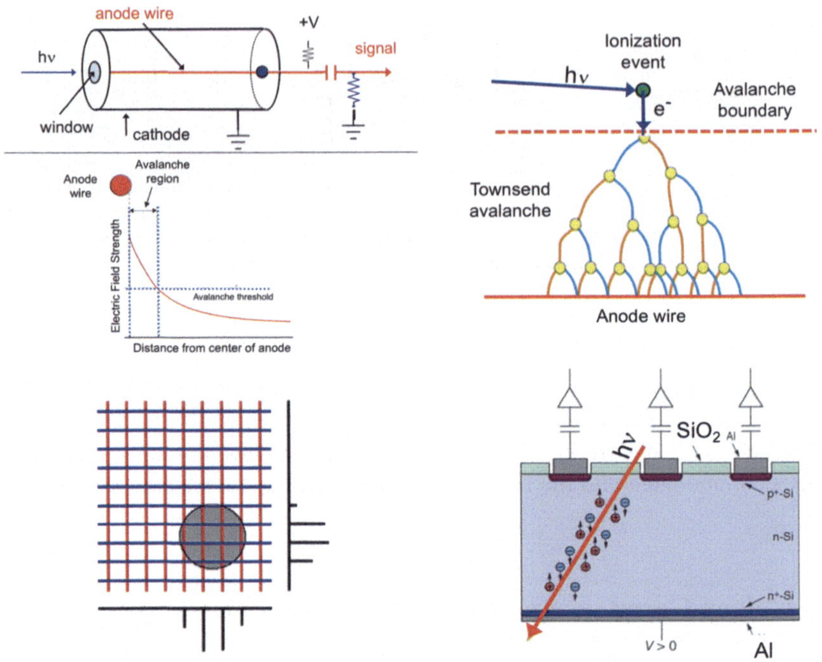

Fig. 5.7 Top left: basic geometry for both a Geiger-Müller tube and a proportional counter and the electric field vs. distance from the wire anode. Top right: in both tubes, the field is strong enough to initiate Townsend avalanches: ionizing electron path (blue line) and liberated electron path (red line). Lower left: a crossed wire MWPC where all wires are read out and position and energy are determined from centroid and integral of charge distribution. Bottom right: charge migration in a Si microstrip detector

the cathode. The resulting dead time limits the maximum linear count rate to ~10 kcps. Since the same signal is produced regardless of the initial photon or particle energy, there is no energy discrimination to this device.

A *gas proportional counter* behaves in between the all-or-nothing Geiger counter and the no-gain ionization chamber. It often employs the same construction as a Geiger-Müller tube (Fig. 5.7), but the Townsend avalanche is moderated by addition of a small fraction of polyatomic "quench gas," such as the popular P-10 mixture that is 90% Ar and 10% CH_4. By limiting the avalanche region to fractions of a millimeter from the anode wire, the device yields a current that is proportional to the initial number of ionizations and hence to the photon energy itself.

Typical gains compared to ion chamber currents are on the order of 10–10,000 with voltages from 200 to 800 V (Fig. 5.4). The extra current produced in a proportional counter makes it easier to detect individual X-rays, and these devices are most often used in photon-counting mode. The energy resolution is limited by the statistics of the electron multiplication process to ~20% at 5.9 keV, while the maximum count rate is restricted to $\sim 10^5\ s^{-1}$ by the ion drift velocities (10^5 cm s^{-1}).

5.8 Multi-Wire Proportional Counters and Silicon Microstrip Detectors

A *multi-wire proportional counter* ("MWPC") consists of sets of thin parallel anode wires placed equidistant between two cathode planes (Fig. 5.7). By giving the wires a significant resistance per unit length, the position of an X-ray absorption event along a given wire can be derived from the ratio of charges collected at each end of the wire. Position information can also be gained by constructing the cathode in perpendicular strips and employing electronic center-of-gravity techniques (Fig. 5.7). Arrays of MWPCs played an important role in tracking particles in high-energy physics experiments, and Charpak received the 1992 Nobel Prize for his development of this device—*"for his invention and development of particle detectors, in particular the multi-wire proportional chamber"*. MWPC detectors were used early on for high-resolution X-ray fluorescence experiments [174], but have since been replaced by Si "microstrip detectors" [175–177]. These devices behave like solid-state analogues of MWPCs, and they have parallel narrow electrodes fabricated on the surface of the silicon (Fig. 5.7).

5.9 Scintillation Detectors

In a scintillation detector, the signal transduction includes an intermediate step—the X-ray energy is converted to visible light, which is in turn detected by a photomultiplier circuit (Fig. 5.8) [178,179]. A common scintillating material for intermediate energy X-rays is NaI doped with a small amount of thallium and referred to as NaI(Tl), yielding a visible fluorescence lifetime of ~250 ns. Other scintillating materials include CsI(Tl) (1000 ns), $LaBr_3$(Ce) (16 ns), and even various plastics with lifetimes of a few ns.

The deadtime of a scintillation detector is determined by the fluorescence lifetime of the scintillator, while the energy resolution depends on the statistics of the light production process. In practical terms, NaI(Tl) detectors can typically run to

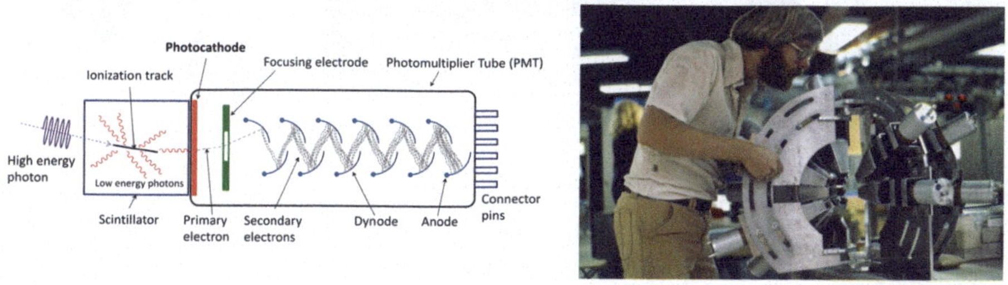

Fig. 5.8 Left: essential components of a scintillation detector. Right: Dr. Robert Scott assembling an array of NaI detectors for fluorescence-detected X-ray absorption experiments [180]

~200 kcps, but with relatively poor energy resolution. For fluorescence-detected X-ray absorption experiments on an element with atomic number Z, filters of element $Z - 1$ can be used to absorb most of the scattered radiation while transmitting the lower energy Kα fluorescence. Arrays of up to 36 NaI(Tl) detectors were employed at one time (Fig. 5.8) [180], but they have generally been replaced by arrays of semiconductor detectors with better energy resolution [181].

5.10 Energy-Dispersive Semiconductor Detectors

In a *semiconductor diode detector* or *solid-state detector*, electron-hole pairs that are created in the main body of the device are transported to the surface contacts under the influence of the applied electric field (Fig. 5.9). If most of the electron-hole pairs created by absorption of an X-ray photon are collected and measured, then the output signal will be proportional to the X-ray energy. In order to do this, an energy-dispersive detector uses a semiconductor crystal, with a bias voltage applied through front and rear contacts, a field-effect transistor coupled to a charge-sensitive preamplifier, and subsequent pulse-processing electronics to shape and quantify the output pulses. A typical construction to accomplish this is shown in Fig. 5.9, and the details of the electronics chain are deferred to Sect. 5.15.

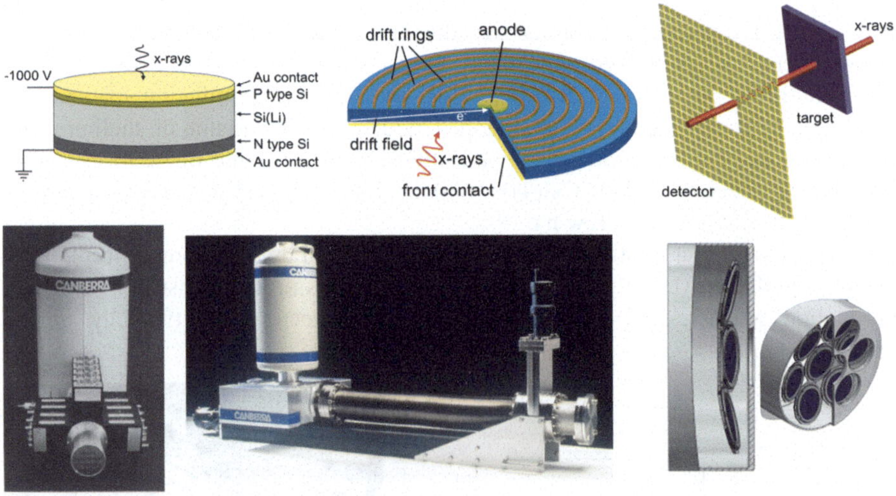

Fig. 5.9 Top left: essentials of a semiconductor detector. Top middle: the anode rings of a drift diode detector direct electrons to a small contact. Top right: a 386-element detector for NSLS-II [182]. Lower left: the first 13-element Canberra Ge detector [181]. Lower middle: an array detector for soft X-ray work—the bellows contracts as detector moves into vacuum chamber [183]. Lower right: a Ge detector with angled elements to better focus on sample

5.10.1 *Energy Resolution*

Unfortunately, not all of the X-ray energy goes into electron-hole pair production—some of it is released to the lattice as heat (phonons). Knowing that the average energy required to create a pair is W (Table 5.3), then for an X-ray energy E, the average number of pairs n is obviously $n = E/W$. The partitioning of energy between heat and pair production results in statistical fluctuations in the number of electron-hole pairs. If the process followed Poisson statistics, then the variance in the number of pairs would be $\sigma_{\mathrm{n}}^2 = <(\mathrm{n}-<n>)^2> = n$, and the fractional energy resolution would be:

$$R = \frac{\Delta E\ (\mathrm{FWHM})}{E} = 2.35\sqrt{\frac{W}{E}} \tag{5.5}$$

It is common to introduce an empirical constant, F, the *Fano factor*, as a measure of how close the actual variance is to Poisson statistics. The result is that the actual variance:

$$\langle n_0 \rangle^2 = F\langle n \rangle^2 = Fn \tag{5.6}$$

and the energy resolution is better than Poisson statistics by:

$$R = 2.35\sqrt{\frac{F\,W}{E}} \tag{5.7}$$

5.10.2 *Count Rates*

To achieve the resolution predicted by Eq. 5.7, one needs to collect a large fraction of the electrons produced by the incident X-ray. The charge collection involves a fast preamplifier and a slower "shaping" amplifier. If another photon arrives during the charge collection time, it will modify the integrated charge and yield an incorrect energy prediction. For conventional HPGe or Si detectors, the ideal shaping time is on the order of 10 μs, which leads to a maximum count rate on the order of 50–100 kcps. Higher total count rates can only be achieved by (a) reducing the necessary collection time and/or (b) spreading the X-ray flux between an array of detectors.

5.10.3 *Drift Diodes*

The ideal shaping time depends in part on the capacitance of the contact at the semiconductor surface. To collect electrons more quickly, one needs to reduce that capacitance—make the contact smaller. The silicon drift diode or SDD accomplishes this using a field gradient applied by ring electrodes on its back surface. The field gradient directs the electron flow to a small region at the center of the ring, where an integral or closely located FET amplifies the initial signal (Fig. 5.9). The lower capacitance allows drift diodes to run at maximum count rates almost 10*x* those of conventional HPGe or Si detectors.

5.10.4 Diode Array Detectors

The count rate limits of individual detectors can also be overcome by creating arrays of devices. In an early incarnation of this idea, Canberra built an instrument with 13 Ge detectors mounted on a common cryostat (Fig. 5.9) [181]. This was followed by devices designed for operation in a vacuum [183] (for soft X-ray work) and up to 36 elements. Since the outer devices might be at less favorable angles with respect to the sample, some vendors have made arrays with more focused individual elements. The dead space between devices can be minimized by employing "monolithic" detectors arrays. Devices with 100 detectors are commercially available, and a 384-element detector has been deployed at NSLS-II (Fig. 5.9) [182]. A beautiful synthesis of SDD and array concepts has been planned—a 384-element array of SDD detectors [184]!

5.10.5 pnCCD Detectors

An even higher level of pixelation is achieved by so-called "pnCCD" detectors [185]. One design, originally built for astronomy (of course), involved a 256 × 256 field of 75 μ × 75 μ pixels. Parallel readout of each 256-pixel channel is achieved in 5 ms, allowing a frame rate of 200 images per second. Thanks to the very low electronic noise, ~123 eV (FWHM) resolution is observed at Mn Kα and 47 eV at C Kα, close to the Fano noise limit. A typical application is as a position-sensitive detector with energy resolution for single-shot experiments at the LCLS [186].

5.11 Avalanche Photodiodes (APDs)

In order to achieve a faster time response, especially in nuclear spectroscopy experiments, many scientists have turned to the *avalanche photodiode or* "APD" detectors [187, 188]. The APD behaves somewhat like a solid-state version of a gas proportional counter. It is a diode constructed so that when a large reverse bias is applied, there is a volume of semiconductor with high enough electric field ($\geq 10^5$ V cm^{-1}) to enable migrating electrons to create secondary ionization events

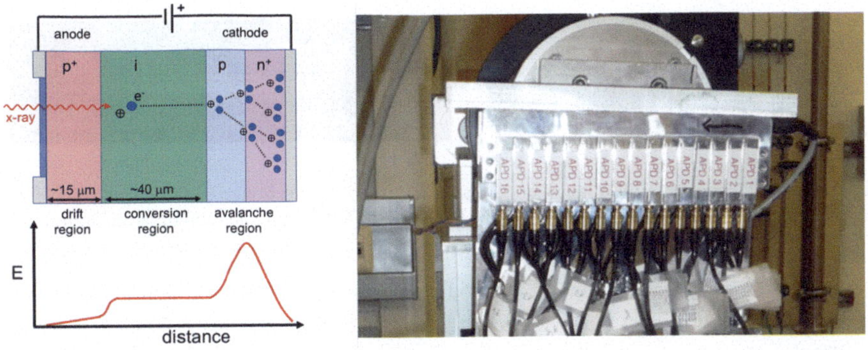

Fig. 5.10 Left: fundamentals of APD operation. Right: a stack of 16 thin Hamamatsu S5344 APDs (3 mm diameter, thickness 30 µm) used at PETRA-III, with overall time resolution of about 0.6 ns (250 ps time resolution for a single element)

before collection (Fig. 5.10). Typical gains achieved by this process are on the order of a few hundred. This higher gain allows shorter amplifier shaping times and contributes to better timing resolution.

For synchrotron applications, APDs are often used where the key issue is speed. Since the high-field drift velocity in Si is about 100 µ ns^{-1}, the leading-edge signal from a 100-µ-thick device will have a rise time on the order of 1 ns. The trailing lifetime will be determined by the device capacitance, but that is less of an issue for many experiments. By using very thin devices, time resolutions of ~120 ps have been achieved [189]. To achieve sufficient stopping power at higher X-ray energies with such thin devices, individual APDs can be *stacked* (Fig. 5.10) and/or *tilted* [190], or the X-ray photons can first be converted to visible light by high-Z scintillators such as Pb-doped plastics [191]. As with SDD detectors, arrays of APDs can be used for higher total count rates—a 10×10 APD array is under development for the high-energy photon source (HEPS) in Beijing [192].

5.12 Streak Cameras

The speed record for detectors is unambiguously held by *streak cameras* or *streak detectors*. A streak camera is a time-resolving detector, which works similar to an old-fashioned TV tube. The key components are a (1) photocathode, (2) a strong electric field for photoelectron acceleration, (3) sweeping plates, (4) an electron lens, and (5) a phosphor screen for imaging the photoelectrons. X-rays that are absorbed in the photocathode release a pulse of photoelectrons, which is both accelerated by a strong electric field and deflected by a fast voltage ramp on two parallel deflector plates. The early electrons see a low electric field and are deflected less than later electrons, which are deflected more. The intensity vs. time information is thus

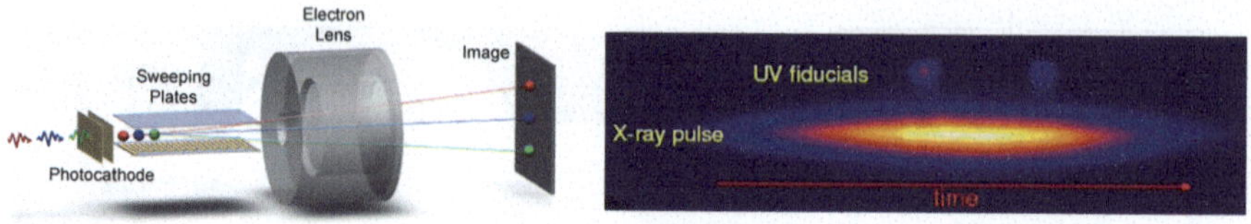

Fig. 5.11 Left: schematic of general streak camera design. Right: image of the X-ray pulses from the ALS. It was obtained by averaging ~10,000 streaks at a 5 kHz rep rate with time resolution ~4 ps

transferred to a spatial dimension—a "streak" on the electron detector. Time resolutions of a few hundred fs have been achieved [193, 194] (Fig. 5.11).

Streak cameras are popular tools for storage ring diagnostics. Why aren't streak cameras used more often for X-ray spectroscopy? They are not single photon-counting devices, and in most cameras there is significant down time between sweeps.

5.13 Superconducting Tunnel Junction (STJ) Detectors

The energy resolution of a semiconductor X-ray detector is intrinsically related to the number of electron-hole pairs produced following X-ray absorption and the statistics of their production. One approach to achieving higher resolution is to create more particles, and this can only be achieved by lowering the energy of individual particle production. The "superconducting tunnel junction" ("STJ") detector achieves improved energy resolution by relying on the disruption of Cooper pairs, which only requires ~0.5 meV (Table 5.4).

The tunnel junction consists of a very thin insulating layer, typically a few nm of Al_2O_3, between two superconducting layers (typically Al) (Fig. 5.12). When an absorbed X-ray releases its energy in the superconductor, it disrupts the Cooper pairs, and the resultant *quasiparticles* are then free to tunnel across the insulating barrier. Measuring the tunneling current then yields a measure of the particle energy. Since the number of quasiparticles produced is ~1000× more than in semiconductor detectors, in principle the resolution improvement is ~30-fold. An expression for the achievable resolution (FWHM) is:

$$\Delta E \cong 2.355 \sqrt{1.7\, \Delta_g\, (F + G)\, E} \tag{5.8}$$

where F is the Fano factor (~0.2), Δ_g is the superconductor energy gap, $G = 1 + 1/\langle n\rangle$, where $\langle n\rangle$ is the mean number of tunneling events per quasiparticle; and E is the X-ray energy. In practice, ~10 eV has been achieved at 5.9 keV [196].

Individual STJs are thin in depth and small in area. For more stopping power, a second superconductor (such as Nb or even Pb) can be layered on top of the first

Table 5.4 Properties of different superconductors for STJ detectors [195]

Material	Energy gap (meV)	T_c (K)	ΔE (@ 1 keV) (eV)[a]	ΔE (@ 6 keV) (eV)[a]
Niobium	1.55	9.3	4.2	10.2
Vanadium	0.82	5.4	3.0	7.5
Tantalum	0.70	4.5	2.8	7
Aluminum	0.18	1.2	1.4	3.5
Molybdenum	0.14	0.915	1.25	3.1
Hafnium	0.019	0.128	0.47	1.15

[a]Theoretical best resolution (FWHM)

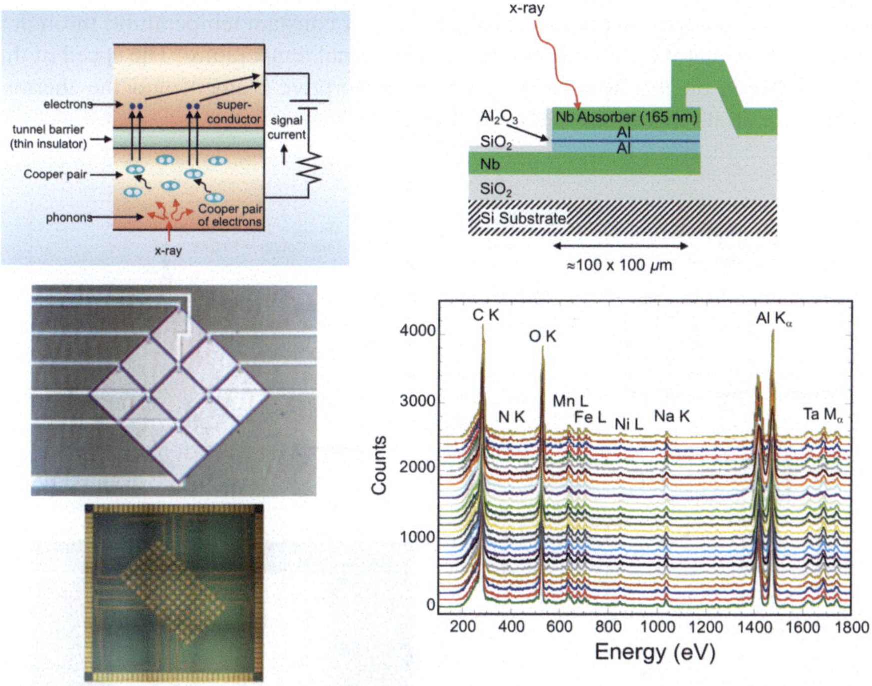

Fig. 5.12 Top left: STJ detector operating principle. Top right: a dual superconductor STJ—most of the X-rays are absorbed in the Nb layer. Middle left: a 9-element STJ array; pixels are 200 μ *x* 200 μ [197]. Lower left: a 112-element STJ array. Individual pixels are 208 μ × 200 μ [200]. Lower right: superposition of signals from 36-element array [199]

superconductor. Since individual STJs are small and limited in count rate, there has been an evolution in arrays of detectors, starting at 9 [197, 198], through 36 [199], then >100 [200, 201], and there are hopes for arrays with >1000 elements [201]. Resolutions on the order of 10 eV at 500 eV and even at 6 keV have been achieved.

5.14 Microcalorimeters and Transition Edge Sensor (TES) Detectors

The ultimate in energy resolution is achieved by measuring the amount of heat produced when an X-ray dissipates all of its energy in a sensor. In this case the energy resolution is determined by thermodynamic fluctuations at the phonon level, as opposed to electron-hole pairs in semiconductor detectors or quasiparticles in STJs. In a conventional *microcalorimeter*, the temperature change in the X-ray absorber is measured by a thermistor in contact with the absorber. (The term *bolometer* is usually reserved for a thermal detector that measures the power of the incident radiation, as opposed to individual particle energies.) The absorptive element is connected to a thermal reservoir (a body of constant temperature) through a thermal link so that it eventually returns to its original temperature. The speed of the device depends on the heat capacity of the absorptive element and the thermal conductance of its link to the reservoir (Fig. 5.13).

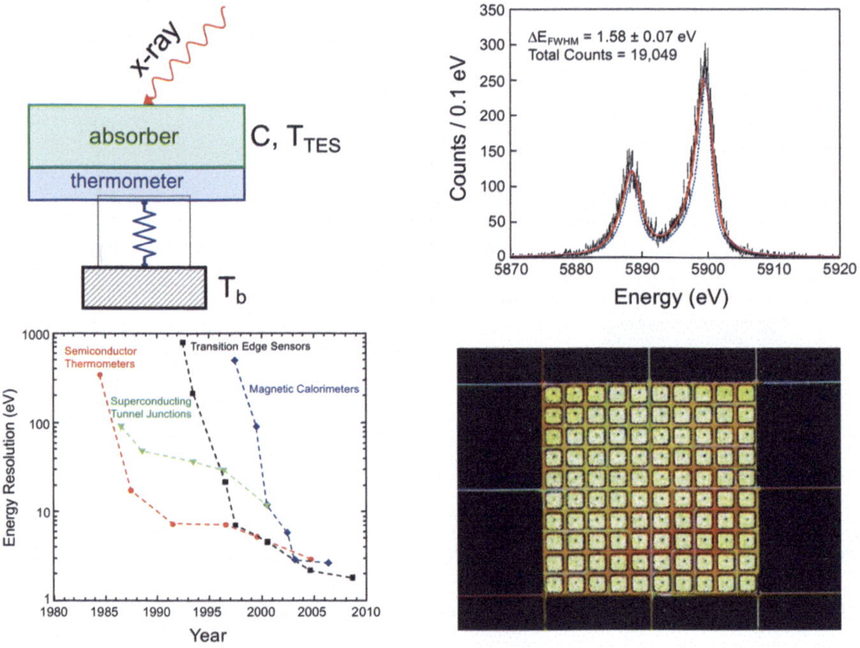

Fig. 5.13 Top left: the general operating principle for thermal X-ray detectors. Absorption of an X-ray changes an absorber temperature, which is then read as a change in resistance or even magnetic properties [202]. A TES detector operates on the edge of the superconducting-normal transition. Top right: ~1.6 eV resolution achieved by a NASA TES [203]. Lower left: the progress in cryogenic detector resolution vs. time [204]. Lower right: a 100-element TES using Au absorbers with 0.05 mm pitch, surrounded by Au/Bi absorbers with 0.25 mm pitch, for the ESA Athena mission. The overall array will have ~4000 elements

A "transition edge sensor" ("TES") is an improved variation on the microcalorimeter concept [203, 205]. In this device, a thin metal film is held just on the rising edge of the transition temperature between a superconducting state and normal conductance. An X-ray photon that strikes the detector raises the temperature enough to alter current through the superconductor, and the current is proportional to the energy of the photon.

The individual X-ray absorbers need to be small (for low heat capacity) and the response is still relatively slow (microseconds). Thus, TES detectors are ideal candidates for array detectors. Devices with 240 detectors have been built and deployed for synchrotron experiments as well as other measurements [202]. Plans for TES arrays with millions of elements are being considered—for astronomers, of course [204].

5.15 Detector Electronics

The electronics that is paired with X-ray detectors has certainly come a long way over the past few decades. In the early days of EXAFS experiments at SSRL, a task as simple as changing the amplifier gain required taking out a soldering iron and swapping resistors. Nowadays, almost every aspect of detection is under computer control, and many of the analogue circuits that we will discuss next are going digital (Fig. 5.14).

Analogue circuits process signals as continuous variables such as voltage or current, whereas digital electronics take the continuous waveforms and convert them into two or more discrete values such as high and low voltages or a range of numbers. Typical waveforms for the various steps are illustrated in Fig. 5.15.

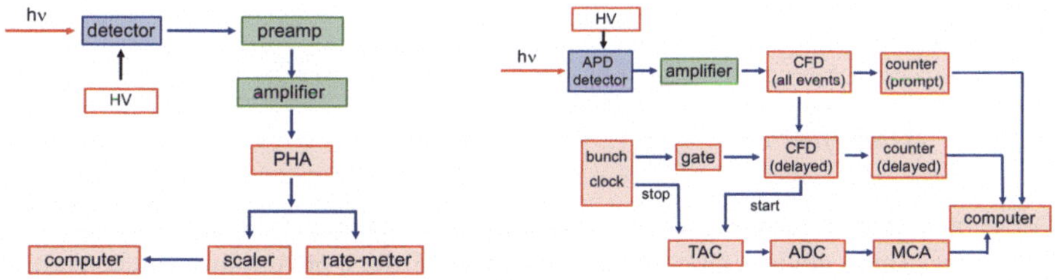

Fig. 5.14 Conventional electronics layouts. Left: electronics for photon-counting X-rays with energy resolution. The detector could be a scintillation detector or semiconductor detector, and the PHA (pulse height analyzer) could be a single-channel analyzer or discriminator. Right: electronics for time-sensitive gating in nuclear spectroscopy experiments at the ESRF

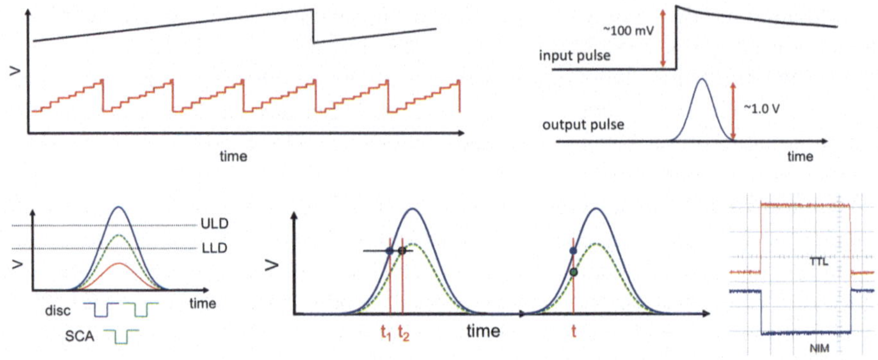

Fig. 5.15 Top left: output from a charge-sensitive preamp in absence and presence of X-ray pulses. The voltage is reset once it reaches a certain limit. Top right: shaping and amplification of discrete pulses by the amplifier. Lower left: a discriminator fires whenever a signal crosses LLD; an SCA fires only when a signal crosses LLD but remains under ULD. Lower middle: a conventional discriminator will fire whenever signal crosses fixed threshold—this time will depend on the pulse height; a CFD fires when the pulse crosses a certain fraction of its overall height, hence at a fixed time relative to pulse arrival. Lower right: typical logic pulses

5.15.1 *Preamplifiers*

The initial signals from most detectors are small currents that require amplification. For example, the currents in an ion chamber might be on the order of nA to μA, and they are usually followed by a current-to-voltage preamplifier with gains on the order of 10^9–10^6 V/A. For semiconductor detectors, two basic types of charge-sensitive preamplifiers are used. One type uses *dynamic charge restoration* (RC feedback), while the other employs *pulsed charge restoration* via pulsed optical or transistor reset methods to discharge the integrator [206]. More details on these and other electronics are in the text by Knoll [206] and also on the commercial websites for various detector companies.

5.15.2 *Amplifiers*

For semiconductor detectors, the next stage in analogue processing involves *shaping amplifiers*. These devices take the step-like output from a preamplifer and convert it into a shaped output pulse. A common output shape is a Gaussian waveform, and the "shaping time" is the standard deviation of that pulse width. The shaping time is usually a compromise—it needs to be long enough to collect most of the charge from the preamplifier, but short enough to avoid significant "pileup" between successive pulses (see **Appendix F**).

5.15.3 Discriminators and Single-Channel-Analyzers

The analogue signal from an amplifier is converted to a logic pulse by these devices. A *discriminator* sends an output pulse every time its input crosses a certain threshold, while a "single-channel analyzer" ("SCA") pulses when the input lies between a lower and an upper level (Fig. 5.15). A typical (NIM-standard) "slow logic pulse" output is nominally +5 V amplitude and 500-ns width. For applications where high speed timing is important, there are also instruments that deliver "fast logic pulses" that are negative with rise times typically on the order of a few nanoseconds.

5.15.4 Multi-Channel Analyzers

A *multi-channel analyzer* or *"MCA"* is an SCA on steroids. Instead of operating with multiple discriminators, an MCA uses a fast ADC to digitize the amplitude of incoming pulses. The MCA then creates a histogram for the number of pulses in different amplitude ranges.

5.15.5 Constant Fraction Discriminators

One downside of discriminators and SCAs is that the time at which they trigger will depend on the size of the input pulse. For applications where the time is more important than the amplitude, spectroscopists employ *constant fraction discriminators* or "CFDs". As illustrated in Fig. 5.15, for pulses that meet a certain threshold, these devices trigger at the same time regardless of the pulse amplitude.

5.15.6 Time-to-Amplitude and Time-to-Digital Converters

In some synchrotron spectroscopy applications, such as nuclear forward scattering and perturbed angular correlation, the arrival time of an X-ray is the important quantity to measure. A "time-to-amplitude converter" ("TAC") measures the time interval between input start and stop pulses, and it then provides an analogue output signal with a voltage proportional to that time. This output can then be processed by an MCA to yield a histogram of events vs. time. A *time-to-digital converter* or "TDC" provides a digital output corresponding to the measured time interval, eliminating the need for an MCA.

5.15.7 Gates

As the name implies, *gates* are electronic devices used to pass or reject signals, which can be in analogue or digital form. Generally there are two inputs, a logic pulse that determines whether the gate should be open or closed for an adjustable duration and the signal itself which is either passed or rejected.

5.15.8 Analogue-to-Digital Converters

An "analogue-to-digital converter" or "ADC" is used to convert analogue signals such as amplifier or TAC outputs into digital form for further processing (Fig. 5.16). As ADCs become faster and cheaper, they are becoming much more frequently used. One application is in digital processing of detector signals as a replacement for shaping amplifiers. The digitized signal can be processed numerically to mimic or supersede the shaping done by an analogue amplifier.

5.15.9 Counters/Scalers

Once a signal is converted into a logic pulse, it can be registered in a *counter*, which adds up the number of pulses over a certain interval and converts the sum into a number that can be transmitted to a computer. It is common to include a number of counters on the same electronic board—this is then called a *scaler*.

5.15.10 NIM Bins and Crates

Where to put all the electronics? A typical beamline will have an assemblage of NIM bins and CAMAC and/or VME crates. The NIM bin provides slots and power for

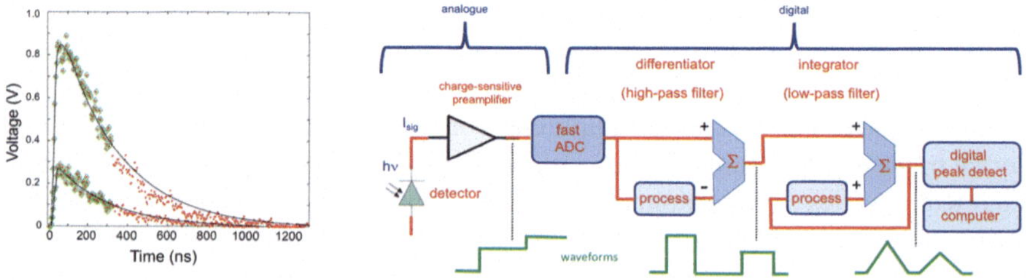

Fig. 5.16 Left: direct digitization of the output from a NaI(Tl) detector, compared with dual exponential fitting functions [207]. Right: replacement of analogue pulse shaping circuits with digital processing electronics

devices that do not need to talk to the beamline computer directly, such as amplifiers, SCAs, gates, and TACs. The crates, such as CAMAC, FASTBUS, VME, and PXI, have backplanes that can send data to and receive commands from the beamline computer. Accordingly, they will contain the devices that convert signals to numbers, such as scalers and ADCs. Good descriptions of all of these electronics are of course on Wikipedia and also in Knoll.

5.15.11 The Trend toward Digital

Just as the signals for our telephones and televisions have gone digital, so has much of the electronics for X-ray detection. A key step has been replacement of analogue shaping amplifiers with numerical processing of digitized detector waveforms, either directly from NaI(Tl) detectors or following preamplification (Fig. 5.16). The benefits of digital processing include (a) better time and energy resolution, (b) better pileup rejection, and (c) higher throughput. As ADCs and processing electronics keep getting faster, this trend is likely to continue for some time.

5.16 Suggested Exercises

1. Calculate the photon flux (I_0) at 20 keV if a 10 cm N_2-filled ion chamber gives a measured current of 1 pA.
2. Calculate the number of electron-hole pairs produced by a 10 keV photon in Ge.
3. Calculate the theoretical best energy resolution for a Si diode detector at 20 keV.
4. Calculate the theoretical maximum number of Cooper pairs disrupted in a Nb STJ.
5. Calculate the temperature change in a TES detector when a 6.4 keV photon is absorbed in a Bi absorber that is 4 μ thick with a cross section that is 320 μ by 305 μ. Assume a heat capacity of 1 pJ K^{-1} [202].
6. Calculate the deadtime losses for a random incident flux of 10^5 photons s^{-1} for a paralyzable detector system with a deadtime of 10 μs. Do the same for a non-paralyzable system.

5.17 Reference Books and Review Articles

1. *Radiation Detection and Measurement*, Glenn Knoll, fourth Edition, Glenn F. Knoll, John Wiley & Sons, New York, 2010 ISBN-10: 0470131489. A classic resource about detectors of ionizing radiation. Updated to include more on cryogenic devices.

2. *Nuclear Electronics: Superconducting Detectors and Processing Techniques*, Vladimir Polushkin, John Wiley & Sons, New York, 2004, ISBN 0470857595. A nice introduction to cryogenic detectors.
3. *Cryogenic Particle Detection*, Christian Enss, Ed. Springer, Berlin, 2005, ISBN 978–3–540-20113-7. Individual edited chapters on all the basic cryogenic detectors and also on applications.
4. Nuclear Electronics, P.W. Nicholson, John Wiley & Sons, New York, 1974, ISBN 0–471–63697-5. Ancient history, but still an accurate description of detector electronics prior to digital age.
5. *Synchrotron Radiation News*, Vol. 31, No. 6, **2018**—"Detectors", an entire issue on some of the more sophisticated detectors.

5.18 Commercial Websites

Company websites, such as for Canberra and Ortec, often contain useful information about detectors and electronics. The links often change, so we recommend starting with the corporate websites:

Canberra (now Mirion): http://Canberra.com
Ortec: https://www.ortec-online.com/

Chapter 6
X-ray Absorption and EXAFS

6.1 Introduction

When X-rays impinge on a sample, they can be transmitted, scattered elastically (Rayleigh scattering and diffraction), scattered inelastically (Raman and Compton scattering), or absorbed. At high X-ray energies, the higher the atomic number of an element, the stronger the absorption, and this contrast mechanism is employed for X-ray imaging of broken bones, at the dentist's office, in mammography, and a myriad of other applications. In this chapter, we will consider the energy dependence of X-ray absorption in more detail.

An X-ray absorption measurement in transmission mode is really quite simple, and the principles are the same as for UV-visible or infrared absorption spectroscopy. For a fixed X-ray energy E, measure the fraction of X-rays absorbed in a sample with *pathlength* t (Fig. 6.1). The ratio of transmitted flux (I) to incident flux (I_0) is called the *transmittance* (T), while the *absorbance* (A) is defined by $A = -\log_{10} T = \log_{10} I_0/I$. By repeating this measurement over a range of energies, one obtains an absorption spectrum $A(E)$ vs. E. For every element, there is a typical energy where there is just sufficient energy to ionize a particular core electron. At this point in the spectrum, there is a rapid rise in the absorption coefficient—the so-called *absorption edge*. The edge energy (E_0) is approximately halfway up the edge. The various edges available for Mo and other elements are illustrated in Fig. 6.2, and Appendix G lists the more commonly used absorption edge energies.

"I already have FT-IR and UV-visible spectrometers in my lab," you object. "Why should I bother going to a synchrotron for yet another absorption measurement?" It turns out that X-ray absorption spectroscopy ("XAS") has many special attributes that will make the trip worthwhile. It is:

- Element specific. One beauty of X-ray absorption is that every element has *absorption edge* features in specific energy ranges. At a given energy, you

S. P. Cramer, *X-Ray Spectroscopy with Synchrotron Radiation*, Biological and Medical Physics, Biomedical Engineering, https://doi.org/10.1007/978-3-030-28551-7_6

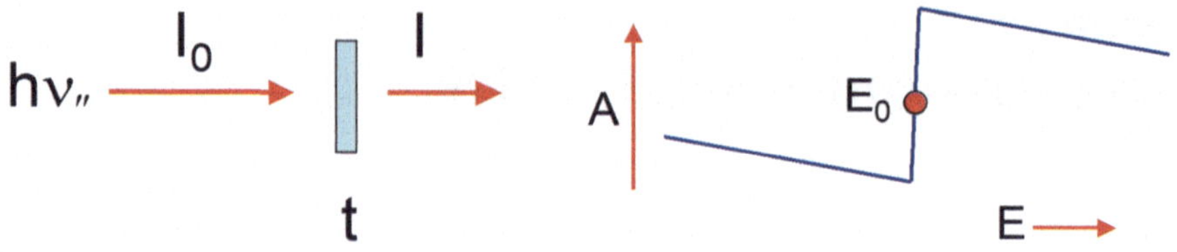

Fig. 6.1 Left: Diagram of an X-ray absorption measurement. Right: a bare bones spectrum

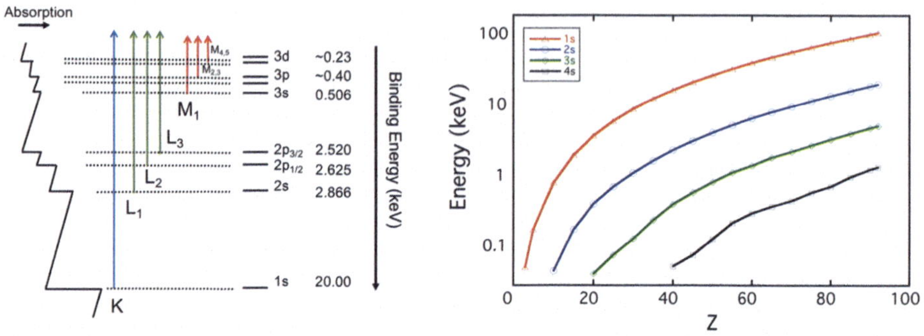

Fig. 6.2 Left: X-ray absorption edges for Mo. Absorption is not to scale. Right: X-ray absorption edge energies as a function of atomic number Z

Table 6.1 Nomenclature for the most commonly used edges

Edge	Strong, dipole-allowed	Weaker, dipole-allowed	Quadrupole
K	1s → (n + 1)p		1s → nd
L_1		2s → 4p[a], 2s → 5p[b]	
$L_{2,3}$	2p → 3d[a], 2p → 4d[b], 2p → 5d[c]		2p → 4f[c]
M_1		3s → 4p[a], 3s → 5p[b]	
$M_{2,3}$	3p → 3d,[a] 3p → 4d[b]		
$M_{4,5}$	3d → 4f[c], 3d → 5f[d]		3d → 5s[c]

[a]First transition metals
[b]Second transition metals
[c]Lanthanides
[d]Actinides

know what you are looking at. Conversely, if you want to study a particular element, you know where to look.

- Sensitive to molecular structure. In particular, the *extended X-ray absorption fine structure* (EXAFS) provides information about the number and types of atoms surrounding an absorber and especially the interatomic distances.
- Responsive to electronic structure and magnetic properties. This information is primarily in the absorption edge region (Table 6.1) and will be treated in the following chapter.

If X-ray absorption is so wonderful, why doesn't everyone have an X-ray spectrometer in their lab? The primary difficulty is obtaining an *intense*, *monochromatic*, and *tunable* source of X-rays. The advent of broadband, collimated synchrotron radiation sources was a major breakthrough for X-ray absorption spectroscopy; to use a term that is overworked but still true, it was a *revolution*.

6.2 The Experiment in More Detail

6.2.1 Detection Modes

There are three common modes for recording an X-ray absorption spectrum: *transmission*, *fluorescence*, and *electron yield*. The transmission measurement is similar to the way one normally records a UV-visible spectrum. One difference is the measurement of the incident intensity—it is usually done with a partially absorbing ion chamber at the same time as a second ion chamber measures the transmitted intensity (Fig. 6.3). A third ion chamber is often placed in line to record the transmission through a reference standard for energy calibration purposes. The use of ion chambers for these measurements means that one does not exactly measure I_0 or I but instead a signal that is linearly proportional to I_0 or I. Since the gas has its own slowly varying spectrum, this needs to be taken into account if one wants the true absorbance (but this is rarely done).

The second approach to recording XAS is fluorescence mode, sometimes called *fluorescence yield* or "FY" (Fig. 6.3). In this measurement, the fluorescence due to the element of interest, I_f, is monitored as the energy is scanned. If a number of assumptions are fulfilled, then the "fluorescence excitation spectrum" will be directly proportional to the absorption spectrum for the edge of interest. X-ray scientists did not invent this approach; fluorescence excitation spectra have been used in optical spectroscopy for over a century.

The relationship between fluorescence and absorption can be derived by integrating the fluorescence received into a solid angle $\Omega/4\pi$ from the entire sample. This

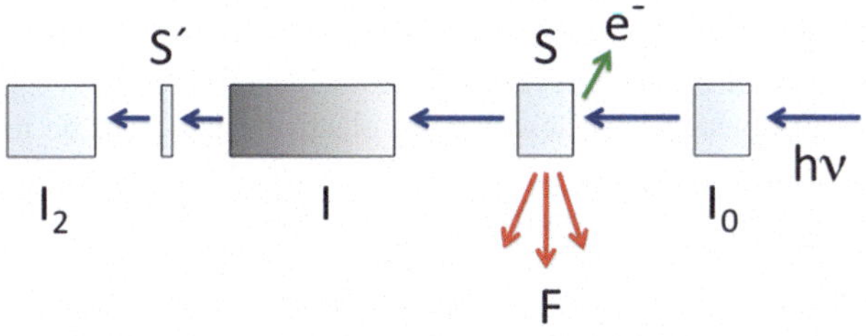

Fig. 6.3 Arrangement for transmission, fluorescence, and electron yield measurements. Symbols represent the following: *S* sample, *S′* calibration foil, *F* fluorescence, e^- electron yield, I_0, incident intensity, *I* transmitted intensity, I_2 transmitted intensity for *S′*. I/I_2 yields transmission measurement for *S′*

yields the resulting "general case" formula, assuming that the incident and fluorescent radiation makes equal angles with respect to the sample normal [208]:

$$I_{\mathrm{f}} = I_0 \frac{\varepsilon_{\mathrm{f}}(E)(\Omega/4\pi)\mu_{\mathrm{s}}(E)}{\mu_{\mathrm{T}}(E) + \mu_{\mathrm{T}}(E_{\mathrm{f}})} \times [1 - \exp\{\mu_{\mathrm{T}}(E) + \mu_{\mathrm{T}}(E_{\mathrm{f}})\}d] \tag{6.1}$$

where E is the excitation energy, E_{f} is the fluorescence energy, ε_{f} is the fluorescence yield, μ_{s} is the absorption coefficient for the sample component of interest, μ_{T} is the total absorption coefficient, and d is the path length.

The general case is unnecessarily complex, and, in most cases, one can work in either of two simplified regimes. In the first regime, one assumes that the sample is sufficiently thick that all of the incident radiation is absorbed. In addition, if the bulk of the absorption is due to the matrix, then $\mu_{\mathrm{T}}(E)$ is slowly varying, and we have the "thick, dilute" limiy:

$$\frac{I_{\mathrm{f}}}{I_0} = \propto \frac{\varepsilon_{\mathrm{f}}(E)(\Omega/4\pi)\mu_{\mathrm{s}}(E)}{\mu_{\mathrm{T}}(E) + \mu_{\mathrm{T}}(E_{\mathrm{f}})} \propto \mu_{\mathrm{s}}(E) \tag{6.2}$$

In the second case, if a sample is thin enough, then the exponential in the last term can be expanded as a Taylor series ($e^{-x} \cong 1 - x + \ldots$) which cancels the denominator of the first term, and the general case simplifies to the "thin, concentrated" limit:

$$\frac{I_{\mathrm{f}}}{I_0} = \propto \varepsilon_{\mathrm{f}}(E)(\Omega/4\pi)\mu_{\mathrm{s}}(E)d \propto \mu_{\mathrm{s}}(E) \tag{6.3}$$

Notice that in simplifying the proportionality, we have assumed that the energy dependence of the fluorescence yield, $\varepsilon_{\mathrm{f}}(E)$, is negligible. This is a good approximation for K-edges, but it turns out to have serious problems for lower energy edges such as the 3d $\rightarrow$ 2p fluorescence at transition metal L-edges, because different $2p^5 3d^{N+1}$ multiplets have different fluorescence yields [209]. In principle, one workaround is to instead use the 3s $\rightarrow$ 2p fluorescence, which has a small but constant yield [210]. This is one example of a *partial fluorescence yield* experiment. There is even an *inverse partial fluorescence yield* experiment, in which one measures the decrease in fluorescence from a different element as the element of interest absorbs more. In practice, there are always experimental reasons for slight differences between all of the detection modes (Fig. 6.4) [211].

A third class of detection modes, usually electron yield, can be more generally described as "non-radiative yield" (Figs. 6.3 and 6.4). After absorption of an X-ray, the core hole is filled by an electron from a higher shell, and if the atom does not emit a photon, the energy difference is instead released as an Auger electron. With an appropriate electron energy analyzer, detection of this Auger electron yields an excitation spectrum in a manner analogous to fluorescence detection. This mode is referred to as "Auger electron yield" or "AEY". However, the mean free path for an Auger electron in a solid ranges from a few Å to perhaps 1000 Å at high energies, and the electron energy analyzer has limited angular acceptance.

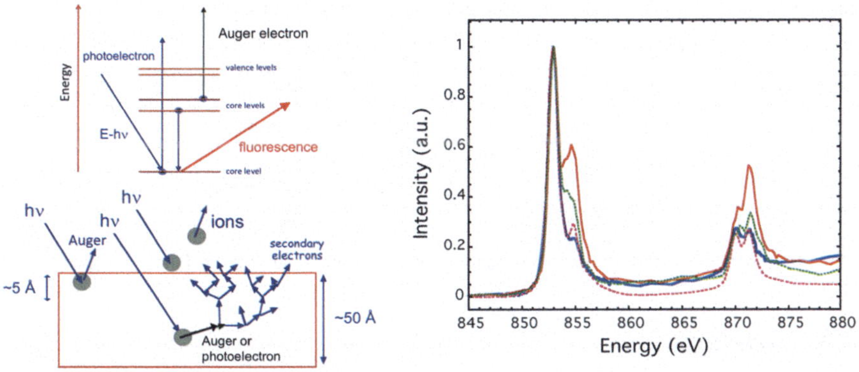

Fig. 6.4 Top left: relaxation following photoabsorption. Lower left: non-radiative detection candidates. Right: L-edges for $Ni(acac)_2$ taken with various detection modes: transmission (blue solid line), total electron yield (red dashed line), Lα (3d → 2p) fluorescence (red solid line), and Lι (3s → 2p) fluorescence (green dashed line) [211]

Most experiments take advantage of the fact that scattering of the Auger electron and photoelectron following X-ray absorption and relaxation results in a cascade of lower energy electrons (Fig. 6.4). Thus, multiple electrons are emitted per absorption event, while typical escape depths are on the order of 50 Å [212]. The detection modes that exploit this are *total electron-yield* ("TEY"), in which electrons emitted from a sample are typically collected by a channeltron electron multiplier, and *sample photocurrent*, in which the current flowing back to the sample from ground is recorded by a high sensitivity electrometer.

The connection between these electron yield measurements and the true absorption spectrum is mathematically identical to that for fluorescence excitation spectra. If the fluorescence yield is f, then the non-radiative yield will be $(1 - f)$. If the total number of electrons emitted from the sample after an absorption event remains constant for different energies, then analogous derivations can be done for both "thin" and "dilute, thick" cases.

6.2.2 *Signal-to-Noise Comparisons*

As the absorption by the component of interest becomes weaker compared to the matrix absorption, fluorescence-detected XAS eventually wins out in terms of sensitivity. In qualitative terms, transmission becomes akin to weighing the captain by weighing the ship with and without him/her aboard. Mathematically, since the S/N for a transmission experiment goes linearly with concentration, while the fluorescence S/N goes as the square root of concentration, there will always be a crossover point where fluorescence wins out. The location of that crossover point depends on the fluorescence yield and the relative cross sections, as summarized in Table 6.2.

Table 6.2 Signal-to-noise comparisons for different detection modes [213]

Mode	Prefactor	Assumptions
Transmission	$0.736\frac{\mu_A(E)}{\mu_T(E)}$	Uniform sample
Fluorescence	$\left(\frac{\varepsilon_f(\Omega_f/4\pi)\mu_A{}^{(E)}}{(1+I_b/I_f)[\mu_T(E)+\mu_T(E_f)]}\right)^{1/2}$	Thick, dilute sample 45° geometry
Non-radiative	$\left(\frac{\varepsilon_n(\Omega_n/4\pi)\mu_A{}^{(E)}}{(1+I_b/I_n)[\mu_T(E)+n(E)]}\right)^{1/2}$	Thick sample compared to electron pathlength

For the complete S/N equation, the above prefactors must all be multiplied by: $\left(\frac{\Delta\mu_A(E)\ I_{inc}^{1/2}}{\mu_A(E)}\right)$

6.2.3 Artefacts

X-ray absorption is a deceptively simple experiment—one will always get "data"—meaning that the beamline computer will always generate an output file. However, the quality of that data and its relationship to the desired quantity, the absorption coefficient $\mu_s(E)$ for a particular 'sample' element, depends on avoiding a host of experimental artefacts. Included among the numerous possible artefacts are:

- Leakage effects.
- Fluorescence saturation effects.
- Detector nonlinearity.
- Crystal Glitches.

6.2.4 Leakage Effects

In transmission experiments, a range of artefacts can be summarized under the heading of "leakage effects" [214, 216]. Suppose that the measurements that purportedly correspond to the incident and transmitted intensities I_0 and I are both contaminated by the addition of a second signal proportional to I_0: αI_0. This additional intensity could come from, among other things, (a) pinholes in the sample that allow beam to pass without attenuation, (b) harmonic wavelengths in the incident beam that have a much lower absorption cross section, (c) electronic noise in the detector itself, and even (d) fluorescence from the sample that is reemitted into the I_0 or I detectors. Regardless of source, leakage results in reduced apparent absorbance and a loss of EXAFS amplitude (Fig. 6.5). The math behind leakage effects has been summarized elsewhere [214, 216].

6.2.5 Fluorescence Saturation Effects

The derivation of Eq. 6.2 from Eq. 6.1 requires that the absorbance contribution from the sample μ_s is small compared to the total absorbance μ_T. On the other hand, if the contribution from the sample μ_s is significant, the fluorescence proportionality to

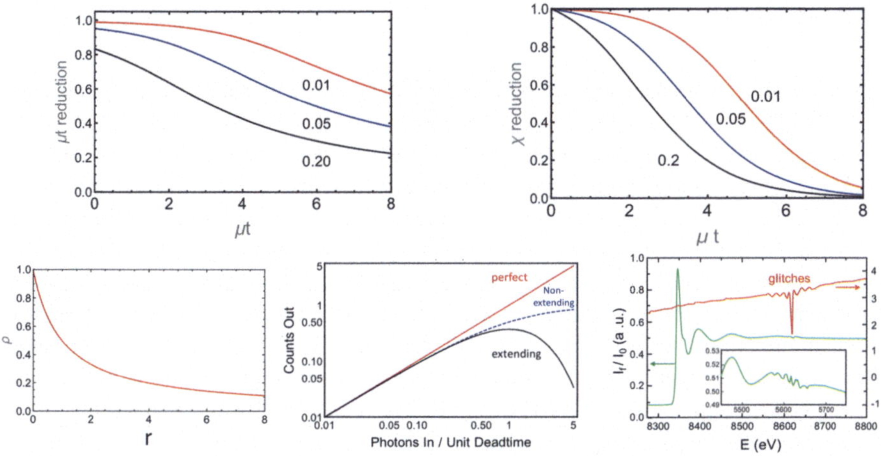

Fig. 6.5 Four different types of EXAFS artefacts. Top left: apparent absorbance reduction for various leakage rates α as a function of sample optical thickness. Top right: reduction in EXAFS amplitude due to same leakage rates. Bottom left: the loss factor (ρ) in fluorescence detected absorption due to saturation as a function of $r = \mu_s/\mu_b$; bottom middle: pileup effects on a perfect detector (red solid line), a non-paralyzable (non-extending) detector (blue dashed line), and a paralyzable (extending) detector (black solid line); bottom right: EXAFS spectrum suffering from glitches, incident intensity (red solid line), and fluorescence detected absorption (green solid line)

absorbance will be reduced by $(F/F_0) = 1/(r + 1)$, where r is the ratio of μ_s to the absorbance of the remaining bulk μ_b (Fig. 6.5) [214–216]. These *saturation effects* can be avoided by using appropriately dilute and/or thin samples. Saturation effects are sometimes unfortunately called *self-absorption*, but that is a quite different process.

6.2.6 Detector Nonlinearity

The fluorescence-detected absorption experiment assumes direct proportionality between the output signal and the amount of fluorescence. However, photon counting detectors have a *dead time*, during which arrival of one photon prevents the processing of the next photon (Chap. 4). Thus, the measured fluorescence signal might no longer be directly proportional to the "real" fluorescence. The input vs. output curves for different alternatives are shown in Fig. 6.5. The bottom line—if the photon flux is too high, the required linearity between measured and true fluorescence breaks down, and the height of an absorption edge will be reduced. In extreme cases, with a paralyzable detector, the edge will even bend over (the detector output will decrease as the fluorescence increases)!

6.2.7 Glitches

Another source of artefacts comes from the crystal monochromator. As one scans the monochromator with the primary reflection obeying Bragg's law, there are usually a few angles at which another crystallographic plane just happens to diffract [217]. When this happens, there is a dip in the intensity of the primary reflection, as some of the beam is directed into the other reflection. Any slight nonlinearity in the measurement will result in a *glitch* in the observed spectrum. (A more sophisticated name is *umweganregung* or "detour radiation" [218]). Even without nonlinearity, there can be changes in the spatial distribution of the beam intensity, the harmonic composition, and even its polarization [219]. Facilities like SSRL keep glitch spectra for different crystal orientations, so you can pick the one with the least severe glitches across your EXAFS region. The expected location of glitches can also be calculated with computer software [220].

6.3 Essential Physics of EXAFS

One of the strengths of X-ray absorption is that each element has an absorption edge with a relatively constant energy and cross section. From the energy of the edge, you immediately know which element you are looking at, and from the size of the edge jump, you can quantitate the concentration of that element. The energies for these edges are summarized in Fig. 6.2 and Appendix G. If X-ray absorption always exactly followed the tables of mass absorption coefficients, it would be a nice analytical tool for elemental composition, but not much else.

The exciting properties of XAS are the deviations from the norm—the fine structure that reveals the electronic and molecular structure of the sample. As a simple example, in Fig. 6.6 we compare the X-ray absorption for the monatomic noble gas krypton with the neighboring diatomic Br_2. Past the absorption edge, the Kr spectrum shows a smooth, featureless falloff with increasing X-ray energy. In contrast, Br_2 shows periodic oscillations past the edge called "EXAFS"—an acronym for "**E**xtended **X-ray** **A**bsorption **F**ine **S**tructure," which reflect the Br-Br distance. It also has a sharp feature at the edge (part of the region called "**X-ray** **A**bsorption **N**ear **E**dge **S**tructure" ("XANES"); see Chap. **7**) that reflects a transition to a vacant π^* molecular orbital. A final acronym, "XAFS", "**X-ray** **A**bsorption **F**ine **S**tructure", is sometimes used to define the overall fine structure patterns including both XANES and EXAFS.

The EXAFS signal is defined as the fractional modulation of X-ray absorption for an atom in a particular environment compared to the free atom X-ray absorption:

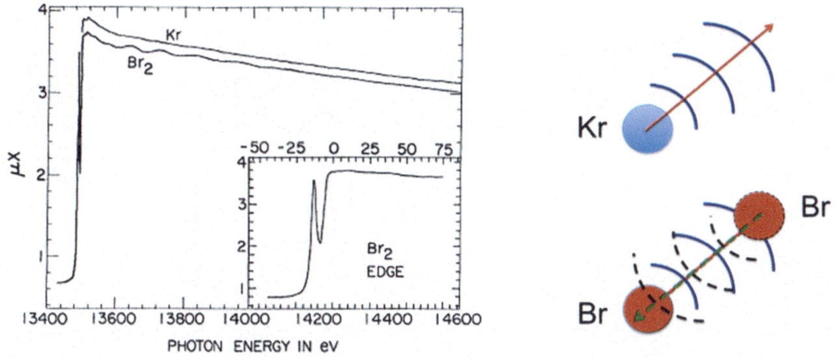

Fig. 6.6 Left: The X-ray absorption spectrum for gaseous Kr compared with that for Br_2. Redrawn from [221]. Right: a photoelectron from Kr propagates in free space, while a photoelectron in Br_2 sometimes scatters from the neighboring Br back to the origin

$$\chi(E) = \frac{\mu(E) - \mu_0(E)}{\mu_0(E)} \tag{6.4}$$

where $\mu(E)$ is the observed absorption coefficient for a particular absorption edge and $\mu_0(E)$ is the structureless, atomic-like background absorption coefficient for that same edge [213,222,223]. In terms of the absorption coefficient, we can rewrite the above as:

$$\mu(E) = \mu_0(E)[1 + \chi(E)] \tag{6.5}$$

All of this is graphically illustrated in Fig. 6.7. In an X-ray absorption event, a photon is absorbed, and its energy is used to promote a core electron (to a valence level or the continuum), with a core hole left on the absorbing atom [222–224]. In the simplest continuum events, the kinetic energy of the photoelectron is given by the difference between the photon energy and the electron binding energy, so as the X-ray photon energy is scanned, the photoelectron kinetic energy also changes. If there is another atom nearby, the outgoing photoelectron can scatter off that neighbor, sometimes directly back to the absorbing atom (Figs. 6.6 and 6.7).

Interference phenomena occur when two or more paths can be travelled by a wave and the observer cannot distinguish those paths (at its simplest, think of the double slit experiment). With EXAFS, the interference is between a simple outgoing photoelectron wave and processes in which the photoelectron wave scatters off a neighbor, back to the absorbing atom, and then scatters out again. (In more complicated cases, the photoelectron can scatter multiple times.) So to describe this interference phenomenon, we need to start by characterizing these photoelectron waves.

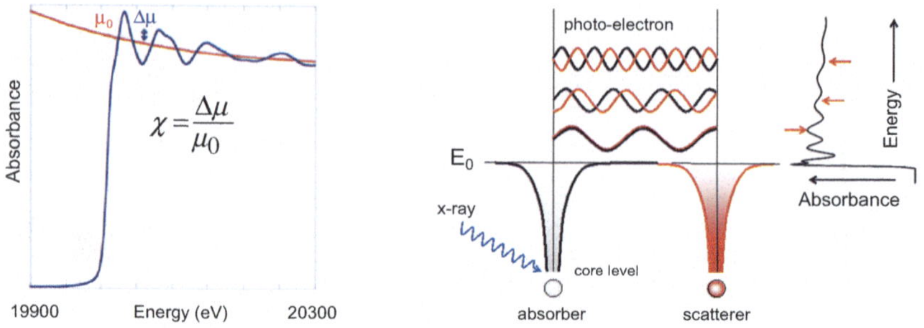

Fig. 6.7 Left: quantitative definition of EXAFS, illustrated with MoS_2 data. Right: The changing photoelectron wavelength and scattering in EXAFS. As the photoelectron wave number changes, the backscattered wave (red line) goes in and out of phase with the outgoing wave (black line) near the nucleus

6.3.1 The Matrix Element

In standard quantum mechanics textbooks (see Resources at end of Chapter), it is derived that the absorption coefficient (or cross section) for the absorption of light is proportional to the square of a matrix element that involves the initial $|i\rangle$ and final $|f\rangle$ state wave functions and the interaction Hamilton $\mathcal{H}_I$:

$$\mu \propto \sigma \propto \left|M_{if}\right|^2 = |\langle f|\mathcal{H}_I|i\rangle|^2 \tag{6.6}$$

We assume for now that only the wave function for a single electron changes. Then, for X-ray absorption, the initial wave function is that of a core electron, and the final-state wave function involves an outgoing electron wave. Following Als-Nielsen and McMorrow [224], we write the final-state wave function as the combination of an outgoing wave $|f_0\rangle$ and a scattered wave $|\Delta f\rangle$, so the matrix element becomes:

$$\begin{aligned}\left|M_{if}\right|^2 &= |\langle f_0 + \Delta f|\mathcal{H}_I|i\rangle|^2 \\ &= |\langle f_0|\mathcal{H}_I|i\rangle|^2\left(1 + \left\{\frac{\langle f_0|\mathcal{H}_I|i\rangle \times \langle \Delta f|\mathcal{H}_I|i\rangle}{|\langle f_0|\mathcal{H}_I|i\rangle|^2} + c.c.\right\}\right)\end{aligned} \tag{6.7}$$

where c.c. means complex conjugate. If we look back to Eq. 6.5, we see that the last part of this equation refers to the absorption coefficient of the free atom, $\mu_0(E)$, while the remainder represents the EXAFS oscillations $\chi(E)$:

$$\chi(E) \propto \langle \Delta f|\mathcal{H}_I|i\rangle \tag{6.8}$$

From these manipulations we see an essential result—the strength of the EXAFS depends on the overlap between the scattered wave $\langle \Delta f|$ and the core wave function $|i\rangle$, as modified by the interaction Hamiltonian $\mathcal{H}_I$. Going forward, we will assume that the Hamiltonian is for a dipole transition, hence the operator looks like position $\vec{r}$. Furthermore, the core electron wavefunction is so localized near the nucleus that it is essentially a δ function at the origin. If we then write the scattered wavefunction in a more conventional form as ψ_s, the matrix element just becomes amplitude of the scattered wave function at the origin:

$$\chi(E) \propto |\langle \Delta f|\mathcal{H}_I|i\rangle| \propto \int \psi_s\left(\vec{r}\right) \cdot \vec{r} \cdot \delta\left(\vec{r}\right) d\vec{r} \propto \psi_s(0) \tag{6.9}$$

Since the scattered wave function has a phase as well as amplitude, the EXAFS signal can be positive or negative, depending on whether there is constructive or destructive interference between outgoing and backscattered waves.

6.3.2 The Scattered Wave function

From the above, we now see that in order to calculate the EXAFS, our task becomes the evaluation of the scattered wavefunction at the origin of the X-ray absorbing atom. At this point, we will skip the math and use Fig. 6.7 to illustrate the result.

So far, we have been describing X-ray absorption in terms of the X-ray energy E, but for EXAFS calculations, we really care about the photoelectron kinetic energy E_{kin} and associated photoelectron wavelength λ and wavenumber k. Above the absorption edge E_0, absorption of an X-ray with energy $E = h\nu$ (where h is Planck's constant) yields an outgoing photoelectron with kinetic energy E_{kin}:

$$E_{kin} = \Delta E = (E - E_0) \tag{6.10}$$

Using $E_{kin} = \frac{1}{2} m_e v^2$ and $p = m_e v$, where m_e, v, and p are the electron mass, velocity, and momentum, leads to the momentum p being given by:

$$p = \sqrt{2m_e\, E_{kin}} = \sqrt{2m_e\, (E - E_0)} \tag{6.11}$$

From the DeBroglie equation, we know that the wavelength of an electron depends on its momentum:

$$\lambda = \frac{h}{p} \tag{6.12}$$

The photoelectron wavelength λ is thus a function of photon energy and is given by:

$$\lambda = \frac{h}{\sqrt{2m_e\, (E - E_0)}} \tag{6.13}$$

We will also need the photoelectron wave number: $k = 2\pi/\lambda$, which is in practical units:

$$k\left[\mathring{A}^{-1}\right] = \frac{2\pi}{\lambda\left[\mathring{A}\right]} = 2\pi\sqrt{\frac{2m_e\,(E - E_0)}{h^2}} = \sqrt{0.2625(E - E_0)[\mathrm{eV}]} \tag{6.14}$$

Thus, when taking a spectrum, as the X-ray energy is increased, the photoelectron wavelength decreases (Fig. 6.7). A key point is that the strength of absorption, given by the absorption coefficient $\mu(E)$, is proportional to the overlap between the initial wave function, a core orbital localized near the nucleus, and the final diffuse outgoing wave function. In the absence of neighboring atoms, the overlap between the initial core electron wave function and the outgoing photoelectron wave function monotonically decreases, and the absorption falls off as the smooth, $\mu_0(E)$ function illustrated for Kr in Fig. 6.6.

When neighboring atoms are present, the outgoing wave from the absorber is scattered in all directions by these neighbors. Back at the absorber, the scattered wave oscillates between adding *in phase*, for *constructive* interference, or *out of phase*, for *destructive* interference. The larger the distance between absorber and scatterer, the higher the frequency the oscillations. In general, the larger the number N or size Z of neighboring atoms, the larger the amplitude of the oscillations. Thus, *EXAFS contains information that can be used to deduce local structure.*

6.4 Single Scattering EXAFS Equation

From the qualitative treatment above, we see that the frequency of the EXAFS depends first and foremost on the distance between the absorbing atom a and the backscattering atom b, while the amplitude will depend on the number and type of backscatterers. A commonly used equation for describing the EXAFS of an absorbing atom a in an environment of backscattering atoms b at distances R_{ab} is Eq. 6.15:

$$\chi(k) = \sum_{\mathrm{b}} S_0^2(k)\overbrace{\frac{N_b\,|f_b(\pi,k)|}{kR_{ab}^2}}^{\text{b.s.amplitude}}\underbrace{\exp\left(-2\sigma_{ab}^2k^2\right)}_{\text{Debye-Waller factor}}\overbrace{\exp\left(\frac{-2R_{ab}}{\lambda}\right)}^{\text{mean free path effect}}\sin\left[2kR_{ab} + \underbrace{2\alpha_a(k)}_{\substack{\text{absorber}\\ \text{p.s.}}} + \overbrace{\beta_b(k)}^{\substack{\text{scatterer}\\ \text{p.s.}}}\right] \tag{6.15}$$

The terms in the above equation are:

- R_{ab}—the distance between absorbing atom a and backscattering atom b. Since the phase of the backscattered wave depends on the distance travelled R_{ab}, absent other terms, the EXAFS would follow $\sin(2kR_{ab})$. The amplitude of the EXAFS diminishes as $\frac{1}{R_{ab}^2}$.

  The EXAFS effect also depends on the type of neighbor. This is because the strength of the scattering depends on the atomic number of the neighbor, through:
- $|f_b(\pi,k)|$—the electron-atom backscattering amplitude

  The type of neighbor also affects the phase of the scattered wave through:
- $\beta_b(k)$—the electron-atom backscattering phase shift

  The type of absorbing atom also contributes a phase shift through:
- $\alpha_a(k)$—the central atom phase shift

  Of course, if there is more than one neighbor at the same distance, then the effect will be multiplied by:
- N_b—the number of scattering atoms at a particular distance.

  If the interatomic distances are slightly different, then the sine waves from different species in the sample will be slightly out of phase. For a Gaussian distribution of such distances with an rms deviation σ_{ab}, this reduces the EXAFS by:
- $\exp(-2\sigma_{ab}{}^2k^2)$—the Debye-Waller factor.

  Finally, there are inelastic effects that can rob EXAFS intensity, including a damped exponential dependence of the photoelectron mean free path λ:
- $\exp(-2R_{ab}/\lambda)$

  and an additional reduction by:
- $S_0{}^2(k)$—a factor that includes losses due to multi-electron excitations.

We now look in detail at the contributions of each of thse terms.

6.4.1 Scattering Functions

In the above equation, which describes EXAFS as a sum of damped sine waves, to first order, there are only three terms that affect the frequency and phase of the oscillations. The most important term is of course the information that we want—the absorber-backscatterer distance R_{ab}. The other two terms are so-called phase shifts that prevent EXAFS analysis from being a slam-dunk procedure. These phase shifts arise as the electron wave propagates across the potentials of the absorbing atom and the backscattering atom. Although one can find published values for phase shifts for different atoms [225,226], nowadays, they are usually calculated as needed within larger software packages such as FEFF [227,228].

6.4.1.1 Phase Shifts

There are two types of phase shifts involved as the photoelectron passes through the potential of (a) the X-ray absorbing atom and (b) the backscattering atom. The absorbing atom phase shifts all look quite similar, as seen in Fig. 6.8.

The backscattering atom phase shifts for relatively light atoms also look quite similar, but, as Z increases, these phase shifts become increasingly structured (Fig. 6.8). Note that the phase shifts for atoms in different rows of the periodic table are shifted (approximately) by an increment of π. This can be helpful in identifying the type of neighboring.

6.4.1.2 Scattering Amplitudes

Very approximately, the strength of electron-atom backscattering increases linearly with the backscatterer Z (Fig. 6.9). Combined with the π difference in phase shifts, this means that the EXAFS contribution from 1 sulfur (S) at a given distance is approximately equivalent to negative two oxygens (O) at the same distance. As with phase shifts, there is increasing complexity in amplitude functions as the backscattering atom Z increases.

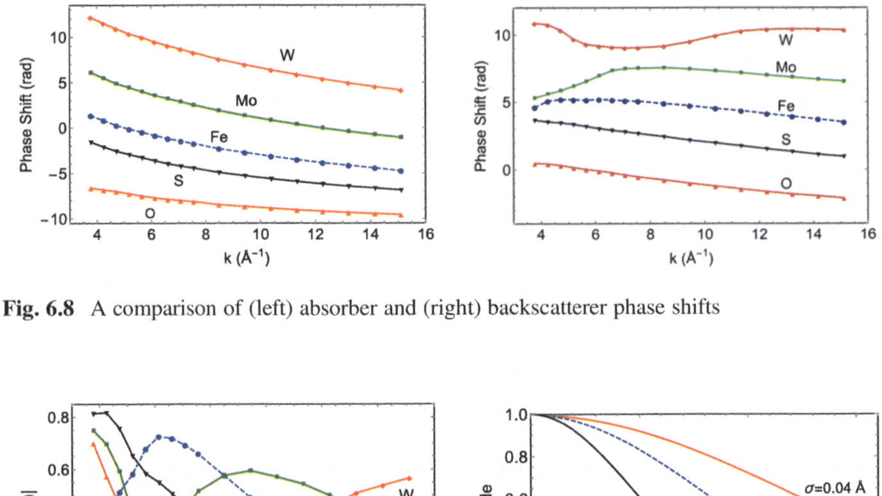

Fig. 6.8 A comparison of (left) absorber and (right) backscatterer phase shifts

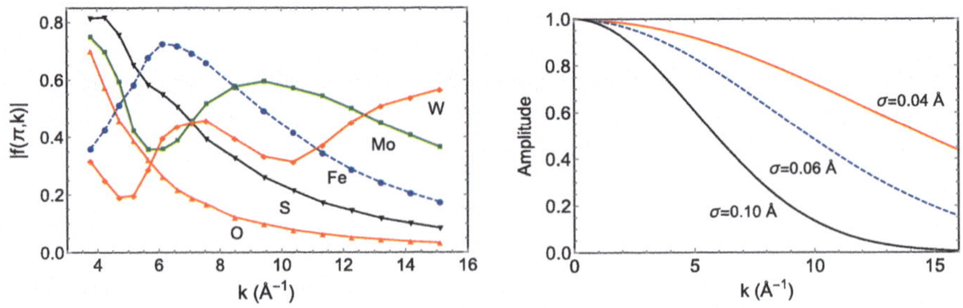

Fig. 6.9 Left: backscattering amplitudes $|f(\pi,k)|$ for different values of Z. Right: Debye-Waller factors vs. k for representative values of σ

6.4.2 Simple Disorder Effects: Debye-Waller Factors

In a real sample, the distances between absorber and backscatterers are not fixed. Instead, there is always a distribution of distances, because of both thermal motion (dynamic disorder) and structural (static) disorder. We can describe this distribution of distances in terms of a *pair distribution function* $P(r)dr$ that defines the probability that the absorber-backscatterer distance lies between r and $r + dr$. The distribution of distances around the centroid R_{AB} means that the observed EXAFS signal will be a sum of sine waves with different frequencies, and the interference between these frequencies will modify the amplitude and (sometimes) phase of the EXAFS.

6.4.2.1 Diatomic Systems

As the simplest case, consider a bimolecular system A–B, with masses m_A and m_B and force constant $k_{A\text{-}B}$. Although in the classical limit the two masses could sit at a fixed distance $R_{A\text{-}B}$, we know from quantum mechanics that this harmonic oscillator will have a "zero-point motion" even at a temperature of absolute zero.

The ground state turns out to have a Gaussian probability distribution around the equilibrium interatomic distance r_0 [229]:

$$P(r) = \frac{1}{\sqrt{2\pi\sigma^2}} \exp\left[-\frac{(r-r_0)^2}{2\sigma^2}\right] \tag{6.16}$$

while the excited states have more complicated distributions involving products of Hermite polynomials and Gaussians (Fig. 6.10). At a given temperature T, an ensemble of oscillators will be found in a Boltzmann distribution of states, and the overall mean square amplitude of vibration is given by [230]:

$$\langle\sigma^2\rangle = \frac{h}{8\pi^2\mu c\omega} \coth\frac{hc\omega}{2kT} \tag{6.17}$$

where μ is the reduced mass and the frequency ω with force constant k is given by $\omega = (k/\mu)^{(1/2)}$ [229]. In a nice illustration of trends within a group down the periodic table, Baran has calculated σ for different dihalogen molecules and different temperatures [231], and we reproduce this work in Fig. 6.11.

6.4.2.2 More Complex Molecules

For nonlinear molecules with N atoms, one has to add the contributions to σ^2 from all of the $3N - 6$ normal modes. In some of the simpler cases, such as tetrahedral AB_4 systems, one can still derive tractable expressions for σ^2 from the observed vibrational frequencies and simplifying assumptions [230]. In the tetrahedral case, for

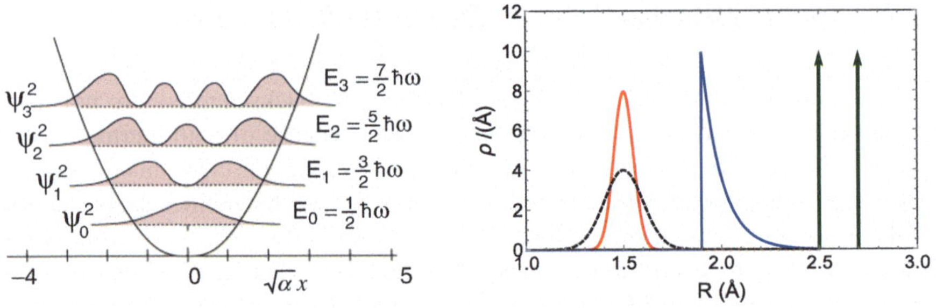

Fig. 6.10 Left: energies and probability distributions for a quantum harmonic oscillator, where $\alpha = m\omega/\hbar$. Right: common distribution functions for EXAFS (left to right) Gaussians, a skewed exponential, and a pair of δ functions (distances are arbitrary)

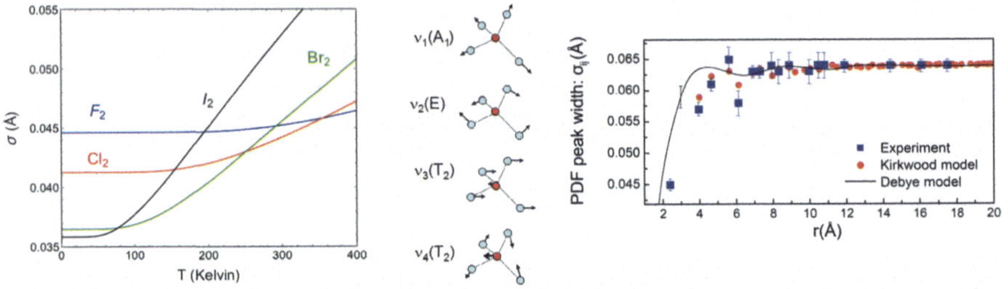

Fig. 6.11 Left: The trend in rms deviations σ for different halogens vs. temperature [231]. Middle: Normal modes of a tetrahedral molecule. The AB σ values can be approximately calculated using the vibrational frequencies for the A_1 and T_2 stretching modes [230]. Right: comparison of neutron PDF with "Kirkwood" model and correlated Debye model calculations for GaAs at 10 K. The latter used a Debye temperature $\theta_D = 250$ K and Debye wave vector $k_D = 1.382$ Å^{-1} [232]. The key point is that in most cases, σ-values rise rapidly and plateau beyond the second coordination sphere

example, only the totally symmetric A_1 and asymmetric T_2 stretching modes give rise to changes in A–B distances, so one can simply add these contributions in quadrature (Fig. 6.11). Of course, for larger molecules and/or lower symmetries, one needs to sum over a much larger number of normal modes. Typical values for σ range from $\sigma = 0.03$ Å for tight metal-oxo or metal-metal bonds to ~0.1 Å for pairs of atoms with virtually no bonding interaction.

6.4.2.3 Einstein Solid and Debye Models

The enormous number of atoms involved in extended solids means that some simplifying assumptions are required to derive the σ^2 values appropriate for EXAFS. An important difference with the molecular treatment is that the set of discrete 3 N-6 normal modes is replaced by a *vibrational density of states*, $\rho(\omega)$, that describes the number of modes $\rho(\omega)d\omega$ between ω and $\omega + d\omega$. For the purposes of

EXAFS analysis, one also needs to define a *projected density of states*, which weights each particular mode at frequency ω by the mean square deviation of the relevant interatomic distance in that mode—$p_j(\omega)$.

The simplest possible assumption for the density of states is the Einstein model, which pretends that all of the vibrational modes for the set of equivalent atoms at distance R have a single frequency ω_E [233]:

$$\rho_R(\omega) = \delta(\omega - \omega_E(R)) \tag{6.18}$$

For a monatomic solid where the atomic mass is M, and using β = 1/kT, the resulting equation for σ^2 is:

$$\sigma_R^2(\omega_E) = \frac{\hbar}{M\omega_E}\coth\left(\frac{\beta\hbar\omega_E}{2}\right) \tag{6.19}$$

Prescriptions for the appropriate ω_E values for different crystal structures and shells of atoms are published [234].

The next step in complexity is the correlated Debye model, where the phonon density of states increases with the square of the frequency ω up to a Debye cutoff frequency $\omega_D = ck_D$, where c is the sound velocity, the Debye wavevector is $k_D = (6\pi^2 N/V)^{1/3}$, and N/V is the number density of the crystal [233]:

$$\begin{aligned}\rho(\omega) &= \frac{9\omega^2}{(\omega_D)^3}, \quad \omega \leq \omega_D\\ \rho(\omega) &= 0, \quad \omega > \omega_D\end{aligned} \tag{6.20}$$

The relevant projected density of states for interatomic distance r_{ij} is [235]:

$$\rho_j(\omega) = \frac{3\omega^2}{\omega_D^3}\left(1 - \frac{\sin(\omega R_j/c)}{(\omega R_j/c)}\right) \tag{6.21}$$

which gives rise to a surprisingly complex equation for the mean square deviation as a function of interatomic distance r_{ij}, Debye temperature $\Theta_D = \hbar\omega_D/k_B$, and actual temperature T [232]:

$$\sigma_D^2(\omega_D) = \frac{6\,\hbar}{M\,\omega_D}\left[\frac{1}{4} + \left(\frac{T}{\Theta_D}\right)^2\Phi_1\right] - \frac{6\,\hbar}{M\,\omega_D} \times \left[\frac{1-\cos(k_D r_{ij})}{2(k_D r_{ij})^2} + \left(\frac{T}{\Theta_D}\right)^2 \int_0^{\Theta_D/T} \frac{\sin\left(\frac{k_D r_{ij} T x}{\Theta_D}\right) / \left(\frac{k_D r_{ij} T}{\Theta_D}\right)}{e^x - 1}\,dx\right] \tag{6.22}$$

where $\Phi_1 = \int_0^{\Theta_D/T} x\,(e^x - 1)^{-1} dx$, and x is a dimensionless integration variable. The bottom line, despite the complexity of the math, is that near room temperature, σ_D, or σ for just about any model, is ~0.06 Å for the first coordination sphere and quickly rises toward an uncorrelated value of ~0.1 Å beyond the second coordination sphere, as illustrated for GaAs in Fig. 6.11.

6.4.3 Multi-Electron Effects: The Amplitude Reduction Factor

The description of the EXAFS effect so far has relied on two implicit assumptions: (1) that all of the extra photon energy ($E - E_0$) is deposited in a single electron, so that the latter has a well-defined wave vector k, and (2) that the photoelectron does not lose energy as it propagates out and back in the medium. However, the creation of a core hole can result in dual or multiple electron excitations, and the resulting losses in EXAFS intensity are traditionally described as *intrinsic losses*. Similarly, the photoelectron can lose energy as it scatters within the medium, and in these cases the losses in EXAFS intensity are assigned as *extrinsic losses*.

To understand these losses, note that so far we have concentrated on the outgoing photoelectron and ignored the effects of the resulting core hole on the remaining electrons. However, the perturbing influence of the core hole in effect makes the remaining electrons see a nucleus with a $Z + 1$ charge, and their orbitals will contract accordingly. As described by Rehr and Albers [223], if one makes the approximation that the final-state wave function can be factored into one-electron orbitals and an outgoing photoelectron:

$$|\Psi_f\rangle = |\Phi'{}_0^{N-1}\rangle\,|\phi_f\rangle \tag{6.23}$$

then the *many body dipole matrix element* is given by:

$$M_{fi} \cong \left\langle \phi'_f|\hat{\varepsilon}\cdot\vec{r}|\phi_c\right\rangle \left\langle \phi_0'^{N-1}|\phi_0^{N-1}\right\rangle^2 \tag{6.24}$$

In the above (and below) expressions, the terms with primes correspond to states calculated in the presence of a core hole. The first part of the above expression contains the EXAFS effects that we have considered so far. The second part is the so-called many-body overlap integral. The absorption in the "primary channel" (that yields EXAFS) is reduced by the square of this integral, leading to the so-called amplitude reduction factor S_0^2:

$$S_0^2 \cong \left\langle \phi_0'^{N-1}|\phi_0^{N-1}\right\rangle^2 \tag{6.25}$$

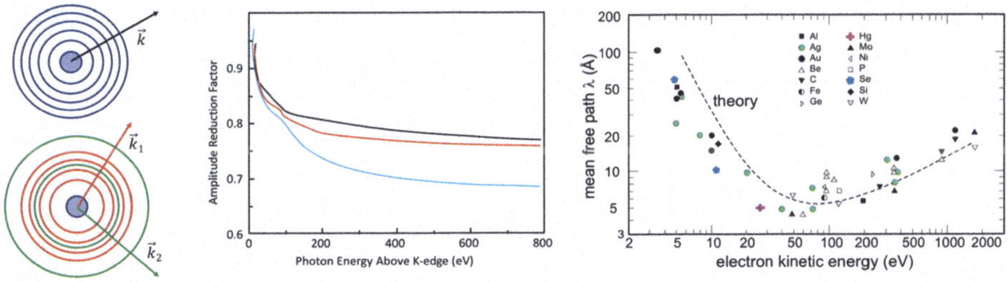

Fig. 6.12 Loss mechanisms. Left: a simplified model for the source of intrinsic losses—two electrons share the photon energy leading to a mix of k-values. Middle: Gurman S_0^2 calculations [236]. Right: experimental values for electron mean free paths for different elements and for different electron kinetic energies

In practice, S_0^2 is on the order of 0.7–0.9, as shown in Fig. 6.12 [236]. Although it is technically k-dependent, most analyses use a fixed value for the entire range of data.

Another way to describe these multi-electron effects is to consider that the absorption events involve two or more electrons. In such cases, the available energy $(E - E_0)$ is shared between the electrons, and the outgoing photoelectron wave vector becomes a distribution of k-values (Fig. 6.12). This will rapidly damp the EXAFS oscillations, similar to the lost intensity we saw for a distribution of R-values.

6.4.4 *Mean Free Paths*

There are other ways to lose intensity for the EXAFS effect. Either the outgoing or backscattered electron wave can be scattered inelastically by an intervening atom, resulting in loss of energy to vibrations or electronic excitations. This energy loss will change the photoelectron wavelength and thus the interference effect that gives rise to EXAFS. Since there are many different ways to scatter the photoelectron, the net effect is a loss of EXAFS intensity.

Another mechanism for loss of EXAFS intensity is collapse of the core hole, with subsequent emission of a fluorescence X-ray or an Auger electron. This typically occurs on the femtosecond time scale. This means that there is only a very short time available for propagation of the photoelectron.

Taken together, both of these effects diminish the EXAFS from longer distance neighbors, and they are usually lumped together into a parameter called the *mean free path*—λ. The inelastic scattering effects depend on the photoelectron energy, as shown in Fig. 6.12. Of course, the photoelectron velocity (and the distances it can probe during the core-hole lifetime) also depends on its energy. Thus, a k-dependent mean free path $\lambda(k)$ is commonly used, resulting in a distance-dependent amplitude reduction term:

$$\chi(k) \propto \exp\left[\frac{-2R}{\lambda(k)}\right] \tag{6.26}$$

6.4.5 Polarization and Orientation Dependence

In our presentation of the original EXAFS equation (Eq. 6.15), we assumed a fluid or powder sample with no favored molecular orientation, and we thus ignored the angular dependence of the outgoing photoelectron. In the case of an oriented sample, this angular dependence can no longer be ignored. For a K-edge or L_1-edge transition, one starts from a spherical *s*-orbital, and the dipole selection rule constrains the outgoing electron to *p*-wave symmetry. In such cases, one can use the result derived by Stern [222]. This says that for a particular absorber-backscatterer A–B pair, the intensity of the oriented EXAFS will be proportional to 3 $\cos^2\theta$, where θ is the angle between the X-ray polarization vector and the A–B axis (Fig. 6.13):

$$\chi(k) \propto 3\cos^2\theta \tag{6.27}$$

The angular dependence of the EXAFS is considerably more complex at L_3 or L_2 edges, which involve a mix of $p \rightarrow s$ and $p \rightarrow d$ transitions. In this case there will be interference between the stronger outgoing *d*-wave and the weaker outgoing *s*-wave, yielding a cross-term in the EXAFS:

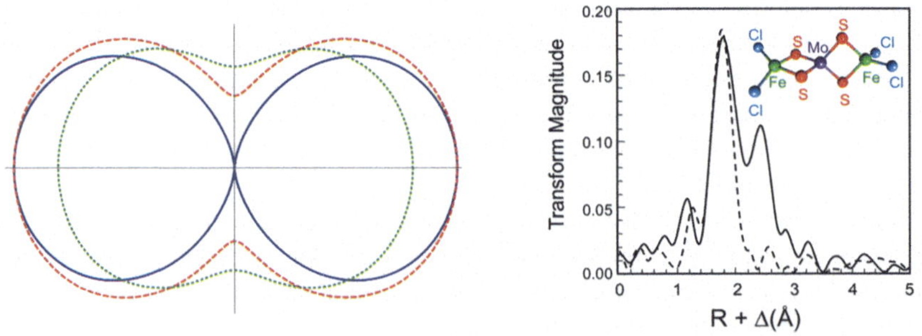

Fig. 6.13 Left: predicted angular dependence for K and $L_{2,3}$-edge EXAFS, $\cos^2\theta$ dependence of K-edge EXAFS (blue line),1 + $3\cos^2\theta$ component of $L_{2,3}$-edge EXAFS (red dashed line), an $L_{2,3}$-edge EXAFS assuming a modest (1 − $3\cos^2\theta$) cross-term component (green dotted line). Right: experimental Mo EXAFS for $(Ph_4P)_2(Cl_2FeS_2MoS_2FeCl_2)$ with electric field ∥ (black solid line) or ⊥ (black dashed line) to the Mo–Fe axis. (inset—structure) [237]

$$\chi_{\text{direct}}(k) \propto \frac{|\langle 2|1\rangle|^2}{2}\left(1+3\cos^2\theta\right)\sin\left[2kR+\delta_{2j}(k)\right] \tag{6.28}$$

$$\chi_{\text{cross-term}}(k) \propto |\langle 1|2\rangle\langle 0|1\rangle|\left(1-3\cos^2\theta\right)\sin\left[2kR+\delta_{02j}(k)\right] \tag{6.29}$$

The $\chi_{\text{direct}}(k)$ term refers to the pure p → d component, while the $\chi_{\text{cross-term}}(k)$ refers to the interference between p → s and p → d transitions. This interference between the two terms leads to a more complicated angular dependence for the amplitude of L_3 or L_2 EXAFS than for K and L_1-edges. Furthermore, because the phase shifts are different for the direct and cross-term components, mistakes in the distance predictions can also result. In case you still dare to attempt such experiments, refer back to the original literature and more recent reviews [238, 239].

6.4.6 More Complex Disorder Effects

Previously (Sect. 6.4.2), we introduced a Debye-Waller factor without derivation to account for the spread in A–B distances due to thermal motion. Of course, there is no a priori reason that the A–B distribution has to be a Gaussian. The source of a non-Gaussian distribution may be dynamic—resulting from an anharmonic potential function. Or, the source might be static, say from structural disorder in an amorphous material. Finally, we note that even if the true distribution function is Gaussian, for EXAFS what matters is the *effective distribution function*, $P(R_b,\lambda)$, which includes all of the R-dependent amplitude terms:

$$P(R_b,\lambda) = P(R_b)\exp\left(\frac{-2R_b}{\lambda(k)}\right)\exp\left(-2\sigma_b^2k^2\right) \tag{6.30}$$

For Gaussian distributions with small σ, the R-dependent terms are nearly equal, but for large σ, the effective distribution function becomes noticeably skewed (Fig. 6.14).

The problems resulting from more complex distribution functions were noted early on by Eisenberger and Brown [240], who pointed out that a conventional analysis of the EXAFS for metallic zinc predicted a nearest neighbor contraction of 0.09 Å between 20 K and room temperature, compared to the known 0.05 Å expansion. Perhaps even more disturbing, the apparent coordination number decreased by an order of magnitude (Fig. 6.14)!

The key point for readers is that whenever the distribution of scatterers strays from a narrow Gaussian function, there are corrections needed to both the phase and the amplitude of the simple EXAFS expression given in Eq. 6.15. More detailed discussion of the issues that arise from non-trivial distribution functions can be found in references at the end of this chapter.

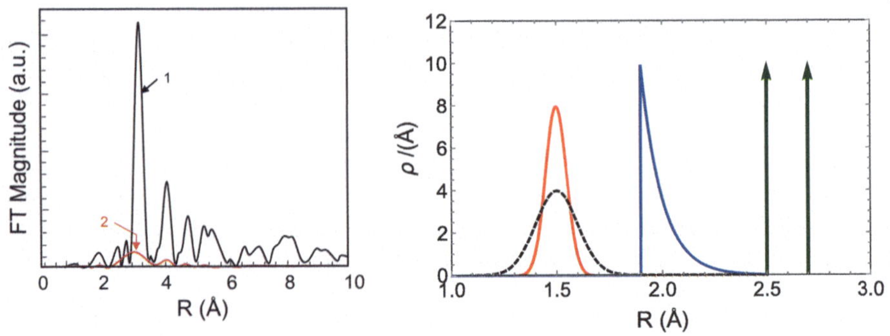

Fig. 6.14 Left: temperature dependence of Zn EXAFS Fourier transforms: low temperature (black solid line) vs. room temperature (red line), redrawn from [240]. Right: some of the distribution functions relevant for EXAFS: left, Gaussians; middle, a skewed exponential, and right, pairwise δ-functions. Distance scale is arbitrary

6.5 Multiple Scattering

So far, we have assumed that EXAFS comes only from single scattering processes, where the photoelectron scatters at 180° from neighboring atoms (Path 1 in Fig. 6.15). Of course, the photoelectron can scatter at other angles and subsequently off additional atoms. Among the plethora of possible multiple-scattering events, the simplest additional process involves scattering off one additional atom, thus yielding triangular paths such as 2a and 2b in Fig. 6.15, with scattering from B → C → A as well as scattering from C → B → A. Thus, this triangular path needs to be counted twice. Finally, there can be contributions from path **3**, with scattering B → C → B → A. We call these cases three-body paths to include the limiting case where the first scattering angle $\beta = 0°$ and hence the second scattering angle $\gamma = 180°$.

The angular dependence of scattering for a carbon atom is illustrated in Fig. 6.16. The scattering amplitude falls off rapidly beyond 20–30° and then rises slightly near 180°. From this we can see that not all three-body paths are equally important. Because scattering is strongest in the forward direction, paths that involve scattering at angles close to 0° or 180° are far more important than those involving 90° scattering. For bound A–B–C systems, multiple scattering is relatively unimportant unless the ABC angle is greater than 150°.

To reiterate, the relative importance of multiple-scattering paths is exquisitely sensitive to the scattering angle β (as defined in Fig. 6.15). Ignoring mean free path and disorder effects for the moment, expressions from Teo for the relative intensities for paths **1**, **2**, and **3** are given by [241]:

$$\text{Path } 1 : \chi_1^{\text{AC}}(k) \propto \frac{F_{\text{C}}(\pi, k)}{R_{\text{AC}}^2} \tag{6.31}$$

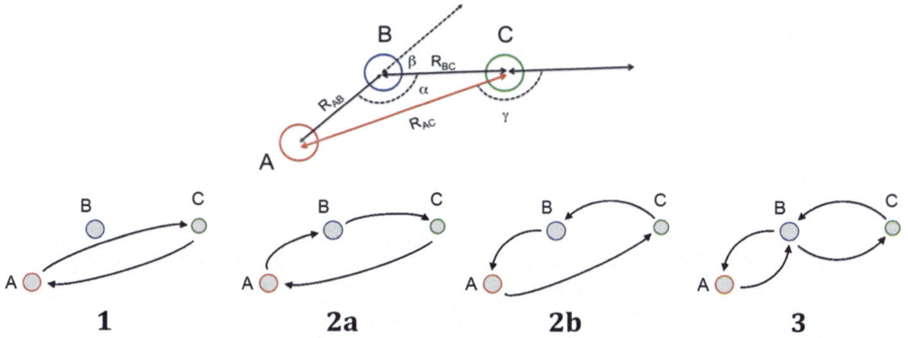

Fig. 6.15 Top: angles and distances in multiple scattering. Bottom (left to right): single scattering (**1**), the simplest multiple-scattering paths (**2a** and **2b**), and a double multiple-scattering path (**3**)

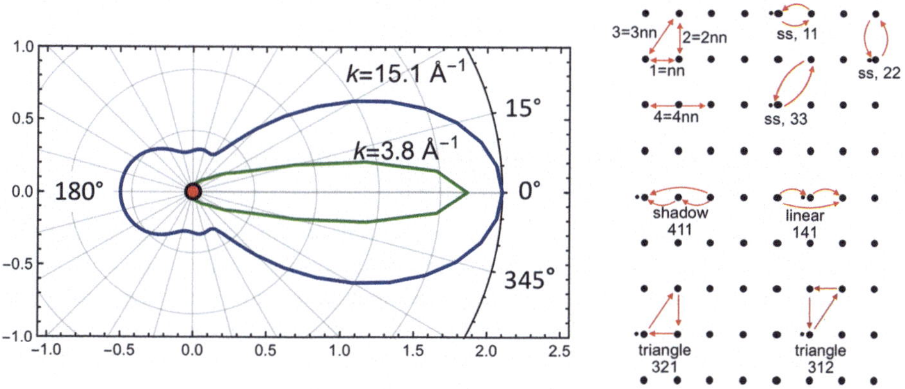

Fig. 6.16 Left: a polar plot of the angular dependence of electron scattering amplitude for carbon, from data in [241]. Right: the variety of scattering paths in Cu metal, redrawn from [223]

$$\text{Path } 2 : \chi_2^{\text{AC}}(k) \propto \frac{2F_\text{B}(\beta, k)F_\text{C}(\pi, k)}{R_\text{AB}R_\text{BC}R_\text{AC}} \tag{6.32}$$

$$\text{Path } 3 : \chi_3^{\text{AC}}(k) \propto \frac{F_B^2(\beta, k)F_\text{C}(\pi, k)}{R_\text{AB}^2 R_\text{BC}^2} \tag{6.33}$$

where $F_\text{B}(\beta,k)$ is the scattering amplitude for atom B at angle β for wavenumber k, and we have assumed that $F_\text{C}(\gamma,k) \cong F_\text{C}(\pi,k)$. The net effect of strong forward scattering from intervening atom B in path **3** means that it can sometimes be much stronger than path **1**; at $\beta = 0°$ it is the only path that needs to be considered. On the other hand, for ABC angles less than 150°, the diminished scattering amplitudes often make multiple scattering unimportant.

In an analysis for fcc Cu, Zabinsky and coworkers found that about half of the EXAFS signal came from single scattering contributions [242]. More than 90% of

Table 6.3 Multiple-scattering analysis for fcc Cu [242]

Type of path	Number of paths	Importance (%)
Single scattering	15	46.2
Shadow	15	24.2
Triangle	17	20.7
Linear	4	4.4

the EXAFS amplitude was then covered by inclusion of two types of multiple scattering—"shadow" paths and "triangle" paths ((Fig. 6.16) and (Table 6.3)). The bottom line—while it might be fun to include more exotic multiple scattering, such as "dogleg" and "quadrilateral" paths, in most cases these are a small contribution to the signal. On the other hand, since multiple scattering is so strong in the forward direction, there are special cases where the symmetry of the molecule or the solid allows long A–B–C–D or even A–B–C–D–E interactions to be observed, as illustrated later in this chapter for MoS_2 in Fig. 6.18.

6.5.1 Distribution Effects on Multiple Scattering

We saw previously that for small amounts of thermal motion, an exponential Debye-Waller factor was a good model for the *k*-dependent amplitude reduction. In part, this is because motion perpendicular to an A–B interaction (to first order) does not change the A–B distance. When dealing with a three-body (or more complex) interaction, this perpendicular motion causes significant changes in both amplitude and phase of the EXAFS interactions. Thus, to properly include the effects of static and vibrational disorder on multiple scattering, one has to include the much more complicated angular dependence of the phase shift and amplitude contributions from multiple scattering. However, this is rarely done.

6.6 Extraction of EXAFS from Experimental Data

Unlike some other spectroscopies, you do not directly measure EXAFS; you measure an X-ray absorption spectrum which contains EXAFS structure. The steps involved in defining the EXAFS include (a) baseline and pre-edge subtraction, (b) spline subtraction, (c) conversion to k-space with amplification and possible smoothing, and (d) optional Fourier transformation. This sequence of steps is illustrated for Na_2MoO_4 in Fig. 6.17. A variety of software packages is available to accomplish these steps (see end of chapter).

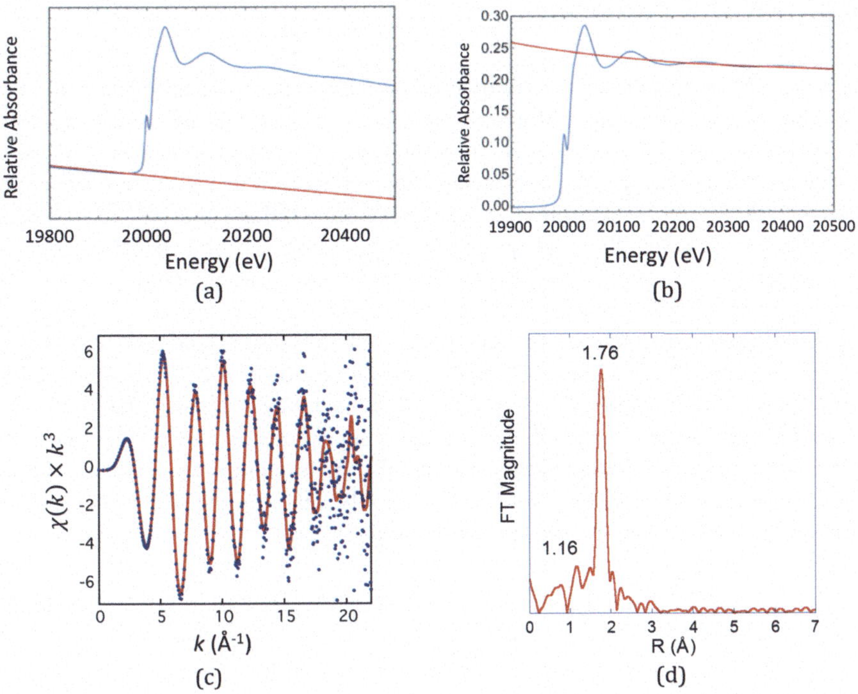

Fig. 6.17 EXAFS data processing, illustrated for $Na_2MoO_4(H_2O)_2$: (**a**) combined baseline and pre-edge subtraction; (**b**) spline subtraction; (**c**) conversion to *k*-space, k^n weighting, and optional smoothing; (**d**) Fourier transformation: $k = 2–22$ Å^{-1}, 2 Å^{-1} damping, with Mo–O phase shift included

6.6.1 Baseline and Pre-edge Subtraction

The first processing step is to remove the contributions from absorption by other elements and by other edges of the element under study. This can be done by extrapolation of a curve fit to the trend below the absorption edge. The functional form of this parameterized curve can be as simple as a straight line, but a more rational form is the so-called *Victoreen function* [243]:

$$\left(\frac{\mu}{\rho}\right)_Z = C\lambda^3 - D\lambda^4 + \sigma_0 N_0 \frac{Z}{A} \tag{6.34}$$

Since extrapolations are notoriously unstable, another approach is to use a pre-edge subtraction to estimate the size of the absorption jump and then to use the Victoreen to estimate the post-edge falloff in intensity of the atomic contribution (see below).

6.6.2 EXAFS Extraction

The next step is to extract the oscillatory part from the total absorption for that particular element and edge. Although various single polynomials and more complex functions have been tried, it is rarely the case that a single function works well over the entire data range. A poor background subtraction results in low-frequency artifacts in the EXAFS spectrum. Experience has shown that cubic splines or related piecewise polynomials do an adequate job. A calculated spline is overlaid with the pre-edge subtracted data in (Fig. 6.17). The gory details are covered in articles by Stern and Bunker [244] and the cited reference books.

Since EXAFS is define as the modulation of the absorbance $(\mu(E) - \mu_0(E))/\mu_0(E) = \Delta\mu(E)/\mu_0(E)$, one has to normalize the amplitude of the oscillations relative to the smooth atomic absorption. Near the edge itself, say at a particular energy E_1, μ_0 can be seen as the value of the interpolating spline. At higher energies, the extrapolated pre-edge subtraction often results in an unreliable estimate of μ_0. Instead, it is safer to record μ_0 close to E_0 and rely again on the Victoreen function in the post-edge region to estimate the denominator as:

$$\chi(E) = [\Delta\mu(E)/\ \mu_0(E_1)]\times[f_{Victoreen}(E)/f_{Victoreen}(E_1)] \tag{6.35}$$

6.6.3 Conversion to k-Space and Amplification

EXAFS is approximately periodic in k-space, so it is general practice to illustrate the data on a k-scale (Å^{-1}) as opposed to an energy (eV) scale. To define the wave number k, you have to know the photoelectron energy, and thus in turn you need to know the X-ray energy E_0 for which $k = 0$. This is usually estimated as approximately half-way up the absorption edge, but a precise definition of E_0 is often elusive.

The EXAFS signal dies rapidly with increasing energy or wave number, and the later oscillations are almost invisible as extracted. It is common practice to enhance the visibility by including a weighting function such as k^2 or k^3, depending on the Z of the backscatterer under study. The goal is to achieve an approximately constant amplitude so that each region of the spectrum is weighted equally in subsequent analysis, as shown for the MoO_4^{2-} example in Fig. 6.17.

6.6.4 Fourier Transforms

The final step in EXAFS extraction is often the first step in interpretation of an EXAFS spectrum. Since the k-space EXAFS for an A–B distance has an approximately sinusoidal dependence on the interatomic distance R_{AB}, an appropriately weighted Fourier transform yields peaks at approximately R_{AB}. However, the phase shifts from the absorber and scatterer are also *k*-dependent functions, and their net effect on the Fourier transform is to shift peaks to shorter distances. The nonlinearity of the phase shifts also skews the Fourier transform peaks and can give rise to additional unphysical peaks.

An approximate solution to this problem is to include the phase shift functions in the Fourier transform itself. If one transforms with respect to $2k + \phi(k)$, where $\phi(k)$ is the total phase shift, then the peaks for the A–B interaction will be approximately in the proper position R_{AB}. Of course, the correction is not perfect when there are multiple types of neighbors, but in practice the resulting pictures are still approximately correct. Note in the Na_2MoO_4 transform example that there are still additional non-physical peaks due to the finite range of the data being transformed. These *truncation effects* are well-known to result when applying a Fourier transform to a signal that has been terminated essentially by convolution with a square wave envelope.

6.7 Interpretation of EXAFS

From the previous discussion, we see that all of the required physics for describing EXAFS is basically understood, and it could be argued that structure determination from EXAFS data should be straightforward. The defect in this assumption is that structure determination from EXAFS is an inverse problem. The structure does not fall out from the data; instead you usually have to follow an iterative procedure involving the following: (a) assuming a structure, (b) calculating the EXAFS, (c) calculating the disagreement between observed and calculated EXAFS, (d) revising the proposed structure, (e) calculating the revised EXAFS, and (f) calculating the new residual and iterating until convergence. As with EXAFS extraction, a variety of software packages is available to accomplish these steps.

Iteration is required because of the nonlinear nature of the EXAFS equation, and finding the optimum simulation is usually accomplished by a nonlinear least squares analysis. To use Eq. 6.15, the nature of the backscatterer has to be assumed, and one can then vary N, R, and σ to achieve the best match to the data. As part of the analysis, you need expressions for the neighbor backscattering amplitude (Fig. 6.9) and both absorber and scatterer phase shift functions (Fig. 6.8). Early on, these functions were obtained empirically from experimental data [245] or interpolated from tables of theoretically calculated values [225]. Nowadays, software packages are available to calculate these functions as needed.

Writing about the things that can go wrong with EXAFS analysis is something of a cottage industry. Indeed, I can think of no other technique that has more review articles about *pitfalls*, *limitations*, or *artefacts*. Sadly, there is a long history of EXAFS analyses gone bad.

6.8 Some Examples

Here we illustrate the strengths and pitfalls of EXAFS analysis with three chemical examples of progressively greater complexity. We will see that a phase-shift-corrected EXAFS Fourier transform can give an approximate radial distribution function but that artefacts persist due to (a) "truncation effects," (b) imperfect background subtraction, (c) missing shells of atoms, and in some cases (d) strong multiple-scattering contributions. We then show how curve-fitting analysis can complement and supersede the Fourier transform interpretations.

6.8.1 *Molybdate: MoO_4^{2-}*

The EXAFS Fourier transform for $Na_2MoO_4(H_2O)_2$ was shown to illustrate data processing in Fig. 6.17. It is also useful as one of the simplest possible cases for EXAFS analysis. In this structure, there is approximate tetrahedral symmetry for four O ligands around Mo at 1.77 ± 0.02 Å [246], and indeed the phase-shift-corrected Fourier transform (FT) gets this average distance almost exactly right. However, this phase-shift-corrected k^3 FT is still not a perfect representation of the true radial distribution function. For example, there are small side lobes symmetrically displaced from the main peak, at ~1.48 and 2.00 Å. These peaks come from the termination of the transform over a finite range—in the FT literature, they are called truncation effects. These very small features persist out to 7 Å and beyond.

Another type of artefact is seen at very short distances, ~1.16 Å and even at 0 Å. These low-frequency features come from an imperfect spline subtraction of the atomic background, leading to small residuals at short distances in the FT. Finally, the FT is missing any evidence for the Na ions in the 3.7 Å range. These ions have a very large thermal motion and also static disorder. Combined with the relatively low Z for Na, their EXAFS contribution becomes invisible. To summarize, the main peak in an EXAFS Fourier transform can do a good job at representing the first coordination sphere, but the FT will also have some peaks that are not real, and some real features will be missing.

6.8.2 *MoS_2*

Molybdenum disulfide, MoS_2, is a highly symmetrical layered solid with each Mo surrounded by a trigonal prismatic array of 6 S ions at 2.37 Å. In this layered structure, each Mo is bridged through S to 6 Mo neighbors with a Mo–Mo distance of 3.16 Å. The symmetry and relatively high Z for these neighbors lead to strong Mo–S and Mo–Mo interactions in the EXAFS, with a clear beat pattern in the k-space data and 2 large peaks in the EXAFS Fourier transform (Fig. 6.18).

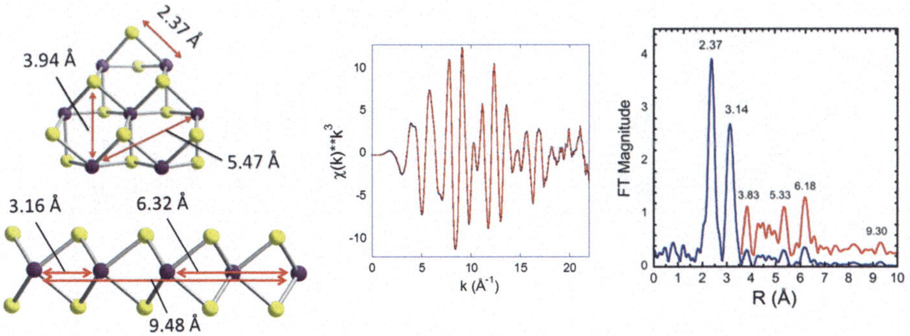

Fig. 6.18 Left: key interatomic distances in MoS_2. Middle: *k*-space EXAFS for MoS_2. Note the strong beat pattern arising from Mo–S and Mo–Mo components of comparable intensity. Right: EXAFS Fourier transform for MoS_2, $k = 2–22$ $Å^{-1}$, 2 $Å^{-1}$ damping, with incorporated Mo–S phase shift

The high symmetry of the MoS_2 structure yields some interesting multiple-scattering features. For example, in the amplified Fourier transform there is a significant feature at 6.18 Å. This arises from multiple scattering along the Mo–Mo–Mo pathway (Fig. 6.18); the multiple scattering yields additional phase shifts that shorten the apparent distance from the true value of 6.32 Å. Additional small multiple-scattering features are also evident, and there is even a small peak at 9.30 Å that corresponds to a Mo–Mo–Mo–Mo pathway. Notice that the multiple-scattering peaks are progressively shorter than the true distances, because of the additional phase shifts.

6.8.3 Nitrogenase

A more complicated example involves nitrogenase (N_2ase)—an enzyme that catalyzes reduction of dinitrogen to ammonia, along with ATP hydrolysis and H_2 evolution [247–249]:

$$N_2 + 8H^+ + 8e^- + 16MgATP \rightarrow 2NH_3 + H_2 + 16MgADP + 16P_i$$

The Mo version of N_2ase contains an $MoFe_7S_9$ cluster knows as the FeMo cofactor (Fig. 6.19). There is a strong beat in the *k*-space EXAFS, in this case caused by Mo–S and Mo–Fe components of comparable intensity. The oblong nature of the cluster is revealed by a modest long distance Mo-Fe peak at 5.0 Å. Although this feature is in the earliest EXAFS data [250], unfortunately it was only understood in light of the crystal structure.

Additional complexity is concealed in the twin peak Fourier transform. The peak at 2.31 Å turns out to be the combination of the Mo–S shell at 2.36 Å and shorter

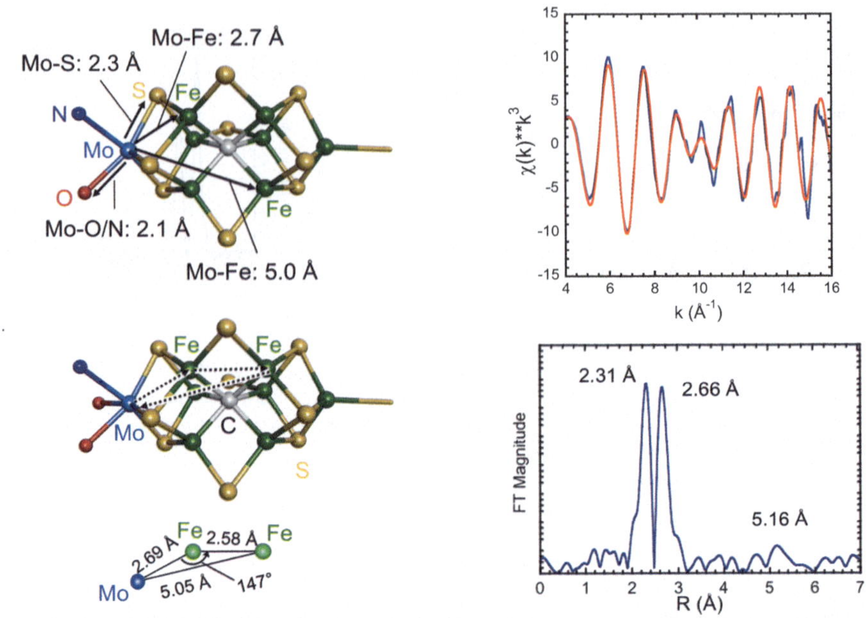

Fig. 6.19 Left: key interatomic distances in the FeMo cofactor. Note the possible complication from multiple scattering in the 5 Å region. Top right: *k*-space Mo EXAFS for the FeMo cofactor, and a representative fit using 3 S @ 2.36 Å, 3 Fe @ 2.70 Å, and 3 N,O @2.26 Å. Note the strong beat pattern arising from comparable Mo–S and Mo–Fe components. Bottom right: *k*-space Mo EXAFS Fourier transform for the FeMo cofactor

distances for homocitrate and histidine Mo–O,N ligands. These are only captured in careful curve-fitting analysis. Here we see the strengths and weaknesses of EXAFS for structure prediction—(a) clear Mo–S and Mo–Fe shells, but—(b) unresolved first shell components and (c) difficult to interpret long distance features.

In summary, every EXAFS Fourier transform is subject to artefacts that can be modest or severe, including (a) the presence of peaks where there are no interatomic distances and (b) the absence of peaks at distances where there are atoms.

6.9 Curve-Fitting

A nonlinear least squares curve-fitting approach is often used for a more quantitative interpretation of EXAFS data, and in Fig. 6.20 we illustrate this process for N_2ase Mo EXAFS. (In this example, we use a wider *k*-range, 2–18.4 $Å^{-1}$, and the transforms are not phase-shift corrected, which explains why the peaks look slightly different.) Looking at the phase-shift-corrected transform of Fig. 6.19, we see that at

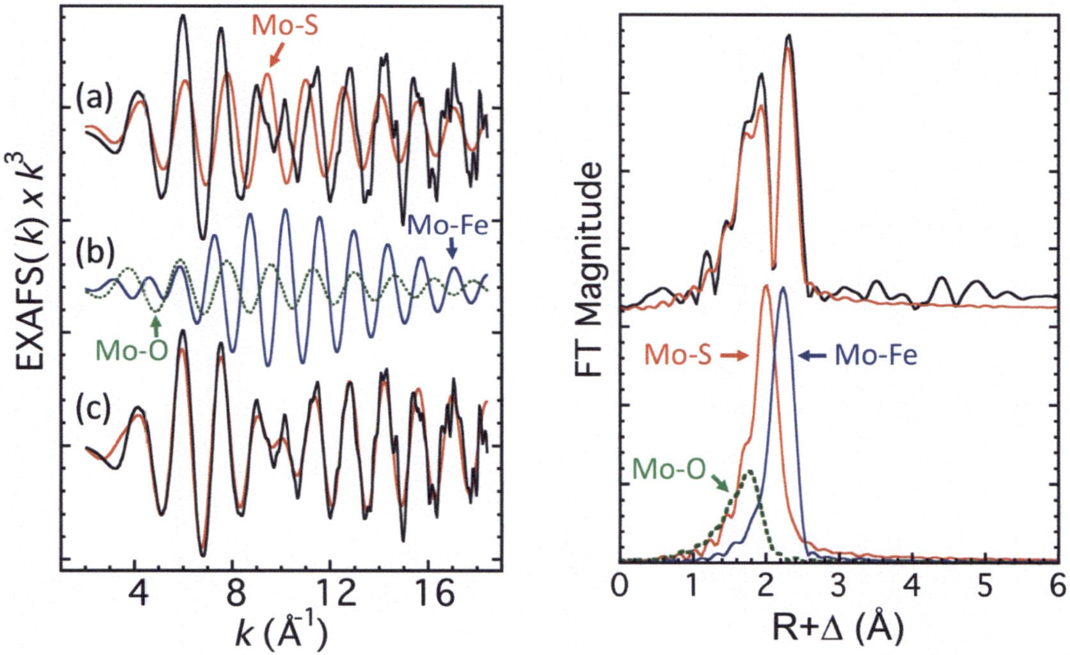

Fig. 6.20 Left: *k*-space fitting. (**a**) A representative fit using 3 S at 2.36 Å, (**b**) additional components added to capture beat pattern 3 Fe at ~2.6 Å, and 3 O at 2.2 Å, and (**c**) final fit with three components. Right: EXAFS Fourier transforms for the curve-fitting process. op: EXAFS Fourier transforms for data (—) and 3-component fit (—). Bottom: Fourier transforms for individual components used in the overall fit

least two components of comparable amplitude will be required to reproduce the *k*-space data. The first attempt using a fixed number of 3 Mo–S refined to a distance of 2.36 Å and σ = 0.045 Å (Fig. 6.20a).

A second component representing 3 Mo–Fe interactions was then added to capture the beat pattern (not shown). Finally, a third component representing 3 Mo–O interactions was added to fill in the largest residual. The Mo–Fe and Mo–O components are shown in Fig. 6.20b, and the final fit is shown in Fig. 6.20c. The results illustrate the strengths and weaknesses of the curve-fitting procedure. Without fitting it would be hard to demonstrate the presence of a weaker shell of Mo–O/N interactions. However, the true distances are Mo–O = 2.16 and 2.19 Å and Mo–N = 2.33 Å, so the fitting has its limits in defining the structure. Furthermore, in this example we have not tried to untangle the multiple scattering contributions to the long Mo–Fe distances shown in Fig. 6.19 and we have also ignored the weak contributions from the outer shell histidine imidazole ligand. EXAFS curve-fitting is often a solid technique for refining structures that have crystallographic data or other evidence for constraining the proposed structure being refined. It has a less than glorious history for prediction of structures out of whole cloth.

6.10 Suggested Exercises

1. **Mass Absorption Coefficients.** Using tabulated mass absorption coefficients (http://www-cxro.lbl.gov), calculate:
   (a) The fraction of X-rays absorbed at 8 keV by Fe for a 10 mM solution of Fe in H_2O.
   (b) The fraction of X-rays absorbed at 800 eV by Fe for a 10 mM solution of Fe in H_2O.
   (c) Which experiments is easier: Fe K-edge or Fe L-edge?
2. **Detection Modes.** Using tabulated mass absorption coefficients and fluorescence yields and assuming a 10% solid angle for fluorescence collection, calculate the concentration where Fe Kα fluorescence has the same S/N as a transmission experiment.
3. **First Shell EXAFS.** Suppose you are planning an EXAFS experiment and you want to estimate how hard you will have to work. Assume you have prepared FeS molecules isolated in a noble gas matrix at 10 K. Assume a bond length of 2.3 Å, a stretching frequency of 500 cm^{-1}, $S_0^2 \sim 0.9$, and a mean free path of 10 Å.
   (a) Estimate the rms σ for the Fe–S distance using: $\sigma_{ab} = 4.106\left(\frac{1}{\mu\bar{\nu}}\coth\frac{1.44\bar{\nu}}{2T}\right)^{1/2}$
   where μ is the reduced mass in atomic mass units, T is the temperature in Kelvin, h is Planck's constant, and $\bar{\nu}$ is the vibrational frequency in wave numbers.

(b) Using the EXAFS equation, calculate the EXAFS signal at $k = 4$ Å^{-1} and at $k = 12$ Å^{-1}. You can use any of the published tables (Teo/Lee, McKale etc.) for phase shifts and amplitudes or do the calculation with FEFF.
(c) If this were a fluorescence measurement, and the noise is just photon statistics, how much longer do you have to count at $k = 12$ Å^{-1} to achieve the same EXAFS S/N as at 4 Å^{-1}?

4. **Mean Free Path.** Using the "universal curve," estimate the amplitude reduction of the EXAFS due to mean free path effects at $k = 4$ Å^{-1} and at $k = 12$ Å^{-1}.
5. **Multiple-Scattering EXAFS.** Consider the O contribution to the Fe EXAFS of an Fe–C–O moiety. Assuming an Fe–C bond length of 1.9 Å and a C–O bond length of 1.43 Å, calculate the relative importance of single, double, and triple scattering paths at $k = 15.1$ Å^{-1}, using the scattering amplitudes illustrated in Fig. 6.16. Assume Fe–CO bond angles of 180°, 165°, or 150°. If you are ambitious, first calculate more modern scattering amplitudes with FEFF.
6. For a more extended multiple-scattering problem, visit: https://bruceravel.github.io/demeter/documents/Artemis/examples/fes2.html

6.11 Reference Books and Review Articles

1. *X-ray Absorption: Principles, Applications, Techniques of EXAFS*, D. C. Koningsberger and R. Prins eds., Wiley-Interscience, New York 1988, ISBN 0-471-87547-3. Somewhat dated but still useful.
2. *Introduction to EXAFS: A Practical Guide to X-ray Absorption Fine Structure Spectroscopy*, Grant Bunker, Cambridge University Press, Cambridge, 2010, ISBN 978-0-521-76775-0. Very practical guide.
3. *XAFS for Everyone*, Scott Calvin, CRC Press, Boca Raton, 2013, ISBN 978-1-4398-78635-7. Loads of practical information. Somewhat quirky style will appeal to some and not to others.
4. *X-ray Absorption and X-ray Emission*, J. A. van Bokhoven and C. Lamberti, eds., Wiley, Singapore, 2016, ISBN 978-1-118-84423-6. Edited volume covers synchrotron radiation and X-ray spectroscopy.
5. *X-ray Studies on Electrochemical Systems*, Artur Braun, De Gruyter, Boca Raton, 2013, ISBN 978-3-11-043750-8. X-ray absorption plus many other X-ray methods. Excellent resource for electrochemistry-related experiments.
6. *X-ray Absorption Spectroscopy for the Chemical and Materials Sciences*, John Evans, Wiley, New York, 2018, ISBN 978-1119990901. Another of the growing number of EXAFS applications books.
7. "A History of X-ray Absorption Fine Structure", R. Stumm von Bordwehr, *Ann. Phys. Fr.* **14**, 377-466, 1989. A good historical review of the early development of EXAFS theory.

6.12 Popular EXAFS Software

Here I have to break my rule about providing links to websites. Several sites maintain lists of popular EXAFS analysis software packages and links to their distribution sites. Two of the best are:

ESRF. XAFS Software page: http://www.esrf.eu/computing/scientific/exafs/intro.html

IXAS (International X-ray Absorption Society). The society web page maintains a variety of resources for X-ray scientists: https://www.ixasportal.net

6.12.1 Theory Packages

In the 1990s, the three most popular packages for theoretical EXAFS predictions were FEFF, GNXAS, and EXCURVE. The latter went commercial, became very expensive, and is now essentially dormant. GNXAS is free and FEFF has a modest licensing fee.

1. FEFF (Full Multiple Scattering). http://feff.phys.washington.edu
2. GNXAS (Full Multiple Scattering) A complete theory package maintained by XAS group at U. Camerino in Italy [251]. The research behind GNXAS was coordinated by C. R. Natoli and the code was mainly written by A. Filipponi. http://gnxas.unicam.it/pag_gnxas.html

6.12.2 Analysis and Fitting Packages

1. IFEFFIT. One of the original analysis programs to interface with FEFF theory package. http://cars9.uchicago.edu/ifeffit/Ifeffit
2. Larch. A Python-based set of EXAFS analysis tools that replaces IFEFFIT. It contains much more. http://cars.uchicago.edu/xraylarch/
3. Demeter. "A comprehensive system for processing and analyzing X-ray Absorption Spectroscopy data". [252] http://bruceravel.github.io/demeter/
4. EXAFSPAK. One of the earliest analysis packages, started as individual subroutines by Tom Eccles in the Hodgson group, transferred to an IBM mainframe by SPC, and subsequently expanded and fully integrated by Graham George. [253] http://www-ssrl.slac.stanford.edu/~george/exafspak/exafs.htm

Chapter 7
XANES and XMCD

In this chapter we will cover the features close to an absorption edge, often called "XANES" (but sometimes NEXAFS). XANES is an acronym for *X*-ray *A*bsorption *N*ear-*E*dge *S*tructure. This is usually considered the region around the absorption edge where there are dramatic changes in the density of states and/or individual transition intensities, so that a simplistic EXAFS interpretation does not apply. As a reminder, the absorption cross section will depend on both the density of states and the transition matrix element for the given transition [254]:

$$\mu \propto \underbrace{\left|\left\langle \psi_f | \hat{\varepsilon} \cdot \vec{r}\, e^{i\vec{k}\cdot\vec{r}} | \psi_i \right\rangle\right|^2}_{\text{dipole+quadrupole+...}} \underbrace{\rho\left(E_f - E_i - \hbar\omega\right)}_{\text{density of states}} \tag{7.1}$$

For the EXAFS region, we were able to assume a smoothly varying density of states, but this is not a good assumption at the edge. In the edge region, there is usually a variety of discrete states, such as molecular orbitals and continuum resonances, and each of these states can have a rather different cross section (Fig. 7.1).

For K-edges, starting at a 1s level, the dipole $\Delta\ell = \pm 1$ selection rule means that transitions to states with *p*-symmetry are dipole-allowed. These edges are therefore mostly 1s $\rightarrow$ np transitions. In certain cases, one can also observe electric quadrupole transitions, where the selection rule is $\Delta\ell = 0, \pm 2$. For first transition element K-edges, this can result in observable 1s$\rightarrow$ 3d features, while for the rare earth L- or M-edges, one can see weak 2p $\rightarrow$ 5s or 3p $\rightarrow$ 5s transitions.

S. P. Cramer, *X-Ray Spectroscopy with Synchrotron Radiation*, Biological and Medical Physics, Biomedical Engineering, https://doi.org/10.1007/978-3-030-28551-7_7

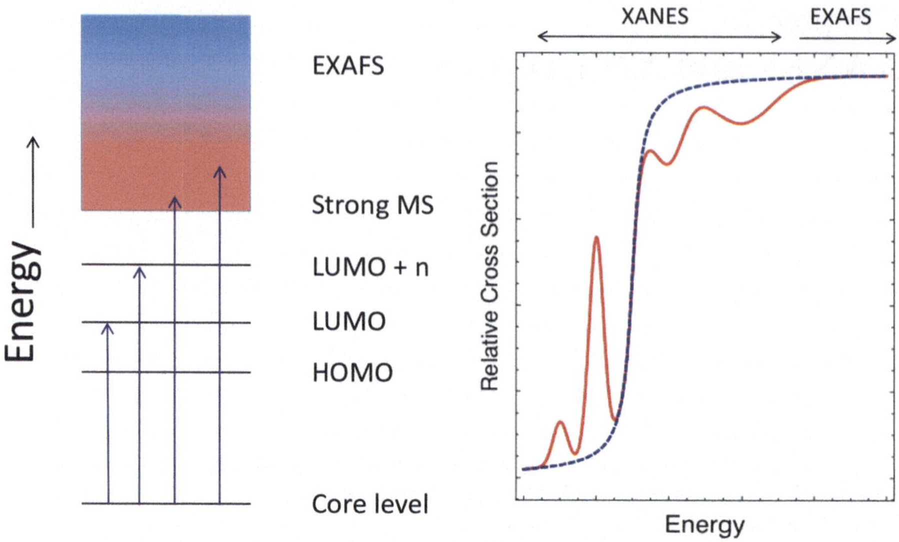

Fig. 7.1 Left: types of transitions in the absorption edge region. Right: a generic absorption edge, showing bound-state transition and multiple-scattering resonances that contribute to XANES, followed by weaker scattering in the EXAFS region. The dashed line represents a broadened step function for the onset of a continuum

7.1 Some Motivation

The absorption edge region is orders of magnitude easier to record than an EXAFS spectrum. First of all, the range is much smaller, perhaps 50 eV instead of 500–1000 eV, so there are fewer points to cover. Furthermore, the magnitude of the features is much larger, with structure sometimes as large as the edge jump itself, as opposed to EXAFS modulations of $\sim<1\%$. Compared to EXAFS analysis, interpretation can also be easy. The XANES region is rich in effects that can be used for chemical information. These include:

- Chemical shifts in the position of edge features.
- Intensity variation that depends on densities of states and selection rules.
- Vibrational fine structure.
- Linear dichroism.
- Magnetic circular dichroism.
- “Natural” circular dichroism.

7.2 Empirical XANES Interpretation

The application of XANES can in many cases be done using empirical trends from standard compounds, which often provide a fingerprint for certain species. Although some might call this scientific “stamp-collecting” mode, in many such cases, this is

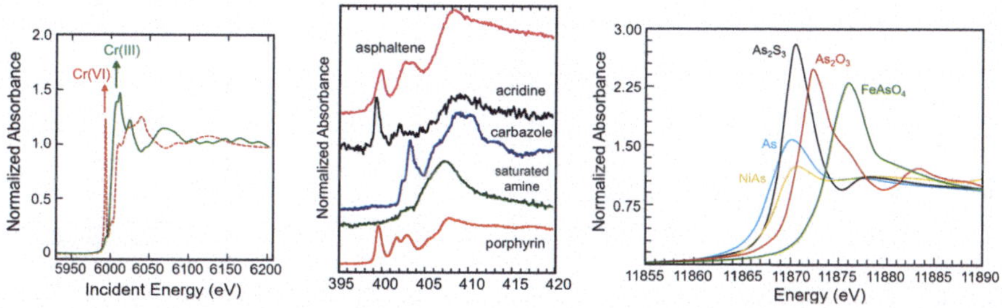

Fig. 7.2 Left: comparison of CrO_4^{2-} K-edge and Cr(III). Middle: K-edges for nitrogen in a variety of chemical states [255]. Right: chemical shifts for As in a variety of oxidation states [256]

all that is needed. For example, if you want to demonstrate the presence of carcinogenic hexavalent Cr, it suffices to know that a very sharp band at 5993 eV comes from a 1s $\rightarrow$ 3d transition in CrO_4^{2-} (Fig. 7.2). Other examples include speciation of different elements in fossil fuels, whether they are major components such as C, N, or S or trace elements such as V and As.

For more theoretical descriptions of the XANES region, interpretation is often based on rather different formalisms, depending on the sample and the absorption edge of interest. *Different systems require different types of models.* The observed features can be described as transitions from core orbitals to:

- Atomic orbitals (atoms).
- Molecular orbitals (molecules).
- Multiple-scattering resonances or band structure (solids).
- Atomic multiplets split by the local environment (inorganic complexes).

Of course, there are no sharp divisions between these approaches, and they all simplify the true nature of the problem. In general, different classes of systems are often described by different models.

7.3 Atomic XANES

K-edges of noble gases are the simplest type of absorption edge to understand, and as examples, spectra for Ar, Ne, and He are shown in Fig. 7.3. The Ar K-edge has been studied since 1939 and is easily explained as a series of lines for 1s $\rightarrow$ 4p, 1s $\rightarrow$ 5p, ... transitions. The Ne K-edge is even sharper and exhibits the expected 1s $\rightarrow$ 3p, 1s $\rightarrow$ 4p, ... transitions. In both cases, successive features are progressively weaker because the 1s-np overlap decreases as n increases. The discrete transitions are followed by an arctangent-like feature for the beginning of transitions into the continuum (the convolution of a step function with a Lorentzian is an arctangent) (Fig. 7.3).

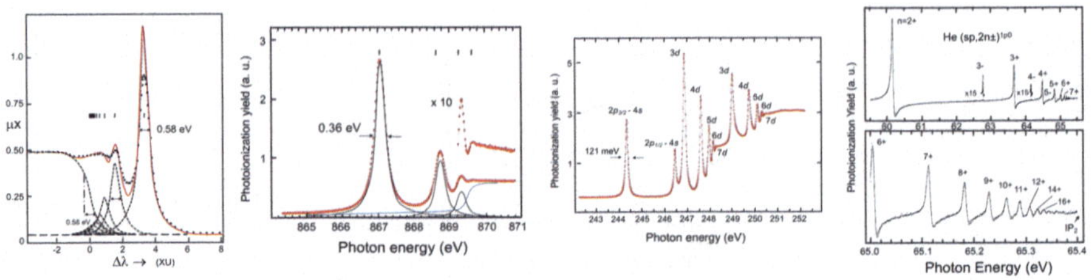

Fig. 7.3 Left to right: historic Ar K-edge spectrum from 1939. Dots are experimental data, and red line is spectrum corrected for instrumental broadening. Dashed lines are fits to individual 1s → np transitions as well as arctangent for continuum. Note abscissa reported in terms of wavelength (redrawn from [257]); Ne K-edge [258]; Ar L-edge [258]. Note split between L_3 and L_2 features, as well as both p → d and p → s transitions; double-excitation features in He absorption edge [259]

A richer spectrum is seen at the Ar L-edge. Here, the $\Delta\ell = \pm 1$ selection rule allows for transitions from the 2p core orbital into final states with s or d symmetry. Also, spin-orbit coupling of the $2p^5$ final-state core-hole splits the spectrum into L_3 ($j = 3/2$) and L_2 ($j = 1/2$) regions. So, every $2p_{3/2} \rightarrow x$ feature has a corresponding $2p_{1/2} \rightarrow x$ at ~2.1 eV higher energy. Here, relative intensities of the $2p_{3/2}$ and $2p_{1/2}$ features are 2:1, which reflects the 4:2 relative degeneracies of the corresponding core holes.

Even noble gas spectra are complex when looked at carefully. For the simplest case, He, the first ionization level involves a 1s → continuum transition at 24.58 eV. However, there are discrete multi-electron events at higher energies, which were first seen using synchrotron radiation by Madden and Codling in 1963 [258]. They were later intensely studied by Shirley and coworkers [259]. The first of these two-electron features is seen at 60.1 eV, and it can be described as simultaneous 1s → 2p and 1s → 2s transitions, thus $1s^2\ {}^1S_0 \rightarrow 2s2p\ {}^1P_1$ (Fig. 7.3). At higher energies there are additional transitions that correspond to mixes of 2s-np and 2p-ns final states. All of these transitions are *Fano resonances* [260]. They exhibit an asymmetric Fano lineshape that results from interference between the transitions to the discrete final states and those to 1s + *continuum* final states [261].

7.4 The Molecular Orbital Approach

In going from atoms to molecules, the lowest-energy absorption edge transitions are now described as from core atomic orbitals to vacant molecular orbitals of the system under investigation. The XANES for N_2 is one of the best-studied molecular X-ray spectra, and a common N_2 molecular orbital (MO) diagram is shown in Fig. 7.4. We see that the lowest unoccupied molecular orbital (LUMO) is the π_g^* orbital made from side-on antisymmetric combination of the $2p_x$ or $2p_y$ orbitals. The first edge feature is thus an intense $1s \rightarrow \pi^*$ transition which is an order of magnitude stronger than subsequent features (Fig. 7.4).

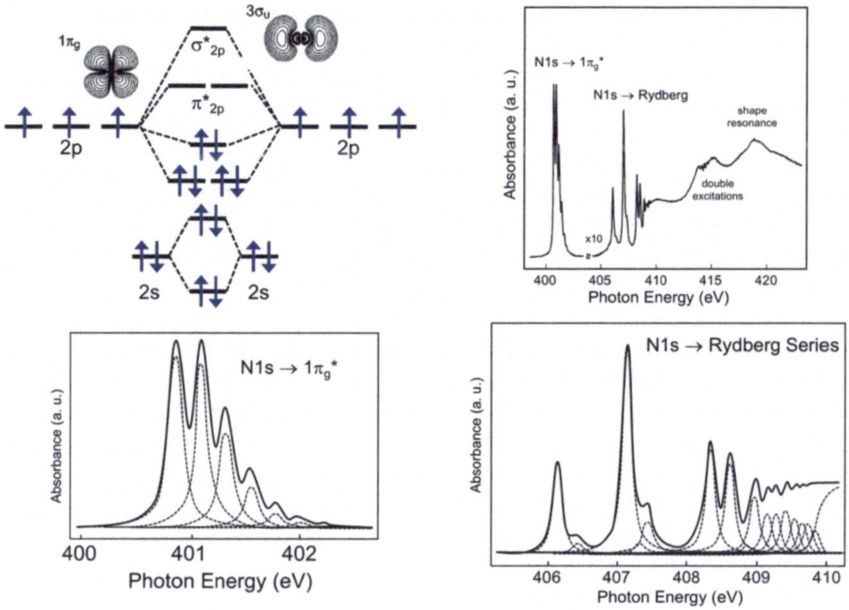

Fig. 7.4 Top left: a molecular orbital diagram for N_2, showing the two empty molecular orbitals that contribute to K-edge structure. Top right: overview of the N_2 K-edge structure. Bottom left: vibrational fine structure in the $1s \rightarrow \pi_g^*$ region. Bottom right: Rydberg transitions in the N_2 K-edge. Latter three figures redrawn from [262]

In the MO diagram, the next available unoccupied orbital is the σ_u^* orbital made from the end-on out-of-phase mixing of the $2p_z$ orbitals. The corresponding $1s \rightarrow \sigma_u^*$ transition occurs at much higher energy, near 420 eV. This is often called a *shape resonance*, because its position depends in a sensitive way on the interatomic distance (see below). In between these two MO-type transitions, there is a so-called Rydberg series, which involves transitions to orbitals that are nearly atomic in character, $1s \rightarrow 3p$, $1s \rightarrow 4p$, $1s \rightarrow 5p$, and so on. And of course, there are the ever-present, pesky two-electron transitions (around 415 eV).

A beautiful (and useful) feature of the N_2 K-edge is the vibrational fine structure in the $1s \rightarrow \pi^*$ transition, with a spacing of about 235 meV or 1880 cm^{-1}. Since the final state has one electron in an antibonding orbital, one expects a weaker N=N bond. For comparison, molecular N_2 has a N≡N stretch at 2331 cm^{-1}, while an Fe-N_2 complex with strong Fe → N_2 backbonding has a stretch at 1719 cm^{-1} cm^{-1} [263]. The 1880 cm^{-1} spacing is thus consistent with an excited state N_2 molecule with a bond order ~2.5.

7.5 Multiple-Scattering and Band Structure Treatments

The 1s $\rightarrow \sigma^*$ transition in N_2 lies above the ionization threshold, so it can also be described as a multiple-scattering resonance, sometimes called a *shape resonance* [264]. In a shape resonance, extensive back and forth scattering between nearest neighbors causes the outgoing photoelectron to be temporarily trapped, resulting in significant density near the absorbing atom (Fig. 7.5). These features occur in all molecular and solid-state spectra, but they are especially clear in light atom K-edge spectra. Sette and coworkers reported that the shape resonance energy δ decreased monotonically with increasing intermolecular distance R: $\delta = E' - \alpha R$ [265]. Shape resonances were proposed as a "molecular ruler" to provide structural information without the need for extensive calculations—"bond lengths with a ruler" (Fig. 7.5). This was criticized as perhaps simplistic [266] [267], and there has been extensive

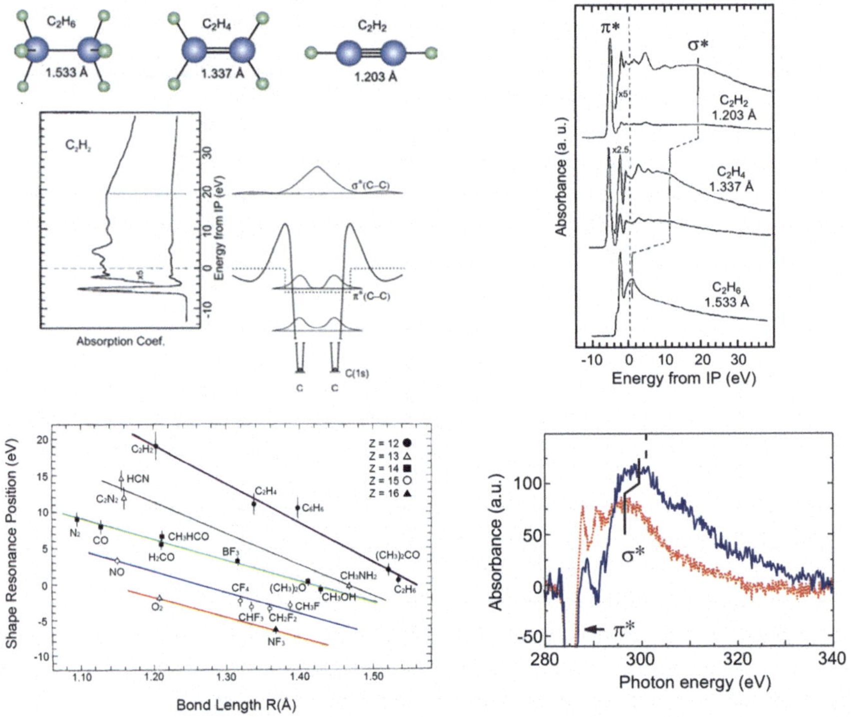

Fig. 7.5 Top left: structures of C_2H_{2n} molecules and an approximate comparison of the differences between π-bonding, π^* antibonding orbitals, and σ^* shape resonances for C_2H_2 [269]. Top right: comparison of C–C σ^* assignments for molecules with progressively shorter C–C bond lengths: C_2H_6, C_2H_4, and C_2H_2 [269]. Bottom left: typical correlations between bond lengths and shape resonance positions, redrawn from [270]. Bottom right: experimental spectra for physisorbed (blue line) and chemisorbed (red line) C_2H_2 on a Cu surface, π^* contributions have been removed (redrawn from [269])

debate in the literature about the validity of the "ruler" approach [268]. The general trend seems reliable within a closely related set of molecules.

Nowadays, shape resonances can be nicely modeled by a full multiple-scattering (FMS) approach using software such as FEFF [228, 269, 271]; they are in fact the same as the DOS that is calculated by DFT. The key difference between XANES analysis and EXAFS analysis is the extent to which high-order multiple-scattering contributions are included. Through its use of Green's function methods, the FEFF approach can include multiple-scattering paths within a modest cluster summed to *infinite order*! In contrast, we saw in Chap. 6 that EXAFS is usually described by summation of a small number of direct and multiple-scattering paths.

Other commonly used "real-space" packages for simulating absorption edges in terms of transitions to molecular orbitals and multiple-scattering resonances are ORCA [272] and StoBe [273]. All of these packages rely on the DFT approximation, so ideally they should yield the same results. In practice, this is not always the case.

A nice application of shape resonances involves monitoring the structural differences between ethylene molecules that have been physisorbed or chemisorbed on a Cu(1 0 0) surface [269]. As seen in Fig. 7.5, the C–C σ^* position moves by ~17 eV as the C–C distances change from 1.2 Å in C_2H_2 to 1.5 Å in C_2H_6. Thus, in Fig. 7.5 a downshift of 3 eV between physisorbed C_2H_4 at 25 K and chemisorbed C_2H_4 at 60 K indicates a modest expansion of the C–C bond upon forming a stronger bond with the surface Cu as electron density flows into the σ^* orbital. A more rigorous FEFF treatment of these spectra does a reasonably good job at reproducing the C–C σ^* features (Fig. 7.5).

7.5.1 *The Water Story*

One of the most controversial applications of XANES (and X-ray spectroscopy in general) involves the structure of liquid water. As seen in Fig. 7.6, there are dramatic differences in the oxygen K-edge for water vapor, liquid water, and ice. The first two gas-phase features are, respectively, assigned as transitions to vacant molecular

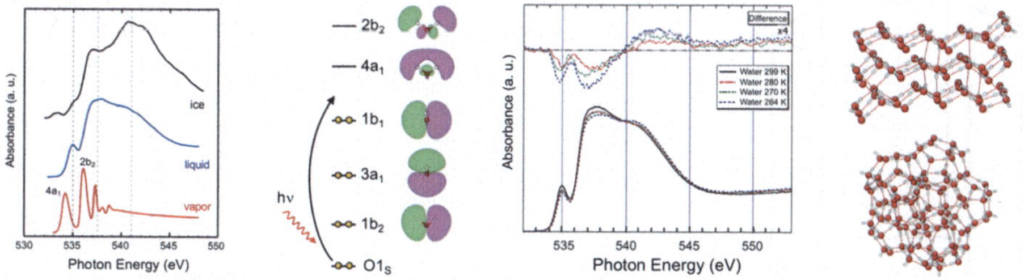

Fig. 7.6 Left to right: spectra of water as vapor, liquid, or solid [271]; molecular orbitals relevant to X-ray absorption by water vapor [274]; spectra for water at different temperatures [275]; and proposed chain and fused dodecahedra structural models, respectively, for high-density and low-density forms of water [276]

orbitals, 1s → $4a_1$ and 1s → $2b_2$, both of which involve antibonding orbitals pointing at the H atoms. In the liquid phase, a single pre-edge feature is observed, and it is argued that the intensity of this feature is related to the number of broken or weakened H-bonds. As expected, as the temperature of water increases, the intensity of this pre-edge feature also increases, while the post-edge intensity decreases (Fig. 7.6).

Based in part on X-ray absorption data, but also involving X-ray emission [277] and X-ray Raman studies [278], a two-phase model for water has been proposed [276]. In this description, most water molecules have strongly distorted hydrogen bonds, leading to denser structures (HDW), while other waters exist in tetrahedrally bonded patches that exist as low-density local fluctuations (LDW). From the same data, others have dismissed the two-phase model, arguing that "X-ray spectra of liquid water in ambient conditions can be understood without a two-structure model" and favoring a continuous distribution model with ~1.7 donated H-bonds per molecule [279]. More than 100 X-ray papers have addressed the pros and cons of different water models. Experts have argued about liquid water for more than a century, so it is clearly beyond the scope of this chapter to settle this controversy.

7.5.2 *Band Structure Approaches*

In going from molecules to solids, the multiple-scattering approach can be continued by conducting DFT or FMS on a fragment of the solid-state structure that is large enough to capture the chemistry and electronic structure of the solid. Alternatively, the analysis can be done in "reciprocal space" via band structure calculations.

In a band structure calculation, the electronic structure is described in terms of periodic electron waves with a certain energy and momentum. Depending on the symmetry of the unit cell, the waves in different directions are given different labels (Fig. 7.7). Calculation of the ground-state band structure follows established

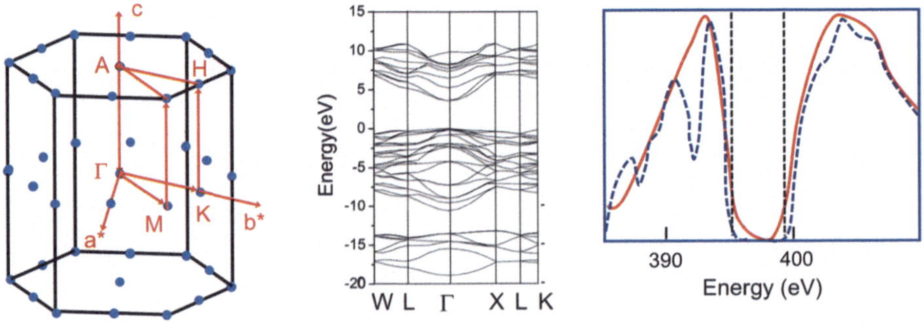

Fig. 7.7 Left to right: the Brillouin zone for α-Si_3N_4 and the labels for the high symmetry points [280]; a typical band structure calculation for α-Si_3N_4 [281]; X-ray absorption at N K-edge for amorphous Si_3N_4 (red line) overlaid with predictions from band structure (blue dashed line), redrawn from [280]

procedures, but for the excited-state band structure, it is essential to include the effects of the core hole. The calculations typically involve a "supercell" consisting of multiple unit cells, including a cell with one excited atom with a core vacancy and an extra valence electron.

The most popular "reciprocal space" package for calculating XANES from band structure is WIEN2k [282]. This software uses a "full-potential (linearized) augmented plane wave and local-orbitals" [FP-(L)APW + lo] basis set to solve the Kohn-Sham equations of density functional theory.

7.6 Charge-Transfer Multiplet Theory

A very different approach to XANES is appropriate for the soft X-ray spectroscopy of transition metal and lanthanide complexes [283]. For these cases, there is a very large interaction between the core vacancy and the final-state valence electrons—this is the key difference from the XANES that has been discussed so far. This breakdown in the one-electron approximation gives rise to additional states that reflect the coupling between the core hole and the valence electrons. These additional interactions are illustrated in Fig. 7.8.

These multiplet effects are very strong for first transition metal L-edge spectra (which measure dipole-allowed 2p → 3d transitions) and lanthanide M-edge spectra (which involve 3d → 4f transitions). Since most of the chemistry of the transition metals involves d-orbitals, L-edges are a natural choice for probing the metal electronic structure. L-edges show multiplet structure that is sensitive to the metal oxidation and spin states. As a bonus, magnetic circular dichroism at transition metal L and lanthanide M-edges is also an excellent probe of magnetic properties.

One of the most successful approaches for interpretation of these spectra has been the *ligand field multiplet theory ("LFMT")* and its extension—*charge-transfer*

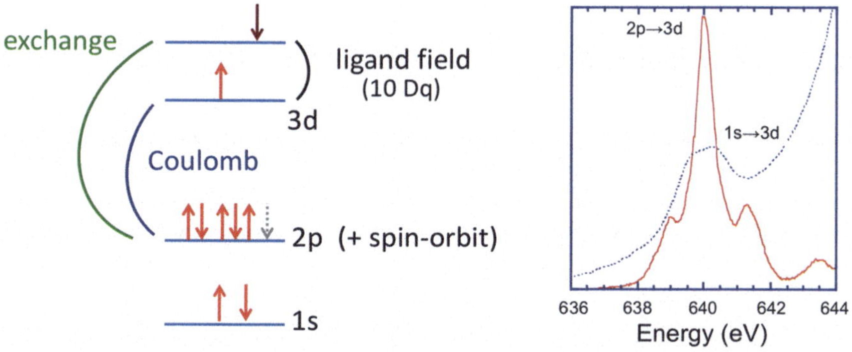

Fig. 7.8 Left: summary of interactions causing splitting of L-edges. Right: comparison of the 1s → 3d region of the $MnCl_2$ K-edge with the L_3-edge [284]

multiplet theory ("CTMT") [285]. These models describe the final states as atomic multiplets that are perturbed by the ligand field of the surrounding atoms. CTMT begins by evaluating the atomic multiplets that arise from a given configuration of core hole and valence electrons. Symmetry and ligand field parameters are then used to calculate how these multiplets are split by the ligand environment. For those who care about the origins of this theory [286,287], there are key papers by Thole, Sawatzky, van der Laan and coworkers [288], who built on group theory by Butler [289] and atomic Hartree-Fock methods from Cowan and coworkers [290].

For example, consider a transition metal ion with *n* 3d electrons. The $L_{2,3}$-edge can then be described as a $2p^63d^n \rightarrow 2p^53d^{n+1}$ transition. The theoretical number of possible final states is given by [283]:

$$6 \times \binom{10}{n} = 6 \times \frac{10!}{(10-n)!n!} \tag{7.2}$$

Similarly, $M_{4,5}$-edges of lanthanides are described as $3d^{10}4f^n \rightarrow 3d^94f^{n+1}$ transitions, and the theoretical number of possible final states is given by:

$$10 \times \binom{14}{n} = 10 \times \frac{14!}{(14-n)!n!} \tag{7.3}$$

In both cases, the actual number of allowed transitions is reduced by the dipole selection rules and possible symmetry of the metal environment. For example, in an extreme case, a $2p^53d^5$ configuration has 1512 possible states, which is reduced to 205 possible term symbols.

Some of these extra multiplet splittings are small, but fortunately, a longer core-hole lifetime usually leads to narrower linewidths in this region, which allows splittings to be seen that are unresolved in broader K-edge spectra. The net results are seen in Fig. 7.8, where K- and L-edges of $MnCl_2$ are compared.

7.6.1 The Hamiltonian

We can illustrate the factors that affect the position, intensity, and shape of L-edge spectra by considering a Hamiltonian that describes the initial and final-state energies. Sequentially turning on various parts of the Hamiltonian allows us to see their effects on the L- or M-edge structure. In the following examples, we use the notation of de Groot for the relevant Hamiltonians [291]. The various interactions that need to be considered are, in order of their usual relative strength:

- Core-hole spin-orbit coupling: ξ. This is the interaction between the spin of the core hole and the magnetic field produced by its orbital motion. It is often the strongest interaction. Across the first transition metals, ξ_{2p} ranges from ~4 eV for Ti to ~14 eV for Cu (Fig. 7.12) [285].

- Core-hole $\leftrightarrow$ valence-electron Coulomb and exchange interactions: F and G. These two-electron interactions are described by so-called Slater integrals. The "Coulomb interactions" are the ordinary repulsions between electrons in overlapping regions of space and are quantified in Slater parameters F. The "exchange interaction" captures the effect of electron-electron correlation that occurs between electrons with the same spin, and it is captured in Slater G parameters.
- Ligand field interactions: 10 Dq, Ds, and Dt. In the ligand field description, the effect of the ligands is to split the atomic multiplets, and this perturbation depends on the symmetry of the ligands around the central ion. In octahedral symmetry the ligand field is described by 10 Dq. In square planar symmetry, additional parameters Ds and Dt are required.

There are also weaker valence electron interactions to consider:

- Valence electron spin-orbit coupling: ξ_{3d}. This is usually on the order of 10's of meV. The final-state effects are usually small, but the ground-state ξ_{3d} can be important for causing splitting of the available ground-state energies that leads to temperature-dependent spectra.
- Valence-electron $\leftrightarrow$ valence-electron Coulomb and exchange interactions: F_{dd} and G_{dd}. There can also be electron-electron interactions in the valence shell.

The Hamiltonian for the ground-state atomic multiplets can be written as:

$$\mathcal{H}_{gs} = \mathcal{H}_{3d} + \mathcal{H}_{mu} + \mathcal{H}_{ls} \tag{7.4}$$

where $\mathcal{H}_{3d}$ is the average energy of the 3d states, $\mathcal{H}_{mu}$ includes all two-electron integrals, and $\mathcal{H}_{ls}$ is the 3d spin-orbit coupling. The Hamiltonian for the final excited state includes extra terms:

$$\mathcal{H}_{es} = \mathcal{H}_{3d} + \mathcal{H}_{mu} + \mathcal{H}_{ls} + \mathcal{H}_{c} + \mathcal{H}_{c.ls} + \mathcal{H}_{c.3d} \tag{7.5}$$

where $\mathcal{H}_c$ is for the core-hole energy, $\mathcal{H}_{c.ls}$ for the core-hole spin-orbit interaction, and $\mathcal{H}_{c.3d}$ for the core-valence Coulomb and exchange interactions.

In Fig. 7.9, we illustrate the sequential turning on of different terms in the Hamiltonian. We use a hypothetical Ti^{4+} L-edge because there are no d-electrons to complicate the ground state. In the absence of any additional interactions, the 2p L-edge is described as a $2p^6 3d^0 \rightarrow 2p^5 3d^1$ transition, and this yields a single line:

$$^1S \rightarrow {}^1P \qquad (7.5)$$

7.6.2 *Core-Hole Spin-Orbit Splitting*

ξ_{2p}. A $2p^5$ configuration can be treated as a single hole with $L = 1$ and $S = 1/2$. The coupling of these angular momenta leads to two possible terms: $J = L + S = 3/2$ and

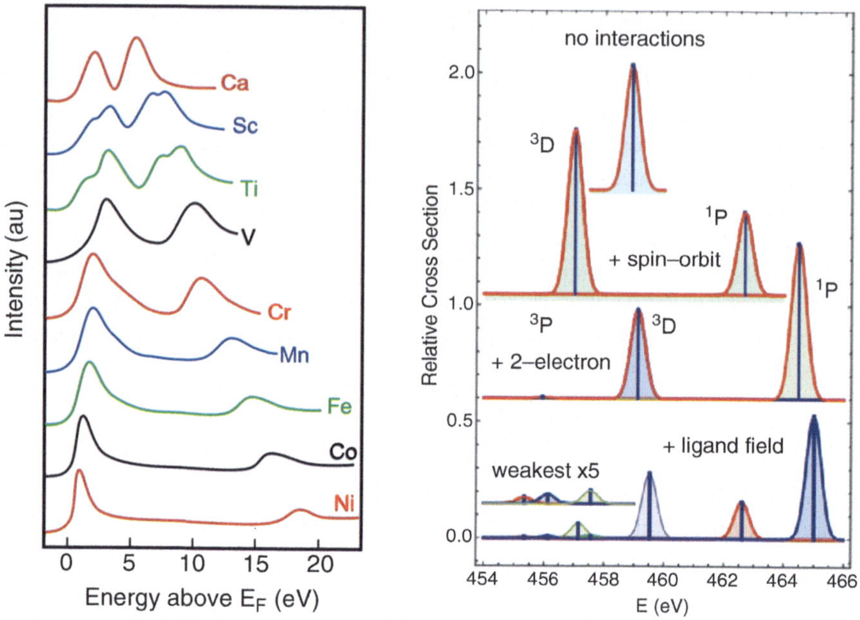

Fig. 7.9 Left: changes in spin-orbit splitting in L-edges across the first transition metals, redrawn from [292]. Right: a series of calculations with different parameters for a hypothetical Ti^{4+} L-edge. Top to bottom: (**a**) no interactions, (**b**) spin-orbit only, $\xi_{2p} = 3.8$ eV, (**c**) with standard two-electron interactions from CTM4XAS, and (**d**) octahedral ligand field added, 10 Dq = 2 eV

$J = L - S = 1/2$. The energies of these terms are split by the "spin-orbit interaction" for the core hole: $\xi_{2p} = H_{c.ls}$. The triplet is shifted by $-(\frac{1}{2})\xi_{2p}$, and the singlet is shifted by $+\xi_{2p}$, so that the overall separation is $(3/2)\xi_{2p}$ (Fig. 7.9), and the transitions are described as:

$$^1S \rightarrow {}^3D \text{ and } {}^1S \rightarrow {}^1P \tag{7.6}$$

The spin-orbit coupling mixes the singlet and triplet terms, so that they both contain $^1P_1$ character in proportions that yield the well-known 3:2 intensity ratio. The conventional nomenclature calls the transitions to the $J = 3/2$ levels the L_3-edge and the transitions to the $J = 1/2$ levels the L_2-edge. ξ_{2p} grows stronger with increasing atomic number, and typical values range from 2.4 eV for Ca [293] to 13.5 eV for Cu (Fig. 7.9).

7.6.3 *Core-Hole ↔ Valence Shell Coulomb and Exchange Interactions:* F *and* G

Next we add two-electron interactions that involve coupling between the final-state valence electron(s) and the $2p^5$ core hole. These two-electron Coulomb and exchange interactions are, respectively, described by so-called Slater integrals *F* and *G*. As seen in Fig. 7.9, the strength of these interactions is such that they can invert the relative intensities of the L_3 and L_2 regions, and the 2p L-edge is then split into three distinct $2p^63d^0 \rightarrow 2p^53d^1$ transitions [283]:

$$^1S_0 \rightarrow {}^1P_1,\quad {}^1S_0 \rightarrow {}^3P_1, \text{and} {}^1S \rightarrow {}^3D_1 \tag{7.7}$$

7.6.4 Ligand Field Splittings: 10Dq

The influence of the crystal field H_{3d} on L-edges is one of the most important reasons why these spectra are chemically useful. For a d^0 system such as Ti^{4+}, the ground state is now described as A_{1g} symmetry. Since the dipole operator behaves as T_{1u} symmetry, only transitions to T_{1u} final states are allowed. As seen in Fig. 7.9, the 2p L-edge is then split into seven distinct $2p^63d^0 \rightarrow 2p^53d^1$ transitions, where we now use the Mulliken nomenclature for the symmetry of the ground-state and excited-state levels to describe the transitions [283]:

$$A_1 \rightarrow {}^1T_1 \tag{7.8}$$

When the splitting of 3d orbital energies is large, its effect can be directly observed in L-edge spectra. For example, a ~4 eV splitting is seen in the L_3-edge of $K_3[Fe(CN)_6]$ (Fig. 11.13). In many other cases, H_{3d} is of the same order as other terms in the Hamiltonian, such as H_{mu} and $H_{c.mu}$, and the complete Hamiltonian has to be considered. It can then be misleading to try to read orbital splittings directly from the spectra (Fig. 7.9).

Another way ligand field splittings can affect the spectra is by changing the initial spin state of the complex. Thole and van der Laan have shown that the ratio of L_3 intensity to total intensity ($L_3 + L_2$), the *branching ratio*, [294] changes with spin state. The so-called "statistical" branching ratio is 4:(4 + 2)—the ratio of the number of electrons assigned to $2p_{3/2}$ and $2p_{1/2}$ levels, respectively. For high-spin complexes, there can be substantial deviations from this ratio of 2/3. For example, a typical branching ratio for high-spin Ni(II) complexes is ~0.77 [295], and even higher branching ratios are seen for high-spin Fe [296] and Mn complexes [284]. In contrast, branching ratios for low-spin Ni(II), Fe(II), and Mn(II) complexes are close to the statistical value.

As an example of spin-state effects on L-edges, we compare the calculated spectra for low-spin and high-spin Ni(II) complexes in Fig. 7.10. In the low-spin case, using parameters typical for a square planar $Ni(II)S_4$ complex, although there are in principal 14 allowed transitions, all of the intensity collapses into two main features with an intensity ratio of 0.123/0.068 = 1.8 and hence a branching ratio of 0.64. In the high-spin case, using parameters typical of NiO, the branching ratio is much higher, ~0.73.

7.6.5 Charge Transfer

For highly ionic systems, such as Mn^{2+}, a single electron configuration such as $3d^5$ suffices to describe the initial state. However, for more covalent systems, one has to recognize the possibility of covalent charge-sharing between the metal and the ligand system. In the charge-transfer multiplet model, this is done by including one or more extra configurations, described as:

$$3d^n + \underline{L}3d^{n+1} \tag{7.9}$$

where $\underline{L}$ denotes a "hole" on the ligands. The energies for the different configurations are described by three parameters, Δ, U_{dd}, and U_{pd}, as illustrated in Fig. 7.10. Δ represents the energy difference between the ground-state $3d^n$ and $\underline{L}\ 3d^{n+1}$ configurations; U_{dd} represents the Hubbard U parameter, the energy difference when an electron is transferred from one metal site to another, thus $3d^n + 3d^n \rightarrow 3d^{n+1} + 3d^{n-1}$ or essentially the correlation energy between two 3d electrons; and U_{pd} is the core-hole potential.

The L-edge transition is then described as:

$$2p^6 3d^n + 2p^6\underline{L}3d^n \rightarrow 2p^5 3d^{n+1} + 2p^5\underline{L}3d^{n+2} \tag{7.10}$$

In extreme cases it might be necessary to include an additional hole on the ligands, yielding a $3d^n + \underline{L}3d^{n+1} + \underline{L}'\underline{L}3d^{n+2}$ configuration, where $\underline{L}'$ denotes the second hole. The result of including one or more alternate configurations is generally the production of "satellite lines" in the X-ray absorption.

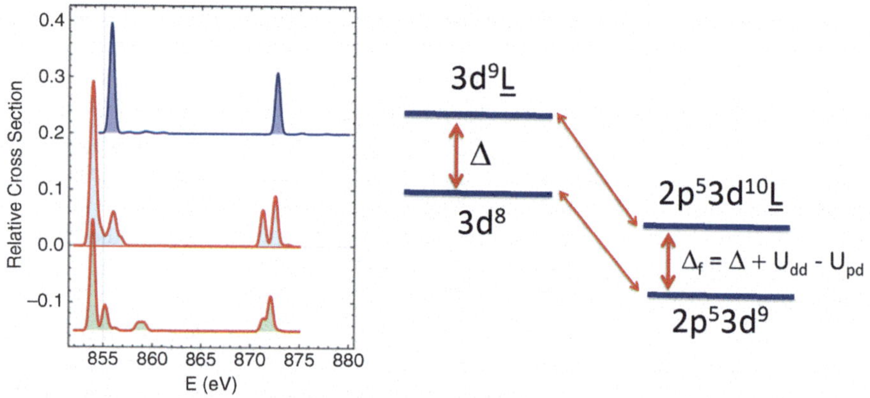

Fig. 7.10 Left: examples of Ni^{2+} CTM4XAS calculations. Top to bottom: a strong tetragonal ligand field yields low-spin Ni^{2+} (10Dq = 5 eV, Ds = Dt = 0.8 eV); an octahedral ligand field yielding high-spin Ni^{2+} (10Dq = 1 eV); high-spin Ni^{2+} with charge transfer (10Dq = 0.7 eV, $\Delta = 3$ eV, $U_{dd} = 6$ eV, $U_{pd} = 8$ eV). Right: the ordering of states for Ni^{2+} in the charge-transfer multiplet model

The contribution of covalency to satellite lines is illustrated in a calculated spectrum for d^8 Ni(II) in Fig. 7.10. For a modest amount of covalency, new bands grow in about 5 eV above the main L_3 and L_2 features. These represent transitions to the higher-energy charge-transfer states. For more covalent systems such as Cu(III), the charge-transfer energy Δ is actually slightly negative, which means that the $3d^9\underline{L}$ configuration actually has a lower energy than the $3d^8$ configuration.

7.7 Simulation-Free Information

It is often useful to analyze edges with quantities that do not depend on a particular simulation procedure. For this purpose, some of the parameters that can be extracted include (a) inflection point or centroid position, (b) branching ratio, and (c) integrated intensity. Some examples are given below.

7.7.1 Inflection Point or Centroid Position

How do you define the position of an absorption edge, when on close inspection, it is often a complex, structured spectrum? For K-edges, especially of metals, a useful feature to report is the first inflection point. For L-edges, which often have rich multiplet structure, a more useful benchmark is the centroid position. Centroid shifts for Ni L-edges are reported in Fig. 7.11. Typical chemical shifts for K-edges of Cr and As compounds were shown in Fig. 7.2, and examples for L-edges of Mn compounds in different oxidation states are illustrated in Fig. 7.12.

7.7.2 Branching Ratio

We saw in the CTM4XAS simulations that the branching ratio was very sensitive to the spin state of Ni(II). The same spin-state sensitivity is seen for other transition metal complexes that can have high-spin or low-spin electronic structures. The combined application of centroid position and branching ratio has been used as an empirical approach to defining the chemical state of Ni species (Fig. 7.11).

7.7.3 Integrated Intensity

Since $L_{2,3}$ edges involve primarily p $\rightarrow$ d transitions, it is reasonable to expect L-edge intensity to reflect the number of *d* vacancies. Similarly, the number of *f* vacancies should be reflected in the intensity of $M_{4,5}$ edges, which are mostly

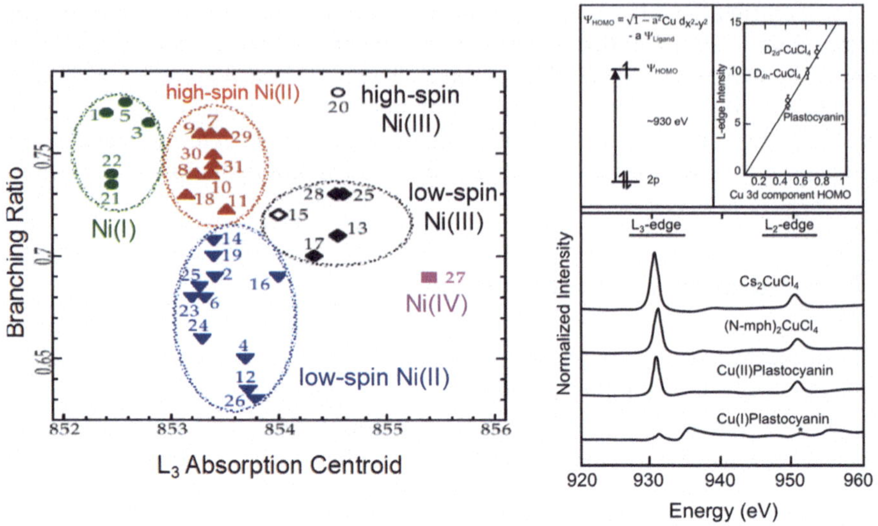

Fig. 7.11 Left: a correlation diagram relating the L_3 centroid position and the branching ratio to the Ni oxidation state and spin state [297]. Right top: correlation of Cu L_3-edge intensity with known 3d vacancies, plus placement of plastocyanin. Right bottom: intensities of 2p → 3d transitions at Cu L-edges after normalization to the 2p → continuum jump [298]

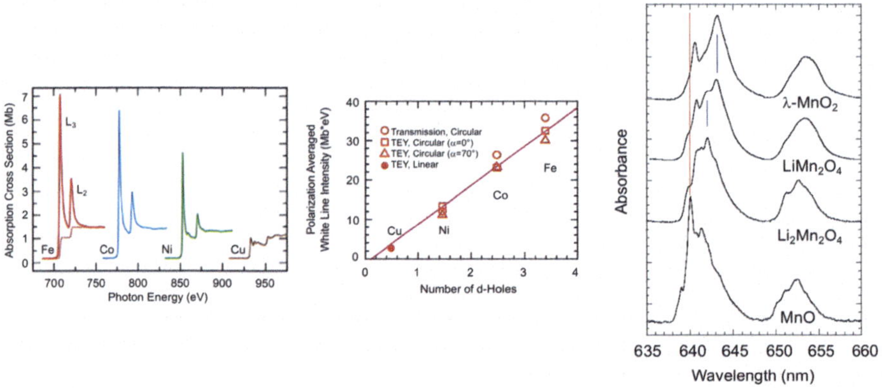

Fig. 7.12 Left: normalized L-edge spectra for Fe, Co, Ni, and Cu, showing the decreasing "white line" intensity as d-electron vacancies are filled [303]. Middle: observed intensity vs. predicted number of holes. Right: the shift in L_3-edge position with oxidation state for a series of Mn oxides. $LiMn_2O_4$ is mixed-valence and shows features from both Mn(III) and Mn(IV)

d → f transitions. This trend can be seen in the strength of the L-edge intensities of the first transition metals as one goes across the periodic table and fills the 3d shell (Fig. 7.12). These expectations were developed as a "vacancy sum rule" by Starace in 1972 [299]. Assuming c and l are the orbital angular momentum quantum numbers for the core and valence shells and n is the number of electrons in the d shell in the ground state, this rule states that the integrated intensity over particular absorption edges reflects the number of empty states with the appropriate symmetry for the transition [299, 300]:

$$\int_{j_+ + j_-} (\mu_0 + \mu_1 + \mu_{-1}) d\omega \propto \frac{4l + 2 - n}{2l + 1} P_{cll}^2 \tag{7.11}$$

where μ_0, μ_1, and μ_{-1} represent absorbance measurements with left, right, and linearly polarized X-rays, with the polarization vector along the magnetic quantization axis.

Sum rule analysis was used to probe the covalency of the blue Cu protein plastocyanin (Fig. 7.11) [298] and the Cu_A site [301], as well as the number of 3d vacancies in compounds with different Ni oxidation states [302].

7.7.4 Some Limitations

There are of course many possible complications in applying the above rules. The absolute energy of an inflection point requires careful energy calibration of the monochromator. Absolute calibration using Bragg angle and *d*-spacing is rarely done; instead, one usually calibrates using a standard compound with a known inflection point or maximum. When comparing results with the literature, it is important to know what standard was used and which energy was assigned. Calculation of a branching ratio depends on background subtraction and a choice for the energy dividing L_2 and L_3 regions. Finally, the calculated integrated intensity for an edge feature depends on the background subtraction as well as the procedure used for removing the continuum (edge jump) features. The devil is in the details.

7.8 X-ray Magnetic Circular Dichroism (XMCD)

In an "*X*-ray *m*agnetic *c*ircular *d*ichroism" or "XMCD" experiment, one measures the difference in absorption of right- and left-circularly polarized X-rays by a magnetized sample:

$$I(\text{XMCD}) \equiv \frac{\sigma^+ - \sigma^-}{\sigma^+ + \sigma^-} \equiv \frac{\mu_{\text{rcp}} - \mu_{\text{lcp}}}{\mu_{\text{rcp}} + \mu_{\text{lcp}}} \tag{7.12}$$

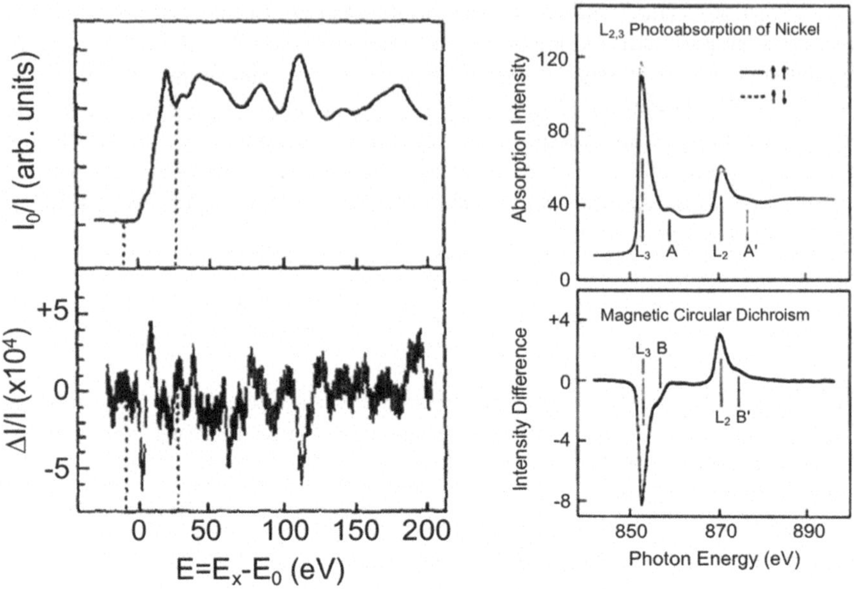

Fig. 7.13 Left: the first Fe K-edge XANES and EXAFS (note different scales) XMCD spectra, redrawn from Schütz et al. [304]. Right: the first soft X-ray MCD, reported for Ni metal, redrawn from Chen and coworkers [305]

Although the first successful XMCD spectra were only recorded in the late 1980s (Fig. 7.13) [304–306], the nature of the measurement is essentially the same as for the UV-visible MCD that has been known since the nineteenth century [307]. Compared to the UV-visible experiment, XMCD has the advantage of elemental specificity that comes with all core electron spectroscopies.

As a complement to conventional XANES results, XMCD can also provide quantitative information about:

- Spin orientations from the sign of the XMCD signal.
- Spin states from magnetization curves.
- From sum rule analysis, quantitation of spin and orbital angular momentum.
- Separate magnetic and nonmagnetic components in heterogeneous samples.

XMCD and the related technique "XMLD" or "*X*-ray *m*agnetic *l*inear *d*ichroism," have proven especially valuable when combined with spatial imaging [308, 309].

7.8.1 The XMCD Experiment

An XMCD experiment requires (a) a source of circularly polarized X-rays and (b) a magnetized sample.

7.8.1.1 Sources of Circularly Polarized X-rays

The bend magnets of a storage ring are the simplest sources of circular polarization. As described in Chap. 3, by viewing the electron beam off-axis, one observes a charge accelerated in an elliptical orbit. A bend magnet beamline can thus have left and right circular polarization by collecting radiation above and below the orbital plane. However, circular polarization comes at the price of lower intensity, and nowadays most soft X-ray XMCD beamlines use elliptically polarized undulators (EPUs) as much brighter sources of variable circular polarization.

In the hard X-ray region, one problem is that crystal monochromators tend to convert circular polarization back to linear polarization, especially when the Bragg angle approaches 45°. For example, on ESRF beamline 12, the degree of circular polarization drops dramatically in the soft X-ray region (Fig. 7.14)—it is 12% at the Mo L_3-edge (2523 eV) and only 4% at the Mo L_2 edge (2629 eV), making Mo L-edge XMCD difficult, but not impossible, as illustrated by the nice spectra for copper octacyanomolybdates reported by Arrio and coworkers (Fig. 7.14) [310]. A better approach is to start with a linearly polarized source and convert to circular polarization (after the monochromator) with an X-ray quarter-wave plate (Chap. 4). This approach is used at APS beamline 4-ID-D.

7.8.1.2 Sample Magnetization

The other requirement for an XMCD experiment is a magnetized sample (Fig. 7.15). For ferromagnetic samples, even at room temperature, a modest field of a few hundred Gauss is often sufficient to overcome any residual hysteresis (Fig. 7.16). In contrast, a paramagnetic sample requires the combination of much higher fields (multiple Tesla) and cryogenic temperatures to approach full magnetization (Fig. 7.16).

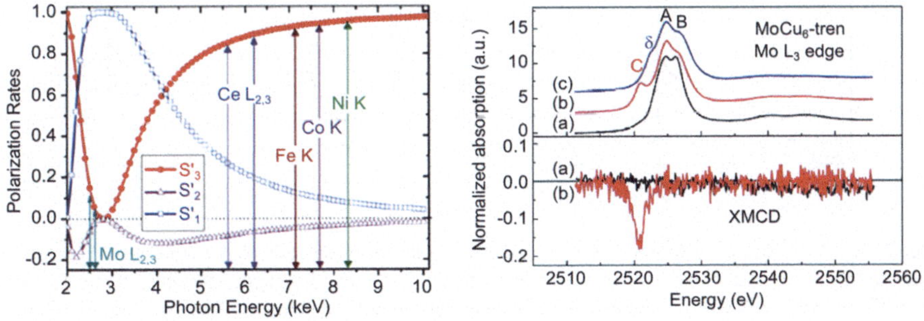

Fig. 7.14 Left: circular polarization rate (S_3) from crystal monochromator at ESRF beamline 12. Right: absorption spectra and XMCD at Mo L_3-edge of $MoCu_6$-tren, recorded at $T = 10$ K and $H = 6$ T, (**a**) before irradiation, (**b**) after exposure to the X-ray beam for 6 h, (**c**) after annealing at 300 K [310]

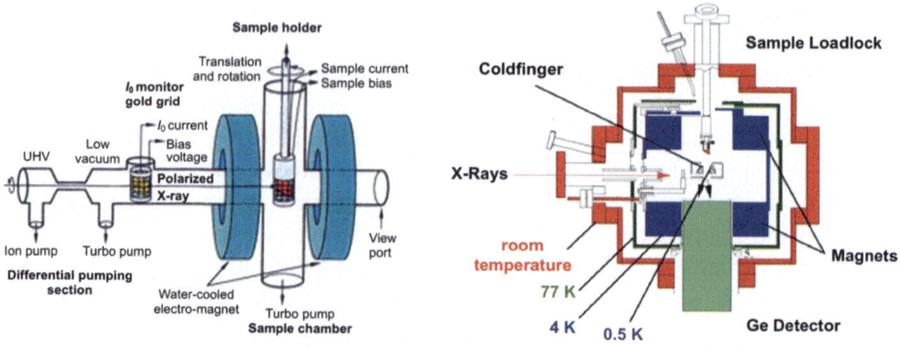

Fig. 7.15 Left: a water-cooled electromagnet XMCD system for electron yield detection [303]. Right: schematic of a low-temperature XMCD instrument for paramagnetic samples

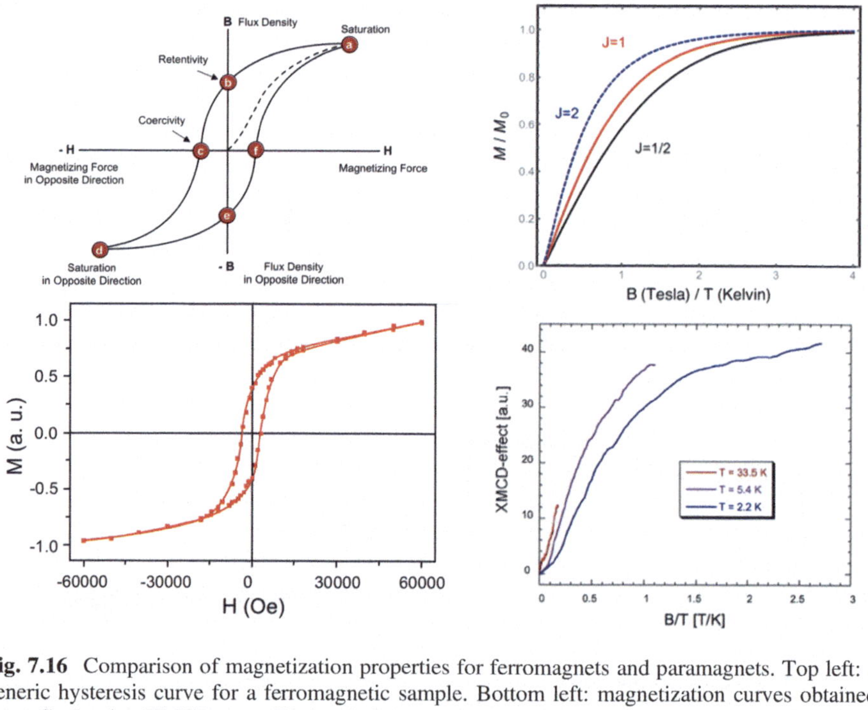

Fig. 7.16 Comparison of magnetization properties for ferromagnets and paramagnets. Top left: a generic hysteresis curve for a ferromagnetic sample. Bottom left: magnetization curves obtained from Co L-edge XMCD on a Fe_3O_4–CoO "nanoaggregate" system [312]. Top right: Brillouin functions for paramagnets with different spins. Lower right: Fe XMCD magnetization curves obtained at different temperatures on an antiferromagnetically coupled Cu–Fe system with $S = 2$ [313]

The simplest model for the magnetization M of N noninteracting atoms or ions with total angular momentum quantum number J in volume V is given by:

$$\frac{M}{M_0} = B(x, J) = \frac{2J+1}{2J}\coth\left(\frac{2J+1}{2J}x\right) - \frac{1}{2J}\coth\left(\frac{1}{2J}x\right) \quad (7.6)$$

where M_0 is the saturation magnetization, $x = (gJ\mu_B H)/(k\mathrm{T})$, and $B(x,J)$ is the *Brillouin function*, which reduces to:

$$\mathrm{XMCD} \propto \tanh x \quad (7.7)$$

for $S = 1/2$ systems with no orbital moment [311]. The curves in Fig. 7.16 illustrate the potential of XMCD magnetization curves as a characterization tool separate from sum rule analysis and multiplet simulations.

7.8.2 XMCD Theory

One approach, the one-electron picture, is commonly applied to metallic samples, while the alternate ligand field multiplet approach, is more often used for transition metal complexes. Below we compare the predictions of both approaches.

7.8.2.1 One-Electron Approach

A large fraction of the XMCD literature, especially papers involving magnetic thin-film and metallic samples, uses the one-electron model along with a two-step approach to XMCD [314]. In this picture, the first step is that with left circular polarization (using our 'optical' definition), at the L_3-edge the atom preferentially (5/8 of the time) emits spin-up electrons, while at the L_2-edge, the atom preferentially (3/4 of the time) emits spin-down electrons (Fig. 7.17). Then, in the two-step model, vacant spin-up and spin-down bands are viewed as spin-sensitive "detectors." For 3d final states, if only spin-down states are available, then the "asymmetry," $(\sigma^+ - \sigma^-)/(\sigma^+ + \sigma^-)$, at the L_2-edge should be a 50% effect, twice as large as at the L_3-edge (and opposite in sign).

7.8.2.2 Multiplet Approach

As mentioned earlier in this chapter, the ligand field multiplet theory (LFMT) is a multi-electron viewpoint that describes the initial and final states as multiplets that are mixed and split by the symmetry of the ligand field [288,293,294,315,316]. In this approach, the XMCD effect emerges naturally as a consequence of angular momentum selection rules. Spectra have been calculated for $3d^1$ through $3d^9$

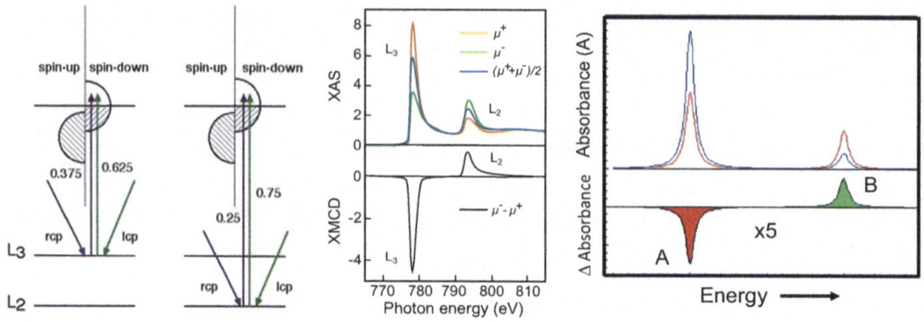

Fig. 7.17 Left: graphic representation of quantities in the two-step model. In the first step, left and right circular polarization, combined with spin-orbit coupling of the $2p^5$ core vacancy, transfer spin-up or spin-down preferences to the excited electron. In the second step, the spin-split valence shell acts as a spin-selective detector [274]. Middle and right: quantities involved in sum rule analysis

systems with a range of crystal field and spin-orbital coupling strengths [227,317], so for many systems, one can check beforehand to see if XMCD will be useful.

7.8.2.3 Sum Rule Analysis

Sum rules are equations based on integrated spectra, and they allow derivation of valuable information without resort to simulation or fitting techniques. For example, most chemists are familiar with the Kuhn-Thomas sum rule, which states that the sum of oscillator strengths for transitions from the ground state to the *k*th excited state f_{0k} is equal to the number of electrons N_e [229].

The first X-ray sum rule was mentioned earlier—the integrated intensity over particular absorption edges reflects the number of empty states with the appropriate symmetry for the transition [299, 300]. For XMCD, there are two new sum rules involving projections of the spin $\langle S_z \rangle$ and orbital $\langle L_z \rangle$ angular momentum of the absorbing species.

As one example, using the nomenclature of van der Laan (see ref. 8 at end of chapter), the "spin sum rule" for $L_{2,3}$-edges of *d*-metals can be written as:

$$\frac{\rho^1_{L_3} - 2\rho^1_{L_2}}{\rho^0_{L_3} + \rho^0_{L_2}} = \frac{2}{3}\frac{\langle S_Z \rangle}{n_h} + \frac{7}{3}\frac{\langle T_Z \rangle}{n_h} \tag{7.8}$$

and the "orbital angular momentum sum rule" can be written as:

$$\frac{\rho^1_{L_3} + \rho^1_{L_2}}{\rho^0_{L_3} + \rho^0_{L_2}} = \frac{1}{2}\frac{\langle L_Z \rangle}{n_h} \tag{7.9}$$

In these expressions, ρ^0 and ρ^1 are the integrated intensities of the isotropic spectrum and XMCD over the L_3 and L_2 edges and n_h is the number of *d* vacancies.

Stöhr and König have argued that $\langle T_z \rangle$ angular averages to zero when combining 3 orthogonal dichroism measurements [318], so the last term in Eq. 7.8 can often be omitted. The quantities involved in sum rule analysis are illustrated graphically in Fig. 7.17.

Using A and B to represent the appropriate integrals, defining C as the square of the p→d radial transition matrix element and assuming an isotropic sample [319], yields the following much simpler expressions for the sum rules for the number of holes n_h and spin and orbital magnetic moments m_S and m_O:

$$n_h \cong \frac{\rho^0_{L_3} + \rho^0_{L_2}}{C} \tag{7.10a}$$

$$m_S \cong \frac{-[A - 2B]}{C} \mu_B \tag{7.10b}$$

$$m_O \cong \frac{-2[A + B]}{3C} \mu_B \tag{7.10c}$$

The sum rules have been tested by comparison with experimental measurements [303,320] and theoretical calculations [321,322]. Analytical calculations for Cu^{2+} in an octahedral ligand field found that at low temperature, the $\langle T_z \rangle$ term makes a large contribution to the spin sum rule and cannot be ignored [323]. Others have shown that $\langle T_z \rangle$ can also be significant for other first transition metals, especially at lower symmetry surface sites [324].

Although the sum rules were generally thought to be accurate within about 10% [321, 322], several papers have pointed to a number of limitations [285, 325, 326, 327]. These include the following:

- The value of $\langle T_z \rangle$ must be known (if it cannot be assumed to be zero).
- The L_3 and L_2 edges must be cleanly separable with a 2:1 ratio.
- The spin-orbit coupling of the valence states must be negligible.

Piamonteze and de Groot found that the "effective spin" sum rule is exactly correct for systems with a filled 3d final state ($3d^9$ Cu^{2+} and Ni^{1+}). However, errors increased from ~5–10% for $3d^8$ Ni^{2+} to ~±50% for $3d^4$ Mn^{3+} transition metals. de Groot and Kotani conclude that "it is usually a difficult, and in some cases a theoretically impossible task, to derive quantitatively correct values of $\langle S_z \rangle$ and $\langle L_z \rangle$ from XMCD experiments." Caveat emptor.

7.8.3 Applications

XMCD has found an enormous number of applications, especially related to characterizing element-specific magnetic properties of materials. For example, Vobornik and coworkers studied the XMCD of a "topological insulator" involving Mn-doped Bi_2Te_3 crystals with a nominal formula $Bi_{1.91}Mn_{0.09}Te_3$ [328]. The samples had a

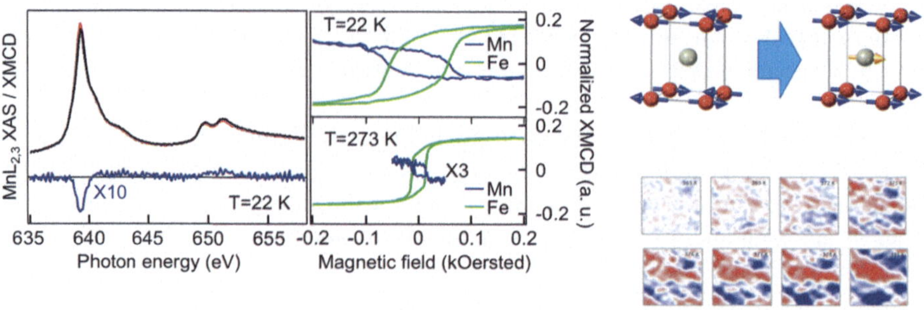

Fig. 7.18 Left: the Mn $L_{2,3}$-edge XAS and XMCD for Mn-doped Bi_2Te_3 with Fe overlayer [328]. Middle: Mn and Fe XMCD magnetization curves for same sample. Right: temperature-dependent PEEM XMCD at the Fe L_3-edge [329]

10 Å overlayer of ferromagnetic Fe, and from the opposite sign of the Mn XMCD compared to the Fe XMCD (Fig. 7.18), it is clear that there is an antiparallel alignment of the Fe and Mn spins. The relatively small size of the Mn XMCD signal suggests that only the topmost layers of the $Bi_{1.91}Mn_{0.09}Te_3$ are magnetic.

Combining XMCD with "*p*hoto*e*lectron *e*lectron *m*icroscopy," or "PEEM," has become a popular tool for imaging the magnetic structure of materials [308]. As one example of so-called "X-PEEM" imaging, Keavney and coworkers examined the structure alloy FeRh [329]. This "multiferroic" material undergoes an antiferromagnetic to ferromagnetic phase transition at ~375 K. The temperature-dependent X-PEEM at the Fe L_3-edge revealed the nucleation and growth of magnetic Fe domains with opposing orientations shown by different signs for the XMCD effect (Fig. 7.18).

7.9 Suggested Exercises

1. **XANES vs. EXAFS.** Suppose a fluorescence-detected absorption experiment yields on average 100 counts/sec with no background, (a) if there are 10 significant points in the XANES region with average deviations from the mean of 10%, how long would it take to record the XANES structure with a S/N of 10? (b) If there are 100 significant points in the EXAFS region with average deviations from the mean of 1%, how long would it take to record the EXAFS structure with a S/N of 10?
2. **Molecular Ruler.** From the data in Fig. 7.5, estimate a value for α for hydrocarbons in the molecular ruler approximation: $\delta = E' - \alpha R$.
3. **Multiplets.** How many multiplets can exist for (a) the $L_{2,3}$-edge of a Co^{3+} ion? (b) the $M_{4,5}$-edge of a Dy^{3+} ion?
4. **Magnetization.** What temperature is required to achieve 95% magnetization of an $S = 1/2$ paramagnet if the applied field is 10 Tesla?

5. **XMCD Sum Rules.** From the XMCD at Mo L_3-edge of $MoCu_6$-tren (Fig. 7.14), what can you say qualitatively about the absence or presence of orbital angular momentum?

7.10 Reference Books and Review Articles

1. *Core Level Spectroscopy of Solids*, Frank de Groot and Akio Kotani, CRC Press, Boca Raton, 2008, ISBN 0-8493-9071-0.—fundamental theory of X-ray absorption and emission, charge transfer multiplet theory, and broad survey of applications to XAS, XPS, XMCD, and RIXS.
2. *NEXAFS Spectroscopy*, Joachim Stöhr, Springer; New York, 2003 ISBN 3642081134.—revised edition of the 1992 classic text, still a valuable source of numerous light element XANES spectra.
3. *Polarized Light in Optics and Spectroscopy*, David Kliger, James W. Lewis, and Cora E. Randall, Academic Press, San Diego, 1990, ISBN 0-12-414975-8.—excellent discussion of circular polarization.
4. *X-Ray Absorption and X-Ray Emission Spectroscopy: Theory and Applications,* Jeroen A. van Bokhoven, Carlo Lamberti, eds. John Wiley, West Sussex, 2016, ISBN 978-1-118-84423-6.—2 chapters on XANES by experts.
5. *Introduction to Magnetic Materials*, B. D. Cullity, Addison-Wesley, Reading, 1972, ISBN 0-201-01218-9. Great background information for magnetic properties.
6. "X-Ray Absorption and Dichroism of Transition Metals and their Compounds", F. M. F. de Groot, *J. Elect. Spec. 67*, 529-622 (1994)—excellent review of the synchrotron XANES up to 1994.
7. "Multiplet Effects in X-Ray Spectroscopy", F. M. F. de Groot, *Coord. Chem. Rev. 249*, 31-63 (2004)—a review including emission spectra as well.
8. "X-ray magnetic circular dichroism—A versatile tool to study magnetism", Gerrit van der Laan and Adriana Figueroa, *Coord. Chem. Rev.* 277-278, 95-129 (2014) — an excellent review of XMCD theory and applications.

7.11 Popular XANES Software

Appendix F of Reference 1 (above) describes the available single-electron approaches to XAS software as (1) real-space multiple-scattering methods, (2) real-space wave function methods, and (3) reciprocal space band structure methods. Appendix F also has more detailed descriptions and references for the following packages.

1. **ORCA** [224]. https://orcaforum.cec.mpg.de. A collection of electronic structure calculation approaches "ranging from semi-empirical methods to DFT to single-

and multireference correlated *ab initio* methods." Designed especially for user friendliness.

2. **FEFF** [228]. http://monalisa.phys.washington.edu/index-feffproject.html. Calculates XAS with real-space multiple-scattering approach based on Green's functions.
3. **StoBe** [273]. "Stockholm-Berlin version of deMon, a Density Functional Theory molecule/cluster package." http://www.fhi-berlin.mpg.de/KHsoftware/StoBe/. A real-space wave function approach based on linear combinations of Gaussian-type orbitals.
4. **WIEN2k** [282]. http://susi.theochem.tuwien.ac.at. A band structure approach to XANES analysis, based on linearized augmented plane wave (LAPW) description of the potential.
5. **TT-Multiplets.** http://www.anorg.chem.uu.nl/people/staff/FrankdeGroot/ttmultiplets.htm. The classic package for calculating XANES, XPS, XMCD, and RIXS within the empirical multiplet approach.
6. **CTM4XAS** [330]. http://www.anorg.chem.uu.nl/CTM4XAS/software.html. A user-friendly PC-based program that implements multiplet calculations for XAS, XES, and XPS.

Chapter 8
Photon-in Photon-out Spectroscopy

8.1 Introduction

A large fraction of the spectroscopy done at synchrotron radiation labs falls under the unwieldy title of "photon-in photon-out" (PIPO) spectroscopy. Although fluorescence-detected X-ray absorption and NRVS are both technically "photon-in photon-out" measurements, for the most part, we reserve the term for those cases where the information content is in the details of the "photon-out" part of the experiment. In this chapter we will consider:

- *H*igh-*e*nergy *r*esolution *X*-*r*ay *f*luorescence (HERXRF).
- *R*esonant *i*nelastic *X*-ray *s*cattering (RIXS).
- *X*-ray *R*aman *s*cattering (XRS) or non-resonant inleastic X-ray scattering (NRIXS).
- *I*nelastic *X*-ray *s*cattering (IXS).

As illustrated in Fig. 8.1, these four PIPO techniques differ in the location of the exciting radiation and the magnitude of the energy shifts that are commonly measured. In broad terms, in (a) conventional XRF, the excitation X-ray energy is far from the fluorescence energy, so ΔE_{XRF} is on the order of hundreds or even thousands of eV; in (b) RIXS experiments, ΔE_{RIXS} can range from hundreds of eV down to hundreds of meV; in (c) most XRS experiments, ΔE_{XRS} corresponds to a light element K-edge, on the order of hundreds eV; and in (d) most IXS experiments, ΔE_{IXS} corresponds to a phonon energy, ranging from tens of meV up to hundreds of meV.

S. P. Cramer, *X-Ray Spectroscopy with Synchrotron Radiation*, Biological and Medical Physics, Biomedical Engineering, https://doi.org/10.1007/978-3-030-28551-7_8

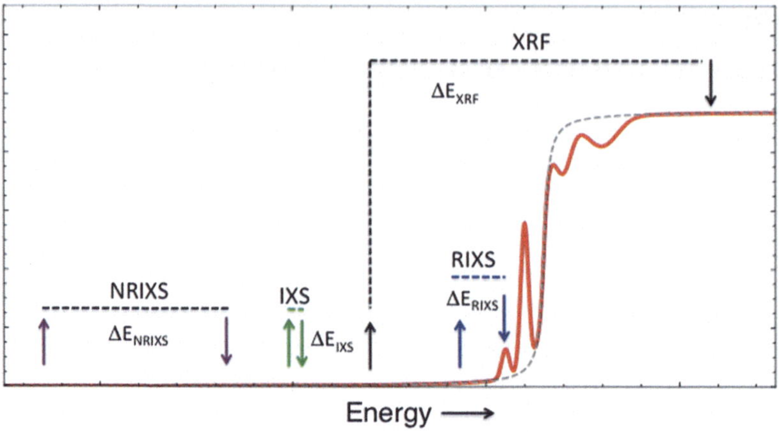

Fig. 8.1 Comparison of different PIPO experiments. The down arrow corresponds to the exciting photon energy and the up arrow is the emitted photon energy

8.2 High-Energy Resolution X-ray Fluorescence (HERXRF)

X-ray fluorescence is the electromagnetic radiation (photons) emitted following creation of an inner-shell vacancy (core hole) in an atom. In some cases, the core hole can be created by electron impact (in an electron microscope), by a proton beam (in a PIXE measurement), or even by radioactive decay (in a K-capture experiment). At a synchrotron, this is a photon-in photon-out process, in which the incoming photon has an energy considerably higher than the absorption edge. As we saw with EXAFS, changing the incoming photon energy will change the outgoing photoelectron energy, but the atom with the core vacancy is the same for different excitation energies. This allows the use of relatively broadband X-ray excitation to increase the flux on sample and the resulting fluorescence intensity.

The relative intensities of different fluorescence lines vary by orders of magnitude, primarily because of the wide variation in overlap integrals between the core vacancy and other orbitals. When comparing the fluorescence intensities between different elements, another important factor is the overall fluorescence yield. This is because the X-ray emission process competes with Auger emission as a means for relaxing the core-hole vacancy (Fig. 8.2):

$$I_f \propto \left|\langle \phi_i | r | \phi_f \rangle\right|^2 \tag{8.1}$$

The general nomenclature, introduced by Siegbahn [331], for the strongest fluorescence lines is summarized in Figs. 8.2 and 8.3. Fluorescence corresponding to 2p → 1s transitions is called Kα, while transitions that fill a 1s vacancy from 3p or higher levels are grouped as Kβ. The labeling is rather archaic and the earliest

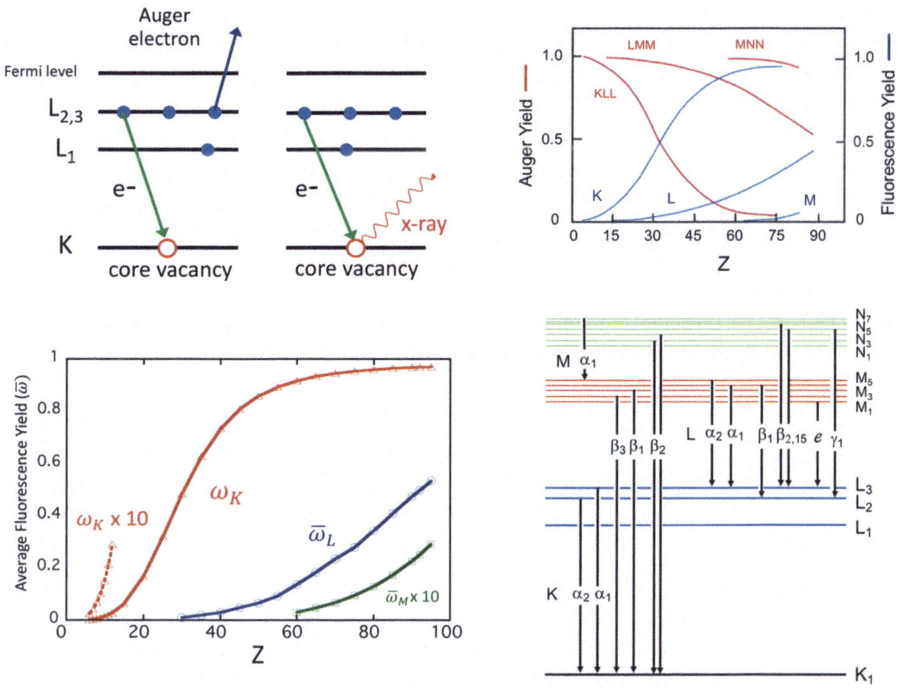

Fig. 8.2 Top left: alternate core-hole relaxation processes. Top right: competition between Auger and fluorescence yields vs. atomic number. Bottom left: average fluorescence yields for different core holes vs. Z. Bottom right: Siegbahn labels commonly applied to X-ray fluorescence lines

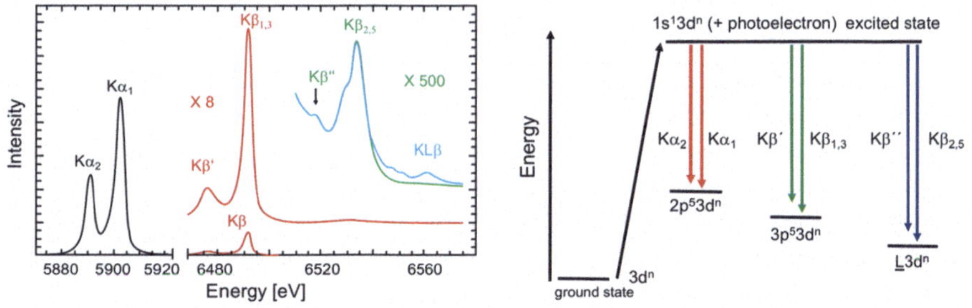

Fig. 8.3 Left: energies and relative intensities of Mn K X-ray fluorescence. Right: origin and nomenclature of K fluorescence lines for first transition metals. The KL emission results from an initial two-electron excitation that yields simultaneous 1s and 2p vacancies in the excited state

literature is not always consistent. In addition to the main lines, there are often "satellite lines" that do not fall on the conventional diagram, and these are labeled with primes, as summarized for transition metal K-edges in Fig. 8.3. (Some researchers abandon the Siegbahn labels and simply specify the levels involved in the fluorescence.)

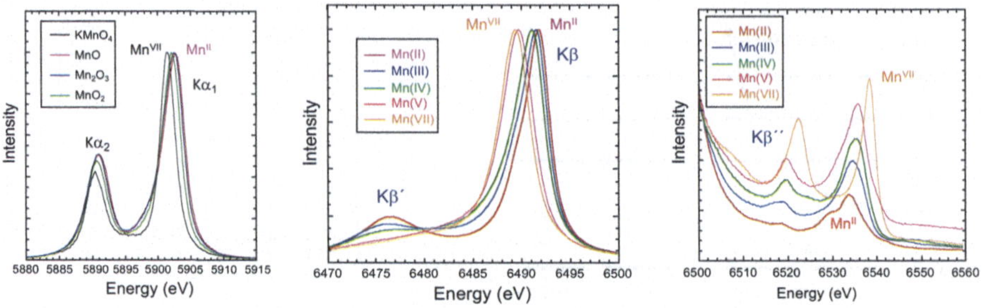

Fig. 8.4. K (np → 1s) fluorescence fine structure for a variety of Mn compounds. Left: Mn $K\alpha_1$ and $K\alpha_2$ lines. Middle: $K\beta_{1,3}$ and $K\beta'$ satellites. Right: $K\beta_{2,5}$ region and $K\beta''$ satellites

The structure and relative intensities of the X-ray emission between two deep core levels are relatively independent of the chemical form of a particular element. For example, to a good approximation, the $2p_{3/2} \rightarrow 1s$ and $2p_{1/2} \rightarrow 1s$ transitions for Mn ($K\alpha1$ and $K\alpha2$ lines) occur in a 2:1 intensity ratio at 5899 and 5888 eV, respectively, with a fluorescence yield of 0.31. As summarized in Appendix J and associated references, there are tables that list the energies, fluorescence yields, and relative intensities of X-ray lines for different elements. This *chemical insensitivity* of X-ray fluorescence is the basis for an entire industry that uses X-ray fluorescence for quantitative elemental analysis [271, 277, 332]. However, as shown in Fig. 8.4, there are sometimes small chemical shifts in fluorescence spectra, and this *chemical sensitivity* is often more significant for the weaker lines in the spectrum.

8.2.1 Why HERXRF?

If the tabulated fluorescence energies and intensities (Appendix F) were fixed in stone, there would be no need to do the HERXRF experiment. However, fluorescence spectra show slight changes with chemistry, and these provide useful information that is sometimes hard to obtain by other measurements. The cases where there are good arguments for HERXRF involve situations where the simpler X-ray absorption experiment does not provide all the answers, or where one wants additional information such as:

- Site selectivity.
- Ligand Z identification.
- Spin-state identification.
- Filled electronic structure information.

8.2.2 The HERXRF Experiment

When conducted as a photon-in photon-out experiment, HERXRF requires (a) a monochromator for the incoming beam (which can be broad-band) and (b) an energy analyzer for the emitted photons. Energy analyzers are more complex than monochromators, because the emission occurs over 4π steradians. As with the other spectroscopies, the optics behind these analyzers were illustrated in Chap. 5.

8.2.3 X-ray Fluorescence Fine Structure

We mentioned earlier that it is the small changes in X-ray fluorescence line energies and intensities that make it a valuable tool for chemical information. The primary sources of fluorescence fine structure are:

- Spin-orbit splitting.
- Multiplet structure.
- Configuration interaction.
- Valence molecular orbitals.
- Magnetic properties.

We now illustrate the effects of these electronic and magnetic properties on the fluorescence.

8.2.4 Spin-Orbit Coupling Structure

The strongest fluorescence lines involve the exchange of one core hole for another. For example, in Kα emission, a 1s core hole is filled by a 2p electron, resulting in a 2p vacancy. In this case, the primary fine structure in the fluorescence spectra comes from spin-orbit coupling in the final-state core holes. For example, coupling of the $S = 1/2$ spin and the $L = 1$ orbital angular momentum splits the $2p^5$ final state into ($J = 1/2$ or $J = 3/2$, respectively) $2p^{1/2}$ and $2p^{3/2}$ final states, yielding the Kα_1 and Kα_2 lines. This is the same spin-orbit coupling that splits L-edge absorption into L_2 and L_3 edges, respectively. This spin-orbit coupling is relatively insensitive to chemistry (Fig. 8.4), and it yields a relatively constant splitting of lines as well as constant intensity ratios. It is this chemical insensitivity that allows creation of tables of emission lines with reasonable accuracy.

8.2.5 Multiplet Structure

For transition metals and lanthanides with partially filled shells of d or f valence electrons, there is fine structure in the emission that can reveal chemical information about the ion under investigation. This fine structure generally comes from the presence of two partially filled shells, which allows for multiple intermediate and final states. This multiplet interpretation is generally favored for early transition metal compounds or "Mott-Hubbard insulators" in the solid-state parlance. These multiplet features in the final state are similar to those discussed earlier with respect to transition metal L- and lanthanide M-edges (Chap. 7). The strongest effects are seen with the largest number of unpaired valence electrons, such as d^5 Mn(II) or f^7 Gd(III).

As an example, consider Mn Kβ fluorescence, and compare it to Mn Kα fluorescence (Fig. 8.4). For Mn(II), in atomic nomenclature, the two allowed terms for the "initial" $1s^1 3d^5$ configuration are ^7S and ^5S. Kβ fluorescence involves filling the 1s vacancy from the 3p shell, so the final configuration is $3p^5 3d^5$, which yields two final terms: ^7P and ^5P. What is different now, compared to the Kα case, is the strength of the 3p–3d exchange interaction. For Mn(II) and Fe(III), this creates a ~15 eV split between final states that have parallel 3p and 3d spins (^7P) and those with antiparallel spins (^5P) (Figs. 8.4 and 8.5).

The dipole selection rules constrain the allowed transitions from ^5S to ^5P and from ^7S to ^7P, with the energy difference between ^5P and ^7P given by the 3p–3d exchange interaction. Thus, for this Kβ fluorescence, the main peak at the highest energy (lowest energy final state) is the $^7S \rightarrow {}^7P$ transition, and the satellite peak is a $^5S \rightarrow {}^5P$ transition. There is also a "spin-flip" shoulder on the main peak, and extra features from 3d–3d multiplet interactions, but that is beyond the current discussion.

Kβ splitting is a useful indicator of oxidation state and spin state, both of which change the 3d shell electronic structure and hence the 3p–3d splitting. As seen in Fig. 8.4, at one extreme, $KMnO_4$ is formally $3d^0$, and as predicted the Kβ splitting essentially disappears. At the other extreme, the same type of exchange interaction effects can be observed in lanthanide emission spectra, where the important interactions are now the 2p–4f and 3d–4f or even 4d–4f exchange interactions. For example, the $4d \rightarrow 2p^{3/2}$ emission for $4f^7$ Gd metal shows a huge (>30 eV) 4d–4f exchange splitting (Fig. 8.5).

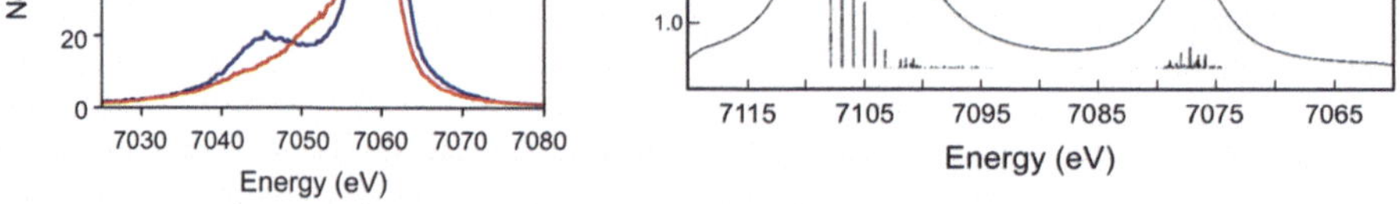

Fig. 8.5 Exchange coupling effects on transition metal and lanthanide emission spectra. Left: experimental Fe Kβ spectra for typical high-spin (blue line) and low-spin (red line) compounds, respectively, $(o\text{-}Me_2smif)_2Fe$ and $(o\text{-}Mesmif)_2Fe$ (courtesy Prof. Serena DeBeer). Right: calculated $4d \rightarrow 2p^{3/2}$ emission for Gd metal with a $4f^7$ ground state [333]. Note reversed energy scale

8.2.6 Configuration Interaction

Another source of fluorescence fine structure is "configuration interaction." For the Mn systems discussed above, we ignored the effects of the core hole on the energy ordering of the valence-electron configurations. However, as one moves to the right in the periodic table, or as the bonding becomes more covalent, situations arise where the ordering of the configurations changes. This is because the 3d electrons are pulled down more by the new potential than are the 4s and 4p electrons (or the ligand electrons). Thus, the intermediate-state configurations will be different from the ground-state configurations.

For the case of Ni(II), it turns out that three configurations have to be considered:

$$1s^1 3d^8 + 1s^1 3d^9 \underline{L} + 1s^1 3d^{10} \underline{L}\,\underline{L}' \rightarrow 3p^5 3d^8 + 3p^5 3d^9 \underline{L} + 3p^5 3d^{10} \underline{L}\,\underline{L}' \quad (8.2)$$

Configuration interaction leads to changes in Ni Kβ spectra that are distinct from multiplet features. For example, in a series of Ni(II) complexes, there are peak shifts and additional features going from covalent $NiBr_2$ to ionic NiF_2 (Fig. 8.6). Although the spectral changes were originally analyzed just in terms of configuration interaction [336], de Groot has shown the essential features are captured by ligand field multiplet theory [335].

8.2.7 Valence Molecular Orbitals

A third source of fluorescence fine structure involves transitions from valence molecular orbitals (or valence bands in solid-state materials) to core vacancies. For the first transition metals, there are molecular orbitals that are primarily ligand in

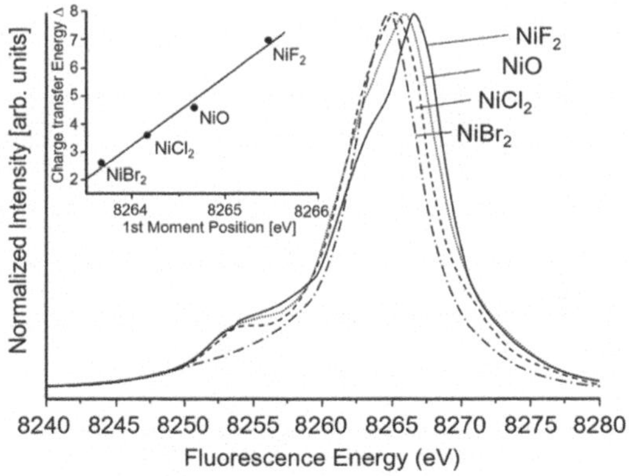

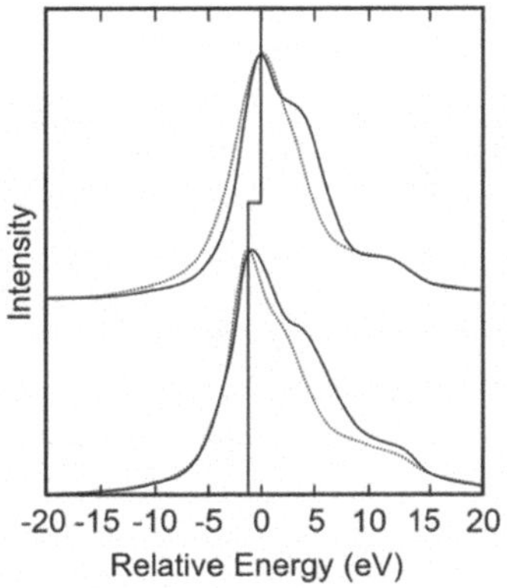

Fig. 8.6 Left: chemical effects on Kβ spectra for a series of Ni(II) complexes [334]. Right: simulation of Kβ spectra for (top) $NiBr_2$ and (bottom) NiF_2 [335]. Note reversed energy direction

character, but mixed with a small amount of metal 4p character. These give rise to weak transitions very close to the absorption edge energy. They are sometimes clumsily labeled K$\beta_{2,5}$ features, but in most cases, they are better described as valence → core transitions.

At lower energies, features sometimes called "crossover transitions" occur from orbitals that are primarily ligand 2s or 3s (or even 4s or 5s) in character, but again with some metal 4p. Since these orbitals are more atomic in character, their energy can be used as an indicator for the type of neighbor. As a nice example, we see in Fig. 8.7 that Cr_2O_3 has the lowest-energy Kβ'' transition, consistent with the fact that the oxygen 2s orbital is the deepest compared to the relevant *s* orbitals of N, Se, or Te.

In some cases valence-to-core transitions are strongly polarized, which can lead to beautiful orientation effects with single crystals. For example, $[Rh(en)_3][Mn(N)(CN)_5]$ has a terminal Mn≡N bond with significant N 2s character. Crossover transitions with N 2s → Mn 1s character are polarized along the Mn–N axis, so that they are maximal in the plane perpendicular to the Mn≡N bond and are not observed when looking along that axis (Fig. 8.7).

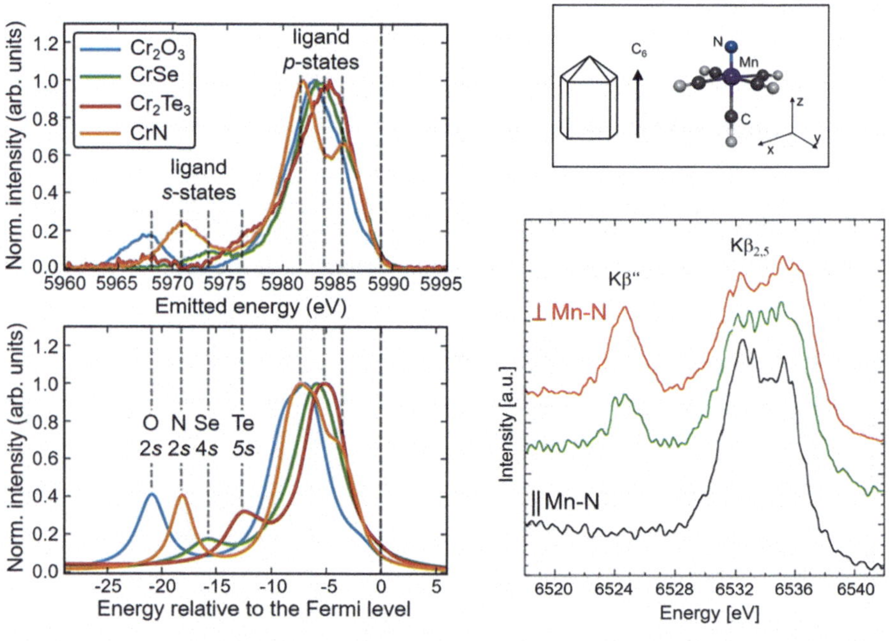

Fig. 8.7 Left: chemical sensitivity of the valence → core region for various Cr compounds. [310]. Right: anisotropy in Kβ'' intensity for a single crystal of $[Rh(en)_3][Mn(N)(CN)_5]$ [337]

8.2.8 *Multi-electron Excitations*

So far we have assumed that core-hole creation results in one-electron excitations. However, a fraction of absorption events results in excitation of a second electron. These are usually valence electrons, and the loss of EXAFS intensity from multi-electron excitations was discussed in Chap. 6. In a more extreme case, a second core electron can be excited. As one example, we show the KL edge and KL emission for MnO_2 [338]. Simultaneous excitation of both 1s and 2p electrons creates a much larger apparent charge and shifts the K fluorescence to an energy actually above the K absorption edge (Fig. 8.8).

8.2.9 *Fluorescence Magnetic Dichroism*

In Chap. 7 we saw that the absorption of magnetized materials could vary between left- and right-circularly polarized X-rays—XMCD. It is thus not surprising that the X-ray emission can also exhibit dichroism—XES-MCD. For K-edge excitation above the absorption edge, the effects are modest, as shown for Co in Fig. 8.9. However, the emission dichroism can be much larger, approaching 10% and even higher on resonance [339]. The on-resonance technique is sometimes given the unwieldy acronyms of RIXS-MCD [340] or MCD-RXES [341].

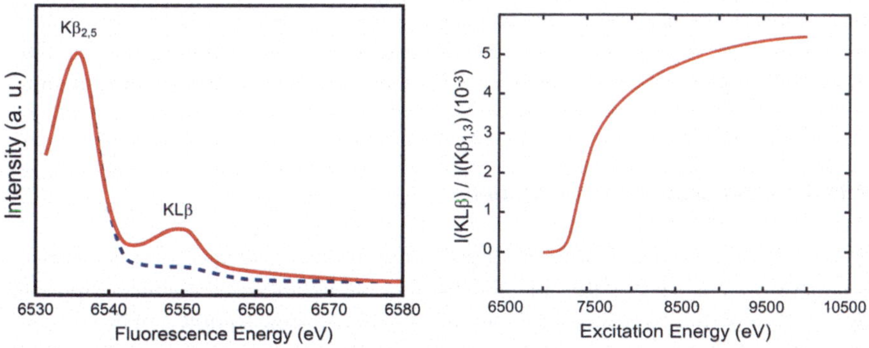

Fig. 8.8 Left: KLβ fluorescence of MnO_2. Excitation energy 7 keV (blue dashed line) or 10 keV (red line). Right: excitation spectrum for the KLβ fluorescence. Both redrawn noise-free from [338]. The onset of KLβ emission occurs at 7246 eV—the combined energies of Mn K-edge and Fe L_3-edge

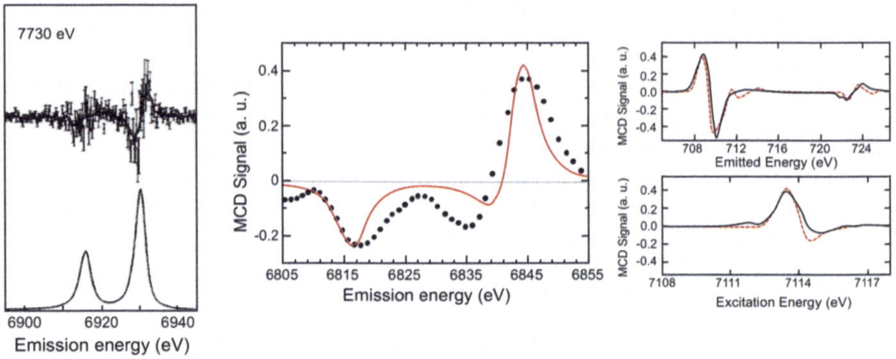

Fig. 8.9 Left: Kα fluorescence dichroism for plate of Co metal, excited at top of the K-edge with left- or right-circularly polarized X-rays [342]. The maximum amplitude is about 0.25%. Middle: $L\beta_{2,15}$ fluorescence ($4d^{10} \rightarrow 2p^{3/2}$) dichroism for 2-mm-thick plate of $4f^7$ EuO at 30 K [343]. Right: RIXS-MCD of Fe_3O_4 (magnetite) powder. Upper panel is a CIE cut and lower panel is a CEE cut, as explained in the next (RIXS) section. Adapted from [340]

8.2.10 Applications

The chemical sensitivity of X-ray fluorescence leads to a variety of applications beyond conventional elemental analysis. In some cases, one can simply examine the fluorescence in detail for (a) chemical characterization. In another application, one can (b) record an excitation spectrum while monitoring the fluorescence with "*h*igh-*e*nergy *r*esolution *f*luorescence *d*etection" in a technique abbreviated as "HERFD" (see below) [344]. The HERFD approach can yield sharper excitation spectra at absorption edges, and it can also allow EXAFS data to be recorded beyond interfering absorption edges, allowing so-called range-extended EXAFS [344]. Finally, one can monitor different parts of the fluorescence while scanning the excitation energy, yielding (c) site-selective and/or (d) spin-selective X-ray absorption data.

8.2.10.1 Chemical Characterization

We have already seen that different parts of the fluorescence spectrum are sensitive to oxidation state, spin state, and the identity of neighboring ligands. Spin-state diagnosis is a particularly nice application because the same sample can be compared with itself at different temperatures or pressures. Rueff and coworkers have used the 1s3p Kβ X-ray emission spectral shape for the analysis of pressure-dependent spectra of FeS [345], FeO [346], and Fe metal [347]. For FeS, the disappearance of the Kβ′ satellite is a nice marker for the transition from high-spin to low-spin Fe (Fig. 8.10). Similarly, the transition between the magnetic (bcc) and the nonmagnetic (hcp) iron phases was seen from the shift of intensity toward the center of gravity.

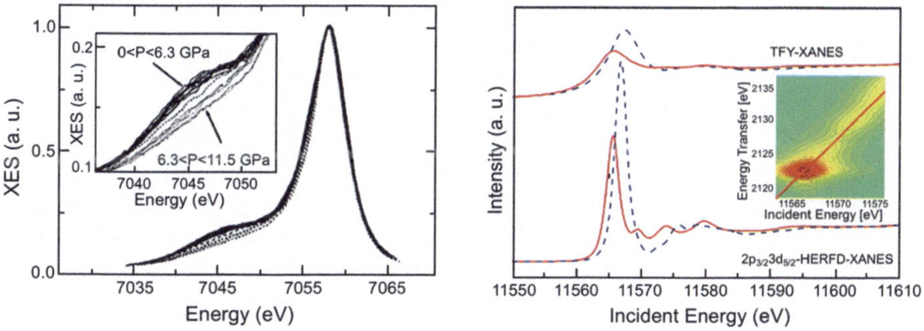

Fig. 8.10 Left: Kβ fluorescence for FeS as a function of pressure [345]. Right: line-sharpening obtained from HERFD spectra for K_2PtCl_6 (red line) and K_2PtCl_4 (blue dashed line) [348]

HERXRF has also seen applications in bioinorganic chemistry. The Kβ spectrum has been used to monitor the oxidation states of Mn in photosystem II (PSII) [349, 350], while the valence → core region was used to identify oxygen ligation in PSII [351] and a carbide ligand for the FeMo cofactor in nitrogenase [352].

8.2.10.2 HERFD: Resolution Sharpening

One limitation of conventional XANES spectroscopy is that absorption edges are significantly broadened by the core-hole lifetime. Hämäläinen and coworkers showed that this broadening could be overcome by measuring both the exciting and emitted radiation with high resolution [353], and in one analysis the new resolution Γ_{exp} is reduced according to the expression below, where Γ_n is the intermediate-state lifetime (Γ_n) and Γ_f is the final-state lifetime [354]:

$$\Gamma_{exp} \cong \frac{1}{\sqrt{(1/\Gamma_n)^2 + (1/\Gamma_f)^2}} \tag{8.3}$$

The technique has the awkward-to-pronounce name of HERFD, but the improvement in spectral quality is dramatic (Fig. 8.10). The experimental linewidth of Pt L_3-edge conducted in transmission mode is normally about 8 eV [354, 355], but in recent measurements this has been reduced to ~3 eV or better eV (Fig. 8.10). The reasons will be clearer once we discuss RIXS.

8.2.10.3 Site-Selective Absorption

The chemical shifts in fluorescence lines can be used to select for the absorption by specific species in multicomponent systems. The application to $Fe_4[Fe(CN)_6]_3$

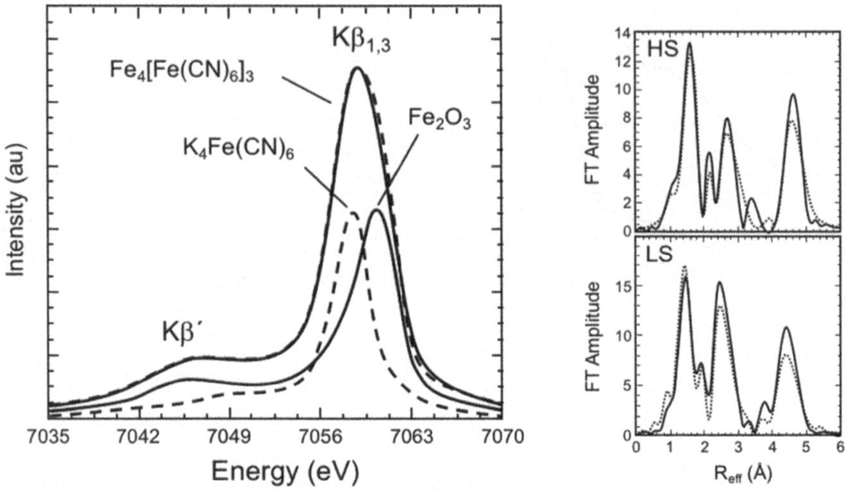

Fig. 8.11 Left: comparison of Kβ emission for high-spin Fe_2O_3 and low-spin $K_4Fe(CN)_6$, along with Prussian blue, $Fe_4[Fe(CN)_6]_3$ [356]. Right: site-selective EXAFS showing Fourier transforms extracted from recording the EXAFS near the peak intensity for high-spin or low-spin Fe [356]

(Prussian blue, which contains a mix of high-spin and low-spin Fe), provides a nice example. Since only high-spin Fe has a strong Kβ´ satellite, by using fluorescence detection in the Kβ´ region, one primarily obtains the EXAFS for the high-spin Fe^{3+} component of this system, while centering detection near the main Fe^{2+} Kβ peak emphasizes the low-spin $Fe(CN)_6$ component. The shorter low-spin Fe–C distances (1.92 Å) compared to the Fe–N distances (2.03 Å) are clearly distinguished in the EXAFS Fourier transforms (Fig. 8.11).

8.2.10.4 Spin-Selective Absorption

A variation on the site-selective theme is to employ the spin sensitivity of Kβ lines to map out the spin-up and spin-down densities of states. As illustrated in Fig. 8.12, if a spin-down 1s electron is excited, then for Kβ fluorescence, only a spin-down 3p electron can fill that vacancy. The resulting final state will have 3p and 3d electrons mostly parallel, and the 3p–3d exchange interaction will make this a lower energy final state. This in turn results in the higher-energy $K\beta_{1,3}$ fluorescence. Conversely, promotion of a spin-up electron will yield lower-energy Kβ´ fluorescence. By monitoring fluorescence-detected absorption on Kβ and Kβ´ features, one can observe this spin-selective absorption. In the case of MnO, there are only spin-down vacancies in the 1s → 3d region; hence these features are only seen with Kβ detection (Fig. 8.12).

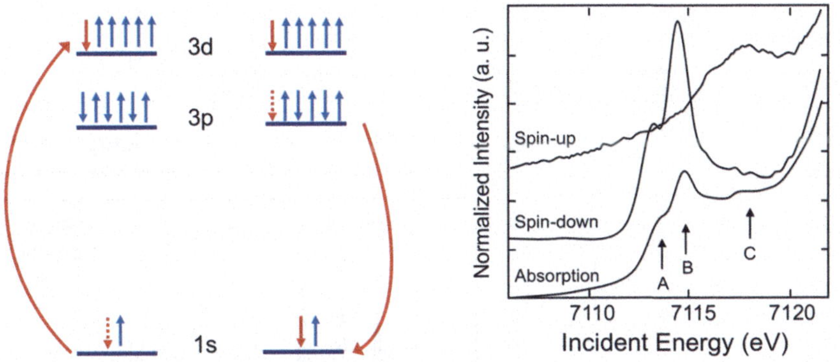

Fig. 8.12 Spin-selective absorption. Left: source of spin-selective fluorescence energy shifts. The 3p–3d exchange interaction lowers final-state energy. Right: comparison of conventional transmission X-ray absorption for MnO with excitation spectrum monitoring the Kβ emission [357]

8.3 Resonant Inelastic X-ray Scattering (RIXS)

*R*esonant *i*nelastic *X*-ray *s*cattering ("RIXS") is a technique similar to conventional resonance Raman spectroscopy, except that the incident and scattered photons are 2–3 orders of magnitude higher in energy. The experiment also resembles the X-ray fluorescence measurements described earlier in this chapter, but in RIXS the core-level electron is excited (with energy/frequency Ω) into a vacant localized state, as opposed to creating a photoelectron wave in the continuum. In RIXS the system decays by emission of the second photon (with energy ω), without dissipation of energy into other channels. The energy loss for the radiation emitted occurs at fixed energies that reveal discrete excitations of the sample (Fig. 8.13).

In Fig. 8.13 we show that there are many different types of excitations that can be seen in a RIXS experiment, and they can be excited from a variety of core levels [358]. With current instruments, it is possible to resolve energy losses from $>$10 eV down to less than 100 meV. This makes the technique sensitive to the charge transfer and d–d transitions that a chemist would normally see in the UV and visible region. One can also see spin-flips (magnons) for systems with relatively large J values. Finally, in favorable cases, molecular vibrations (phonons) can been seen by RIXS. As new RIXS instruments push toward meV resolution, there will be many more experiments using RIXS in the manner of a conventional resonance Raman experiment.

8.3.1 Why RIXS?

A quick look at Fig. 8.13 reveals that the range of energy transfers involved in typical RIXS experiments corresponds to the energies of infrared through visible to

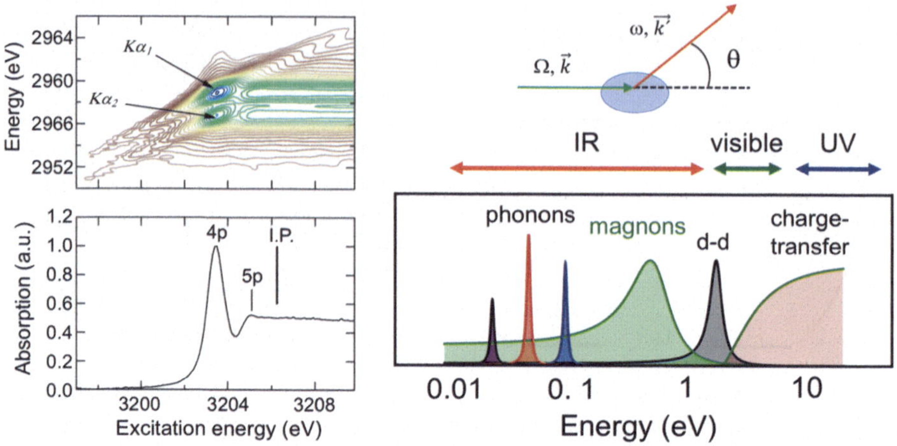

Fig. 8.13 Left: the evolution from RIXS excitations to conventional X-ray fluorescence at the Ar K-edge. The energy loss is constant over a range below the edge. The emission energy then becomes fixed at the Kα values above the edge. Top right: quantities involved in a RIXS experiment. Note: other photon-in photon-out techniques such as X-ray Raman and IXS often label incoming and outgoing photon energies differently. For those sections we will relabel the diagram to be consistent with most of that literature. Bottom right: typical low-energy excitations observed in high-resolution RIXS experiments

ultraviolet photons. Since such radiation is available in a typical chemistry lab, why bother with RIXS and the synchrotron? The cases where there are good arguments for RIXS generally involve situations where the simpler experiment either is not feasible or does not provide the required information, such as:

- A need for element selectivity.
- Interest in the q (momentum) dependence of excitations.
- Transitions forbidden by conventional selection rules.
- Experimental constraints such as windows or pressure cells.

8.3.2 *The RIXS Experiment*

One of the beautiful attributes of the RIXS technique is that the intrinsic resolution only depends on the final-state lifetime. This compares with absorption and conventional fluorescence where the linewidths depend on the lifetimes of the core hole(s) involved. To observe this enhanced resolution of a RIXS measurement, one needs two devices with comparable resolution, a monochromator for the incident radiation and an analyzer for the scattered radiation. Since the scatter comes out in all directions, the design of the analyzer is more difficult, especially when one is trying to achieve both high efficiency and good energy resolution. In Chap. 4 we saw the different solutions used for soft and hard X-rays.

The fundamental point to understand about RIXS is that it is a two-photon process. As shown in Fig. 8.13, the incoming photon has a certain energy ω and momentum $\hbar\vec{k}$, and a given polarization, and the outgoing photon will have energy ω, momentum $\hbar\vec{k}'$, and perhaps a different polarization. The difference in energy, $\Delta E = \omega - \Omega$, and the difference in momentum, $\hbar\vec{Q} = \hbar\vec{k}' - \hbar\vec{k}$, reflect transfer to the sample as an excitation that can propagate through a material. These excitations can involve vibrations (phonons), magnetic fluctuations (magnons), or d–d or charge-transfer electronic excitations. The fact that X-rays can have significant *momentum* as well as energy makes RIXS somewhat different from Raman spectroscopy with visible light, where the photons have so little momentum that for most cases it can be ignored. Below we give a very cursory introduction to the theory of how RIXS works. Our explanation is rudimentary because compared to X-ray absorption, resonant inelastic X-ray scattering is *complicated*.

8.3.3 Direct RIXS

The simplest RIXS process to explain is the so-called "direct RIXS" process for electronic excitations (Fig. 8.14), where the incoming X-ray excites an electron from a core level into an empty orbital in the valence band. The core vacancy is then filled by an electron from an occupied orbital, with the emission of a second X-ray of lower energy. The energy loss between the two photons corresponds to a net excitation of the sample. In previous treatments for the single step X-ray absorption and X-ray fluorescence processes, we could use matrix elements that ultimately relate to Fermi's golden rule. However, the two-step RIXS process requires a higher-order treatment, known as the "Kramers-Heisenberg equation" [359]. About the simplest possible formula that one can extract from such treatments is [334,360]:

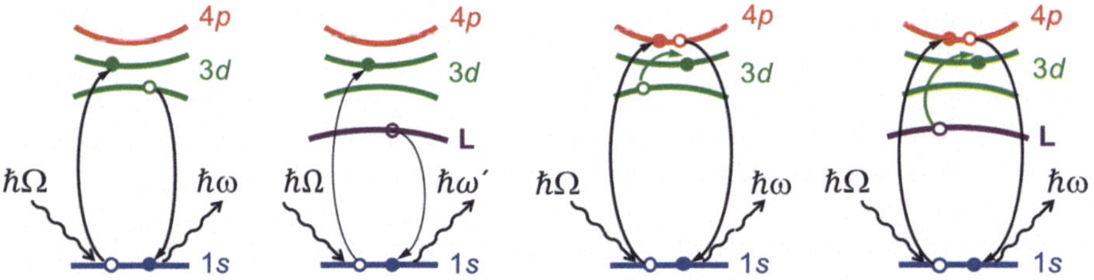

Fig. 8.14 Left: schematic illustration of the direct RIXS process for d–d and ligand → metal transitions. Right: illustration of the indirect RIXS process, showing d–d and ligand → metal excitation promoted by intermediate core-hole perturbation

$$F(\Omega,\omega)=\sum_f\left|\sum_i\frac{\langle f|T_2|i\rangle\langle i|T_1|g\rangle}{E_g-E_i+\Omega-i\Gamma_i/2}\right|^2\times\frac{\Gamma_f/2\pi}{\left(E_g-E_f+\Omega-\omega\right)^2+\Gamma_f^2/4}\tag{8.4}$$

In the above equation, $|g\rangle$, $|i\rangle$, and $|f\rangle$ are the wave functions for the initial, intermediate, and final states of the system under study; E_g, E_i, and E_f are the energies for those states; and Γ_i and Γ_f are the lifetime broadenings for the intermediate and final states. Finally, T_1 and T_2 are the optical transition operators for the excitation and emission events. Ordinarily, these can be either the electric dipole operator, $T\propto\vec{e}\cdot\vec{r}$, or the electric quadrupole operator, $T\propto i(\vec{e}\cdot\vec{r})(\vec{k}\cdot\vec{r})$.

Note that each of the matrix elements inside Eq. 8.4 looks like a dipole or quadrupole transition matrix element that we saw before in the X-ray absorption or emission discussions. Thus, for direct RIXS to have any intensity, both transitions must be "allowed," with nonzero electric dipole or electric quadrupole matrix elements.

8.3.4 *Indirect RIXS*

Although the direct process can explain many RIXS spectra, there are some surprising cases where theorists distinguish a second type of process: *indirect RIXS*. In this process, one excites the sample a few eV above the lowest empty valence orbitals. The core hole acts as an additional perturbation to the system, causing excitation of the valence electrons. Indirect RIXS is thus due to the shake-up excitations described in Chap. 6 (multi-electron excitations that robbed intensity from EXAFS oscillations). We will see examples of indirect RIXS at both K- and L-edges.

8.3.5 *The RIXS Plane*

From Eq. 8.4 we see that (like resonance Raman) the RIXS intensity depends on both excitation and emission energies. But instead of emission energy ω, the quantity of interest is usually the energy loss, $\Omega-\omega$, since that is the energy of the excitation in the sample. Thus, RIXS spectra are frequently presented as 3d representations or the corresponding contour plots vs. excitation energy Ω and energy loss $\Omega-\omega$ (Fig. 8.15).

There are several important "slices" through the RIXS plane. As shown in Fig. 8.15, the most common are (a) signal vs. Ω for fixed $\Omega-\omega$, "constant energy transfer" or CET; (b) signal vs. $\Omega-\omega$ for fixed Ω, "constant incident energy" or CIT; and (c) signal vs. Ω for fixed ω, "constant emission energy" or CEE. The last of these corresponds to the HERFD experiment previously discussed.

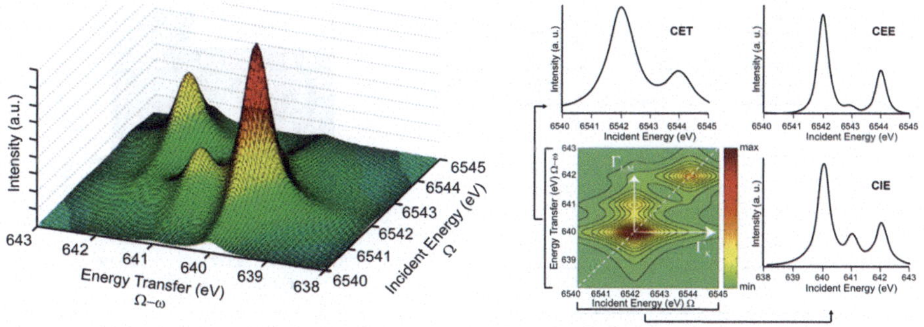

Fig. 8.15 Left: the overall RIXS plane for MnO. Right: a RIXS contour plot (lower left) can be sliced to yield 2-d CET, CEE, and CIE plots (redrawn from [334])

8.3.6 K-L RIXS

In our treatment of X-ray fluorescence, one of the processes we described involved knocking out a 1s core electron and then filling that hole with an electron from a 2p or 3p orbital, to yield Kα or Kβ fluorescence, respectively. What happens if instead of ejecting the 1s electron into the continuum, one instead promotes the electron into a vacant 3d orbital? If the core hole is then filled by a 2p or 3p electron, the final state will then be the same as that achieved by the soft X-ray absorption at the L- or M-edges discussed in Chap. 6!

$$1s^2 3d^n \rightarrow 1s^1 3d^{n+1} \rightarrow 1s^2 2p^5 3d^{n+1} \quad \text{or} \quad 1s^2 3d^n \rightarrow 1s^1 3d^{n+1} \rightarrow 1s^2 3p^5 3d^{n+1} \tag{8.5}$$

Given all the difficulties of working in the soft X-ray region, K-L or K-M RIXS is an attractive way to obtain comparable information. An example of such spectra for NiF_2 using the Ni K-edge is shown in Fig. 8.16 [361].

8.3.7 Charge-Transfer RIXS

Earlier in this chapter, we discussed "valence-to-core" fluorescence involving bands or orbitals that are mostly ligand in character. These transitions can also occur when exciting into the d-orbitals at a transition metal edge, and the resulting final state is the same as a "ligand-to-metal charge transfer" ("LMCT") transition seen in the UV-visible region:

$$1s^2 3d^n \rightarrow 1s^1 3d^{n+1} \rightarrow 1s^2 \underline{L} 3d^{n+1} \tag{8.6}$$

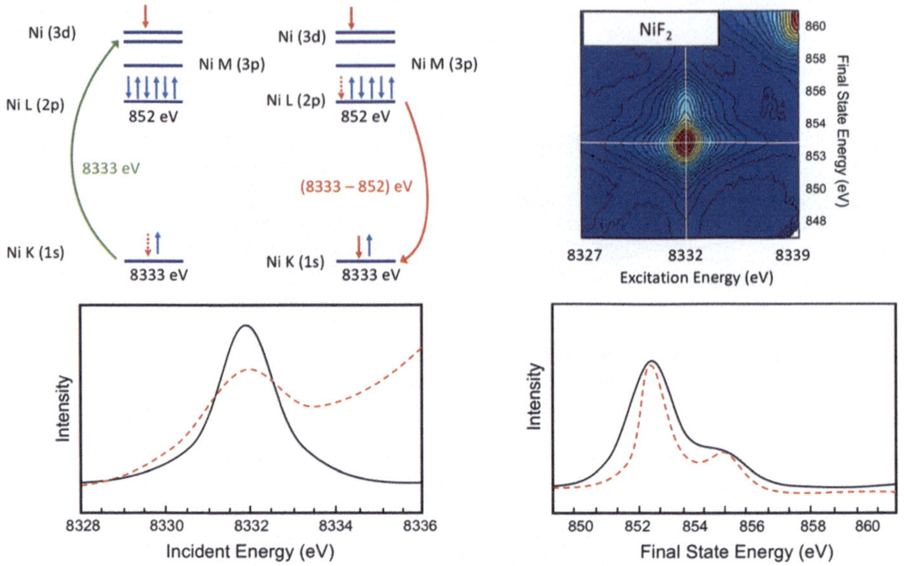

Fig. 8.16 Top left: schematic representation of K-L RIXS. Top right: a RIXS plane for high-spin Ni(II) in NiF_2. Bottom left: conventional K pre-edge absorption spectra (red dashed line) and RIXS constant final-state energy plots (black solid line). Bottom right: conventional L absorption spectra (red dashed line) compared to RIXS line plots (black solid line) [361]

Of course, it's easier to use a UV-visible spectrometer. But there are many situations where something else in the sample might obscure the transitions of interest. Or, there might be something about the experimental apparatus itself that obscures the region of interest. In such cases a RIXS experiment would allow you to get the same information.

One example is high-pressure studies using diamond anvil cells. Since pure diamond has a band gap of 5.5 eV, it is impossible to obtain optical spectra in these devices for samples at shorter wavelengths than about 225 nm. However, by using Ni K-edge RIXS, the charge-transfer excitations of NiO samples could be studied from standard conditions all the way to 100 GPa (almost one million atm) [362]. The normal atmospheric pressure spectra showed features from 4 to 12 eV that agreed well with UV reflectance data (Fig. 8.17). For example, in both cases the onset of the O → Ni charge-transfer band appears just below ~4 eV. This edge stayed relatively constant with increasing pressure, whereas the maximum at ~8.5 eV shifted continuously to higher energies. Data like this could not be obtained in a conventional UV transmission or reflectance experiment.

Charge-transfer excitations are also visible in L-edge RIXS. Remember that in discussing L-edge or Kβ features with the charge-transfer model, the ground state was written as a configuration interaction between a purely ionic configuration and a configuration with an electron transferred from the ligand: $\alpha 3d^8 + \beta 3d^9\underline{L}$ (Fig. 8.17).

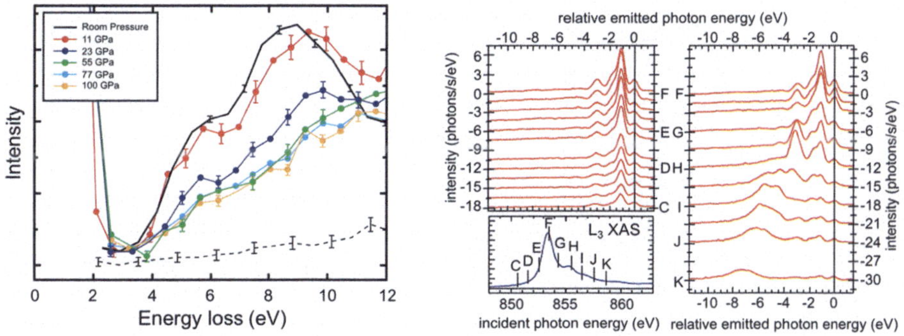

Fig. 8.17 Charge-transfer RIXS in NiO. Left: Pressure dependence of NiO K-edge RIXS charge-transfer bands [362]. Right: excitation-dependent charge-transfer RIXS at the NiO L_3-edge

L-edge transitions can occur to excited states with that same ligand hole, which can then emit a photon to yield a final state that is the same as an optical LMCT:

$$\alpha 2p^6 3d^8 + \beta 2p^6 3d^9 \underline{L} \rightarrow 2p^5 3d^{10} \underline{L} \rightarrow \alpha' 2p^6 3d^8 + \beta' 2p^6 3d^9 \underline{L} \tag{8.7}$$

At the NiO L_3-edge, these charge-transfer transitions emerge when exciting several eV above the L-edge maximum. Initially, the main intensity occurs at an energy loss of ~5 eV, and the intensity extends out to ~10 eV when exciting at even higher energies. These are the same charge-transfer final states that were seen at the K-edge (Fig. 8.17).

8.3.8 d–d RIXS

Transitions between d-orbitals are "forbidden" (yet still weakly observed) in optical spectroscopy. In contrast, net d–d excitations are allowed and frequently observed by RIXS. For example, suppose we excite the same 1s → 3d absorption in NiO that was discussed above, but now examine the scattered radiation with better resolution. As shown in Fig. 8.18, three new distinct features appear—these are excitations to triplet d–d excited states. The overall process can be described as:

$$1s^2 3d^8 \rightarrow 1s^1 3d^9 \rightarrow 1s^2 3d^{8*} \tag{8.8}$$

where the final state $3d^{8*}$ indicates transitions where the 3d electron configuration remains $3d^8$ but the electrons have been rearranged to yield a different energy level.

If we assume octahedral symmetry around the Ni(II) ion in NiO, the available excited states are all multiplets arising from a net $t_{2g}{}^6 e_g{}^2 \rightarrow t_{2g}{}^5 e_g{}^3$ transition, as opposed to the $p^5 d^{N+1}$ multiplets discussed in reference to L-edges. For a 10DQ value of 1.05 eV, we find that the order and energies of the triplet excited states are

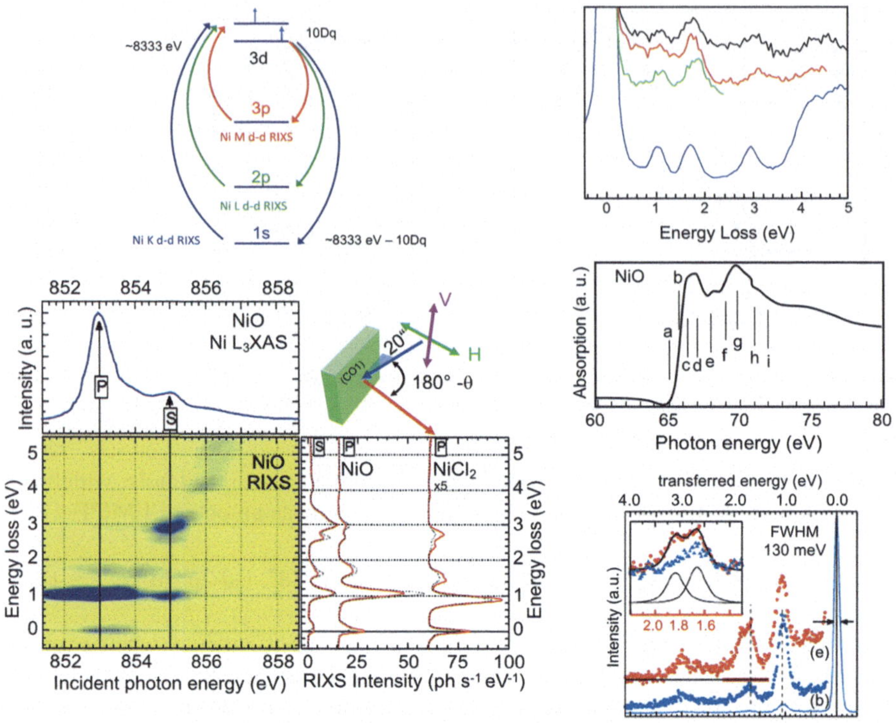

Fig. 8.18 Observation of d–d excitations in NiO RIXS at different edges. Top left: levels involved in direct RIXS at K, L, or M-edge for d–d transitions. Top right: Ni–O d–d excitations in RIXS at using different excitation energies K-edge. Lower left: NiO L-edge RIXS. Lower right: M-edge RIXS (130 meV) [363, 364]

$^3T_{2g}$ (1.05 eV), $^3T_{1g}$ (1.7 eV), and another $^3T_{1g}$ (3.0 eV), and this is approximately what is observed.

The RIXS experiment on NiO can also been done at the L_3 edge, where the *d-d* excitation process can be described as:

$$2p^6 3d^8 \rightarrow 2p^5 3d^9 \rightarrow 2p^6 3d^{8*} \tag{8.9}$$

where, as before, $3d^{8*}$ indicates transitions in which the 3d electrons have been rearranged to yield a different level. In the resulting L_3 RIXS, the elastic line from the $3d^8 \rightarrow 2p^5 3d^9 \rightarrow 3d^8$ process is the surprisingly weak feature at 0 eV (Fig. 8.18). When exciting on the main L_3 feature, we again see the same triplet excited states as at the K-edge.

Continuing to lower energies, the RIXS experiment on NiO can even be done at the M_3 edge, where the different events can be described as:

$$3p^6 3d^8 \rightarrow 3p^5 3d^9 \rightarrow 3p^6 \left[3d^8, 3d^{8*}, 3d^9 \underline{L}\right] \tag{8.10}$$

In the resulting M-edge RIXS (Fig. 8.18), the same three d–d excitations are clearly visible, and the elastic line from the $3p^63d^8 \rightarrow 3p^53d^9 \rightarrow 3p^63d^8$ process is also relatively strong.

The ability to excite at different energies is a powerful tool in RIXS. For example, part of the L-edge absorption corresponds to a "spin-flip" transition to an intermediate state with primarily singlet character. This causes the final states of RIXS excited at this energy to be dominated by the singlet final states. Thus, by changing the excitation region, one can selectively enhance triplet or singlet final states, a feature not available to conventional optical spectroscopy. Refer to [365] for details.

8.3.9 Magnetic Excitations Via RIXS

In a binuclear complex with two magnetic metals (Fig. 8.19), the energy of the system depends in part on the relative spin orientations on the two metal sites, and this "exchange energy" is given by the Heisenberg spin Hamiltonian [366]:

$$H_{ex} = -2J_{12}\vec{S}_1 \cdot \vec{S}_2 \tag{8.11}$$

where $\vec{S}_1$ and $\vec{S}_2$ are quantum numbers for the spins on sites 1 and 2 and J_{12} is the exchange constant. For copper acetate, $-2J_{12}$ is positive, and the lowest-energy state of the molecule has the two $S = 1/2$ Cu(II) ions antiferromagnetically coupled with opposite spin orientations and a total spin $S' = 0$. The first excited state has spins in parallel alignment such that $S' = 1$, and this triplet excited state is $-2J_{12} \cong 300\ \text{cm}^{-1}$ above the ground state. The value for J_{12} can be deduced from the magnetic

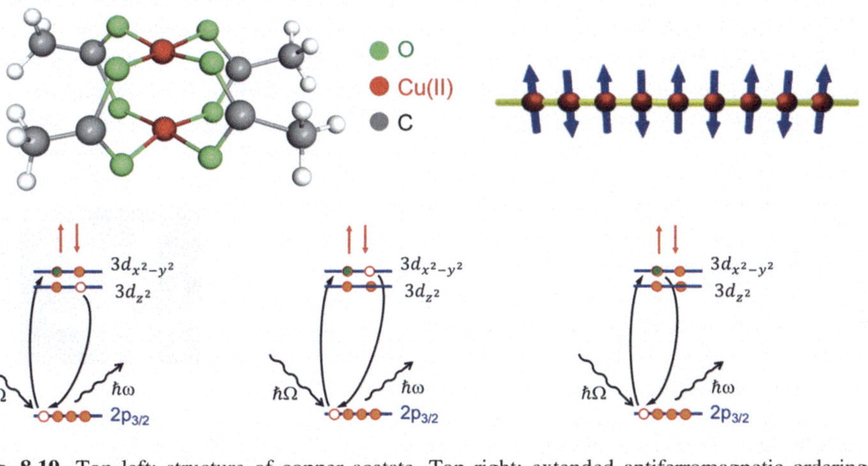

Fig. 8.19 Top left: structure of copper acetate. Top right: extended antiferromagnetic ordering along chains in quasi-1D Sr_2CuO_3. Bottom left to right: comparison of mechanisms for d–d excitations and spin-flip transitions at Cu L_3 edge

susceptibility as a function of temperature [367]. Although invisible by IR and Raman spectroscopies [368], the singlet-triplet excitation can be seen directly by inelastic neutron scattering [369].

RIXS provides another approach for direct observation of spin-flip transitions. The processes by which L- or M-edge RIXS can cause spin flips at single sites or at a collection of sites are sketched in Fig. 8.19. The key point is that thanks to spin-orbit coupling, spin itself is not a "good quantum number" for the 2p or 3p vacancy. This hole can be filled by an electron of either spin, with either $\Delta S = 0$ conservation of spin or an overall $\Delta S = 1$ transition.

So far, magnetic RIXS has been applied primarily to solid-state systems with extended chains of magnetic metals. In such systems, such as CuO and NiO, a simple spin flip as described above for copper acetate will propagate through the lattice as a wave [370]. This spin wave is often described as a quasiparticle called a *magnon*. Like other quasiparticles, a magnon carries a fixed amount of energy, lattice momentum, and angular momentum, in this case a spin of $\hbar$.

The energies of magnetic excitations are frequently in the 1–100 meV range, quite a bit lower than charge-transfer and d–d transitions. Magnetic RIXS thus makes exquisite demands on the instrumental energy resolution $E/\Delta E$, and it is still early going for this technique. Fortunately, some of the most interesting systems such as cuprates have large J values that make them accessible at current beamlines.

Thanks to the interest in high T_c superconductors, there has been an enormous amount of work on the spin distributions of copper oxides [371]. These can be arranged into quasi-1D systems such as Sr_2CuO_3, quasi-2D systems such as La_2CuO_4, and of course full 3D systems. A spectrum for a single magnon excitation in the La_2CuO_4 system (LCO to physicists) is shown for a single energy in Fig. 8.20. Transitions involving multiple magnons can also be observed at Cu edges [358, 372, 373] and even at the oxygen K-edge.

Like phonons, magnons exhibit dispersion, meaning that their energy can depend on their wavelength and momentum. You can control the momentum transfer in a RIXS experiment by varying the scattering angle (Fig. 8.13), and the dispersion data from an INS experiment is compared with RIXS results in Fig. 8.20. The agreement is excellent. One advantage of RIXS is clear—the data were collected on a 100-nm-

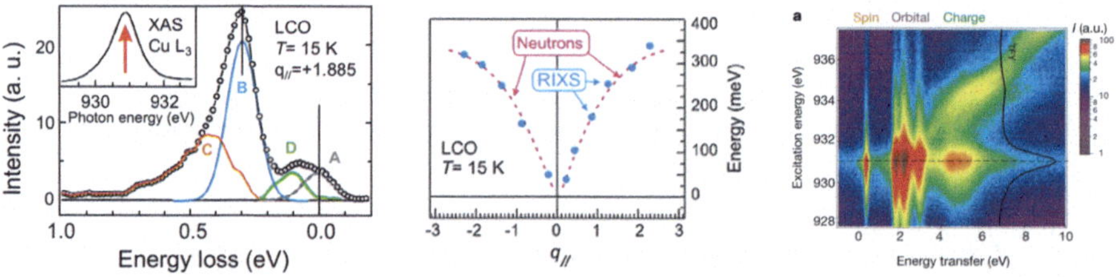

Fig. 8.20 Left: evidence for single magnon transitions in RIXS data (peak B) at the copper L_3 edge of La_2CuO_4 [374]. Peak C is identified as multiple magnons, and peak D is attributed to phonons. Middle: comparison of single magnon INS and RIXS measurements for La_2CuO_4 [375]. Right: using RIXS to observe spin, orbital, and charge excitations in Sr_2CuO_3 [376]

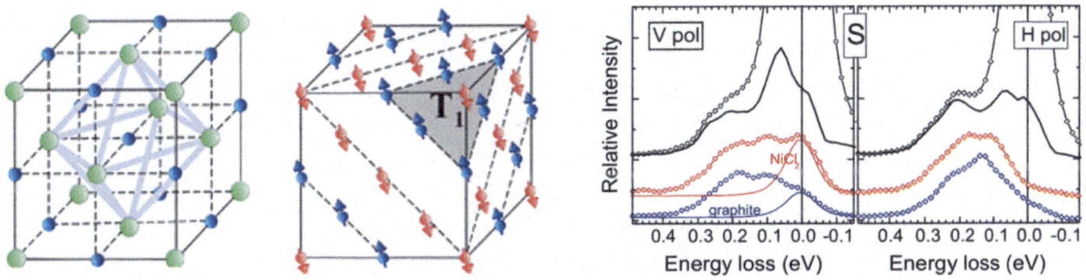

Fig. 8.21 Left: a schematic description of the NiO magnetic structure. Right: magnon transitions in L_3 RIXS of NiO compared with $NiCl_2$ and graphite as controls observed with excitation on P (primary) peak or S (secondary, spin-flip) peak from [364]

thick film of La_2CuO_4, while the INS measurement used multiple crystals with a total mass of about 50 grams!

Turning to 3D systems, a prototypical 3D antiferromagnet is NiO. There is considerable interest in its magnetic properties because NiO is used in so-called spin valves and magnetic data storage. In the NiO lattice, the spins on the linear Ni–O–Ni chains are antiferromagnetically coupled. Each Ni has six such interactions, and a single ion model predicts spin-flip peaks around $6J$ and $12J$, where J is the interatomic superexchange interaction. In fact, since J is $\sim$20 meV for NiO, evidence can be seen in the L_3 RIXS (Fig. 8.21) [364]. This is a case where better resolution would help. For future studies on less strongly coupled antiferromagnets such as $NiCl_2$ ($J \sim$3.8 meV), the technique clearly needs another order of magnitude improvement in resolution.

8.3.10 Vibrational RIXS: Phonons

If it is early days for the study of magnetic excitations by RIXS, the situation is almost pre-Cambrian for study of molecular vibrations and phonons by this technique. The resolution of the best current RIXS instruments is typically 50 meV or 400 cm^{-1}, which makes it possible to see only well-separated and relatively high-energy molecular vibrations. Thus, a beautiful example is the RIXS of gaseous O_2, as shown in Fig. 8.22. By exciting into the $1s \rightarrow \pi*$ resonance, investigators were able to see a strong progression of vibrational excitations of the O–O bond. A similar progression for the C=O stretch with a fundamental frequency of 210 meV was seen in oxygen K-edge RIXS of acetone (Fig. 8.22) [377].

Phonons are lattice vibrations of a periodic solid. They are quantized like the normal modes of molecules that are familiar to chemists. However, unlike molecular normal modes, where the center of mass is not displaced, acoustic phonon modes carry momentum. Since phonons typically have energies only up to ~100 meV, they require exquisite resolution to separate them from the elastic scatter. An attempt to

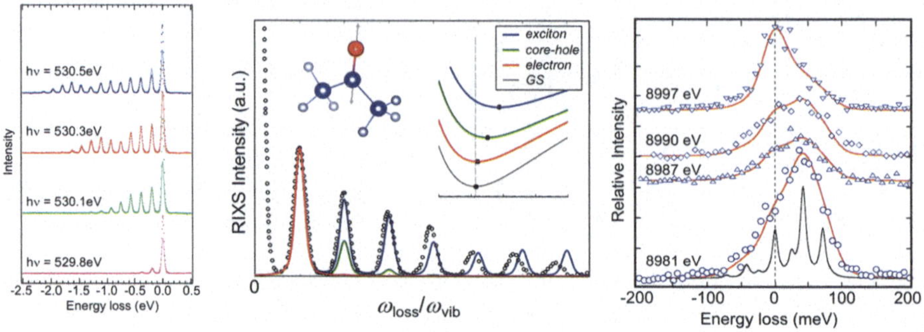

Fig. 8.22 Left: stretching vibrations in O_2 by RIXS, excited at different energies peaking at the 1s → π∗ resonance [378]. Middle: RIXS data (∘∘∘∘) and assorted calculations for the C=O stretch in acetone, for which $\omega_{\mathrm{vib}} = 210$ meV [377]. The model with phonon-exciton coupling clearly gives the best results. Right: partially resolved phonon contributions at Cu K-edge of CuO [379]

resolve phonons in CuO is shown in Fig. 8.22 [379]. Clearly, the study of phonons by RIXS is just in its infancy.

8.3.11 Polarization and Magnetic Effects in RIXS

For samples with spatial or magnetic orientation, there are additional effects that employ variable linear or circular polarizations. Both of these are complex subjects, perhaps appropriate for the second edition.

8.4 X-ray Raman Scattering (XRS)

X-rays can scatter inelastically from a sample even when their energy is not in resonance with the absorption edge of a particular element (Fig. 8.23). If the X-rays scatter off individual electrons, the process is known as Compton scattering. If the scattering process excites an electronic transition in the sample, it is called "*X*-ray *R*aman *s*cattering" or "XRS." This is essentially the same scattering that was discovered by Raman in Kolkata in the 1920s [380] [381]. The same experiment also goes by other names and abbreviations, "*n*on-*r*esonant *i*nelastic *X*-ray *s*cattering" ("NRIXS" or just "NIXS"). Experimental observation of XRS goes back more than 50 years [382], but progress in synchrotron radiation sources has made for enormous improvements in the quality of these measurements.

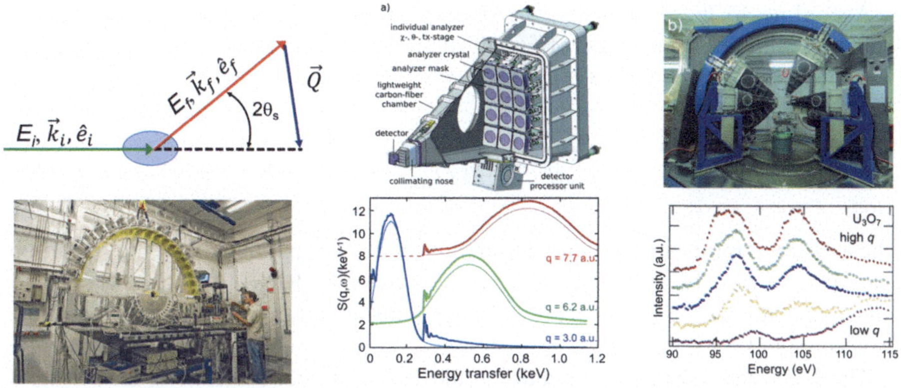

Fig. 8.23 Top left to right: terms used for X-ray Raman and inelastic X-ray scattering; analyzer module hosting 12 analyzer crystals on a 1 m Rowland circle at XRS end-station at ESRF; 6 modules surrounding the sample region [383]. Bottom left to right: LERIX spectrometer at APS [384]; scattering from polycrystalline diamond, with scattered energy fixed at 13 or 16 keV and incident energy scanned for appropriate energy and momentum transfer [385]. Smooth curves underneath represent the Compton scattering; U_3O_7 XRS excited at 10 keV [386]. The q values were 3.1, 5.3, 7.7, 8.9, and 10.0 Å^{-1}. Transitions are primarily $l = 1$ at low q, through $l = 3$ to $l = 5$ at high q

8.4.1 The XRS Experiment

Since X-ray Raman scattering is a weak effect, most measurements employ an array of spherically bent crystals similar to those used for high-energy resolution X-ray fluorescence. Until recently, most of these devices had a moderate resolution of ~1 eV. A recently upgraded end-station at ESRF can achieve ~0.3 eV overall resolution [383] while providing a phenomenal solid angle (at lower resolution) approaching 10% of 4π steradians by using 72 individual spherically bent analyzer crystals! The device uses the same Rowland circle geometry discussed in Chap. 4 (Fig. 8.23).

8.4.2 XRS Theory

The theory of X-ray Raman scattering was summarized by Tohji and Udagawa [387], based on earlier work [388], while more recent summaries are in [385, 389] and the book by Schulke [390]. XRS is the first spectroscopy that we discuss that exploits the $\vec{A}^2$ part of the interaction of radiation with matter (Appendix H.1). As put by Schulke, this scattering results from "time-dependent electron density fluctuations" [390]. When the charge fluctuations and excitations are associated with core electrons, we have XRS, while fluctuations associated with phonons yield the IXS discussed in the next section.

According to these theorists, the double differential cross section for photon scattering into solid angle Ω, originally derived by van Hove [391], is given by (in atomic units) [385, 392–394]:

$$\frac{d^2\sigma}{d\Omega dE} = \left(\frac{d\sigma}{d\Omega}\right)_{\text{Th}} S\left(\vec{Q}, E\right) \tag{8.12}$$

where $\vec{Q}$ and E represent the momentum transfer and energy transfer in the scattering event (Fig. 8.23). (Note the different symbols from the RIXS literature, where Ω is often used for the incident energy. Some XRS literature simply uses θ for the scattering angle).

The first term in the above equation is the Thomson cross section for scattering of electromagnetic radiation by electrons, and this just depends on the experimental geometry:

$$\left(\frac{d\sigma}{d\Omega}\right)_{\text{Th}} = r_0^2 \frac{E_f}{E_i} \left(\widehat{e}_i \cdot \widehat{e}_f\right)^2 \tag{8.13}$$

where E_{i} and E_{f} are the incoming and scattered photon energies, $\widehat{e}_i$ and $\widehat{e}_f$ are the corresponding polarization vectors, and r_0 is the classical electron radius.

The second term in Eq. 8.12 is the dynamic structure factor $S(\vec{Q},\omega)$ which can be expressed via Fermi's golden rule as:

$$S\left(\vec{Q}, \omega\right) = \sum_f \left| \langle f \mid \sum_j \exp\left(i\vec{Q} \cdot \vec{r}_j\right) | i \rangle \right|^2 \delta(\omega + E_{\text{i}} - E_{\text{f}}) \tag{8.14}$$

Here $|i\rangle$ and $|f\rangle$ are initial and final-state wave functions, ω is the energy transferred to the sample, and the internal summation is over electrons at positions $\vec{r}_j$. This expression can be simplified by using a Taylor series expansion for the exponential:

$$\exp\left(i\vec{Q} \cdot \vec{r}\right) \cong 1 + i\vec{Q} \cdot \vec{r} - \frac{\left(\vec{Q} \cdot \vec{r}\right)^2}{2} + \ldots \tag{8.15}$$

The first term does not contribute because the wave functions are orthogonal, so for small values of Q, the dynamic structure factor becomes equivalent to a sum over the same dipole matrix elements we have seen for X-ray absorption, with the direction of $\vec{Q}$ taking the role of the photon polarization in XAS. The two important points to take home from the above equations are the following: (a) at small scattering angles, the XRS effect is proportional to the same electric dipole matrix element seen in conventional X-ray absorption, and (b) at higher angles, there can be interesting non-dipole components to the intensity. The gradually increasing importance of non-dipole terms is illustrated for U_2O_3 XRS in Fig. 8.23.

8.4.3 An Intensity Estimate

How strong is the XRS signal? Sahle and coworkers did a nice calculation using Eq. 8.16 [389]:

$$I = I_0 \frac{d^2\sigma}{d\Omega\, d\omega_2} \Delta\Omega\, \Delta\omega_2\, \rho\, d\, a_{r/t} RD \tag{8.16}$$

where I_0 is the incident flux, ω_2 is the analyzer energy, $\Delta\Omega$ is the solid angle collected, $\Delta\omega_2$ is the energy resolution, ρ is the number density of scatterers in the interaction volume, d is the sample thickness, R is the analyzer crystal reflectivity, D is the detector efficiency, $a_{r/t}$ is an absorption factor to account for losses of the incoming and outgoing beams, and $d^2\sigma/d\Omega d\omega_2$ is the double differential scattering cross section from Eq. 8.12 (using ω_2 instead of E to describe the energy transfer). Assuming an XRS instrument with 12 spherically bent 10 cm diameter Si(6 6 0) crystals and an incident flux of 10^{13} photons s^{-1} at ~10 keV, they estimated a signal count rate for the C K-edge of better than 10 counts s^{-1} for a 10% acetic acid solution. They conclude that samples at ~0.4 mol% are feasible over an 8-h shift. XRS is orders of magnitude less sensitive than conventional X-ray absorption, but it has become quite powerful for cases where it is the only feasible approach.

8.4.4 X-ray Raman Applications

If the signal is so weak, compared to X-ray absorption, then why do an X-ray Raman experiment? As with the RIXS technique, the cases where there are good arguments for XRS generally involve situations where a simpler XAS experiment either is not feasible or does not provide the required information, such as:

- Light element K-edges without UHV conditions.
- Low-energy transition metal M- and L-edges using windows or pressure cells.
- Low-energy edges without artifacts from fluorescence detection.
- Interest in alternate selection rules.
- A need for bulk sensitivity.

8.4.4.1 Dissecting Absorption Edges

As we saw in Chap. 7, conventional X-ray absorption is dominated by electric dipole transitions. One strength of XRS is the ability to enhance higher Δl transitions by changing the scattering angle and Q value. Bradley and coworkers have used this feature to observe $\Delta l = 3$ (octupole) and even $\Delta l = 5$ (triakontadipole) features in the $O_{4,5}$ (5d) edges of actinides such as U_2O_3 (Fig. 8.23).

8.4.4.2 Hydrocarbons and Fuels

Fossil fuels contain an enormous variety of carbon and nitrogen functionalities that could benefit from C and N K-edge analysis. Asphaltenes are a particularly nasty fossil fuel fraction—they are an asphalt-like residue from the final stages of petroleum distillation, and UHV scientists are loath to put petroleum residues in their pristine vacuum chambers. Using X-ray Raman, Bergmann, and coworkers collected C K-edge spectra on hydrocarbon models and asphaltenes (Fig. 8.24). They quantified the relative amounts of aliphatic and aromatic C through the strength 1s $\rightarrow \pi^*$ feature [395].

8.4.4.3 Water

High-temperature (~90 °C) water is another sample that is hardly compatible with UHV chambers. Hence, water ice and liquid water at both 25 and 90 °C have instead been compared by XRS (Fig. 8.24). For the high-temperature water sample, an increase in intensity of pre-edge features was attributed to a larger fraction of water molecules with one uncoordinated or weakly coordinated O–H group [396].

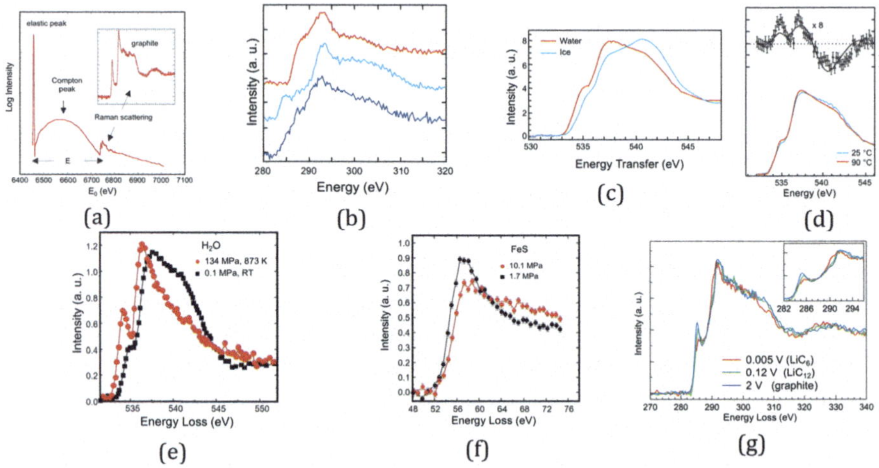

Fig. 8.24 Applications of X-ray Raman spectroscopy: (**a**) wide energy scan of X-ray scattering by graphite, showing relative intensities of elastic peak, Compton scattering, and X-ray Raman [395]; (**b**) comparison of C K-edges for a saturated hydrocarbon (red line) (paraffin) with an aromatic carbon (blue line) (coronene) and with asphaltene (black line) showing intermediate properties [395]; (**c**) XRS of water and ice; (**d**) XRS of liquid water at 25 °C (blue line) and 90 °C (red line), with difference spectrum above [396]; (**e**) XRS of water under ambient conditions and at high temperature and pressure; (**f**) FeS $M_{2,3}$-edges obtained via XRS, demonstrating transition from high-spin to low-spin Fe [397]; (**g**) C K-edge XRS of the graphite electrode at different voltages in a lithium-ion battery. For details see [398]

8.4.5 High-Pressure Samples

High-pressure conditions are another obvious situation where soft X-rays are out of the question. Diamond anvil cells are the most common way to achieve high pressures, and the walls can be penetrated by hard (but not soft) X-rays. One illustration of their use is shown in Fig. 8.24, where Nyrow and coworkers used NRIXS-derived Fe M-edges to observe the conversion of FeS from high-spin Fe to low-spin Fe as pressure is increased to 10.1 GPa [397].

8.4.6 In Situ Batteries

Batteries under operating conditions are another example where the elements of interest are low Z but soft X-ray absorption is not feasible. Nonaka and coworkers observed beautiful changes in the C K-edge of a graphite electrode at different voltages corresponding to different amounts of Li intercalation (Fig. 8.24) [398].

8.5 Inelastic X-ray Scattering (IXS)

There is another type of non-resonant inelastic X-ray scattering, often called just "*i*nelastic *X*-ray *s*cattering" and abbreviated "IXS" [394], which is also quite different from X-ray fluorescence or RIXS. As with XRS, in IXS, a photon scatters inelastically by interacting with charge density fluctuations of the system—in this case fluctuations produced by phonons. Early on, this process was observed as "*T*hermal *D*iffuse *S*cattering" ("TDS") in X-ray diffraction patterns as seen in Fig. 8.25, and TDS analysis is an entire field of its own [399]. However, before the advent of synchrotron radiation sources, there were no instruments with sufficient energy resolution to resolve the small energy shifts from IXS underlying the TDS features.

In the IXS experiment, one again measures a double differential scattering cross section:

$$\frac{d^2\sigma}{d\Omega\, dE} = \frac{\text{photons s}^{-1}\ [\Omega, d\Omega][E, dE]}{\text{incident photons s}^{-1} \times d\Omega \times dE} \tag{8.17}$$

As with XRS, the IXS cross section is sensitive to the collective motions of particles in the system via the "dynamic structure factor" $S(\vec{Q},\omega)$, and we rewrite the cross section in a form similar to that used by Baron [394, 401]:

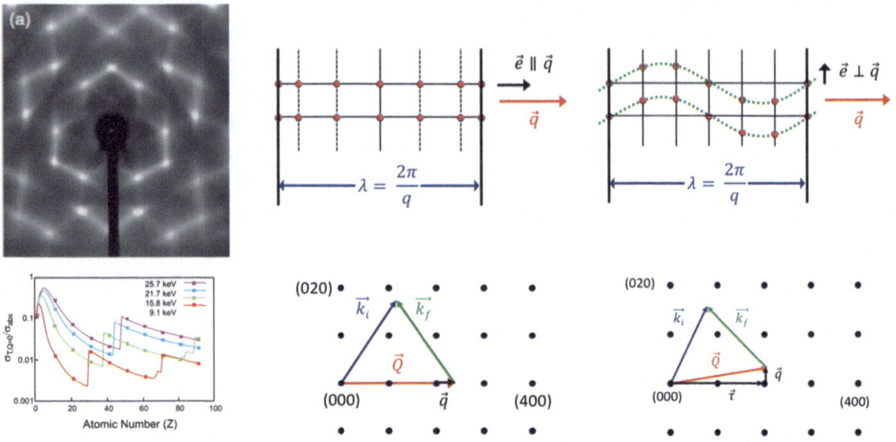

Fig. 8.25 Top left: thermal diffuse scattering from Si along Si(1 1 1) direction [399] [400]. Lower left: ratio of cross section for Thomson scattering at $Q = 0$ to absorption cross section at various energies, redrawn from [394]. Middle and right: atomic displacements for a longitudinal phonon in real space and geometry for an IXS measurement in reciprocal space vs. a transverse phonon in real and reciprocal space, redrawn from [393]. $\vec{Q}$ is the scattering vector, $\vec{q}$ is the phonon wavevector, defined with respect to the nearest lattice point $\vec{\tau}$

$$\left(\frac{d^2\sigma}{d\Omega dE}\right)_{\vec{k}_{\mathrm{i}}\hat{\varepsilon}_{\mathrm{i}}\to\vec{k}_{\mathrm{f}}\hat{\varepsilon}_{\mathrm{f}}} = r_0^2\frac{k_{\mathrm{f}}}{k_{\mathrm{i}}}\left|\hat{\varepsilon}_{\mathrm{f}}\cdot\hat{\varepsilon}_{\mathrm{i}}\right|^2 S\left(\vec{Q},\omega\right) = r_0^2\frac{E_{\mathrm{f}}}{E_{\mathrm{i}}}\left|\hat{\varepsilon}_{\mathrm{f}}\cdot\hat{\varepsilon}_{\mathrm{i}}\right|^2 S\left(\vec{Q},E\right) \tag{8.18}$$

where $\hbar\vec{k}_{\mathrm{i}}$ and $\hbar\vec{k}_{\mathrm{f}}$ are the incident and scattered photon momenta, $\hat{\boldsymbol{e}}_{\mathrm{i}}$ and $\hat{\boldsymbol{e}}_{\mathrm{f}}$ are polarization unit vectors, and r_0 is the classical electron radius. The momentum and energy transfers are, respectively, given by:

$$E = \hbar\omega \equiv \left(E_i - E_f\right) \tag{8.19}$$

and

$$\hbar\vec{Q} \equiv \hbar\left(\vec{k}_i - \vec{k}_f\right) \tag{8.20}$$

Notice that the convention is to write the dynamic structure factor as a function of ω and to define positive energy transfer as corresponding to the sample gaining energy—the Stokes side of the spectrum.

8.5.1 IXS and Phonons

With a known crystal structure and force field, all of the observed changes are in principle calculable from the well-developed theory behind IXS. Thus, many IXS investigations are designed to test agreement between proposed force fields and the experimental data. This theory has been extensively reviewed elsewhere [393, 402–404], so here we merely summarize the key factors that govern IXS intensities.

As with XRS, the single phonon cross-section for scattering, $\partial^2\sigma/\partial\Omega\partial E$, is proportional to $S\left(\vec{Q},\omega\right)$, the "dynamic structure factor," but in this case, $S\left(\vec{Q},\omega\right)$ is sensitive to fluctuations in atomic positions (phonons). Along with terms accounting for phonon population and polarization effects, $S\left(\vec{Q},\omega\right)$ is in turn proportional to the "inelastic structure factor", $F_{in}\left(\vec{Q}\right)$, which involves a summation over all the atoms in the primitive cell and thus for a particular phonon mode [393, 394, 405]:

$$\frac{\partial^2\sigma}{\partial\Omega\partial E}\propto S\left(\vec{Q},E\right)\propto F_{in}\left(\vec{Q}\right)$$
$$=\left|\sum_m \mathrm{M}_m^{-1/2} f_m\left(\vec{Q}\right)\left[\vec{e}_m^{\,n}\left(\vec{q}\right)\cdot\vec{Q}\right]\exp\left(i\vec{Q}\cdot\vec{r}_m\right)\exp\left(-w_m\right)\right|^2 \quad (8.21)$$

In the above expression, atom m with mass M_{m} is located at $\vec{r}_m$ and has X-ray form factor $f_m(Q)$ and Debye-Waller factor $\exp(-w_{\mathrm{m}})$. The term $\vec{e}_m^{\,n}\left(\vec{q}\right)$ is the phonon polarization eigenvector for the direction of motion of atom m in phonon mode n with wavevector $\vec{q}$ [401]. Since as $Q\rightarrow 0, f_m(Q)\rightarrow Z$, at small angles the strength of scattering from a particular atom in a sample will vary approximately as Z^2. The overall signal strength depends on the product of three terms: $f_m(Q)$, Q^2, and a polarization factor that goes as $\cos^2\theta$. In practice, this product maximizes at $Q\cong 10$ Å^{-1} for many elements.

A key term in Eq. 8.21 is the product $\vec{e}_m^{\,n}\left(\vec{q}\right)\cdot\vec{Q}$. This is the projection of the atomic motion onto the total momentum transfer vector $\vec{Q}$, and it plays a critical role in the observed intensities. Thus the intensities of the IXS signal can be directly related to the motions of particular atoms in a given normal mode [393, 402–404] projected onto $\vec{Q}$. Since one can choose different $\vec{Q}$ by controlling the scattering geometry, IXS can be made more sensitive to particular phonons in a sample. The appearance of the sum inside the magnitude signs means that the intensity is sensitive to the relative phase of the motions within one primitive cell. Baron provides very approximate rules of thumb: low-frequency acoustic modes (larger displacements) tend to be stronger than high-frequency optic modes, and these long wavelength acoustic modes tend to be stronger near strong diffraction peaks [394].

8.5.2 The IXS Experiment

The IXS experiment can be considered as an NRIXS experiment on steroids (Fig. 8.26). In both cases, the goal is to measure the energy difference E between incoming and outgoing photons, but since IXS attempts to measure phonons, the

Fig. 8.26 The high-resolution IXS spectrometer at BL35XU [123] of SPring-8, showing (left) main components and (**b**) staff for scale (left to right: A. Baron, D. Miwa, D. Ishikawa, and Y. Tanaka)

desired resolution is 1 meV or better, about 10^3-fold better than common IXS instruments. This is achieved by working at high energies with crystals in an extreme backscattering geometry, as described in Chap. 4. For the best sensitivity, an IXS experiment should optimize the ratio of the Thomson-scattering cross section to the absorption cross, as shown in Fig. 8.25, and this also favors high energies.

The quantities that are measured in an IXS experiment are as before in Fig. 8.23. Apart from the energy change E, the momentum transfer $\vec{Q}$ is also critical, because one goal is to map the energy of a particular vibration as a function of momentum—a so-called dispersion curve. A simplifying aspect of the IXS experiment is that energy transfer is small compared to incident and scattered photon energies. From this it can be shown that the magnitude of the momentum transfer, Q, is completely determined by the photon energy (hence momentum) and the scattering angle:

$$\hbar Q = 2\hbar \cdot k_i \cdot \sin\theta \tag{8.22}$$

To summarize, an IXS experiment consists in measuring the intensity of scattering $S\left(\vec{Q}, E\right)$ or $S\left(\vec{Q}, \omega\right)$ as a function of energy loss E and as a function of momentum transfer $\vec{Q}$, which in turn is determined by the scattering angle 2θ.

8.5.3 IXS Applications

IXS has become one of the best techniques for characterizing the dynamic properties of liquids and the phonon spectra of solid materials. IXS can visualize phonons in samples that are orders of magnitude smaller than the multi-gram quantities required for inelastic neutron scattering. It is admittedly a photon-hungry experiment that requires the absolute state of the art in storage ring, undulator, and X-ray optics. However, it is gradually becoming more routine as the brightness of sources continues to improve.

8.5.3.1 Dynamics of Liquids

One of the simplest applications of IXS was a study of liquid N_2 [406], illustrated in Fig. 8.27. The N≡N stretch (previously mentioned in Chap. 7.4) is obvious in all the spectra at ~289 meV (2331 cm^{-1}). There is a clear Q dependence to the intensity of this band, which was used to deduce a N≡N distance of 1.11 ± 0.02 Å, in agreement with the 1.100 ± 0.011 Å X-ray diffraction value.

The IXS spectra of liquid water and ice have also been thoroughly studied [407–411]. In ice, there are two clear components, a slowly dispersing component for a transverse phonon and a more rapidly dispersing component for a longitudinal wave. From the slope of the Q dependence, one can deduce two sound velocities, respectively, 1500 m s^{-1} and 3200 m s^{-1} (Fig. 8.27). The liquid water data also exhibited a component that dispersed with a slope of 3200 m s^{-1}, and this was attributed to "fast sound" that travels twice as fast as a conventional sound wave in water [407].

8.5.3.2 Phonons in Solids

The quality of phonon spectra that can be obtained via IXS especially stands out when applied to single crystals of solids. By changing the scattering angle, one can obtain the same dispersion curves seen by RIXS in Fig. 8.20, but with more than an order of magnitude better resolution. We first give examples of applications relevant to superconductivity.

The discovery of superconductivity in LaOFeP in 2006 opened up a new direction for high-T_c research based on layered "iron pnictide" compounds. The basic

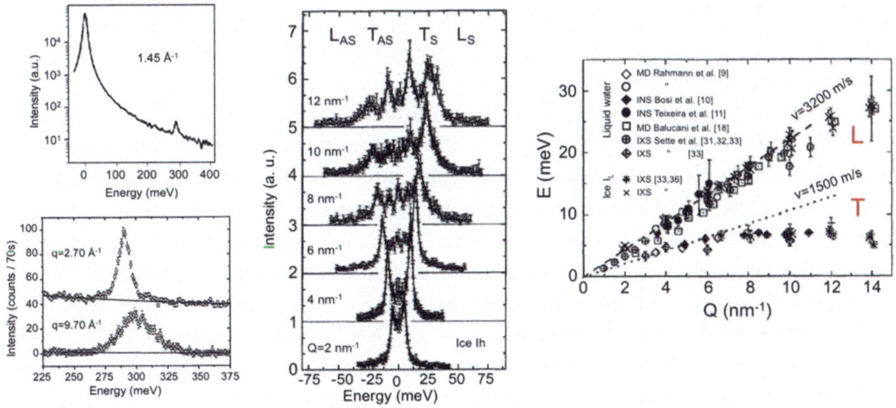

Fig. 8.27 IXS of simple systems. Top left: a representative IXS spectrum for liquid N_2 with incident energy 13.840 keV at $k = 1.45$ $Å^{-1}$ momentum transfer [406]. Lower left: close-up of the feature for stretching mode. Middle: IXS spectrum for polycrystalline I_h ice at different momentum transfers [409]. Right: dispersion curves for ice and liquid water with deduced speeds of sound; L and T correspond approximately to longitudinal and transverse modes [409]

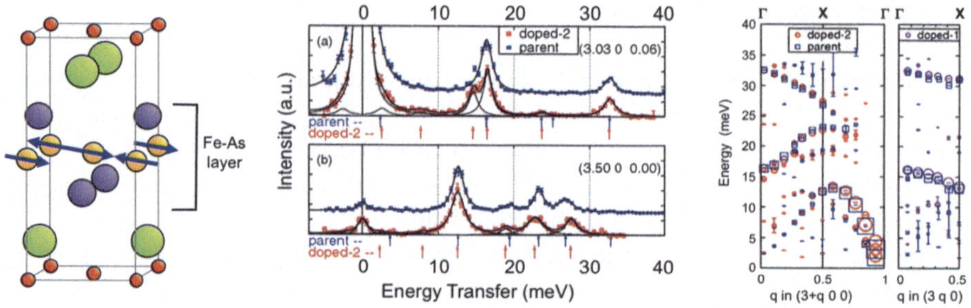

Fig. 8.28 Left: structure of $PrFeAsO_{0.7}$. Scheme: O (red dot), As (blue dot), Fe (yellow dot), Pr (green dot); arrows on the Fe atoms indicate the spin orientations when $T < T_N$. Middle: IXS for doped $PrFeAsO_{0.7}$ (open red circle) and parent PrFeAsO (filled blue circle) at room temperature near Γ point ($Q = (3.03\ 0\ 0.06)$) and at Brillouin zone boundary ($Q = (3.50\ 0\ 0.00)$). Right: dispersion relations for parent PrFeAsO vs. doped sample [412]

structure of the ReFeAsO subset is illustrated in Fig. 8.28. Fukuda and coworkers obtained beautiful spectra on the parent compound PrFeAsO and slightly O-deficient $PrFeAsO_{0.7}$ with a superconducting transition temperature T_c of 42 K [412]. Agreement between theory and experiment required reducing the strength of the predicted Fe–As bond by 30%.

Interest in another class of materials was sparked by the finding of superconductivity in MgB_2 with a $T_c = 39$ K [413]. Related intermetallic compounds with honeycomb layered structures such as ternary silicides MAlSi, M = Ca, Sr, Ba were also found to be superconducting, with $T_c = 8$ K for the Ca version [414]. The IXS for CaAlSi was studied to gain insight into its mechanism of superconductivity (Fig. 8.29) [415]. There is a "soft mode" involving out-of-plane Al/Si vibrations, and they argue that the low frequency of the soft mode enhances its coupling to the electronic system and leads to a relatively high T_c.

We also mention a family of HTS superconductors with Tc up to 43 K that results from doping $Ca_2CuO_2Cl_2$ oxychloride systems (Fig. 8.29) [416]. In this system the Cu–O bond stretching modes are of particular interest, where it has been suggested that the dispersion is different for different directions. For details, see the original work [416] (Fig. 8.30).

As a final example, we show a geophysical application. The elastic properties of minerals under the high-pressure conditions found at the lower mantle are important for interpretation of seismic observations [417]. IXS data at pressures up to 41.2 GPa revealed strong anisotropy in shear wave velocities (Fig. 8.31). Again, for details, see the original work [417].

In summary, IXS has become one of the best techniques for characterizing the phonon spectra of solid materials. IXS can visualize phonons in samples that are orders of magnitude smaller than those required for inelastic neutron scattering. It is admittedly a photon-hungry experiment that requires the absolute state of the art in storage ring, undulator, and X-ray optics. As with other synchrotron methods, it will become even more powerful as sources and optics inexorably improve.

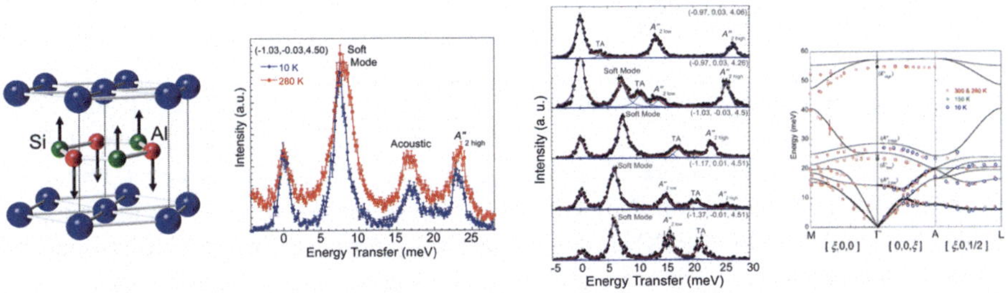

Fig. 8.29 Left to right: structure of CaAlSi. Scheme: Al (red dot), Ca (blue dot), Si (green dot); IXS spectra at 10 K and 280 K [415]; IXS spectra for different $\vec{Q}$-values and fits from a theoretical model; dispersion curves

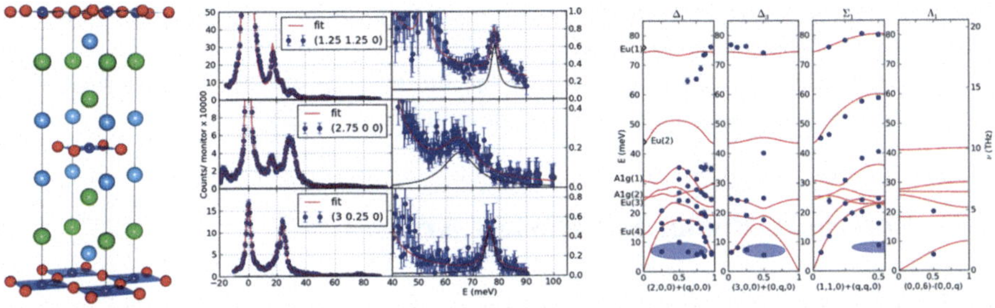

Fig. 8.30 Left to right: layered structure of $Ca_2CuO_2Cl_2$. Scheme: O (red dot), Cu (blue dot), Ca (skyblue dot), Cl (green dot); IXS spectra, with close-ups highlighting Cu–O stretching region; dispersion curves [416]

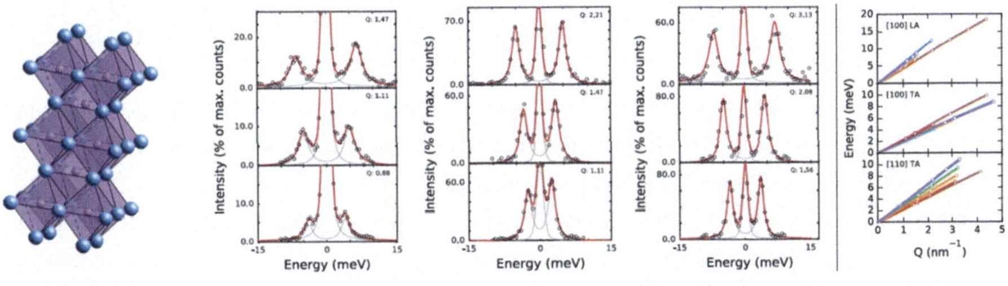

Fig. 8.31 Left: at low pressure magnesiowüstite (Fe-rich ferropericlase, (Mg,Fe)O) has a rock salt structure with octahedral Fe and disordered Mg. Middle: IXS spectra for different directions. Right: dispersion curves [417]

8.6 Suggested Exercises

1. Estimate the count rate for a fluorescence experiment on the Fe Kβ line of a 1 micron Fe foil at a 45° angle, excited at 8 keV, with an incident flux of 10^{12} photons s^{-1} and an analyzer perpendicular to the beam that accepts 1% solid angle with 1 eV resolution. Neglect absorption of the fluorescence in the sample.
2. RIXS—Outline which RIXS processes are fully dipole allowed or fully quadrupole allowed for seeing d–d transitions using the L_1- or $L_{2,3}$-edges of Ni.
3. XRS—Using the Taylor series expansion, estimate where the second and third terms have comparable amplitudes for a carbon XRS experiment at 20 keV: $\exp\left(i\vec{q}\cdot\vec{r}\right) \approx 1 + i\vec{q}\cdot\vec{r} - \frac{(\vec{q}\cdot\vec{r})^2}{2} + \cdots$
4. XRS—Estimate the count rate for a pure diamond XRS experiment at 20 keV. Justify all of the required assumptions.
5. IXS—Derive the relationship between scattering angle and momentum transfer for an IXS experiment (Eq. 8.22):
6. IXS—Calculate the absolute value of the momentum transfer Q if an incoming 24.797 keV photon is scattered through an angle $\theta = 15°$.

8.7 Reference Books and Review Articles

1. *Electron Dynamics by Inelastic X-ray Scattering*, Winfried Schülke, Oxford University Press, New York, 2007, ISBN 978-0-19-851017-8—a deep treatment of inelastic X-ray scattering.
2. *Core Level Spectroscopy of Solids*, Frank de Groot and Akio Kotani, CRC Press, Boca Raton, 2008, ISBN 978-0-8493-9071-5.
3. *X-ray Spectroscopy*, B. K. Agarwal, Springer-Verlag, New York, 1979, ISBN 0-387-09268-4—a nice resource for much of the pre-synchrotron fluorescence literature.
4. *X-ray Absorption and X-ray Emission*, J. A. van Bokhoven and C. Lamberti, eds., Wiley, Singapore, 2016, ISBN 978-1-118-84423-6. Edited volume covers synchrotron radiation and X-ray spectroscopy.
5. P. Glatzel and U. Bergmann, "High resolution 1*s* core hole X-ray spectroscopy in 3*d* transition metal complexes—electronic and structural information", *Coord. Chem. Rev.* **249**, 65–95, 2005.—a nice review.
6. *Synchrotron Radiation News*, Vol. 31, No. 2, **2018**—"*Resonant Inelastic X-ray Scattering*", an update on ultra-high resolution RIXS.
7. A. Q. R. Baron, "High-Resolution Inelastic X-ray Scattering I & II" in *Synchrotron Light Sources and Free-Electron Lasers: Accelerator Physics, Instrumentation and Science*, edited by E. Jaeschke, S. Khan, J. R. Schneider, & J. B. Hastings, pp. 1643–1757. Springer International Publishing, 2016. ISBN 978-3-319-14395-8. see also https://arxiv.org/abs/1504.01—an excellent introduction to IXS.

Chapter 9
Nuclear Hyperfine Techniques

The interaction of X-rays with electrons is so strong that we can usually ignore nuclear effects, dismissing the nucleus as a point charge too massive and slow to respond to the electromagnetic field of the X-ray. However, the ability to produce bright X-ray beams with narrow linewidths now permits observation of nuclear transitions that are normally observed with radioisotope sources. In this chapter we first describe some very general properties of nuclear transitions and nuclear excited states, and we then describe the conventional way of doing things. We then present the most popular synchrotron nuclear techniques for capturing the same information.

"I am a chemist (biologist, materials scientist . . .)," you object. Why do I care about nuclear properties? It turns out that nuclear spectroscopy can reveal a wealth of information about:

- Electronic structure.
- Magnetic properties.
- Vibrational modes, and all of this with:
- Elemental and isotopic sensitivity for the nucleus under study.

How is this possible? It turns out that small shifts and splittings in nuclear energy levels can report back about *electronic structure* and hence the chemistry of the sample. This information is so valuable that NASA included a nuclear instrument (a Mössbauer spectrometer) on the Rover mission to Mars (Fig. 9.1). In addition, the motion of the nucleus can be detected by synchrotron techniques, allowing characterization of the dynamics and normal modes around that nucleus (Chap. 10).

To be specific, there are three techniques that reveal the nuclear energy levels and hence electronic structure and magnetic properties:

- *S*ynchrotron *M*össbauer *s*pectroscopy (SMS).
- *N*uclear *f*orward *s*cattering (NFS), and.
- *S*ynchrotron *r*adiation *p*erturbed *a*ngular *c*orrelation (SRPAC).

S. P. Cramer, *X-Ray Spectroscopy with Synchrotron Radiation*, Biological and Medical Physics, Biomedical Engineering, https://doi.org/10.1007/978-3-030-28551-7_9

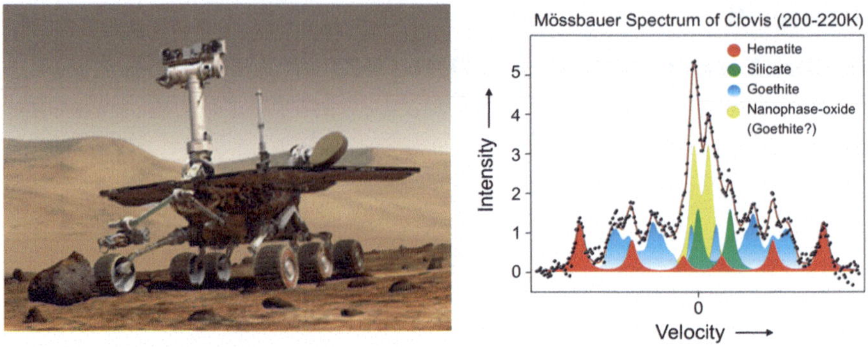

Fig. 9.1 Left: Artist's rendering of rover in action [418]. Right: mineral identification on Mars by fitting Mössbauer spectrum with a set of standard spectra

Finally, there is another technique that probes the vibrational motions associated with the nucleus under study, which we save for the next chapter:

- Nuclear resonance vibrational spectroscopy (NRVS).

All of these measurements have become vastly improved by the high brightness of modern synchrotron radiation sources.

9.1 Nuclear Properties and Nuclear Transitions

From a chemist's point of view, a nucleus can be considered as a bag of nucleons (protons and neutrons) with a particular size and shape. Just as electronic transitions involve a rearrangement of an electron's probability distribution (wave function) within an atom or molecule, nuclear transitions involve a rearrangement of the nucleons. The possible nuclear wave functions and their associated energies are quantized by a nuclear Hamiltonian operating on a nuclear wave function:

$$\mathcal{H}\psi_N = E_N\psi_N \tag{9.1}$$

9.1.1 Energy Levels, Spins, Lifetimes, and Linewidths

Nucleons are held together primarily by the so-called strong interaction. Just as electrons in an atom can be promoted to excited states, the nucleus can also be excited to higher levels and make transitions back to lower levels, absorbing or emitting photons in the process. Nuclear states have characteristic energies E_N and total angular momenta characterized by nuclear spin quantum number I. Note that when speaking about nuclear properties, "spin" often refers to the total angular

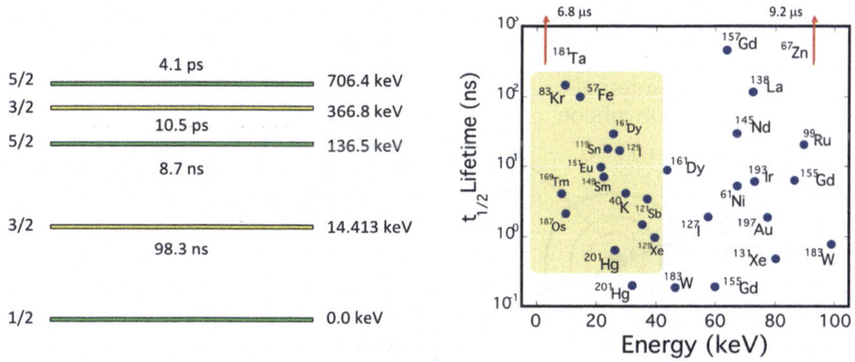

Fig. 9.2 Left: energy levels for ^{57}Fe. Spins are on the left and lifetimes center. Right: other nuclear energies and lifetimes. Shaded area highlights isotopes conducive to NRVS (next chapter)

momentum, rather than the spin of an individual particle. A typical nuclear energy level scheme, for the ground and first several excited states of ^{57}Fe, is shown in Fig. 9.2. For any particular isotope, these quantities can be found on a variety of websites [419].

Nuclear transitions can be exquisitely sharp, especially when compared to X-rays. As with X-rays, the levels above the ground state have a characteristic lifetime τ_0, and this determines the linewidths via the Heisenberg uncertainty relation:

$$\Delta E \cdot \Delta t \geq \hbar \tag{9.2}$$

which in turn leads to:

$$\Gamma \cdot \tau_0 = \hbar \tag{9.3}$$

where $\Gamma = \Delta E$ is the full width half maximum of the Lorentzian and τ_0 is the $1/e$ lifetime of the exponentially decaying excited state. For the first ^{57}Fe excited state shown in Fig. 9.2, the $t_{1/2} = 98$ ns half-life (144 ns $1/e$ lifetime) corresponds to a linewidth $\Gamma = 4.6 \times 10^{-9}$ eV. This presents extraordinary challenges for the design of monochromators, but it also creates exciting opportunities for the observation of very small chemical effects on those levels.

Most "γ-ray" photons involved in nuclear transitions are higher in energy than the X-rays associated with core electron excitation, extending up to the MeV region. However, there are still quite a few nuclear transitions below 100 keV that are accessible to synchrotron radiation (Fig. 9.2). Once in an excited state, a nucleus can relax by photon emission ("nuclear fluorescence") or by the *internal conversion* process, in which the excitation energy is transferred to an emitted core electron (Fig. 9.3). The parameter α is called the "internal conversion coefficient" and represents the ratio of events decaying by electron emission to events decaying by nuclear fluorescence.

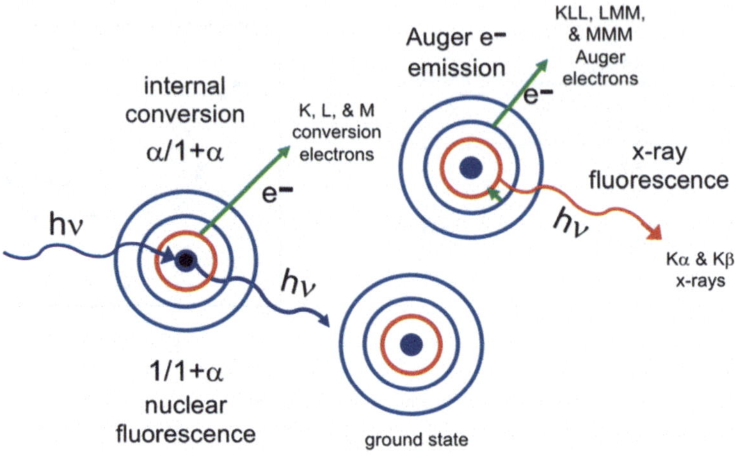

Fig. 9.3 Some excitation and relaxation events in nuclear transitions

Internal conversion leads to creation of a core hole, similar to that created by X-ray absorption. As in that case, this vacancy can relax by X-ray fluorescence or Auger emission, and the probabilities for both events are the same as with core holes created by photons in X-ray absorption.

9.1.2 Nuclear Sizes, Shapes, and Magnetic Moments

The spatial scale of a nucleus is about five orders of magnitude smaller than typical atomic sizes (1 Å). A common analogy is that if an atom were the size of a barn, the nucleus would be the size of a bee flying around inside the barn. The nuclear charge distribution is not necessarily spherical, and, furthermore, it changes from one nuclear energy level to another. For many nuclei, an ellipsoid is a good approximation to the true distribution, yielding cigar-shaped (prolate) or pancake-like (oblate) ellipsoids.

In the most general case, the first-order deviations of the nuclear charge distribution $\rho_n(\vec{r})$ from spherical symmetry are described by a 3 × 3 matrix known as the quadrupole tensor $\widetilde{Q}$. However, if one aligns the z-axis along the axis of cylindrical symmetry, then the tensor simplifies to a scalar quadrupole moment Q given by:

$$Q \equiv \frac{1}{e} \int \rho_n(r) r^2 \left(3 \cos^2\theta - 1\right) d\tau \tag{9.4}$$

The dimensions of Q correspond to an area, and the common unit for Q is a "barn," b, where 1 b $= 10^{-24}$ cm$^2 = 10^{-8}$ Å^2. Only nuclei with $I > 1/2$ have nonzero quadrupole moments. For ^{57}Fe, the ground state is $I_g = 1/2$, and hence the quadrupole moment $Q_g = 0$, while for the first excited state, $I_e = 3/2$ and $Q_e = 0.16$b.

Quadrupole moments and other properties for important nuclei can be found in [420].

If a nucleus has a nonzero spin quantum number, then the motion of charged protons within that nucleus is like a solenoid that leads to a magnetic dipole moment, $\vec{\mu}$, with magnitude μ. In such cases, the moment in the ground state, μ_g, will differ from the excited-state moment μ_e. Magnetic moments are usually tabulated in units of the nuclear magneton μ_N ($\mu_N = eh/4\pi m_p = 5.05 \times 10^{-27}$ J T^{-1}), and values for important nuclei are also listed in [421].

9.2 Hyperfine Interactions

Hyperfine interactions are the result of interactions in an atom between the electrons and the nuclear moments (Fig. 9.4). *Hyperfine structure* was first observed in atomic spectra by Michaelson in the early 1890s [422,423]; it is called *hyperfine* because the splittings are much smaller than in the *fine structure* that arises from electron-electron interactions [424]. Two important hyperfine interactions are between (1) the nuclear electric quadrupole moment and the electronic electric field gradient

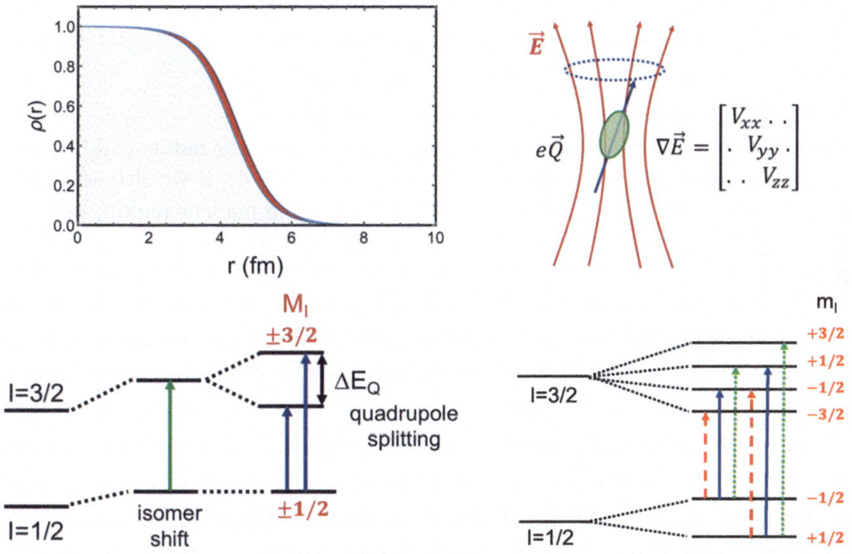

Fig. 9.4 Sources of hyperfine effects. Top left: changes in nuclear charge distribution for ^{57}Fe [425–427] give rise to isomer shifts. The change in nuclear radius between ground and excited states is highly exaggerated to allow visibility. Estimates of $\Delta R/R$ for ^{57}Fe are ~−4 × 10^{-4} [428]. Top right: different angular momentum levels of a nucleus with a quadrupole moment are split in an electric field gradient. Lower left: the combination of an isomer shift and quadrupole splitting on nuclear energy levels. Lower right: a magnetic field will split the levels of a nucleus with spin I into $2I + 1$ different energies

and (2) the nuclear magnetic dipole with the electronic magnetic field. The first of these interactions gives rise to the nuclear *quadrupole splitting*, while the second yields the nuclear *Zeeman splitting*. A final chemical effect involves sensitivity to the electron density at the nucleus, which yields the *isomer shift*.

Why do we care about hyperfine interactions? The details will come later, but the bottom line is that these *perturbations of nuclear energy levels tell us about the ground-state electronic structure of the molecule containing the isotope under study*. This can provide a fingerprint for properties such as oxidation state, or an insight into the magnetic properties.

9.2.1 Electric Monopole Interactions: The Isomer Shift

The simplest hyperfine interaction is the electric monopole interaction between the nuclear charge, which is spread over a finite volume, and the electron density within that nuclear region. Suppose that the nuclear charge distribution is uniform over a sphere with radius R. Then, outside the nucleus, the potential energy at a distance r from the center of the nucleus is given by $V_0(r) = -Ze^2/r$. However, inside the nucleus the potential will depend on the radius R via:

$$V(r) = \left(\frac{Ze^2}{r}\right)\left(-\frac{3}{2}+\frac{r^2}{2R^2}\right), \quad 0 \leq r \leq R \tag{9.5}$$

This distinction is of interest because the effective nuclear radius is different for the ground state and each excited state of a nucleus (Fig. 9.4). If we also assume that the electronic wave function is a constant $\psi(0)$ over the nuclear region, then there will be a shift in energy caused by the finite nuclear volume. The energy of a nuclear transition is modified by the *difference in shifts*, $\delta(\Delta E)$, between two nuclear states with different nuclear volumes. Finally, what is actually measured in nuclear spectroscopy is the *difference in difference of shifts*, between a reference standard and the given absorber, $\delta(\delta(\Delta E))$. This quantity is known as the *isomer shift* and is given by:

$$\delta(\delta(\Delta E)) = \frac{4}{5}\pi Ze^2\left\{|\psi(0)|_A^2 - |\psi(0)|_S^2\right\}R^2\frac{\delta R}{R} \tag{9.6}$$

The nuclear factor $\Delta R/R$, which gives the relative change in radius going from excited to ground state, can be either positive or negative. In particular, $\Delta R/R$ for ^{57}Fe is negative. For this case, this in turns means that a positive isomer shift with respect to a reference absorber means a relative *increase* in electron density at the nucleus (Fig. 9.5).

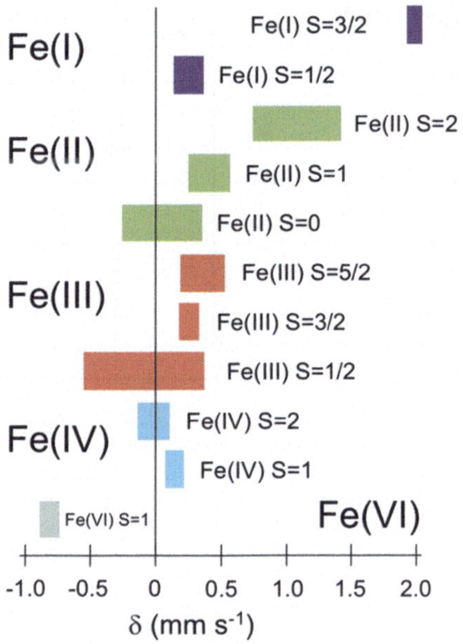

Fig. 9.5 Representative isomer shifts for different spin states and oxidation states of Fe [429]

9.2.2 *Electric Quadrupole Interactions*

A nucleus with an electric quadrupole moment will interact with the local electric field gradient (EFG) to yield a quantized set of energy levels (Fig. 9.4). As an example, for the ^{57}Fe excited state with $I = 3/2$, in an axial EFG, the $m_I = \pm 3/2$ levels are raised by $eQV_{zz}/4$, while the $m_I = \pm 1/2$ levels are lowered by $eQV_{zz}/4$. Typical values for this splitting between $\pm 3/2$ and $\pm$ 1/2 levels range from 0 in octahedral complex to >6 mm s^{-1} (2.9×10^{-7} eV) in Fe complexes with terminal nitride ligands (Fig. 9.6) [430].

9.2.3 *Magnetic Dipole Interactions*

The final important hyperfine interaction is between the nuclear magnetic dipole moment and the local magnetic field. Suppose one has an isotope with μ as the nuclear magnetic dipole moment, *g* is the nuclear magnetogyric ratio, in magnetic field H, *I* is the nuclear spin operation and MI is the magnetic spin quantum number. Using β_N as the nuclear magneton (3.15×10^{-8} eV/Tesla), this yields the following first-order energy levels:

$$E_M = -\frac{\mu H m_I}{I} = -g_N \beta_N H m_I \tag{9.7}$$

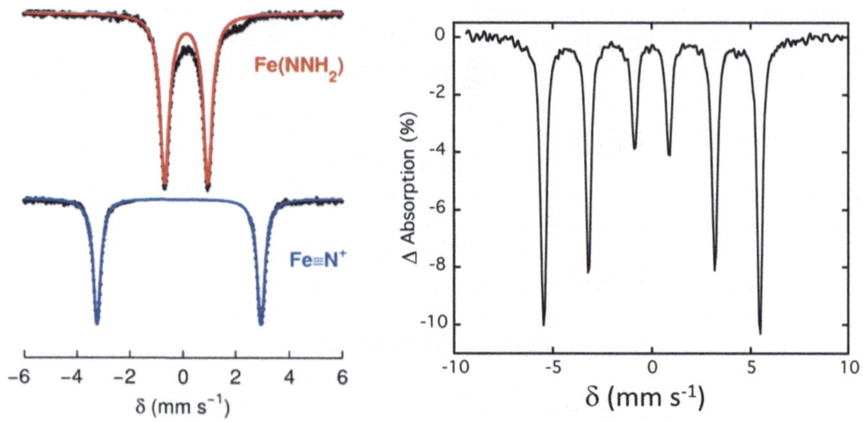

Fig. 9.6 Left: example of structure dependence of the quadrupole splitting, redrawn from [430]. Right: the classic Zeeman splitting of Fe metal Mössbauer spectrum (courtesy Yisong Guo)

Continuing with our ^{57}Fe example, the ground state with $I = 1/2$ and $\mu_g = +0.09$ will split into two substates with $m_I = +1/2$ lower, while the excited state with $I = 3/2$ and $\mu_e = -0.15$ will be split into four substates with $m_I = -3/2$ lowest (Fig. 9.4). In Fe metal, the internal magnetic field is ~33 T, larger than any applied fields that are commercially available (~15 T). This internal field results in an overall splitting between excited-state levels of about (~2×10^{-7} eV) and an overall spectral range of about 10^{-6} eV (Fig. 9.6).

9.2.4 Combined Quadrupole and Zeeman Interactions

In the most general case, there can be significant quadrupole and Zeeman interactions, in which case one must resort to numerical solutions for the level splittings and intensities.

9.3 Conventional Mössbauer Spectroscopy

From the discussion so far, we have shown that nuclear energy levels are sensitive to the electron density, density gradients, and magnetic fields surrounding the nucleus. In other words, the nucleus can report on *chemistry*. But consider the magnitude of these effects—on the order of tens of *nano*electron volts. We have already shown spectra that illustrate such splittings. But, how is it possible to resolve *nano*-eV features on top of transitions with energies that are tens or hundreds of *kilo*-eV?

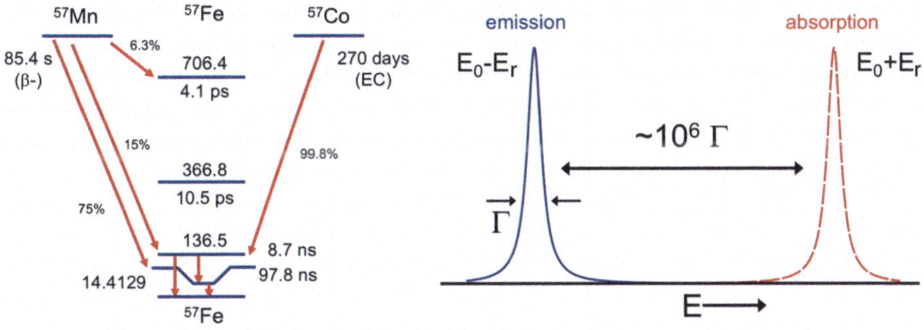

Fig. 9.7 Left: major decays and nuclear energy levels for ^{57}Mn, ^{57}Co, and ^{57}Fe. Right: recoil effects on absorption and emission energies, illustrating their scale compared to the natural linewidth

We consider first a commonly used nuclear transition, from the ground state to the first excited state of ^{57}Fe. The conventional method for observing this is to use a radioactive source, in which a parent nucleus decays to an excited state of a daughter nucleus, which in turn decays to the ground state with emission of nuclear fluorescence (γ-rays). In the case of ^{57}Fe, one could in principle use either ^{57}Mn or ^{57}Co as the parent nucleus (Fig. 9.7). Since the initial states and final states for fluorescence and absorption are reversed, it would appear at first that source and sample are perfectly matched. Things are never so simple.

9.3.1 *Recoil and Doppler Shifts*

The outgoing γ-ray carries momentum E_γ/c. If the nucleus of interest is in the gas phase, then to conserve momentum, upon emitting a photon, it will recoil with momentum $-E_\gamma/c$. Ignoring relativistic effects, the kinetic energy of the recoiling nucleus of is:

$$E_R = \frac{{p_n}^2}{2m} = \frac{E_\gamma^2}{2mc^2} = \frac{5.37 \times 10^{-4} E_0^2}{A}\,[\mathrm{eV}] \tag{9.8}$$

In the final term of the above expression, A is the dimensionless relative atomic mass for the given isotope, the energy E_0 is given in keV, and we have assumed that $E_0 \sim E_\gamma$. If we again use ^{57}Fe as an example, we find the recoil energy E_R is 1.95 meV. The emitted photon will be shifted to $E_0 - E_R$, and since the argument works in reverse for absorption, the absorption energy will be shifted to $E_0 + E_R$. The total separation of $2E_R$ is about a million times larger than the natural linewidth. Things look hopeless (Fig. 9.7).

So far, we have assumed that the emitting and absorbing nuclei are at rest with respect to each other. Suppose now that the emitter is approaching with velocity v. Then the energy of the γ-ray will be shifted by the Doppler energy $E_D = v/c$ to $E_\gamma = E_0 - E_R + E_D$. In the gas phase, there will be a statistical distribution of emitter velocities. The mean value for E_D is related to the mean kinetic energy $\overline{E}_k$ in a particular direction by:

$$\overline{E}_D = 2\sqrt{\overline{E}_k E_R} = 2\sqrt{\frac{k_B T}{2} E_R} \tag{9.9}$$

Thus, in the gas phase, as kT approaches the recoil energy, the Doppler shift and broadening can actually compensate for the recoil energy and bring a small fraction of nuclei into resonance. However, because the gas atoms in fact have a Maxwellian distribution of velocities, only a small fraction matches the resonance condition.

9.3.2 Recoilless Nuclear Absorption: The Mössbauer Effect

Suppose our radioactive source atom is in a solid-state lattice instead of the gas phase. In some cases, the absorption of recoil momentum is by the entire lattice of the solid. Since the lattice is much more massive than the nucleus, its recoil kinetic energy is effectively zero. In this case the gamma-ray carries away exactly the energy of the transition, and the emission and absorption lines overlap, centered about the transition energy. *This is the essence of the Mössbauer effect.*

Suppose a nucleus emits a gamma-ray (without recoil) of energy E_γ in making the transition from state 2 to state 1. The overlap between emission and absorption lines allows another nucleus to absorb that gamma-ray in going from state 1 to state 2. Such resonant absorption is the key to Mössbauer spectroscopy. The probability of recoil-free emission or absorption can be calculated using quantum mechanics.

9.3.3 Nuclear Absorption with Recoil: The Lamb-Mössbauer Factor

Atoms in solids behave differently. Whereas in a gas an atom can move freely; in a solid an atom vibrates around an average position. The lattice vibrational motion is quantized into phonons, which have discrete energies. Along with the recoil-free transitions discussed above, there are also events in which the recoil energy from the γ-ray emission is transferred to the lattice vibrations in units of one of these phonon energies. In such cases the emitted gamma ray will be shifted (typically by tens of meV), and these events will steal intensity from the Mössbauer effect.

The fraction of recoil-free events is termed the Lamb-Mossbauer factor *f*, or alternatively as the *recoil-free fraction* or "recoilless fraction". It is similar to the Debye-Waller factor that appears in EXAFS. The same simple models that were used to describe lattice vibrational motion for EXAFS analysis (Chap. 6) can be used to estimate this loss of intensity. For example, assuming an *Einstein solid* with vibrational frequency ω_E, incident photon wave vector k_γ, and recoil energy E_R yields Lamb-Mössbauer factor:

$$f = \exp\left(-\frac{E_R}{\hbar\omega_E}\right) = \exp\left(-k_\gamma^2\langle X\rangle^2\right) \tag{9.10}$$

where $\langle X\rangle^2$ is the mean square motion of the resonant nucleus.

9.3.4 The Conventional Mössbauer Experiment

A modern conventional Mössbauer spectrometer is a surprisingly simple instrument, consisting essentially of (a) a radioactive source, (b) a velocity transducer, (c) a γ-ray detector, (d) source and sample environmental controls, and (e) associated computers and electronics. Since our focus in this book is on the synchrotron experiment, we only briefly summarize the conventional state-of-the-art experiment. For more details on other approaches, there are excellent experimental chapters in the reference texts cited at the end of this chapter [431].

For a conventional Mössbauer experiment, one first needs to choose a particular radioisotope as the radioactive source. For example, the ^{57}Fe resonance can be excited by decay from either ^{57}Mn or ^{57}Co (Fig. 9.7). One also needs to choose a chemical environment for the radioisotope that provides a single emission line and a large Lamb-Mössbauer factor. After nearly a half century of research, the best matrices for particular isotopes have generally been worked out. For example, for ^{57}Fe work, ^{57}Co in a Rh matrix can provide a strong unsplit source with close to the natural linewidth.

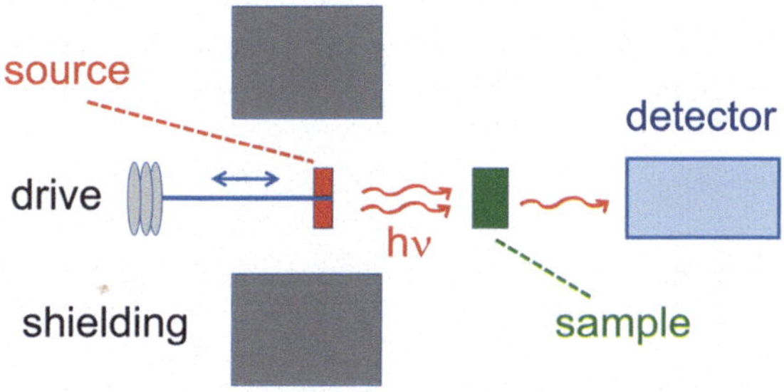

Fig. 9.8 Diagram of a conventional Mössbauer experiment

The essence of the laboratory experiment is to create and vary a Doppler shift between the sample and source. Nowadays, most instruments use an electronically controlled voltage feeding a drive coil similar to a loudspeaker system. The drive coil causes motion of the drive tube, upon which is mounted the Mössbauer source. By varying the voltage input, the devices can run in constant acceleration mode (velocity sawtooth), in sinusoidal motion, or any other desired pattern. For low-energy Mössbauer experiments such as ^{57}Fe work, gas-filled proportional counters have good efficiency and an additional advantage—they do not absorb much of the background 122 and 136 keV radiation from the typical ^{57}Co source.

9.4 Synchrotron Mössbauer Spectroscopy: Energy Domain Approaches

The energy scales for hyperfine splittings, tens of nano-eV, are vastly different from those normally encountered with X-rays monochromators. For example, the EXAFS monochromators discussed in Chap. 4 had typical resolutions on the order of 1 eV, and even the high-resolution monochromators employed for RIXS experiments only achieve ~1 meV. How can one hope to achieve the extra 6 orders of magnitude required for Mössbauer spectroscopy? The three most commonly employed approaches are:

- Build a monochromator based on *nuclear diffraction*.
- Excite a nuclear resonance in a single-line standard and employ the re-radiated photons.
- Switch to time domain detection of *nuclear forward scattering*.

The first approach uses nuclear Bragg diffraction, as illustrated in Fig. 9.9. This method employs multiple clever tricks to achieve a single line. First of all, as noted 50 years ago by Smirnov, one has to pick a reflection for which electronic diffraction is strictly forbidden but for which nuclear diffraction is still allowed [432]. In the case of $FeBO_3$, the (3 3 3) reflection is an appropriate choice. Then, to achieve a single-line nuclear source, one raises the crystal temperature close to the Néel point (~348 K) to convert from antiferromagnetic to paramagnetic state, and additionally employing a small (150 Oe) magnetic field, a single-line source is achieved [433]. Finally, by vibrating the crystal back and forth at about 10 Hz, a Doppler-shifted beam is achieved with an energy given by:

$$E_{\text{ref}} = E_0[1 + (v/c)\cos\theta_{\text{B}}] \tag{9.11}$$

where E_0 is the nuclear resonance energy for the crystal at rest, v is the crystal speed, and θ_{B} is the Bragg angle.

Thanks to the high brightness of synchrotron radiation, the monochromatic beam can also be focused to a small spot—something impossible to do efficiently with a

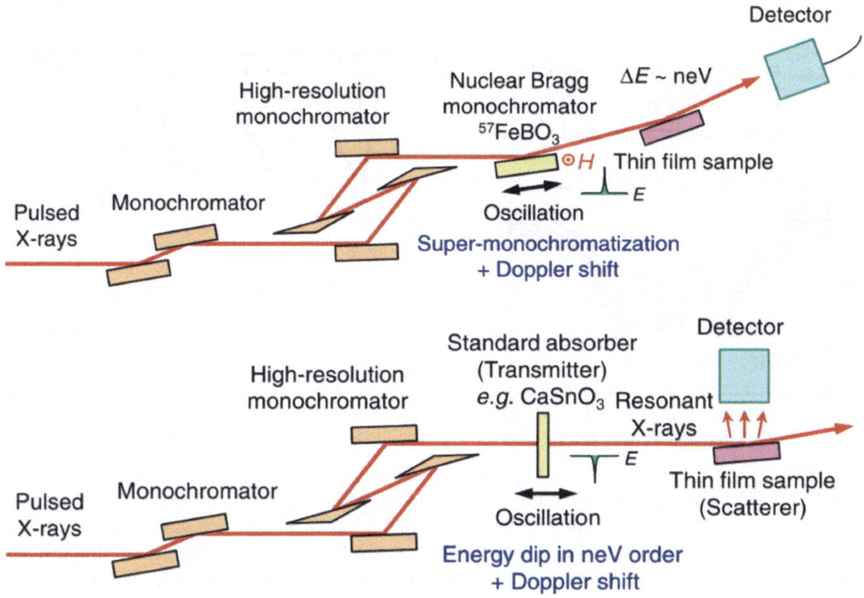

Fig. 9.9 Energy domain Mössbauer spectroscopy using synchrotron radiation. Top: the nuclear diffraction approach, using an isotopically enriched crystal, a purely nuclear reflection, and Doppler shifting [434]. Bottom: an alternative approach, using a moving single-line standard to serve as an energy analyzer

radioactive source emitting into 4π steradians. This is particularly useful for surface studies. As an example, Mibu and coworkers obtained a beautiful energy-domain spectrum for a monatomic layer of ^{57}Fe at a the Fe/Cr interface in a Fe/Cr bilayer (Fig. 9.10) [434].

The second energy-domain approach uses a scatterer containing the same Mössbauer isotope and placed behind the sample on the path of the synchrotron beam (Fig. 9.9) [435, 436]. The sample or scatterer is then Doppler-shifted using a standard Mössbauer velocity transducer, and the delayed emission from the scatterer is measured as a function of the relative velocity using a detector placed below and/or above the scatterer. A spectrum taken with this approach for $NiCr_2O_4$ is shown in Fig. 9.10.

9.5 The Time Domain Approach: Nuclear Forward Scattering

Another approach to observing the ~nano-eV splittings of nuclear levels is to *coherently excite some or all of the resonances at the same time* and then to observe the beating between their different frequencies. In this synchrotron "*n*uclear *f*orward

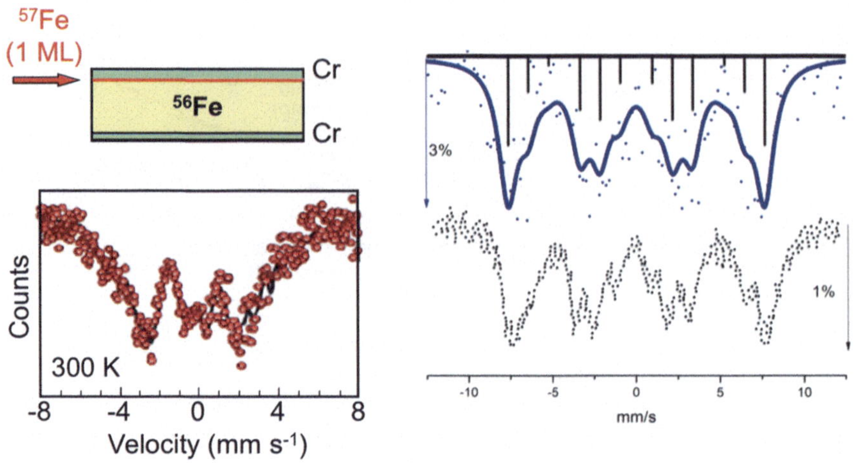

Fig. 9.10 Applications of energy domain synchrotron Mössbauer spectroscopy. Left: spectrum from a single layer of ^{57}Fe deposited at the interface of a Fe/Cr bilayer, obtained using a $^{57}FeBO_3$ nuclear Bragg monochromator [434]. Right: spectrum of $NiCr_2O_4$ using single-line nuclear analyzer (blue line) compared with conventional spectrum using radioactive sources [437]

scattering" or "NFS" experiment, the sample is excited by a very short synchrotron pulse ($<$100 ps), and the forward-scattered intensity is monitored as a function of time. The excitation is *coherent* because the sample is in the same state before and after the scattering process. Since the particular atoms involved in the scattering cannot be identified, one has to calculate the response as the sum over all possible scattering paths.

9.5.1 NFS Methodology: The Synchrotron Experiment

Everything in a synchrotron nuclear experiment is designed to separate the small nuclear signal from the overwhelming electronic background. Although peak nuclear cross sections are large, they are also very narrow. For these experiments to be successful, one requires (a) *a high spectral brightness X-ray source* (b) *a very high-resolution incident beam monochromator* and (c) *a high-speed detection system* to separate nuclear events from ordinary X-ray scattering and fluorescence. Typical arrangements are shown in Fig. 9.11.

Nowadays, the source is invariably an undulator on a third-generation synchrotron radiation facility or even an FEL. To reduce the vertical divergence, some beamlines employ X-ray compound refractive lenses (Chap. 4). In either case, a high heat load monochromator is used to select a narrow energy slice from the undulator beam—otherwise the thermal load would overload the next stage, a high-resolution

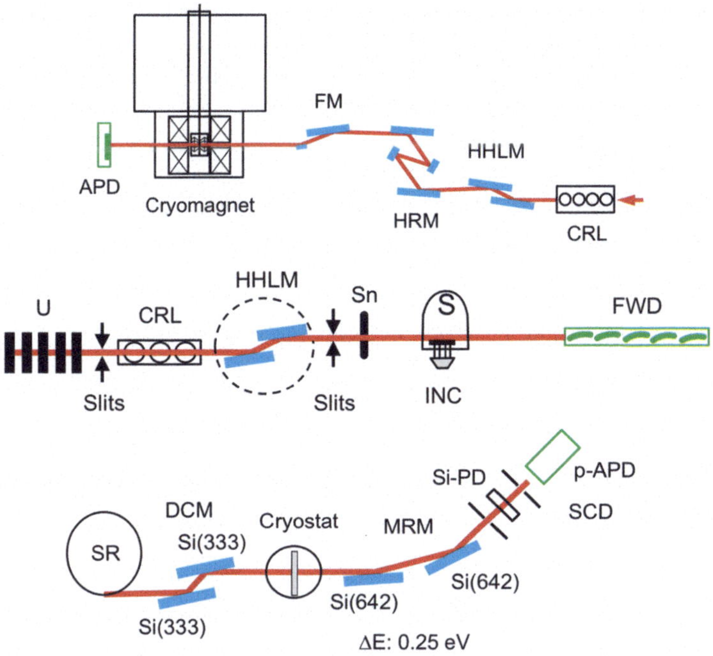

Fig. 9.11 Typical beamlines for synchrotron nuclear experiments. Top: arrangement for low-energy (~14.4 keV) nuclear forward scattering at ESRF ID18. Middle: arrangement for high-energy (~90 keV) nuclear forward scattering at ESRF ID18 [438]. Bottom: arrangement for ^{61}Ni NFS (67 keV) at Spring-8 BL09 [191]. Abbreviations: *CRL* compound refractive lens, *HHLM* high heat load monochromator, *HRM* high-resolution monochromator, *FM* focusing mirror, *APD* avalanche photodiode, *U* undulator, *FWD* detector, *DCM* double-crystal monochromator, *MRM* medium-resolution monochromator, *PD* photodiode, *p-APD* proportional mode Si-APD

asymmetrically cut crystal monochromator (Chap. 4). The output from the monochromator can then be refocused onto the sample with KB mirrors or other optics.

The sample, usually at cryogenic temperatures, may also be exposed to a high magnetic field or other conditions. Finally, the scattered radiation is recorded with a high-speed detector. Because it has excellent (<1 ns) time resolution, as well as the ability to quickly recover from saturating overloads, the detector of choice has been the avalanche photodiode [187,439] (Chap. 5). Both single elements and large arrays [440] have been used.

9.5.2 *The Refractive Index Model*

There are a variety of approaches to describing nuclear forward scattering, including the microscopic nuclear exciton model [441] and the macroscopic complex refractive index model [442]. Some of the mathematics for the latter approach are included in work by Smirnov [442] and discussed in theses by Baron [443] and Lübbers [444] and in a book by Röhlsberger [445]. In this model, the incident X-ray beam is described as an electromagnetic wave packet with "radiation amplitude" A which consists of a band of spectral components broader than the nuclear linewidth [442]. At first we will describe the sample as having a single nuclear resonance, described by a complex index of refraction $\tilde{n}$.

Then, as a function of distance z through the sample:

$$A(z) = A_0 e^{ik_0 \tilde{n} z} \tag{9.12}$$

For a nuclear transition, the index of refraction $\tilde{n}$ is given by:

$$\tilde{n}(\omega) \cong 1 + \frac{2\pi\rho}{k^2} f_{\rm n} \tag{9.13}$$

where for simplicity we omit the electronic contribution. In turn, the scattering amplitude $f_{\rm n}$ can be written as:

$$f_n = f_n{}' + if_n{}'' = \frac{f_0 x}{x^2 + 1} + i\frac{f_0}{x^2 + 1} \tag{9.14}$$

where $x = 2(E - E_0)/\Gamma_0$ is the deviation from exact resonance in units of the natural linewidth Γ_0 and f_0 is proportional to the oscillator strength of the transition:

$$f_0 = \frac{f_{\rm LM}}{2k_0}\frac{2I_{\rm e} + 1}{2I_{\rm g} + 1}\frac{1}{1 + \alpha} \tag{9.15}$$

Here, $f_{\rm LM}$ is the Lamb-Mössbauer factor, $I_{\rm g}$ and $I_{\rm e}$ are respectively the nuclear spins for the ground state and excited state, and α is the internal conversion coefficient.

Exactly on resonance, the scattering amplitude is purely imaginary, and the scattered wave interferes destructively with the incident wave. This gives rise to the normal absorption term. Off-resonance, there is a real component to the scattering amplitude. What is particularly important is that as one goes off-resonance, this real component f' diminishes only in proportion to ΔE, while the absorptive term f'' falls off as ΔE^2 (Fig. 9.12 and Eq. 9.14).

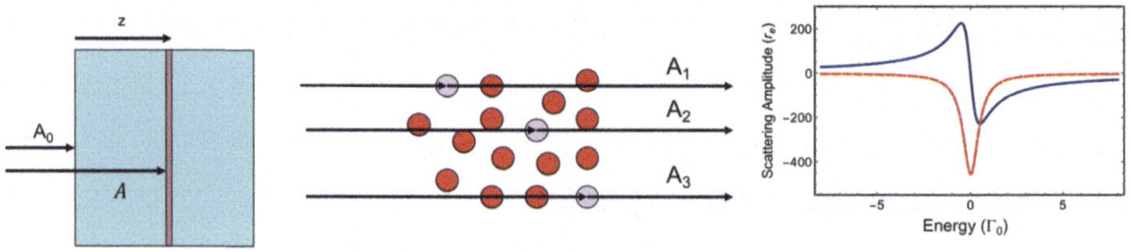

Fig. 9.12 Left: incident and transmitted waves in a forward scattering sample. Middle: forward scattering of the incident radiation by individual scatterers at different positions in the sample. Right: the real (blue) and imaginary (red) nuclear parts of the complex refractive index for a single ^{57}Fe

In a typical synchrotron experiment, the spectral width of the incident beam is much larger than the width of the resonance, so a number of propagating waves with different group velocities can be excited. The interference between these waves leads to a complex integro-differential equation that was solved by Shvyd'ko [446], and the derivation is further described elsewhere [420]. The net effect is a complicated, time-dependent interference pattern between different frequencies and different spatial regions of the sample, yielding a beat pattern that can be described as a Bessel function (Appendix D):

$$I(\tau) \propto \mathrm{e}^{-\tau} \frac{\chi}{\tau} J_1^2\left(\sqrt{\chi\tau}\right) \tag{9.16}$$

The previous equation has been simplified by using a key parameter, χ—the "effective thickness." In the nomenclature of Lübbers, this parameter is defined as [444]:

$$\chi = \frac{1}{4}\rho\sigma_0 f_{\mathrm{LM}} d \tag{9.17}$$

where f_{LM} is the Lamb-Mössbauer factor of the nuclear resonance and the other terms have been defined above. (In the Mössbauer literature, $\chi = \rho\sigma_0 f_{\mathrm{LM}} d$). The overall effects on the time-dependent scattering lead to some non-intuitive effects, such as "lifetime shortening" and "dynamical beats."

9.5.2.1 Lifetime Shortening and Coherent Enhancement

For thin samples or short times after excitation, Eq. 9.16 can be further simplified to:

$$I(\tau) \propto \mathrm{e}^{-(1+\chi)\tau} \tag{9.18}$$

Thus, for thin samples, there is a speedup of the exponential decay by a factor of (1/1 + χ) in the apparent lifetime (Fig. 9.13). The coherence of the scattering also leads to an enhancement of the signal strength in the forward direction.

9.5.2.2 Dynamical Beats

For describing thicker samples and/or longer times, one needs to use the entire Bessel function in Eq. 9.16. One observes an aperiodic beat pattern with nodes determined by the zeroes of the Bessel function (Fig. 9.13).

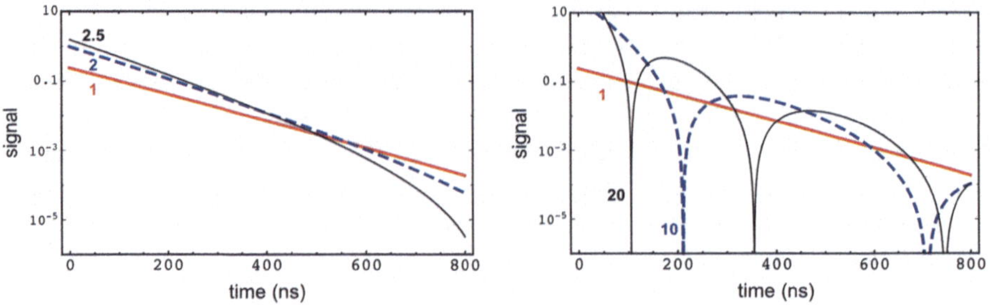

Fig. 9.13 Theoretical curves for the nuclear forward scattering from a collection of single resonance ^{57}Fe nuclei for different values of the effective thickness. Left: small values of $\chi = 1, 2, 2.5$, showing "speedup" as increase in initial slope of decay. Right: χ and time-dependent NFS for larger values of χ, showing evolution of quantum beats for $\chi = 10$ and $\chi = 20$ compared to normal decay

9.5.3 *Quantum Beats 1: Quadrupole Splittings and Isomer Shifts*

For most experiments, it is the chemical properties of the Mössbauer nucleus that are most of interest. For the remainder of the discussion, we assume a thin sample with moderate dynamical beats. If the sample under consideration contains two resonance lines, say from quadrupole splitting, then the interaction between those two frequencies is observed as "quantum beats." If the separation $\Delta\omega$ is large compared to their effective linewidth, the resulting time spectrum can be approximated as the product of the dynamical beat pattern and a quantum beat pattern reflecting the interference between the two frequencies:

$$I(\tau) \propto \mathrm{e}^{-\tau}\frac{\chi}{\tau} J_1^2\left(\sqrt{\frac{\chi\tau}{\tau}}\right) \cos^2\left(\frac{\Delta\omega}{2}t + \frac{\chi\Gamma_0}{8\Delta E}\right) \tag{9.19}$$

A nice example of such quantum beats involves the NFS from ferrocene (Fig. 9.14). The observed beat period of 35.8 ns yields a quadrupole splitting of 2.39 mm s^{-1}.

Another chemically useful quantity that can be obtained from NFS is the isomer shift. The absolute measurement of the γ-ray frequency with sufficient precision to define shifts is obviously impossible, but the relative frequency and energy can be obtained by observing beats between signals from the sample and a single-line reference material [447]. As illustrated in Fig. 9.14, if one places a single-line reference material in the beam before or after the sample, then there will be interference between the forward scattering from this reference and from the sample. The difference in isomer shifts can then be directly extracted from the beat frequency in the forward scattering signal. For our ferrocene example, introduction of a stainless steel reference introduced an extra beat in the NFS that was used to calculate an isomer shift of 0.44 mm s^{-1} (Fig. 9.14).

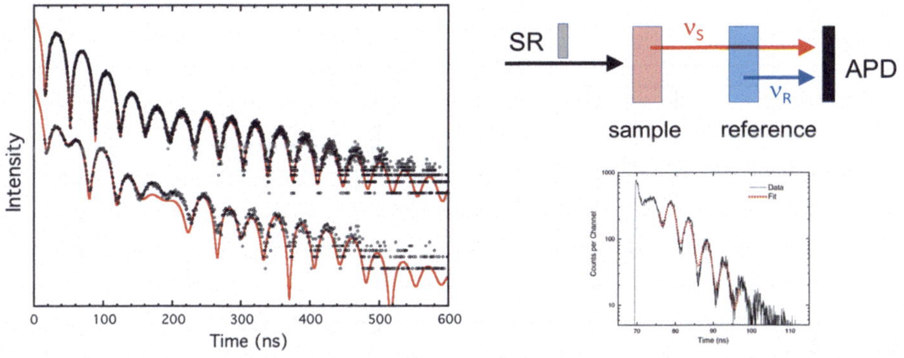

Fig. 9.14 Left: (top) NFS data (∘∘∘∘) for ferrocene and simulation (red line) with a quadrupole splitting of 2.39 mm s^{-1}, (bottom) ferrocene NFS data with a stainless steel reference inserted (∘∘∘∘) and simulation (red line). Both ferrocene simulations also modeled sample thickness distribution. Top right: one geometry used for isomer shift measurements. Lower right: NFS data allowing deduction of isomer shift between MgS:Eu and Eu_2O_3 [447,448]

In fact, the sample and standard need not be physically separated. By making a mix of MgS/Eu and Eu_2O_3, Konjhodzic and coworkers observed a beat pattern corresponding to an energy difference of ~12.8 mm s^{-1} between Eu^{2+} and Eu^{3+} at the 21.54 keV ^{151}Eu nuclear resonance (Fig. 9.14) [447,448].

9.5.4 Quantum Beats 2: Magnetic Splittings and Polarization Effects

For a sample without electric quadrupole splittings, we saw that a magnetic field splits the levels of a nucleus with spin I into $2I + 1$ Zeeman states. In the specific case of ^{57}Fe, this leads to two possible initial states, four possible final states, and a familiar six-line spectrum (Fig. 9.6). The interaction of six frequencies in the NFS yields a very complicated spectrum, but considerable simplification can be achieved using the source polarization properties and selection rules (Fig. 9.15).

For example, by aligning the magnetic field perpendicular to the beam direction, only the $\Delta M = 0$ transitions are allowed, and the observed NFS shows the beating between these two different frequencies (Fig. 9.15). In contrast, with the magnetic field along the beam direction, the four different $\Delta M \pm 1$ transitions are allowed, and a more complicated beat pattern appears (Fig. 9.15). As expected, the beat pattern becomes quite complicated if there are a large number of allowed transitions. For example, a magnetized Fe foil without any polarization selectivity yields the complex pattern shown in Fig. 9.15.

9.5.4.1 Applications

The properties of metals and minerals under extremes of pressure and/or temperature are critical for understanding the interior structure of the earth and other planets

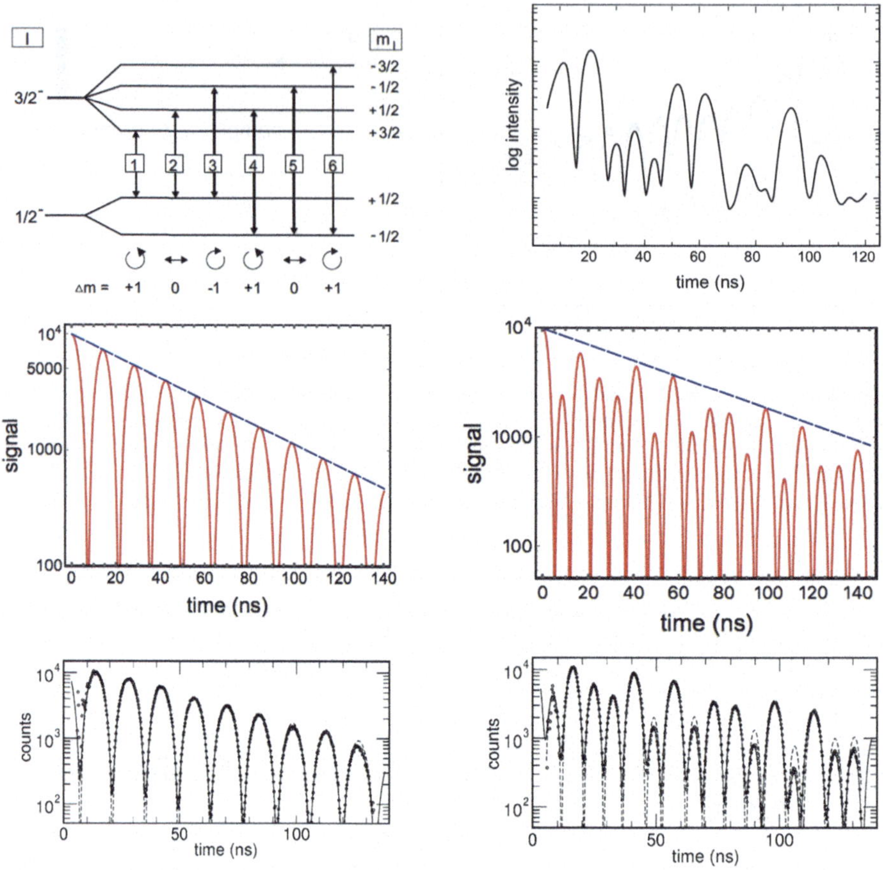

Fig. 9.15 Top left: the transitions possible for a magnetized ^{57}Fe foil. Top right: experimental spectrum without polarization selectivity, corresponding to beating of all six transitions [442]. Middle left: calculated beat pattern associated with two Mössbauer transitions of equal intensity. Values are representative for $\Delta m = 0$ transitions in Fe, with splitting of 6 mm s^{-1} and a frequency difference of $2\pi/28$ ns, superimposed on the 144 ns $1/e$ lifetime of the ^{57}Fe excited state. Middle right: calculated beat pattern from two different beat frequencies of $\omega_1 = 2\pi/16.4$ ns and $\omega_2 = 2\pi/$ 103 ns with a respective 3:1 intensity ratio. Lower left: experimental NFS for magnetic field applied perpendicular to beam direction and polarization, corresponding to the 2 $\Delta m = 0$ transitions. Lower right: experimental NFS for magnetic field applied parallel to beam. The beat pattern arises from the 4 $\Delta m = \pm 1$ transitions [449]

(Fig. 10.13). Such extremes of pressure can only be achieved over very small volumes, so the ability to focus SR to micron scale spots is critical for such experiments. As one example, NFS was used to demonstrate a high-spin to low-spin magnetic transition in $(Mg_{.16}Fe_{.84})O$ between 85 and 121 GPa (Fig. 9.16).

NFS also shows great potential for investigating properties of high-energy Mössbauer transitions, some of which do not have convenient radioactive sources. A recent demonstration was for ^{99}Ru at 89.6 keV (Fig. 9.16), and other resonances under 100 keV seem likely candidates, including ^{156}Gd at 88.97 keV through ^{183}W at 99.1 keV.

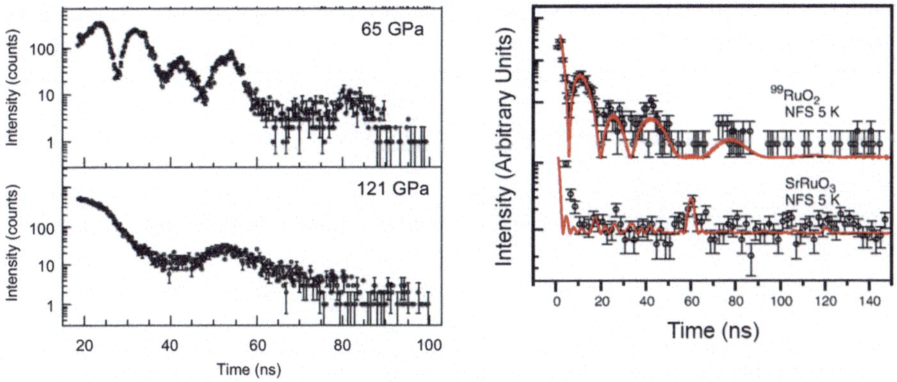

Fig. 9.16 Left: NFS for $(Mg_{.16}Fe_{.84})O$ at different pressure [450]. The loss of rapid oscillations in the 121 GPa spectrum indicates a transition from magnetic high-spin Fe to non-magnetic low-spin Fe as the pressure is increased. Right: NFS patterns for the ^{99}Ru resonance at 89.6 keV, adapted from [438]. The RuO_2 data were fitted with a quadrupole splitting of 0.44 mm s^{-1} as well as dynamical beats. The $SrRuO_3$ data were fit with a hyperfine field of 33.9 T, redrawn from [438]

9.6 Perturbed Angular Correlation

Angular correlation (AC) experiments depend on the production and decay of oriented nuclei. In contrast with NMR, in which a strong magnetic field creates a difference in spin populations, in the angular correlation measurement the oriented nuclei are selected "after the fact." In a conventional *t*ime-*d*ependent *p*erturbed *a*ngular *c*orrelation ("TDPAC") experiment, the "fact" is emission of a γ-ray, while in the synchrotron experiment, *s*ynchrotron *r*adiation *p*erturbed *a*ngular *c*orrelation or "SRPAC", the "fact" is an absorption event.

9.6.1 γ-γ Angular Correlations

In both TDPAC and SRPAC, the experiments work because the emission and absorption of γ-rays are usually anisotropic with respect to the nuclear spin axis. The general case for a nucleus decaying with successive emission of two γ-rays is developed in a classic article by Frauenfelder and Steffen [451]. The equations take an entire page, but for randomly oriented samples, the complicated angular momentum algebra can be simplified to the following equation for the time and angle correlation between two emitted photons:

$$W(\theta, t) \propto e^{-t/\tau_0} \sum_k A_k(\gamma_1) A_k(\gamma_2) G_{kk}(t) P_k(\cos\theta) \tag{9.20}$$

where τ_0 is the intermediate state lifetime, θ and t are the angle and time between two successive γ-rays, $A_k(\gamma_1)$ and $A_k(\gamma_2)$ are, respectively, "orientation coefficients" and "correlation coefficients," $P_k(\cos\theta)$ is a Legendre polynomial, and $G_{kk}(t)$ are "perturbation functions":

$$G_{kk}(t) = \frac{1}{2k+1}\sum_{N=-k}^{N=k} G_{kk}^{NN}(t) \tag{9.21}$$

where the $G_{kk}(t)$ contains terms in cos ωt, with $\omega = \Delta E/\eta$. In this case:

$$W(\theta,t) = \mathrm{e}^{-t/\tau_0}\left[1 + A_2(\gamma_1)G_{22}(t)\left(\frac{3}{2}\cos^2\theta - \frac{1}{2}\right)\right] \tag{9.22}$$

The perturbation function depends on the spin of the intermediate nucleus, and for the most common cases, $I = 3/2$ and $I = 5/2$, in the absence of other effects:

$$I = \frac{3}{2}:\ G_{22}(t) = \frac{1}{5} + \frac{4}{5}\cos\omega_0 t \quad \omega_0 = 6\omega_Q \tag{9.23}$$

$$I = \frac{5}{2}:\ G_{22}(t) = \frac{7}{35} + \frac{13}{35}\cos\omega_0 t + \frac{10}{35}\cos 2\omega_0 t$$

$$+\frac{5}{35}\cos 3\omega_0 t \ \ \omega_0 = 6\omega_Q \tag{9.24}$$

where $\omega_Q = -eQ\,V_{zz}/(4I(2I-1)\hbar)$

Equations for other intermediate spins as well as a vastly deeper treatment can be found in an article by Butz [452].

9.6.2 TDPAC: The Conventional Experiment

TDPAC is a marvelous technique, when it works. Unfortunately, a large number of conditions have to be met simultaneously. These include (a) an intermediate nuclear state with lifetime between ~5 ns and 1 μs, (b) emission of strong γ-rays upon populating and depopulating the intermediate state, (c) a significant anisotropy in the gamma-ray angular correlation, and (d) a "mother isotope" with a usable lifetime. Fortunately, there are some isotopes that make it through this sieve, and the most popular ones are summarized in Table 9.1. Typical applications are reviewed elsewhere [453].

9.6.3 The Synchrotron Experiment

There is another way to obtain the same information as in a TDPAC experiment—using synchrotron radiation instead of radioisotopes. In a *synchrotron radiation perturbed angular correlation* (SRPAC) measurement, the sample is excited by a

Table 9.1. Popular TDPAC isotopes [453]

Isotope	Half-life	Decay	Isomer	Half-life (ns)	$E(\gamma_1)$ (keV)	$E(\gamma_2)$ (keV)
^{62}Zn	9.2 h	EC /β⁺	^{62}Cu	4.57	597	41
^{99}Mo	66 h	β⁻	^{99}Tc	3.61	739	181
^{111m}Cd	49 months	IT	^{111}Cd	85.0	151	245
^{111}In	2.8 days	EC	^{111}Cd	85.0	171	245
^{111}Ag	7.5 days	β⁻	^{111}Cd	85.0	97	245
^{133}Ba	10.5 years	EC	^{133}Cs	6.27	356	81
^{160}Tb	72 days	β⁻	^{160}Dy	2.02	197	87
^{181}Hf	42 days	β⁻	^{181}Ta	10.8	133	482
^{199m}Hg	43 months	IT	^{199}Hg	2.45	374	158
^{204m}Pb	67 months	IT	^{204}Pb	265	912	374

EC electron capture, β^+ positron emission β decay, β^- electron emission β decay, *IT* isomeric transition

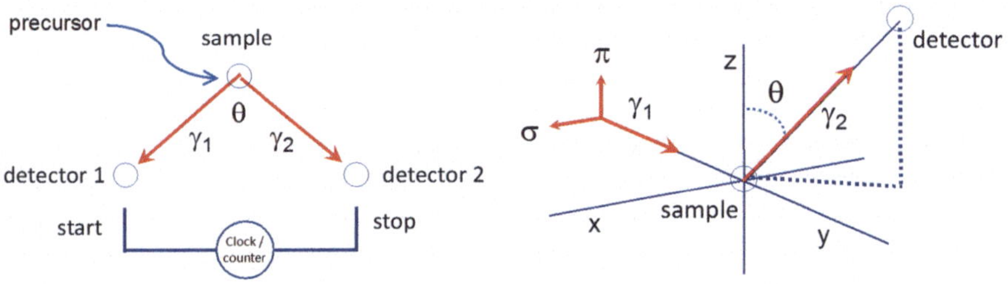

Fig. 9.17 Left: geometry of a TDPAC experiment. Right: geometry of the SRPAC experiment

pulsed synchrotron source (Figs. 9.17 and 9.18), which starts the "clock" for the experiment. The incoming beam direction defines the quantization axis, and the decay back to the ground state is typically monitored by one or more detectors perpendicular to this axis. The basic physics is thus the same for both radioisotope and synchrotron experiments—a time-dependent angular correlation results from the nucleus precessing in the local chemical and magnetic environment. Alternatively, one can view the results as quantum beats between two or more nuclear states excited by the synchrotron beam.

As derived by Sergueev in the case of M_1 magnetic dipole transitions, SRPAC for a powder sample with only a quadrupole splitting can be described by [454]:

$$I(t,\theta) = I_0 e^{-t/\tau_0}\left[1 - 2 \cdot A_{22}^{\exp} \cdot P_2(\cos\theta) \cdot G_{22}(t)\right] \tag{9.25}$$

with "anisotropy coefficient A_{22}," "perturbation factor $G_{22}(t)$," excited-state lifetime τ_0, and θ as the angle between the direction of the magnetic polarization $\vec{h}_1$ of the incoming photon and the direction of scattering for outgoing photon $\vec{k}_2$. Theoretical anisotropy coefficients are given in Table 9.2. The experimental anisotropy coefficient will be reduced from the theoretical value because of the finite solid angle covered by the scattered photon detector.

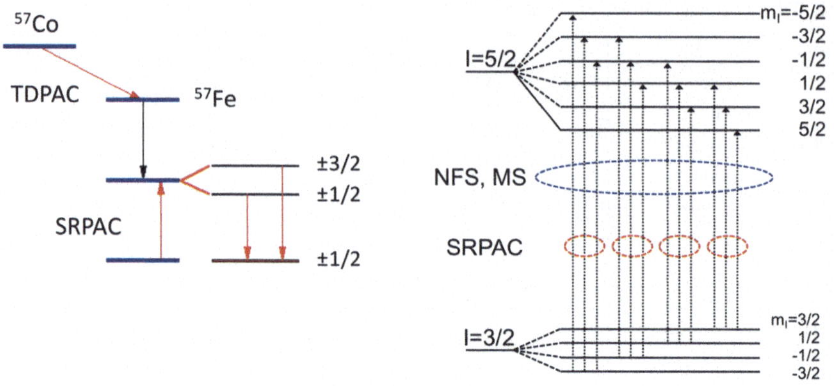

Fig. 9.18 Left: contrast between radioisotope TDPAC and synchrotron SRPAC experiments, illustrated for quadrupole split ^{57}Fe. Right: the simplification of SRPAC compared to NFS, illustrated for magnetically split ^{61}Ni. In SRPAC, the quantum beats come from individual atoms and the interferences are only between transitions with a common initial state. NFS involves interference between multiple atoms and hence all possible transitions

Table 9.2 Theoretical anisotropy coefficients [454]

Isotope	I_g	I_e	A_{22}
^{57}Fe, ^{119}Sn	1/2	3/2	¼ = 0.25
^{61}Ni	3/2	5/2	7/50 = 0.14
^{121}Sb, ^{151}Eu	5/2	7/2	3/28 ≅ 0.11
^{149}Sm	7/2	5/2	1/56 ≅ 0.018

For example, in the case of ^{119}Sn, there is a useful M_1 ½ → 3/2 → ½ transition where the excited state has a lifetime of 26 ns (Fig. 9.19) [455], for which the perturbation factor is:

$$G_{22}(t) = \frac{1}{5} + \frac{4}{5} \cdot \cos(\omega t). \tag{9.26}$$

and the angular frequency ω corresponds to $\omega = \Delta E/\eta = eQ_e V_{zz}/2$.

A slightly more complicated pattern occurs for ^{61}Ni SRPAC, where the excitations are M_1 3/2 → 5/2 → 3/2 transitions. For a Zeeman-split excited state with magnetic field B and nuclear g-factor g, a random magnetic orientation and Larmor frequency $\omega_B = -g\mu_N B/\hbar$, where μ_N is the nuclear magneton, the perturbation factor is given by:

$$G_{22}(t) = \frac{1}{5}(1 + 2\cos\omega_B t + 2\cos 2\omega_B t) \tag{9.27}$$

Although there are now two beat frequencies, there is still a tremendous simplification over the NFS experiment which involves beating between 12 different frequencies. An example is shown later in Fig. 9.21.

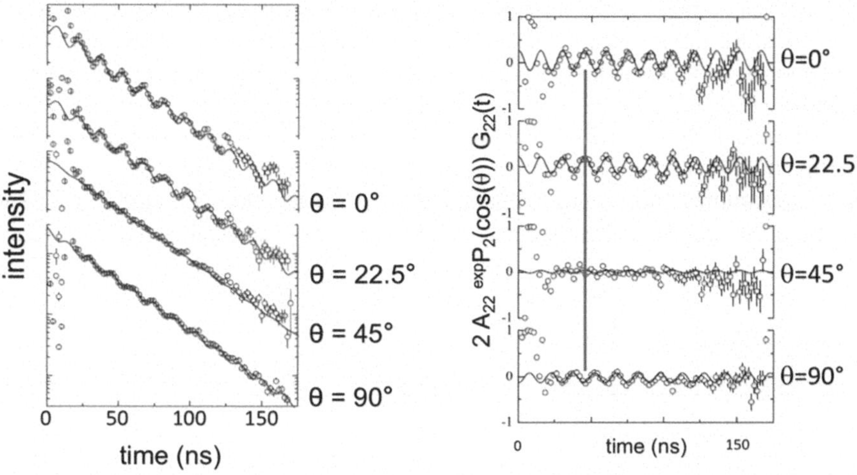

Fig. 9.19 Left: SRPAC intensity from tributyltin-fluoride polymer for different detector angles in plane perpendicular to the incident beam. The quantum beat period of 15.14 ns corresponds to quadrupole splitting of 3.43 mm/s [455]. Right: perturbation functions extracted from data on left. Note 90° data is opposite in phase to 0° and 22.5° data, and almost no beating is observed at 45°

9.6.4 Applications to Dynamics

SRPAC has special merits as a technique when the Lamb-Mössbauer factor becomes very small or where simplification of hyperfine structure has merits. This is often the case when a sample is fluctuating over time and one wants to understand the dynamics. Suppose the nucleus under study is embedded in an environment that rotates randomly in a sort of Brownian motion with jump or relaxation rate λ. Then the normal precession of the nucleus around the electric field gradient (EFG) at frequency Ω will be modified by the stochastic rotation of the EFG itself. In an analysis of this effect, Sergueev distinguished two extreme cases that depend on the relative rates λ and Ω [454].

In the slow relaxation case, $\lambda \ll \Omega$, the strength of the SRPAC perturbation factor G_{22} is exponentially damped by the random motions:

$$G_{22}(t) \cong \mathrm{e}^{-\lambda t} \cdot G_{22}^{(0)}(t) \tag{9.28}$$

In the fast relaxation case, $\lambda \gg \Omega$, the SRPAC oscillations disappear, and the perturbation factor becomes a simple exponential:

$$G_{22}(t) \cong \exp\left(-4\Omega^2 t/5\lambda\right) \tag{9.29}$$

This sensitivity to angular motion allows SRPAC to characterize rotational dynamics with timescales of picoseconds to microseconds. As one example, ^{57}Fe ferrocene was doped into the molecular glass-former dibutyl phthalate (DBP). Over the temperature range from <180 to 328 K, the time evolution of the anisotropy varied enormously, allowing the relaxation rate to be determined over ~4 orders of magnitude (Fig. 9.20).

In a similar fashion, SRPAC can be used to study the average magnitude of the hyperfine field as a function of temperature. In one example, the SRPAC for metallic ^{61}Ni was examined from room temperature to the Curie temperature of 630 K. The intensity of the signal gradually decayed and eventually vanished at T_C (Fig. 9.21).

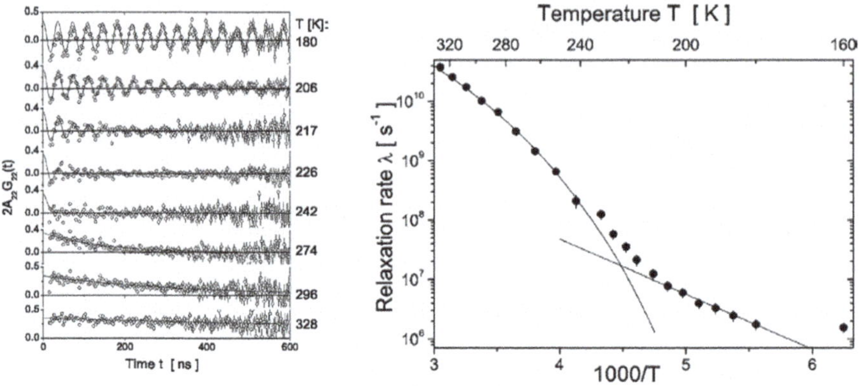

Fig. 9.20 Left: temperature-dependent ^{57}Fe SRPAC for ferrocene in DBP [454]. Right: derived relaxation rates from simulation of same SRPAC data

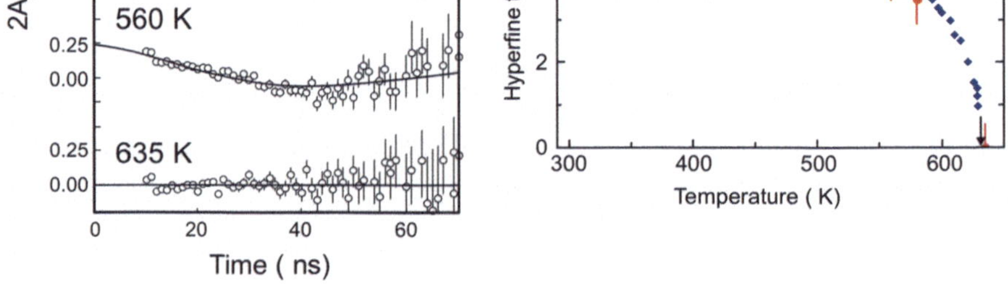

Fig. 9.21 Left: temperature-dependent ^{61}Ni SRPAC data (δ) and simulation (solid line) for Ni metal at key temperatures. Right: derived hyperfine field from broader set of same SRPAC data (red dot) compared with results from NMR measurements (blue diamond), both redrawn from [456]

9.7 Some Nuclear History

Nuclear spectroscopy began, of course, with the discovery of the nucleus, described by Rutherford in his 1911 paper on the structure of the atom [457]. In an experiment outlined in many freshman chemistry books, it was observed that α-particles impinging on a thin gold foil were occasionally scattered back at large angles. From this it could be deduced that the atom was not a "plum-pudding" of electrons immersed in positive charge, but, instead, the positive charge was concentrated in an extremely dense "nucleus." In fact, the density was such that a common analogy portrayed the nucleus in the atom like a "fly in a cathedral" [458]. What is not commonly appreciated is that the person who actually did the experiments was a 20-year-old undergraduate student, Ernest Marsden, who observed the backscattering events with his own eyes, one at a time in a dark room, as flashes from a ZnSe screen [458,459].

Over the next 40 years, the properties of the nucleus as a composite of protons and neutrons were gradually defined, and a number of experiments observed resonant scattering of γ-rays from nuclear excited states, analogous to the scattering of visible light by electronic excited states [461]. However, because of the much higher energy of γ-rays, the recoil energy for atoms in the gas phase or solution is significant, and it was necessary to use high-velocity radioactive sources to achieve the appropriate Doppler shifts. Success by this method came in 1950, when Moon used a centrifuge rotor plated with a 411 keV ^{198}Au source spinning at ~800 m s^{-1} to observe nuclear resonance scattering in liquid ^{198}Hg [462]. Since most scientists hesitate to have radioactive sources flying in their labs faster than the speed of sound, something had to change to increase the popularity of this experiment.

Mossbauer Effect. In 1957 Rudolf Mössbauer, a graduate student under Professor Maier-Leibnitz at the Max-Planck Institute in Heidelberg, set out to investigate the nuclear resonance scattering from ^{191}Ir. He began at room temperature, and since the recoil energy of a free ^{191}Ir nucleus is $E_R = 0.047$ eV, while Doppler broadening at room temperature is ~0.1 eV, he did observe some resonant scattering. Mössbauer then cooled *both* his source and absorber, to reduce the Doppler broadening, and he expected a sharp decrease in signal. Instead, he observed a rapid increase in effective cross section (Fig. 9.22). He had discovered the *recoil-free* emission and absorption of γ-rays from nuclear transitions [463–465], for which he was awarded the Nobel Prize in Physics 3 years later at the age of 32 [460]. Within a year of Mössbauer's first publication, efforts were under way at Harvard, Harwell, Los Alamos, and Argonne National Lab

(continued)

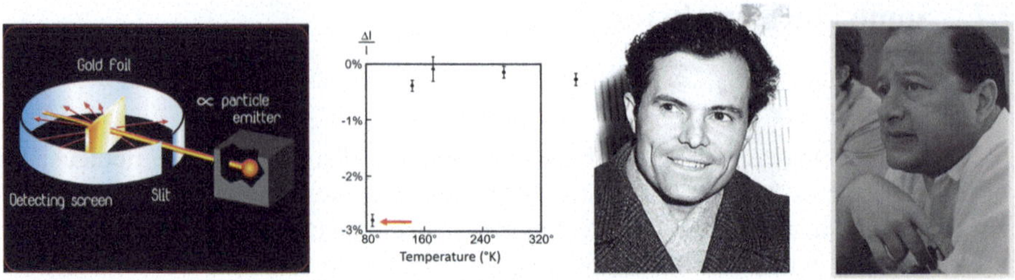

Fig. 9.22 Left to right: Geiger/Marsden experiment done in Rutherford's lab; temperature dependence for ^{191}Ir nuclear resonance observed by Mössbauer, highlighting the increase in absorption at low temperature, redrawn from [460], Mössbauer; Stan Ruby

to repeat and expand on this discovery. With the discovery of a strong Mössbauer effect in ^{57}Fe [466–467], the field took off.

Perturbed Angular Correlation. The history of perturbed angular correlation [453] goes back to 1940, when Hamilton published the theory of directional correlation between successive γ-ray emissions in a radiative cascade [468]. Experimental confirmation came after the war, when Brady and Deutsch reported a pronounced anisotropy in the successive gamma-ray emission from both ^{60}Co and ^{46}Sc [469]. In 1946, Goertzel analyzed the effect of extranuclear fields on the nuclear emissions, thus the "perturbed angular correlation." The time dependence of the perturbed angular correlation (TDPAC) was first observed in 1955 [470].

Synchrotron Nuclear Experiments. In 1974, Stan Ruby at SSRL made the rather outrageous suggestion that synchrotron radiation could be used for a resonant excitation of nuclei [471,472]. Although a number of attempts were made at SSRL and elsewhere to observe the predicted effects, as described in Gerdau's historical overview, these initial attempts were "doomed to failure" [473]. Improvements in source brightness, monochromator resolution, detector speed, and samples were needed. A decade later, Gerdau and coworkers accomplished the first successful nuclear Bragg scattering (NBS) experiments with synchrotron radiation, using an isotopically enriched ^{57}Fe single crystal and the DORIS storage ring source at HASYLAB [474]. The first nuclear forward scattering (NFS) measurements were reported in 1991 by Hastings and coworkers [475], followed rapidly by Seto's nuclear inelastic scattering measurements [476] and the first synchrotron radiation perturbed angular correlation experiments by Baron and coworkers [477]. By now, nuclear experiments with synchrotron radiation are routinely done at many facilities around the world.

9.8 Suggested Exercises

1. ^{61}Ni has a nuclear transition at 67 keV. Calculate the recoil energy.
2. ^{125}Te has an excited state at 35.48 keV with a 1/*e* lifetime of 2.14 ns. Calculate the linewidth (FWHM) in (a) neV and (b) mm s^{-1}.
3. A typical value for the Debye temperature of Fe metal is ~400 K. Use this to calculate the predicted Lamb-Mössbauer factor at absolute zero.
4. For an NFS experiment, calculate the position of the first dynamical beat node for a modestly thick Fe sample with an effective thickness $\chi = 10$.
5. For an NFS experiment, calculate the quantum beat frequency for a very thin Fe sample with a quadrupolar splitting of 4 mm s^{-1}.
6. For ^{57}Fe SRPAC, calculate the theoretical amplitude of the quantum beats at $\theta = 10°$ and $\theta = 100°$. Estimate the diminished amplitudes at both angles if the collected solid angle is $\pi/2$ steradians.

9.9 Reference Books and Review Articles

1. *Mössbauer Spectroscopy and Transition Metal Chemistry: Fundamentals and Applications*, Philipp Gütlich, Eckhard Bill, Alfred X. Trautwein, Springer, New York 2011, ISBN 3540884270.
2. *Nuclear Condensed Matter Physics with Synchrotron Radiation—Basic Principles, Methodology & Applications*, Ralf Röhlsberger, Springer-Verlag, Berlin, 2004, ISBN 3-540-23244-3.
3. *Mössbauer Spectroscopy—Applications in Chemistry, Biology, and Nanotechnology*, Virendar K. Sharma, Göstar Klingelhöfer, and Tetsuaki Nishida, eds., Wiley, Hoboken, 2013, ISBN 978-1-118-05724-7.
4. *Principles of Mössbauer Spectroscopy*, T. C. Gibb, Halsted Press, 1976, ISBN 978-0470297438. An old book but the theory is presented simply and does not go out of date.
5. *The Mössbauer Effect—A Review with a Collection of Reprints*, Hans Frauenfelder, W. A. Benjamin, New York, 1963. A very old book, but a nice collection of papers documenting the early history of this technique.

9.10 Nuclear Software

CONUSS: W. Sturhahn. CONUSS and PHOENIX: Evaluation of nuclear resonant scattering data. *Hyp. Int.*, 125(1):149–172, 2000. ISSN 1572-9540. doi: 10.1023/A:1012681503686. URL https://doi.org/10.1023/A:1012681503686

MOTIF: Y. V. Shvyd'ko. MOTIF: Evaluation of time spectra for nuclear forward scattering.
Hyp. Int., 125(1):173–188, 2000. ISSN 1572-9540. doi: https://doi.org/10.1023/A:1012633620524.
URL https://doi.org/10.1023/A:1012633620524

Chapter 10
Nuclear Resonaynce Vibrational Spectroscopy

We learned in Chap. 9 that Mössbauer spectroscopy and nuclear forward scattering rely on the zero-phonon, recoil-free fraction, as determined by the Lamb-Mössbauer factor—f_{LM}. In the Mössbauer experiment, the intensity is reduced by this factor, so that the observed cross section is reduced from the total cross section according to $\sigma(0) \sim (\pi/2)\, \sigma_N\, f_{LM}$. However, there is a "sum rule" that states that the integrated cross section for a nuclear transition remains the same [478]. So, where does this missing intensity go?

The missing intensity is in the "recoil fraction," $1 - f_{LM}$, in which nuclear transitions couple to vibrations, resulting in the creation or destruction of phonons (Fig. 10.1). It turns out that this fraction contains useful and detailed information about the vibrational properties of a sample. Before the advent of synchrotron radiation sources, most experiments probed the recoil fraction indirectly, by observing the loss of Mössbauer intensity. (There were also a few heroic experiments with radioactive sources spinning around in ultracentrifuges to achieve the requisite Doppler shifts.) With high brightness third-generation sources, we can now probe the recoil fraction directly by the technique known as "*N*uclear *R*esonance *V*ibrational *S*pectroscopy" ("NRVS") or sometimes called "*N*uclear *R*esonant *I*nelastic *X*-Ray *S*cattering" ("NRIXS").

Why is NRVS of interest to chemists, biologists, geologists, and materials scientists? Among other things, it is an exciting tool because:

- NRVS yields a vibrational spectrum for a specific isotope of a specific element.
- The resulting vibrational spectrum is easily calculated and interpreted.
- The isotopic sensitivity allows labeling experiments, say of surfaces or of specific sites in metalloenzymes.
- NRVS provides quantities such as the speed of sound and average kinetic energy.

Before going further, we present a pair of examples to show what all the excitement is about (Fig. 10.2). At one extreme is the spectrum of Fe metal, in which the vibrations are described as propagating phonons with a continuous density

S. P. Cramer, *X-Ray Spectroscopy with Synchrotron Radiation*, Biological and Medical Physics, Biomedical Engineering, https://doi.org/10.1007/978-3-030-28551-7_10

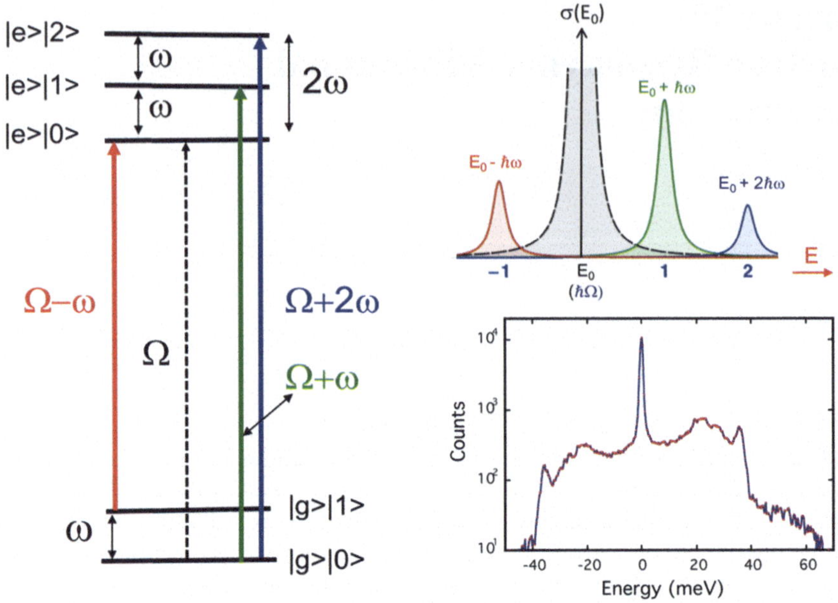

Fig. 10.1 Left: schematic of zero-phonon and recoil events for a system with a single fundamental vibrational frequency. The nuclear ground state is $|g\rangle$, and the excited state is $|e\rangle$, while the number of phonons is $|n\rangle$. The recoil-free Mössbauer line is thus the $|g\rangle|0\rangle \rightarrow |e\rangle|0\rangle$ transition. Upper right: comparison of energies and intensities for the same system. The phonon creation or "Stokes" region is to higher energy, with single-phonon and multiple phonon events. The phonon annihilation or "anti-Stokes" region is at lower energy and has its intensity reduced by a Boltzmann factor. Lower right: raw nuclear spectrum for Fe metal at 14.4 keV—note log scale for intensity [479]

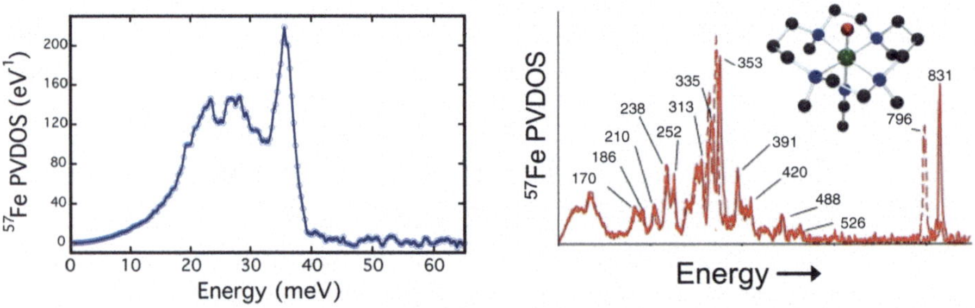

Fig. 10.2 NRVS examples. Left: the "partial vibrational density of states" or "PVDOS" for Fe metal, derived from the previous raw spectrum [479]. Right: spectra for a complex with a terminal Fe=O group, using ^{16}O (solid line) or ^{18}O (dashed line) [480]

of states. Analysis of such spectra can provide important geophysical properties such as the velocity of sound, specific heat, and average force constant. At the other extreme is the spectrum of an Fe coordination complex with a terminal Fe=O ligand. In this case the analysis is much more straightforward, and the frequency of the

relatively isolated Fe=O stretching mode in the 800 cm^{-1} region reveals information about the Fe=O force constant and thus the type of bonding.
Despite these differences in modes of interpretation, the experiment is always done in about the same manner.

10.1 The NRVS Experiment

Although the recoil fraction might have an integrated cross section comparable to the recoil-free fraction, the intensity is spread over tens or hundreds of meV, compared to tens of neV for the Mössbauer effect. Thus, the absorption cross section at any particular energy is effectively a million times weaker. So, how is it ever possible to observe the NRVS effect? The key to the synchrotron experiment is to *exploit the time delay* for the subsequent emission of nuclear fluorescence and internal conversion X-ray fluorescence. Of course, having a high-brightness source and a high-resolution monochromator system is also essential. A typical NRVS experimental setup is illustrated in Fig. 10.3.

In a synchrotron NRVS experiment, the sample is excited by a very short synchrotron pulse (<100 ps), and the delayed emission of photons (or sometimes electrons) is monitored by a fast avalanche photodiode (APD) detector (Fig. 10.3). Currently, the source of X-rays is invariably an undulator on a high-energy third-generation storage ring. In order to lower the power density, the beam is first passed

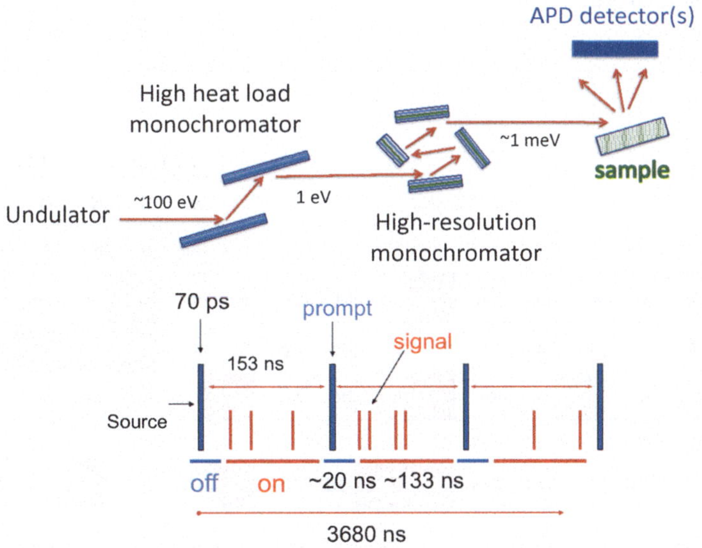

Fig. 10.3. Top: schematic illustration of the NRVS experiment. Bottom: a representative timing scheme for an NRVS experiment. The specific times are for typical APS operating conditions

through a ~ 1 eV resolution cryogenically cooled diamond or Si pre-monochromator (Chap. 4). A high-resolution monochromator using asymmetrically cut crystals then narrows the bandpass to ~1 meV. Finally, the beam impinges on a sample that is usually maintained at low temperature in a liquid nitrogen or liquid He cryostat.

The APD detector usually has a time resolution on the order ~1 ns, but it takes multiple time constants to relax to baseline after the strong prompt pulse. A time gate is set on the output from the APD, so that counts are only recorded after a predetermined delay with respect to the prompt radiation (Fig. 10.3). In this manner, the NRVS signal, consisting of delayed nuclear fluorescence and X-rays (or electrons) following relaxation by internal conversion, is separated from purely electronic events such as Rayleigh and Compton scattering and normal X-ray fluorescence.

Not every Mössbauer isotope is suitable for an NRVS experiment. First of all, one needs a monochromator with sufficiently narrow bandpass to resolve vibrational features, generally ~meV resolution. At the moment, this limits the nuclear resonance energy to <40 keV. Second, the nuclear excited-state lifetime has to be long enough (a few hundred ps) to allow the detector to distinguish nuclear events from prompt electronic events. The isotopes that have made the cut so far are illustrated in Fig. 10.4. Fortunately, one of the best isotopes is ^{57}Fe.

Despite the difficulty of the experiment, the combination of third-generation synchrotron sources, long undulators, asymmetrically cut monochromators, and fast APD detectors has made NRVS a viable technique for samples down to millimolar concentrations. At the moment, there are about 13 isotopes accessible (Fig. 10.4) and 4 storage ring facilities worldwide (Fig. 1.4 and Appendix K) where the measurement can be done.

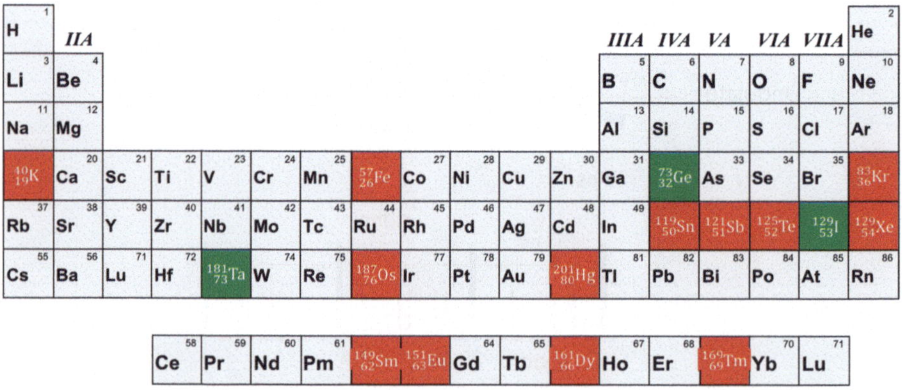

Fig. 10.4 Periodic table of isotopes accessible by NRVS. Red boxes represent isotopes already demonstrated; green boxes are isotopes that should be feasible but have not yet been employed for NRVS. Energies and lifetimes of isotopes that make the NRVS cut were shown in Fig. 9.2

10.2 NRVS Intensities for Discrete Normal Modes

How are vibrations and phonons with energies of meV coupled to nuclear transitions at the tens of keV level? The occurrence of vibrational sidebands (analogous to those in AM radio communication) is one way to understand the NRVS effect. A nucleus vibrating at frequency Ω and illuminated by a monochromatic beam at $\omega_N + \Omega$ will experience a negative sideband at $(\omega_N + \Omega) - \Omega = \omega_N$ and hence be in resonance for a nuclear transition. In addition, a nucleus that is part of a vibrational excited state will be in resonance via the positive sideband: $(\omega_N - \Omega) + \Omega = \omega_N$. Finally, multi-phonon resonances will also occur as higher-order sidebands, but their intensity will be reduced.

The overall cross section $\sigma(E)$ for nuclear resonant absorption of a photon with energy E can be factored into two terms, one of which depends on the properties of the nucleus and one that depends on the environment in which that nucleus is placed:

$$\sigma(E) = \frac{\pi}{2}\sigma(E_0)\Gamma_0 S(E - E_0) \tag{10.1}$$

and if one assumes a harmonic lattice or molecule:

$$S(E) = f_{\mathrm{LM}} \underbrace{\delta_\Gamma(E)}_{\text{Mössbauer}} + \underbrace{\sum_{n=1}^{\infty} S_n(E)}_{\text{NRVS}} \tag{10.2}$$

where f_{LM} is the Lamb-Mössbauer factor, $\delta_\Gamma(E)$ is a Lorentzian of width Γ, and the function $S_n(E)$ refers to events involving n phonons.

10.2.1 Stokes Fundamentals

Suppose we have a harmonic oscillator molecular system with a set of normal modes labeled α that are described by the displacements $\vec{r}_{k\alpha}$ for atom k and normal mode α. We assume a lineshape function L to account for the experimental resolution and lifetime broadening, and we convert the energy scale for our normal mode α to a frequency $\bar{\nu}_\alpha$ in wave numbers. As a specific example, we use the $FeCl_4^-$ ion, which in T_d symmetry will have normal modes with four types of symmetry species. The particular frequencies are labeled ν_1, corresponding to the totally symmetric A_1 Fe–Cl stretch, ν_2, for the doubly degenerate E Fe–Cl bend; ν_3, for the triply degenerate T_2 Fe–Cl stretch; and ν_4, for the triply degenerate T_2 Fe–Cl bend (Fig. 10.5).

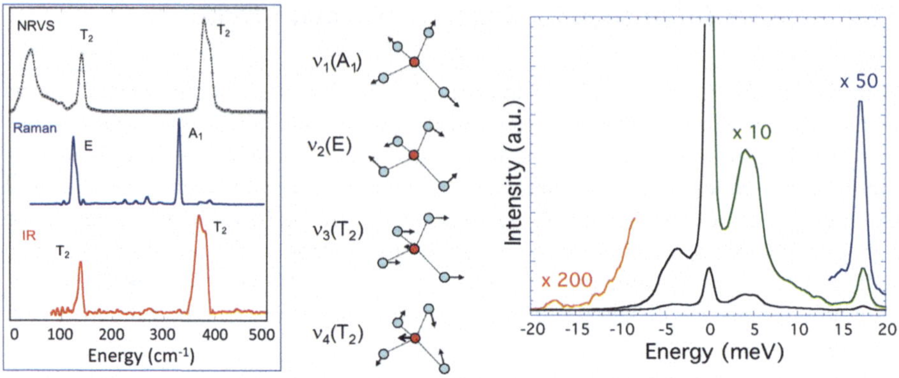

Fig. 10.5. Spectra for $(NEt_4)(FeCl_4)$. Left: comparison of ^{57}Fe PVDOS with Raman and IR spectra, illustrating the dependence of NRVS mode strength on the amount of Fe motion. Middle: descriptions of atomic motion in different modes (not to scale). Right: expansion of the NRVS PVDOS showing the relative amplitudes of the Stokes and anti-Stokes features

The normalized excitation probability can then be rewritten as [481]:

$$S(\overline{\nu}) = f_{\mathrm{LM}}\mathcal{L}_0(\overline{\nu}) + \sum\nolimits_{\{\Delta n_\alpha\}} \phi_{\{\Delta n_\alpha\}} \mathcal{L}_\nu \left(\overline{\nu} - \sum_\alpha \Delta n_\alpha \overline{\nu}_\alpha \right) \tag{10.3}$$

In the above expression, $\phi\{\Delta n_\alpha\}$ refers to the fractional area (integrated probability) corresponding to a transition from initial population n_α to final population $n_\alpha + \Delta n_\alpha$. In the particular case of the $FeCl_4^-$ ion, α would range from 1 → 4, with appropriate factors to account for the degeneracies.

The NRVS effect depends on the amount of nuclear motion in a particular normal mode or phonon. The critical term that captures this is the "mode composition factor," $e_{j\alpha}^2$, which is the fraction of kinetic energy in normal mode α associated with motion of nucleus j with mass m_j and mean square displacements $r_{j\alpha}^2$ [482]:

$$e_{j\alpha}^2 = \frac{m_j r_{j\alpha}^2}{\sum_k m_k r_{k\alpha}^2} \tag{10.4}$$

For a randomly oriented sample, and a nuclear transition with recoil energy $\overline{\nu}_{\mathrm{R}}$ and Lamb-Mössbauer factor f_{LM}, an expression that captures the properties that govern NRVS intensity for a fundamental transition from $n_\alpha \rightarrow n_\alpha+1$ is [481]:

$$\phi_\alpha = \frac{1}{3}\left(\frac{\overline{\nu}_{\mathrm{R}}}{\overline{\nu}_\alpha}\right)(\overline{n}_\alpha + 1) f_{\mathrm{LM}} e_{j\alpha}^2 \tag{10.5}$$

The term $\overline{n}_\alpha$ is the mean occupation number for mode α and is given by Boltzmann statistics:

$$\overline{n}_\alpha = \frac{1}{\exp\left(hc\overline{\nu}_\alpha/k_\mathrm{B}T\right) - 1} \tag{10.6}$$

The mode composition factor is the key term for describing the NRVS intensity of a given normal mode. It thus plays a role similar to the transition dipole moment in IR spectroscopy or the polarizability tensor in Raman spectroscopy. One thing that makes NRVS so special is that once you have a normal mode description, it is *easy* to calculate the mode composition factor. In contrast, IR or Raman calculations require knowledge about molecular properties such as transition dipole moments or polarizabilities, which in turn require quantum mechanical calculations.

In summary, there are four terms that govern NRVS intensities:

- The Lamb-Mössbauer factor f_LM.
- A general $1/E$ dependence that reduces intensity of higher-energy transitions.
- The temperature T, which governs the distribution of occupied vibrational levels.
- Most importantly, the mode composition factor $e_{j\alpha}^2$.

Because the NRVS signal depends on motion of the nucleus of interest, some modes will be strictly forbidden by symmetry. For our $FeCl_4^-$ example, both the totally symmetric (A_1) stretch and the symmetric (E) Fe–Cl bend lack Fe motion and hence are invisible to NRVS (Fig. 10.5). As another example, for Fe hydride complexes, Fe–H stretches are extremely difficult to observe because most of the motion is in the light atom. Higher-frequency modes also suffer from the $1/E$ dependence. Weak modes can be salvaged to some extent by preferentially weighting the acquisition time (as done with EXAFS scans).

10.2.2 Anti-Stokes Intensity

As with Raman spectroscopy, at photon energies E_0-$\overline{\nu}_\alpha$, there are transitions involving "annihilation" of phonons. The relative intensity of these "anti-Stokes features" is given by:

$$\phi_\alpha = \frac{1}{3}\frac{\overline{\nu}_\mathrm{R}}{\overline{\nu}_\alpha}(\overline{n}_\alpha) f_\mathrm{LM} e_{j\alpha}^2 \tag{10.7}$$

In these transitions, $n_\mathrm{a} + 1 \rightarrow n_\mathrm{a}$, and the intensity is strongly temperature dependent because these transitions start from vibrational excited states. As illustrated in Fig. 10.5, the relative strength of Stokes and anti-Stokes features depends on the temperature via the Boltzmann factor:

$$\frac{\phi_\text{anti-Stokes}}{\phi_\text{Stokes}} = \frac{\overline{n}_\alpha}{\overline{n}_\alpha + 1} = \exp\left(-hc\overline{\nu}_\alpha/k_\mathrm{B}T\right) \tag{10.8}$$

By rearranging this equation, the relative strengths of the anti-Stokes and Stokes transitions at frequencies $\pm\overline{\nu}_\alpha$ can be used to determine the sample temperature.

10.2.3 Multi-Phonon Events: Overtone and Combination Bands

Again like Raman and IR spectroscopy, an NRVS spectrum also exhibits overtone bands and combination bands involving changes of two or more phonons. How significant are they? It all depends on the Lamb-Mössbauer factor. By integrating the individual $S_n(E)$ curves, the n-phonon probabilities are obtained:

$$P_n = \int S_n(E)\,dE = f_{\mathrm{LM}} \frac{(-\ln f_{\mathrm{LM}})^n}{n!} \tag{10.9}$$

From this, Sturhahn gives a very simple expression for the average number of phonons excited over the nuclear excitation spectrum:

$$\langle n \rangle = -\ln f_{\mathrm{LM}} \tag{10.10}$$

As shown in Fig. 10.6, this means that multi-phonon contributions are less than 10% of the single-phonon contribution as long as $f_{\mathrm{LM}} < 0.83$.

Sage and coworkers have investigated the properties of two phonon contributions in some detail [481,487]. In the same low-temperature (high-frequency)

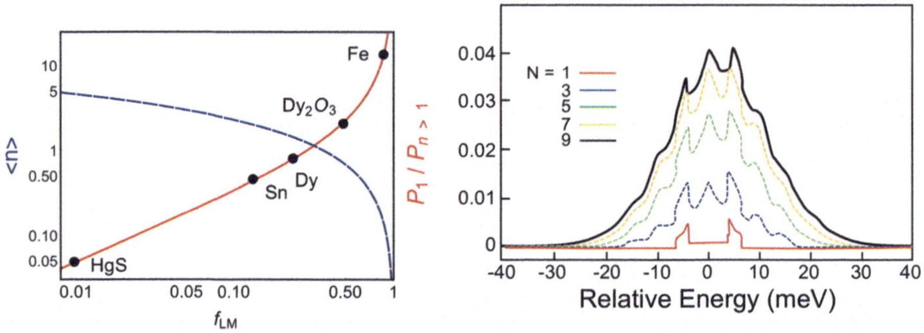

Fig. 10.6. Left: the average number of phonons $\langle n \rangle$ excited (blue dashed line) (integrated over all transition probabilities) and the ratio of single-phonon to multi-phonon events (red solid line) as a function of the Lamb-Mössbauer factors f_{LM}, illustrated for a variety of typical samples and conditions, HgS at 300 K [483], β-Sn at 100 K [484,485], Dy and Dy_2O_3 [486], Fe metal [449]. Right: NRVS spectrum for HgS, an extreme case of multi-phonon events dominating the nuclear spectrum. Redrawn from [483] with error bars removed for clarity

approximation used for the single-phonon approximation, the contribution $\phi_{\alpha\alpha}$ of the overtone transition, $n_\alpha \rightarrow n_\alpha + 2$, at energy $2\overline{\nu}_\alpha$ is given by [487]:

$$\phi_{\alpha\alpha} = \frac{9}{10}\frac{\phi_\alpha^2}{f_{\mathrm{LM}}} \tag{10.11}$$

To assess the relative intensity of the first overtone band, divide the above by ϕ_α, yielding a relative intensity of $0.9\phi_\alpha/f_{\mathrm{LM}}$. Assuming an arbitrary value of $\phi_\alpha = 0.002$ and a favorable $f_{\mathrm{LM}} = 0.5$, we see the overtone intensity is less than 1% of the fundamental intensity.

With the same approximations, a similar expression can be derived for the strength $\phi_{\alpha\beta}$ of the combination band, $n_\alpha \rightarrow n_\alpha + 1$, $n_\beta \rightarrow n_\beta + 1$, at energy $\overline{\nu}_\alpha + \overline{\nu}_\beta$, with angle $\xi_{\alpha\beta}$ between displacements for the resonant nucleus in the given normal modes [487]:

$$\phi_{\alpha\beta} = \frac{3}{5}\left(\frac{\phi_\alpha \cdot \phi_\beta}{f_{\mathrm{LM}}}\right)\left(1 + 2\cos^2\xi_{\alpha\beta}\right) \tag{10.12}$$

10.2.4 Orientation Dependence

Although most NRVS experiments are conducted on powder or solution samples, there is extra information to be gained when oriented or single-crystal samples are available. In the most general case, the NRVS will be different for an incident beam along three perpendicular directions in the sample: $\widehat{x}$, $\widehat{y}$, and $\widehat{z}$. There will then be three distinct mode composition factors, corresponding to the projection of the nuclear motion in those three directions:

$$e_{j\alpha,x}^2 = \frac{m_j\left(\vec{r}\cdot\widehat{x}\right)_{j\alpha}^2}{\sum\limits_k m_k r_{k\alpha}^2} \quad e_{j\alpha,y}^2 = \frac{m_j\left(\vec{r}\cdot\widehat{y}\right)_{j\alpha}^2}{\sum\limits_k m_k r_{k\alpha}^2} \quad e_{j\alpha,z}^2 = \frac{m_j\left(\vec{r}\cdot\widehat{z}\right)_{j\alpha}^2}{\sum\limits_k m_k r_{k\alpha}^2} \tag{10.13}$$

As illustrated in Fig. 10.7, for a crystal with isotope I and neighbor J, the intensity of the I–J stretching mode in the NRVS will vary as $\cos^2\theta$, where θ is the angle between the photon direction and the interatomic axis. This contrasts with the $\cos^2\theta'$ angular dependence of EXAFS, where the θ' refers to the angle between the electric field $\vec{E}$ vector and the interatomic axis.

In one example of using this orientation dependence, the NRVS for a single crystal of FeOEP(NO) was examined, with the photon beam oriented in three perpendicular directions. The observed intensities were then used to deduce the direction of Fe motion in particular normal modes, such as the Fe–NO bending motions [488] (Fig. 10.8).

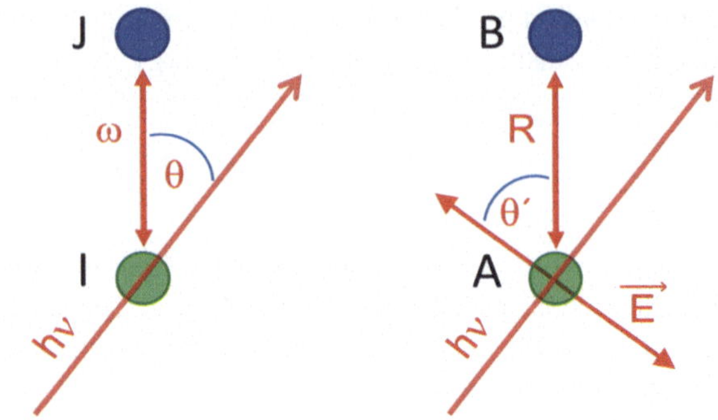

Fig. 10.7. Relevant angles for NRVS intensity (left) of a stretching mode and EXAFS intensity (right) from a neighboring atom

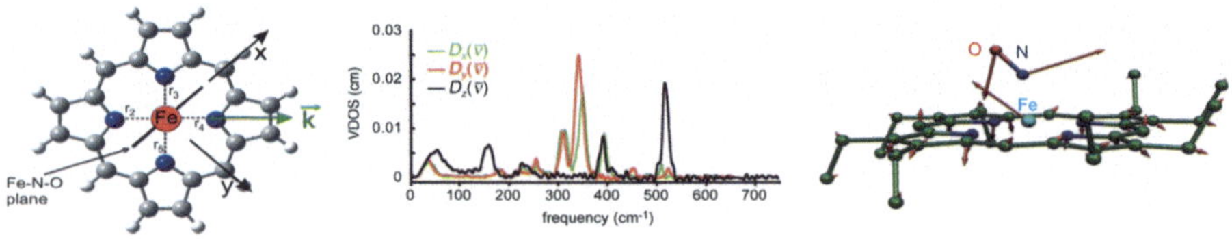

Fig. 10.8. Left: orientations used for NRVS on a single crystal of Fe(OEP)NO. Middle: observed PVDOS along *x*, *y*, and *z* directions. Right: deduced motion for the Fe–NO bending mode

10.3 Partial Vibrational Density of States (PVDOS) Treatment

The discrete, isolated NRVS features seen in the $(NEt_4)(FeCl_4)$ spectrum are the exception. Instead, there are often a large number of unresolved modes. In addition, along with localized "optical" modes, all samples will also have a set of "acoustic" modes that involve motion of the entire unit cell and form a low-energy continuum. For $(NEt_4)(FeCl_4)$ such modes are seen below 100 cm^{-1}. Although for many chemical and biological applications the acoustic modes are not of keen interest, they contain information about properties such as the speed of sound in a sample that is of critical importance to geophysicists [489]. We thus need a way to describe the NRVS for the more general case where there is a continuous distribution of phonon energies.

In the NRVS literature, this is done using a continuous "*p*artial *v*ibrational *d*ensity *o*f *s*tates" or "PVDOS", commonly written as $D(E)$ or $g(E)$ and $D\left(E, \widehat{k}\right)$ or $g\left(E, \widehat{k}\right)$ if the directional dependence of the PVDOS is retained. This weighted vibrational density of states can be written as:

$$D_j\left(E, \widehat{k}\right) = \sum_{\alpha} \left(\widehat{k} \cdot \vec{e}_{j\alpha}\right)^2 \mathcal{L}(E - E_\alpha) \tag{10.14}$$

In this formulation, the one phonon term becomes:

$$S_1\left(E,\hat{k}\right) = \frac{E_\mathrm{R}}{E(1 - \mathrm{e}^{-\beta E})} D\left(|E|,\hat{k}\right) \qquad (10.15)$$

and the multi-phonon terms for $n > 1$ are obtained from successive convolutions with the one-phonon probability function:

$$S_n\left(E,\hat{k}\right) = \frac{1}{n f_{LM}} \int S_{n-1}\left(E,\hat{k}\right) S_1\left(E - E',\hat{k}\right) dE' \qquad (10.16)$$

Although an exact vibrational density of states need not follow any simple expression, a useful approximation for NRVS analysis is the Debye model that was discussed in Chap. 6. As we saw there, the Debye PVDOS increases as the square of the frequency (or energy) until it cuts off abruptly at a certain threshold ω_D:

$$g(\omega) = \frac{9\omega^2}{\omega_\mathrm{D}^3} \quad \text{or} \quad D(E) = \frac{9E^2}{E_\mathrm{D}^3} \qquad (10.17)$$

By incorporating typical numerical values for a Debye function into the above single-phonon formula, we obtain the single-phonon excitation spectra illustrated in Fig. 10.9.

10.4 Other Quantities from NRVS Analysis: Sum Rules

As we mentioned earlier, there are a number of physical properties that can be extracted from NRVS spectra, and a short list is included in Table 10.1. Many of the useful quantities are obtained from so-called sum rule analysis, where the sum relies on integrals yielding various moments of the excitation spectrum. Specifically, the nth moment is defined as $W_n = \int E^n S(E) dE$. (We use W_n instead of S_n to avoid confusion with the n-photon absorption probability.)

The first rule involving the zeroth moment is trivial—the sum of probabilities over all possible events is unity, something has to happen. Thus, by definition, the zeroth moment W_0 is the integrated transition probability and hence unity:

$$W_0 = \int S(E)\, dE = 1 \qquad (10.18)$$

The first moment is useful because it turns out to be independent of the chemical environment of the nucleus under study—it just depends on the recoil energy E_R:

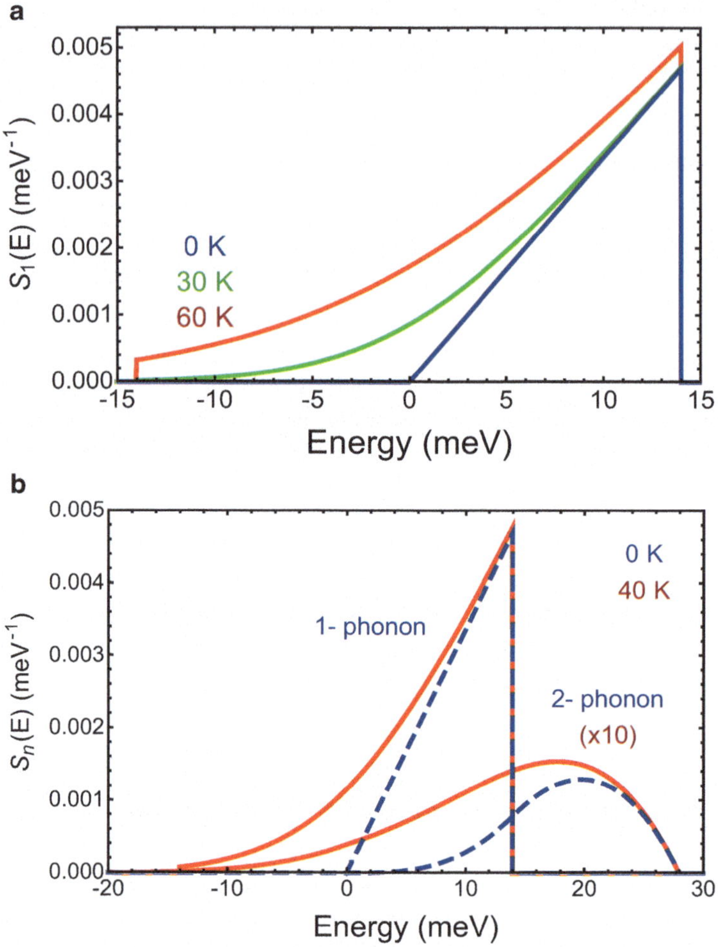

Fig. 10.9. Left: single-phonon excitation probabilities, $S_1(E)$, for a Debye model PVDOS with a cutoff energy of 14 meV at different temperatures. Right: comparison of one-phonon and two-phonon excitation probabilities for a Debye model as a function of temperature

Table 10.1 Quantities derived from sum rule analysis

Quantity	Fe metal (300 K) [449]	Goethite [490]
Lamb-Mössbauer factor f_{LM}	0.796	0.78
Mean square displacement $\langle z^2 \rangle$ 10^{-3} (Å^2)	4.13	4.66
Average kinetic energy, T_{av} (meV)	14.16	15.58
Average force constant, k_{av} (N/m)	180	107.1

$$W_1 = \int ES(E)\,dE = E_R \tag{10.19}$$

This expresses a satisfying result: the average energy transfer to the lattice is equal to the recoil energy of the free atom. Since the recoil energy is already known from $E_R = E_0^2/2mc^2$, the above equation turns out to be a convenient tool for normalization of the overall spectrum.

The next two sum rules are simpler if one uses "centered moments" defined as:

$$R_n = \int (E - E_R)^n S(E)\,dE \tag{10.20}$$

The centered second moment provides the average kinetic energy T_{av} of the nucleus under study:

$$R_2 = \int (E - E_R)^2 S(E)\,dE = 4E_R T_{av} \tag{10.21}$$

Finally, the centered third moment turns out to be proportional to the average force constant k_{av} holding an atom in its position along the average photon direction $\widehat{k}$:

$$R_3 = \int (E - E_R)^3 S(E)\,dE = \frac{\hbar^2}{m} E_R k_{av} \tag{10.22}$$

One obvious nicety about sum rule analysis is that it directly yields important information with fewer presuppositions than force-field analysis. As concrete examples, in Table 10.1 we present results from sum rule analyses for Fe metal [449] and goethite [490].

Sum rule analysis can also be done using the processed PVDOS $D(E)$ instead of the raw spectrum $S(E)$ [490, 491]. When using DOS moments, normal moments $g_1(k)$ and thermally averaged moments $\widetilde{g}_1(k)$ are employed, as described in [490]. Calculations using both procedures give reasonably close agreement (Table 10.2).

Table 10.2 Relations between moments from NRVS $S(k,E)$ and PVDOS

$R_0(k)$	1	1—by definition of probability
$R_1(k)$	0	0—From centering at E_R
$R_2(k)$	$2E_R\,\widetilde{g}_1(k)$	Yield T_{av}—Average kinetic energy
$R_3(k)$	$E_R\,g_2(k)$	Yield k_{av}—average force constant

10.4.1 Speed of Sound

The speed of sound provides some of our best information about the inner structure of the earth. It is also relevant to the properties of many synthetic materials, such as thermoelectrics used for direct conversion of heat to electricity. NRVS turns out to be an excellent tool for obtaining the speed of sound on small samples and (if necessary) under extreme conditions such as the high pressures and high temperatures relevant to the earth's interior. Furthermore, in the acoustic modes involved in sound propagation, the unit cell moves as a whole, and all of the atoms have the same displacements. This also allows one to determine the speed of sound for a material that does not have a natural NRVS isotope by using a dilute probe nucleus.

To relate the NRVS-derived PVDOS to the Debye model total DOS, we make the approximation that the intermolecular and intramolecular modes are decoupled. In the simplest case of an isotropic and/or polycrystalline solid, the Debye model yields for the speed of sound v_{D}:

$$D(E) = \left(\frac{\widetilde{m}}{m}\right) \frac{E^2}{2\pi^2 n \hbar^3 v_{\mathrm{D}}^3} \tag{10.23}$$

where $\widetilde{m}$ is the mass of the "probe" nucleus, m is the average atomic mass in the material, and n is the number density of atoms. This can be rearranged to give an expression for the Debye speed of sound v_{D}:

$$v_D = \left[\left(\frac{\widetilde{m}}{m}\right) \frac{E^2}{2\pi^2 n \hbar^3 D(E)}\right]^{1/3} \tag{10.24}$$

or in terms of wave numbers:

$$v_D = \left[\frac{4\pi \mathrm{m}_X \mathrm{c}^3}{\rho} \frac{\overline{\mathrm{n}}^2}{D(\overline{\mathrm{n}})}\right]^{1/3} \tag{10.25}$$

Although one could in principle use Eq. 10.24 or Eq. 10.25 at a single energy to derive the speed of sound, in practice it is better to use a range of data to check for consistency. One approach is to fit the low-energy portion of the spectrum with a parabola according to Eq. 10.23 and use the region over which there is a good match. In the same vein, one can make a plot of D(E) vs. E^2 and extract ν_{D} from the slope. Finally, one can even plot D(E)/E^2 vs. E and use the range over which this appears constant. In any case, it is important to only use the range of data over which the density of states D(E) exhibits Debye-like behavior (Fig. 10.10 and Table 10.3).

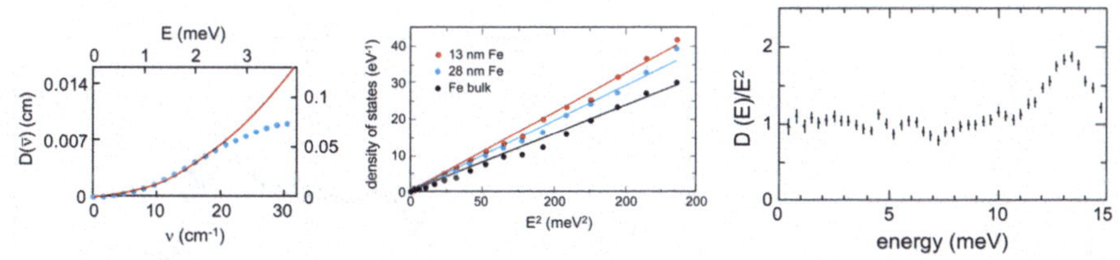

Fig. 10.10 Alternate approaches to extracting the Debye speed of sound. Left: a parabolic fit to low-frequency data for FeOEP [492]. Middle: linear fits to *D*(*E*) vs. E^2, for different thickness of Fe on ZERODUR® [420]. Right: data for 1% ^{119}Sn in Pd from Hu et al. [489]

Table 10.3 Speeds of sound measured by NRVS on different systems

Sample	Nucleus	Speed (m/s)	Reference
$Mg_{0.65}Fe_{0.33}Ti_{0.02}O$ (140 GPa)	^{57}Fe	6565	[493]
Pd	^{119}Sb	2193	[489]
Fe_2O_3	^{57}Fe	4279	[489]
Fe	^{57}Fe	3488	[489]
Fe(OEP)Cl	^{57}Fe	1145	[492]
Bi_2Te_3	^{125}Te	1750	[494]
$Eu_{14}MnSb_{11}$	^{151}Eu	1690	[495]

10.5 Data Processing

To compare theory and experiment, we often need to extract the PVDOS from the experimental data. A variety of programs are available for the analysis of NRVS spectra, including DOS [496] and PHOENIX [479]. As illustrated schematically in Fig. 10.11, the key steps in the analysis of NRVS data are:

- Determination of the resolution function *R(E − E′)*
- Normalization using sum rules
- Subtraction of the elastic component
- Decomposition into *n*-phonon contributions
- Derivation of the PVDOS

Determination of the Resolution Function R(E). The raw NRVS spectrum consists of a measurement of intensity *I*(*E*) vs. energy *E*. The spectrum in the region of the Mössbauer (zero-phonon) resonance is effectively a δ-function convolved with the monochromator resolution function *R*(*E*), while the wings of the spectrum depend on the probability for vibrational excitation, *S*(*E*). One can thus express the raw experimental spectrum as:

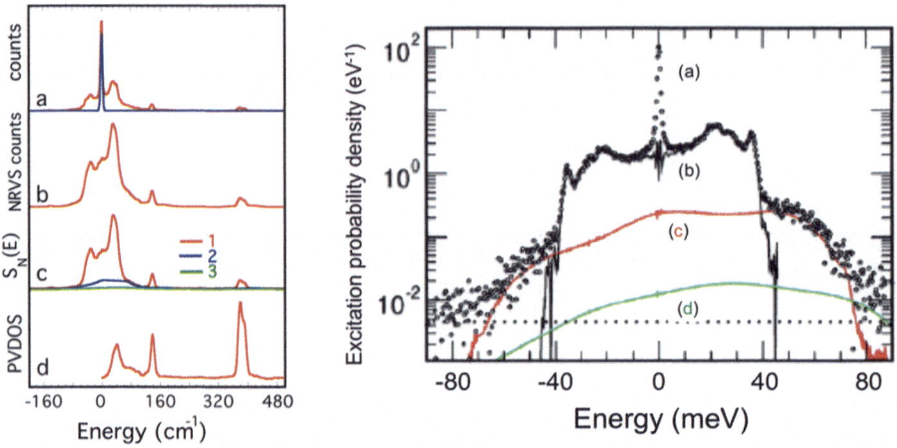

Fig. 10.11 Left: stepwise NRVS data processing for $(NEt_4)(FeCl_4)$—(**a**) normalized raw data (**red line**) and resolution function (blue line); (**b**) after removal of Mössbauer peak; (**c**) removal of multi-phonon contributions, (**d**) conversion to PVDOS. Right: data processing for Fe metal—(**a**) normalized raw data (dotted line); (**b**) one-phonon term $S_1(E)$ (solid line); (**c**) two-phonon term $S_2(E)$ (red line); (**d**) sum of higher-order terms (green line). Dotted line indicates noise level. Revised and reproduced from [449]

$$I(E) = \int_{-\infty}^{\infty} \underbrace{R(E-E')}_{\text{resolution}}\{\underbrace{aS(E')}_{\text{NRVS}} - b\underbrace{\delta(E')}_{\text{Mössbauer}}\}dE' \tag{10.26}$$

where parameter a is a normalization constant and b is a value that accounts for saturation effects when on the Mössbauer resonance. The shape of $R(E)$ is obtained by adjusting a parameterized function or by using a tabulated data file (Fig. 10.11).

$$\int_{-\infty}^{\infty} E\,S(E)dE = E_R \tag{10.27}$$

Normalization Using Sum Rules. Once the resolution function has been determined, the zeroth and first moments of (Eq. 10.26) can be determined, and a and b can be adjusted so that $I(E)$ satisfies the first two sum rules: $W_0 = 1$ and $W_1 = E_R$.

Subtraction of Elastic Component. Once the form of $R(E)$ has been determined, the elastic component can be removed from the spectrum (Fig. 10.11), yielding an experimental curve that reflects single- and multi-phonon contributions:

$$I'(E) = I(E) - c_1R(E) = af\sum_{n=1}^{\infty}\int_{-\infty}^{\infty} G_n(E')R(E-E')dE' \tag{10.28}$$

Decomposition into n-Phonon Contributions. The normalized spectrum still represents the sum of single- and multi-phonon contributions, as well as the convolution with the resolution function. In PHOENIX, the single-phonon contribution is obtained by a so-called Fourier log method. This exploits the so-called convolution theorem—the Fourier transform of a convolution is the product of the individual Fourier transforms.

Derivation of the PVDOS. The single-phonon contribution still contains the Stokes and anti-Stokes regions, as well as the $1/E$ energy dependence of the intensity. Finally, the single-phonon contribution is converted to the PVDOS by employing:

$$g(E) = \frac{E}{E_R} \tanh \frac{\beta E}{2} (S_1(E) + S_1(-E)), \quad \text{for}\, E \geq 0 \tag{10.29}$$

The various steps in this extraction process are illustrated both for $(NEt_4)(FeCl_4)$ and for Fe metal in Fig. 10.11.

10.6 Applications and Interpretation

Once the NRVS-derived PVDOS is in hand, the interpretation depends a lot on the problem of interest. Among the approaches employed are:

- Single-mode observations combined with isotope shifts (Fig. 10.2)
- Graphical analysis for speed of sound (Fig. 10.10)
- Sum rule analysis for quantities of interest (Table 10.1)
- Optimization via empirical force fields
- Interpretation via DFT calculations

10.6.1 Empirical vs. DFT Force Fields

The tetrahedral $(FeCl_4)^-$ ion was a good system for illustrating NRVS selection rules in comparison with other spectroscopies (Fig. 10.5). As shown in Fig. 10.12, the PVDOS for this system can be modeled well by an empirical Urey-Bradley force field [497], which captures the intramolecular modes.

At the other end of complexity are applications to metalloproteins such as hydrogenase (H_2ase). In this case, DFT calculations were used to reproduce the observed Fe–H bending modes (Fig. 10.12) [498]. Nowadays, DFT calculations are the most common approach for interpreting the spectra of molecular systems. Packages such as ORCA and Gaussian are used to calculate normal modes, and the isotope-specific PVDOS is then extracted either within the DFT program or by use of external Python scripts.

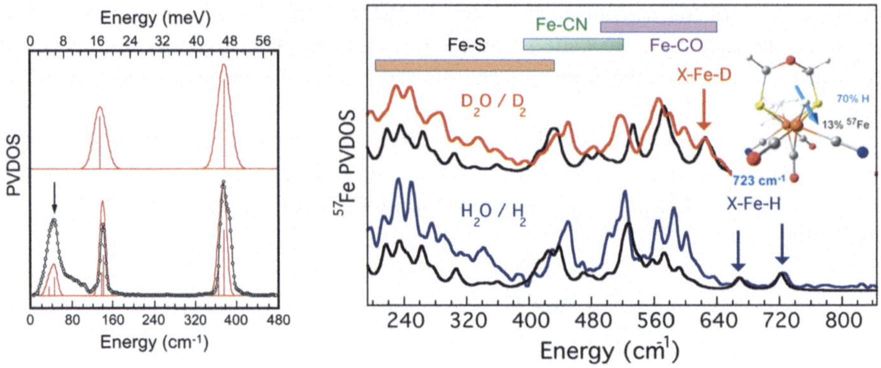

Fig. 10.12. Top left: reproduction of NRVS PVDOS spectra for $FeCl_4^-$ ion with empirical force field in tetrahedral symmetry [497]. Lower left: effects of inclusion of $(NEt_4)^+$ counterion. Right: DFT simulation of Fe–H bending modes in H_2ase [498]

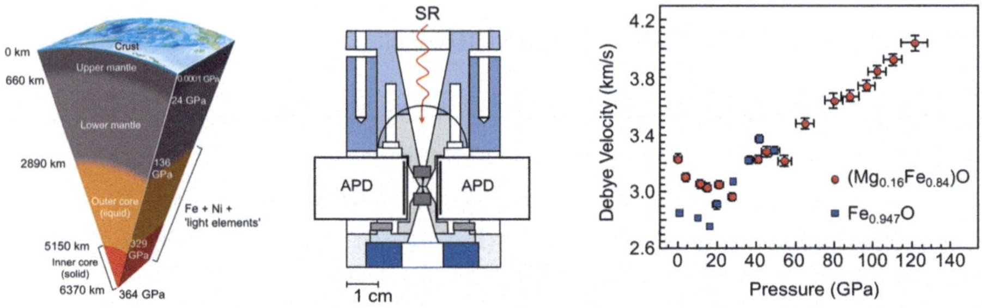

Fig. 10.13. Left to right: a model for the interior of the earth; an apparatus for high-pressure NRVS; sound velocities deduced from NRVS for an Fe-rich $(Mg_{0.16}Fe_{0.84})O$ sample relevant to the core-mantle boundary [450]

10.6.2 Minerals under Pressure

The detailed composition of the earth's core and core-mantle boundary is still the subject of debates (Fig. 10.13) [499]. Geophysicists have struggled to determine which Fe alloys with other elements will give the right densities and sound velocities to match the observed data. Similar questions exist about the other layers of the earth including the lower mantle. In experiments to mimic these conditions, "diamond anvil cells" are used to create pressures up to and exceeding 150 GPa. The construction of these cells limits the observation cross section to small fractions of a mm (Fig. 10.13). Since the synchrotron beam for NRVS experiments can be easily be focused down to dimensions on the order of $50 \times 50\ \mu^2$ or less, NRVS is well adapted to the study of materials at high pressure.

The results from a typical high-pressure experiment on an $(Mg_{0.16}Fe_{0.84})O$ system are illustrated in Fig. 10.13. There is a decrease in the Debye velocity in the region around 28 GPa, consistent with other evidence for transition to a magnetically ordered state in that region. The NFS data (Fig. 9.16) provides evidence for an additional high-spin to low-spin transition around 121 GPa.

10.6.3 Use of Multiple NRVS Centers: Thermoelectric Materials

Thermoelectric materials can use a temperature gradient to generate electricity or vice versa. A common figure of merit for such devices at temperature T is $ZT = (\sigma\Sigma^2 T)/\kappa$, where σ is the electrical conductivity, S is the "Seebeck coefficient," and κ is the thermal conductivity. Since thermal conductivity is intimately related to lattice vibrations, acoustic phonons, and the speed of sound, NRVS can play a valuable role in characterizing these materials.

One promising class of thermoelectrics is skutterudites, compounds or minerals with the composition LM_4X_{12}, where M is a transition metal, X is a group V element, and L is often a lanthanide. These materials have a large unit cell with voids that can accommodate a variety of L ions, and there is a wide range of combinations which can be explored and tuned for desirable properties. Skutterudites are interesting thermoelectric candidates in part because they can operate at higher temperatures than conventional Bi–Te-based materials.

One particular skutterudite, $SmFe_4Sb_{12}$, is an ideal system for NRVS studies because each of its constituent elements has an NRVS isotope: ^{149}Sm, ^{57}Fe, or ^{121}Sb (Fig. 10.14). The sharp peak in the ^{149}Sm spectrum is assigned to a localized rare earth "shake mode" that is decoupled from the lattice acoustic modes and presumably diminishes thermal conductivity.

10.6.4 Difficult Cases: Multi-Phonon Problems and Anharmonic Samples

We have seen above that the probability of multi-phonon events rises as the Lamb-Mössbauer factor decreases (Fig. 10.6 and Eq. 10.10). We know from Eq. 9.10 that f_{LM} decreases as the Mössbauer resonance increases in energy (higher recoil) and as the material gets softer (larger thermal motion). The NRVS of HgS (Fig. 10.6) is a nice example of how bad things can get—this material has a relatively high-energy resonance (26.3 keV) and a very low Debye temperature (~90 K), and the high atomic weights of the HgS pair yield a number of optical modes at about 4 meV [483]. Ishikawa estimated a Lamb-Mössbauer factor of ~0.005–0.01, which yields an average number of phonons of about 5 [483]. Simulation of the spectrum thus requires inclusion of contributions up to $S^9(E)$ (Fig. 10.6).

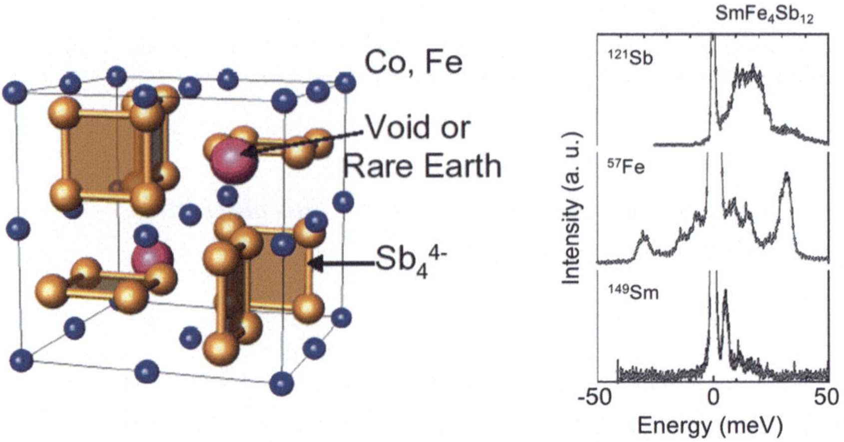

Fig. 10.14. Left: structure of the mineral skutterudite. Right: the raw spectra for $SmFe_4Sb_{12}$ from three different elemental points of view [500]

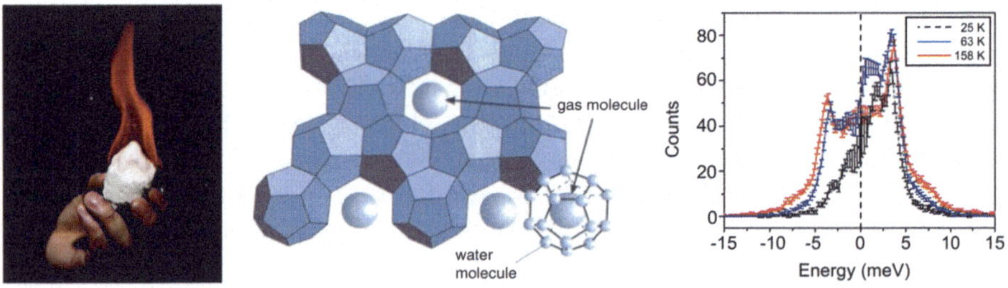

Fig. 10.15 Left: burning methane from a "fire ice" gas hydrate or "clathrate." Middle: molecular structure of a gas hydrate. Right: raw NRVS spectrum for a Kr hydrate [501]. The peak at ~4 meV is assigned to "rattling" of Kr in the smaller cages of the clathrate lattice

Another problematic case for NRVS analysis involves samples with anharmonic motion. The Kr clathrate hydrate, where Kr replaces the more common CH_4 molecule, is of interest in part because it exhibits anomalous glass-like thermal conductivity. This has been ascribed to "rattling" of the Kr guest, with some similarities to "rattling" of rare earths in the previous skutterudite example (Fig. 10.14). Tse and coworkers analyzed the ^{83}Kr NRVS for Kr in a structure II clathrate hydrate (Fig. 10.15) [501]. They found that the usual analysis procedure, based on the harmonic approximation, was no longer valid. The anharmonicity of the Kr motion was linked to the special thermal properties of these materials. A similar study was done on a Xe clathrate hydrate, using the 39.581 keV resonance of ^{129}Xe [502]. This stands as the highest-energy NRVS study to date (Fig. 10.15).

10.7 Prognosis

Shenoy and collaborators have speculated on the future of nuclear spectroscopies at synchrotron sources [503]. A recent workshop report on the potential of XFELO sources predicted enormous gains in sensitivity and range of isotopes (see Chap. 12) [504]. Improved sources will clearly provide more flux, allowing for better data, study of more dilute samples, weaker modes, and so on. These sources will also provide more high-energy flux and shorter pulses, perhaps quite a few resonances in the 40–100 keV region for NRVS experiments (Fig. 10.4). It remains to be seen if future monochromator developments with novel materials will allow meV resolution above 40 keV [117]. It would also be valuable to have more detectors with time resolution on the 100 ps scale, as recently demonstrated by Masuda and coworkers [189]. The combination of these three improvements could open up at least another dozen resonances, including important elements such as W and Ni. Stay tuned.

10.8 Suggested Exercises

1. Estimate the temperature of the sample in Fig. 10.5 from the ratio of (a) peaks at $\sim\pm 3$ meV and (b) peaks at $\pm 17\ \text{cm}^{-1}$. (A more accurate calculation would use the areas.)
2. Calculate the overall one-, two-, and three-phonon excitation probabilities, P_1, P_2, and P_3, assuming a Lamb-Mössbauer factor f_{LM} of (a) 0.85 and (b) 0.5.
3. Calculate the average number (over the whole spectrum) of phonons excited, $<n>$, assuming a Lamb-Mössbauer factor f_{LM} of (a) 0.85 and (b) 0.5.
4. Compare the relative cross sections for an ^{57}Fe vs. a ^{119}Sn nucleus in a powder sample, assuming $T = 50$ K and in both cases a mode frequency 150 cm^{-1}, $e^2_{Fe} = e^2_{Sn} = 0.2$, and $f_{LM} = 0.5$.
5. Suppose you have an ^{57}Fe system with fundamental frequency 150 cm^{-1}, $e^2_{Fe} = 0.2$, and $f_{LM} = 0.9$. What is the ratio of two-phonon intensity ($\phi_{\alpha\alpha}$) to one-phonon intensity ($\phi_{\alpha\alpha}$) in the low-temperature limit of $T \sim 0$ K?

10.9 Reference Books and Review Articles

1. Ralf Röhlsberger, Nuclear Condensed Matter Physics with Synchrotron Radiation - Basic Principles, Methodology and Applications, Springer, Berlin/Heidelberg, 2004. ISBN 3-540-23244-3.
2. E. E. Alp, W. Sturhahn, T. S. Toellner, J. Zhoa, M.Hu, D. E. Brown. "Vibrational Dynamics Studies by Nuclear Resonant Inelastic X-ray Scattering", *Hyp. Int.* **144/145**: 3–20, 2002.

3. W. R. Scheidt, S. M. Durbin, J. T. Sage, "Nuclear resonance vibrational spectroscopy– NRVS", *J. Inorg. Biochem.* **99**, 60–71, 2005.
4. H. Wang, E. E. Alp, Y. Yoda, & S. P. Cramer, "A Practical Guide for Nuclear Resonance Vibrational Spectroscopy (NRVS) of Biochemical Samples & Model Compounds", Metalloproteins: Methods & Protocols, J. C. Fontecilla-Camps & Y. Nicolet, eds. Springer, 2014.

10.10 NRVS Software

1. PHOENIX—a free data analysis program developed by Wolfgang Sturhahn and now distributed by NRIXS Scientific Software (NRIXS.com). Significantly improved since early write-up: W. Sturhahn, "CONUSS and PHOENIX: evaluation of nuclear resonant scattering data", *Hyp. Int.* **125**: 149-172, 2000.
2. DOS—the European analysis program, developed by Kohn and Chumakov: V. Kohn and A. I. Chumakov "DOS: Evaluation of phonon density of states from nuclear resonant inelastic absorption", *Hyp. Int.* **125**: 205-221, 2000
3. SciPHON—a *Mathematica*-based analysis package developed by scientists at the APS, N. Dauphas, M. Y. Hu, E. M. Baker, J. Hu, F. L. H. Tissot, E. E. Alp, M. Roskosz, J. Zhao, W. Bi, J. Liu, J.-F. Lin, N. X. Niea and A. Heard, "SciPhon: a data analysis software for nuclear resonant inelastic X-ray scattering with applications to Fe, Kr, Sn, Eu and Dy", *J. Syn. Rad.* **25**: 1581–1599, 2018 http://originslab.uchicago.edu/Software-and-Facilities

Chapter 11
Photon-in Electron-out Spectroscopies

11.1 Introduction

A variety of X-ray spectroscopies can be classified as "photon-in electron-out" techniques. Analogous to the "photon-in photon-out" methods discussed in Chap. 8, these electron-based methods involve measuring both the incident photon energy and the outgoing electron energy and in some cases the electron momentum and/or spin. Depending on the energy of the incoming photon, the type of electron analysis, and the experimental conditions, an inevitable slew of acronyms has been applied:

- *U*ltraviolet *p*hotoelectron *s*pectroscopy (UPS).
- *X*-ray *p*hotoelectron *s*pectroscopy (XPS).
- *Ha*rd *X-ray* *p*hotoelectron *s*pectroscopy (HAXPS).
- *A*ngle-*r*esolved *p*hoto*e*lectron *s*pectroscopy (ARPES).
- *S*pin-*p*olarized *X-ray* *p*hotoelectron *s*pectroscopy (SPXPS).
- *S*pin-*p*olarized and *a*ngle-*r*esolved *p*hoto*e*lectron *s*pectroscopy (SP-ARPES).
- *A*mbient *p*ressure *X-ray* *p*hotoelectron *s*pectroscopy (APXPS).
- *A*uger *e*lectron *s*pectroscopy (AES).

Another important "photon-in electron-out" technique for structure determination has been "photoelectron diffraction," where photoelectrons emitted from a surface are imaged. As a diffraction (instead of spectroscopy) method, for now we will refer to excellent review articles [505–508].

The first four of these methods are illustrated schematically in Fig. 11.1. The essence of the experiment is to illuminate a sample with monochromatic X-rays and to record the kinetic energies of the emitted photoelectrons. A basic equation that describes the conservation of energy in this experiment is [509]:

S. P. Cramer, *X-Ray Spectroscopy with Synchrotron Radiation*, Biological and Medical Physics, Biomedical Engineering, https://doi.org/10.1007/978-3-030-28551-7_11

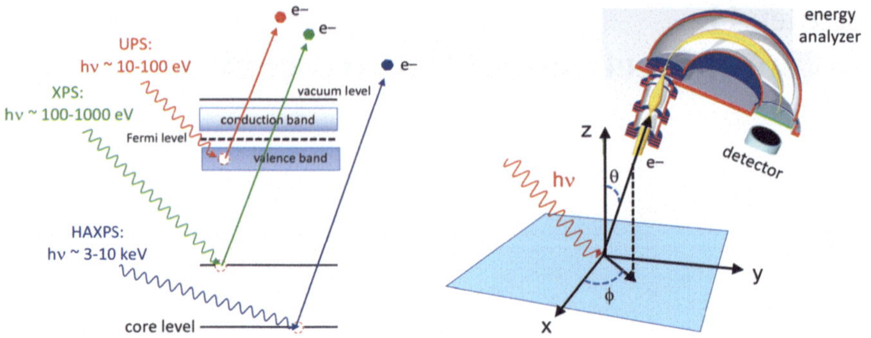

Fig. 11.1 Left: typical energies involved in different photoelectron spectroscopies. Right: a generic PES, ARPES, or SPXPS experiment. In ARPES, PES data are recorded as a function of θ and ϕ. In SPXPS, a spin analyzer is added to the detector

$$h\nu = E_{\text{binding}}^{\text{vacuum}} + E'_{\text{kinetic}} = E_{\text{binding}}^{\text{Fermi}} + \varphi_{\text{spectrometer}} + E_{\text{kinetic}} + \ldots \tag{11.1}$$

where $E_{\text{binding}}^{\text{vacuum}}$ is the binding energy relative to the vacuum level, E'_{kinetic} is the kinetic energy of the photoelectron as it leaves the sample, $E_{\text{binding}}^{\text{Fermi}}$ is the binding energy relative to the Fermi level, E_{kinetic} is the kinetic energy as measured in the spectrometer, and $\varphi_{\text{spectrometer}}$ is the work function of the spectrometer [509]. In some cases, there are complications due to sample charging, and an additional potential V_{charge} needs to be added.

Photoelectron spectroscopy has a long history. Heinrich Hertz demonstrated that light could cause an electrical current to flow from a sample in 1877 [510], and in 1905 Einstein's explanation of the photoelectric effect was a key step toward quantum theory [511], and it led to his 1921 Nobel Prize, "for his discovery of the law of the photoelectric effect" (although his Nobel lecture was about relativity). In the mid-1950s, Kai Siegbahn built electron energy analyzers with a precision of 10^{-5}–10^{-6}, and in multiple papers, he showed that resolvable chemical shifts occur for atoms in different environments (Fig. 11.2), leading to the acronym "ESCA"—"*E*lectron *S*pectroscopy for *C*hemical *A*nalysis" [512, 513]. Siegbahn received his own Nobel prize in physics in 1981, "for his contribution to the development of high-resolution electron spectroscopy [514]."

The first synchrotron XPS experiments were done by Pianetta and Lindau in 1974 at SSRL. They were forced to use hard X-rays because the beamline was cut off from ring vacuum by Be windows. With 8 keV radiation, they observed the *4f* spin-orbit splitting for Au metal with ~0.35 eV instrumental resolution (Fig. 11.4) [517].

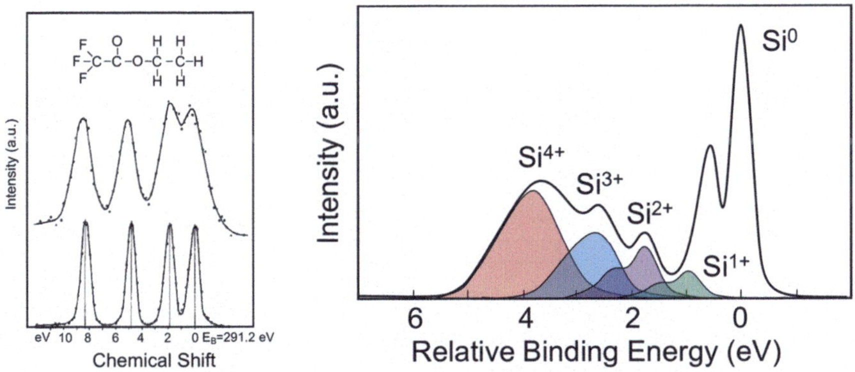

Fig. 11.2 Left: chemical shifts in the 1s XPS for the four different types of carbon in ethyl trifluoroacetate. Top spectrum is from early work and bottom with improved X-ray monochromatization in 1974 [514, 515]. Right: high-resolution Si 2p spectrum from a Si(0 0 1) surface (solid line) that has been partially oxidized at 600 °C and at an ambient pressure of 5×10^{-7} torr (redrawn with smoothing from [516]), revealing the variety of Si oxidation levels (filled curves)

11.2 X-ray Photoelectron Spectroscopy (XPS)

With high-resolution X-ray monochromators and electron analyzers, XPS spectra were found to exhibit a wealth of information, including (1) chemical shifts, (2) multiplet splittings, and even (3) vibrational fine structure. These are features we have already seen in XANES and X-ray fluorescence spectra, so it should not be surprising that they are observed when we analyze systems with core vacancies using electrons. A nice review by Bagus summarizes the theory behind chemical effects on XPS spectra, including multiplet splittings, covalent mixing, satellite peaks, and core-level binding energy shifts [518].

XPS began primarily as a surface science technique. We previously saw (Fig. 6.12) that the mean free paths for electrons in solids are often in the 10–20 Å range—a few to a half dozen atomic layers. This surface sensitivity was seen as a special benefit of XPS, because there was an enormous scientific and technological interest in the properties of surfaces. For examples, witness the numerous papers describing the surface properties of Si—the key component of modern electronics. With the scaling down of device structures, leading semiconductor companies used the surface-sensitive XPS techniques to probe the Si–SiO_2 interface on an atomic scale, for the Si itself as well as for metallic impurities [519].

More recently, characterization of silicon wafers for solar energy applications is another field that has benefited from the surface sensitivity of XPS. A typical application involves understanding the processing of silicon heterojunction solar cells, which consist of thin amorphous silicon layers deposited on crystalline silicon

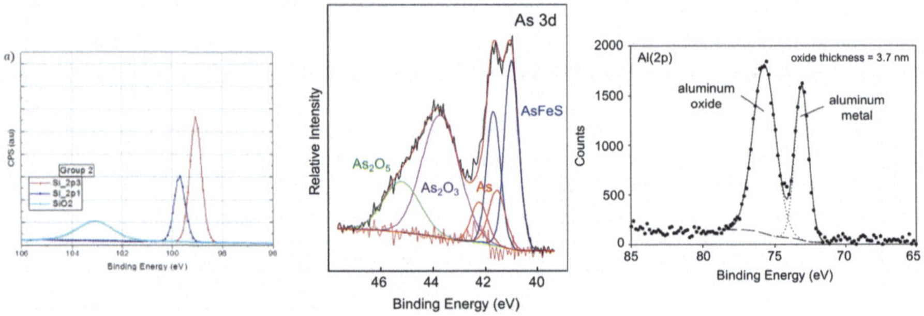

Fig. 11.3 Left: Si XPS region for an Si surface before and after HNA etching, redrawn from [521]. Note loss of 103.5 eV peak. Middle: As 3d XPS spectrum for arsenopyrite. Although technically FeAsS in the bulk, the surface sensitivity of XPS reveals oxidized species such as As_2O_3 and As_2O_5, presumably from air exposure. $FeAsO_4$ might also be present. Right: Al 2p XPS spectrum from a metal surface. The 3.7 nm surface Al_2O_3 thickness was deduced from the relative intensities

wafers [520]. XPS was used to demonstrate that an oxide component on the surface was effectively removed by an HNA (a mix of HF and HNO_3) etching treatment, as illustrated in Fig. 11.3 [521].

11.2.1 Chemical Shifts

We have already seen that the energies of absorption edges are sensitive to the oxidation state of the element under study (Figs. 7.2 and 7.12), and similar trends are seen in XPS spectra. Just as the absorption edge shifts to higher energy for higher oxidation states, the XPS binding energy also shifts in that direction. The shifts are large enough that four or five different chemical species can be disentangled in the spectra of $CF_3COCH_2CH_3$, an Si interface, or a complex As mineral surface (Figs. 11.2 and 11.3). In addition, combined with the surface sensitivity of the technique, the relative strength of Al_2O_3 and Al metal lines can be used to estimate the thickness of the oxide layer on the metal surface (Fig. 11.3).

11.2.2 Multiplet and Spin-Orbit Structure

We saw with transition metal L-edge XANES and, for example, Kβ fluorescence that the combination of a core-hole vacancy and a partially filled valence shell can lead to characteristic multiplet structure. The same effects can be seen in XPS spectra. For direct comparison, consider the XPS of MnF_2 vs. the Kβ spectrum (Fig. 11.4). Ejection of a 3p electron yields a $3p^5 3d^5$ final state—the same final state reached

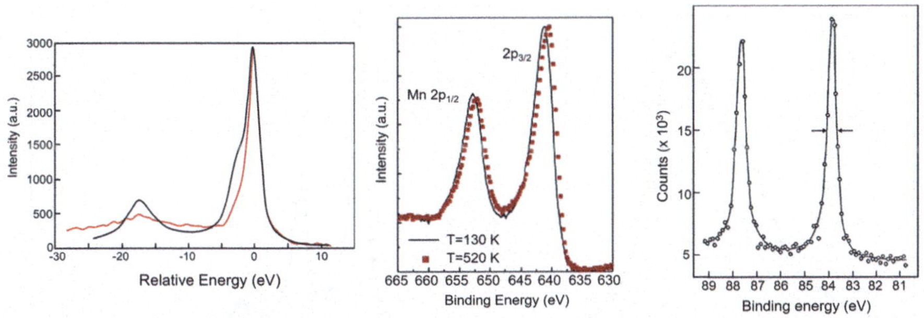

Fig. 11.4 Left: $3p^53d^5$ multiplet structure in 3p XPS of MnF_2 (red line) vs. Kβ fluorescence (black line), redrawn from [357]. Middle: spin-orbit splitting between $2p_{1/2}$ and $2p_{3/2}$ levels for $La_{0.7}Sr_{0.3}MnO_3$, redrawn from [516]. Right: 4f spin-orbit splitting for Au metal, redrawn from [517]

after the 3p→1s transition observed in the Kβ spectrum. As expected, the same 17 eV splitting between ^7P and ^5P levels is seen in both spectra.

Spin-orbit coupling is also prominent in XPS. The $2p_{3/2}$ vs. $2p_{1/2}$ splitting that contributes to separate L_3 and L_2 edges, as well as Kα1 and Kα2 fluorescence lines, is seen in transition metal 3p XPS (Fig. 11.4). For heavier elements such as Au, spin-orbit splitting of $4f_{7/2}$ and $4f_{5/2}$ levels is visible in the XPS (Fig. 11.4). Thus, the same spin-orbit and core-valence multiplet effects are seen across a variety of X-ray spectroscopies.

11.2.3 Vibrational Fine Structure

Vibrational fine structure can be seen in gas-phase XPS of small molecules (just as we saw in the XANES of N_2). The spectrum of methane provided the first evidence for such structure in a core photoelectron spectrum. With ~0.25 eV resolution, in 1974 Siegbahn's group was the first to see this fine structure in XPS [515]. Twenty-five years later, dramatic improvements could be seen with synchrotron radiation, approaching 40 meV resolution [522] (Fig. 11.5).

The observed value of 397 meV for the totally symmetric stretch of $(CH_4)^+$ is higher than the ground-state value of 378 meV for CH_4, but it is quite close to the symmetric stretch of 401 meV for $(NH_4)^+$. Thus, with a core hole, a C atom behaves like a Z + 1 N atom. Note the difference with the XANES for N_2, where the promoted electron went into an antibonding orbital, leading to a lower frequency for the excited-state N≡N stretch.

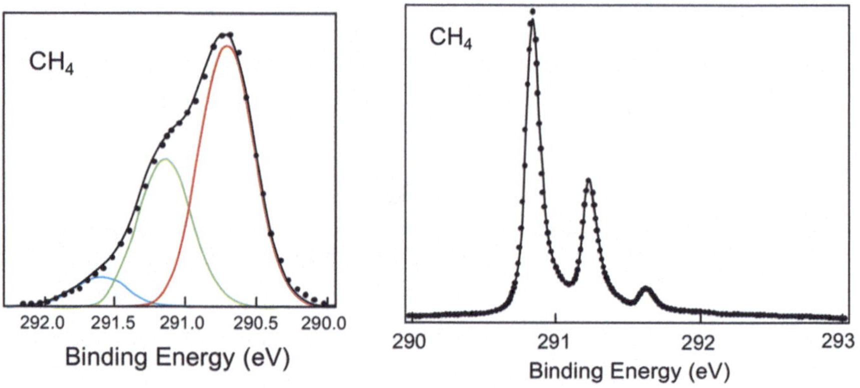

Fig. 11.5 Left: an XPS spectrum for CH_4 in 1974 with a resolution of ~250 meV [515]. Right: improved synchrotron XPS spectrum from the ALS in 1999 with a combined Gaussian resolution of ~42 meV [522]. A fourth peak is just above 292 eV. Note reversed directions of energy scales

A Selective History of Photoemission and Auger Spectroscopies

1877—Hertz observes photoemission [510].

1905—Einstein explains photoelectric effect in terms of quantized photons [511].

1922—Meitner and Auger discover core-hole relaxation via electron emission [523] [524].

1957—Siegbahn publishes first "modern" XPS (for Cu) showing discrete lines [525].

1969—Siegmann reports first spin-resolved photoemission spectrum for Gd metal [526].

1974—Lindau and Pianetta conduct first synchrotron XPS for Au using SSRP [517].

2017—Takagi et al. demonstrate ambient pressure HAXPS at 1 atm at Spring-8 [527].

11.3 Hard X-ray Photoelectron Spectroscopy (HAXPS)

The higher flux of high-energy photons available from third-generation synchrotron sources has contributed toward XPS becoming a bulk-sensitive technique. This is achieved by using incident photon energies of 3–10 keV (or higher). Another key development for using higher energies was commercialization of electron spectrometers capable of reliable operation with retarding voltages in the multi-keV range. The use of "*Ha*rd *X*-ray *P*hotoelectron *S*pectroscopy" ("HAXPS") brings two benefits. First, the photon penetration is much deeper than with soft X-rays.

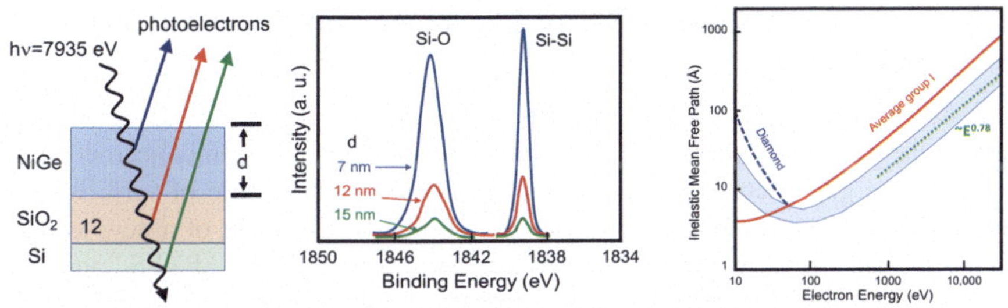

Fig. 11.6 Left: geometry of an SiO_2/Si interface under a NiGe overlay of variable thickness d. Middle: resulting Si 1s HAXPS as function of overlay thickness d [528]. Right: blue band is approximate range of inelastic mean free paths for 41 elements, calculated using the TPP-2 M formula [531]. The dashed green line corresponds to $\lambda \propto E^{0.78}$. Outliers from the general trend are group 1 elements (red solid line) and diamond (dashed blue line). Revised from [516]

Furthermore, the escape depth of the higher-energy photoelectrons is larger, allowing a good approximation to bulk sensitivity. As documented in Appendix K, there are now more than 24 HAXPES beamlines in operation or under construction around the world.

How deep can HAXPS probe? For some samples, it is useful to employ an "information depth," defined by Kobayashi as the thickness of the layer from which 90% of the photoelectron signal originates [528]. But for buried interfaces, the attenuation length λ of photoelectrons is more useful. Sacchi and coworkers used various overlayer thicknesses on top of a Si substrate to experimentally determine λ for different materials and photoelectron energies [529]. As defined in Eq. 11.2 and Fig. 11.6, where I_0^{s} is the expected intensity without an overlayer and d is the overlayer thickness, they found λ on the order of 65 Å at 6 keV for Co, Cu, and Ge, and theory predicts values of 200–1000 Å as one approaches 30 keV (Fig. 11.6) [530]. As seen in Fig. 11.6, with ~8 keV X-rays, signals from a Si substrate can be observed even with 27 nm of intervening NiGe and SiO_2:

$$I = I_0^{\mathrm{s}}\, \mathrm{e}^{-\frac{d}{\lambda}} \tag{11.2}$$

A popular algorithm for estimating the mean free path λ is TTM-2 [532], where the parameters are given in the referenced paper:

$$\lambda = \frac{E}{E_{\mathrm{p}}^2\left[\beta \ln(\gamma E) - \left(\frac{C}{E}\right) + \left(\frac{D}{E^2}\right)\right]} \tag{11.3}$$

while to simply estimate relative energy dependence, a rule of thumb was noted by Fadley:

$$\lambda \propto E^{0.78} \tag{11.4}$$

11.4 Ambient Pressure X-ray Photoelectron Spectroscopy (APXPS)

Despite the enormous power of XPS, scientists have always been concerned about the "pressure gap" between UHV surface science and the nature of surface under real-world reaction conditions. This has driven the development of a new technology, "*A*mbient *P*ressure *X*-ray *P*hotoelectron *S*pectroscopy" or "APXPS," that uses barriers and/or differential pumping to separate an ambient pressure sample from a UHV electron analyzer (Fig. 11.7).

11.5 Spin-Resolved Photoemission Spectroscopies

Adding electron spin detection to the energy measurement can provide additional information that is especially important for characterization of magnetic materials [534]. This approach, variously known as "*S*pin-resolved *P*hoto*e*mission *S*pectroscopy" or "spin-polarized photoemission," as well as a plethora of acronyms such as "SPES," "SR-PES," "SPPES," "SRXPS," and so on, requires measuring the electron spin as well as the electron energy. The first spin-resolved photoemission experiment was done by Siegmann and coworkers [526]. Since then it has become an important tool in XPS research.

The most widely used device for measuring the electron spin is the notoriously inefficient Mott polarimeter. In this instrument, the spin-orbit interaction in a heavy metal such as Au introduces an asymmetry, dependent on the electron spin, in the

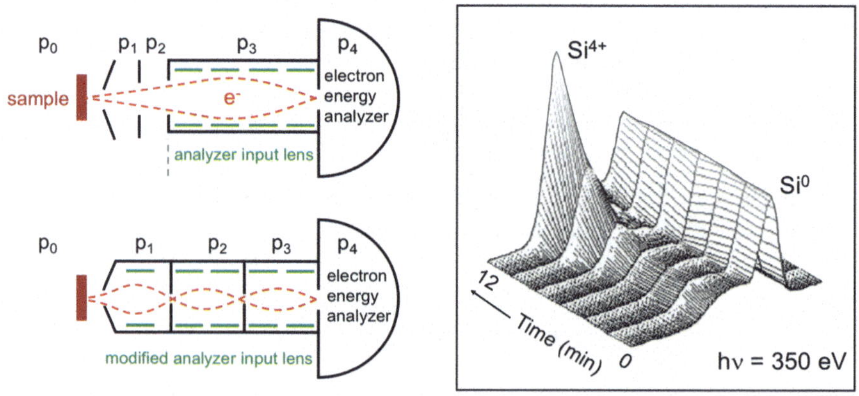

Fig. 11.7 Left: two approaches to APXPS. In the top case, differential pumping occurs between the sample and a standard electron spectrometer. In the bottom case, differential pumping occurs wholly within a modified analyzer. Intermediate approaches are of course also possible [533]. Right: Time-dependent oxidation of a Si surface observed by APXPS. Spectra shown here were recorded every 8 s (redrawn from [516])

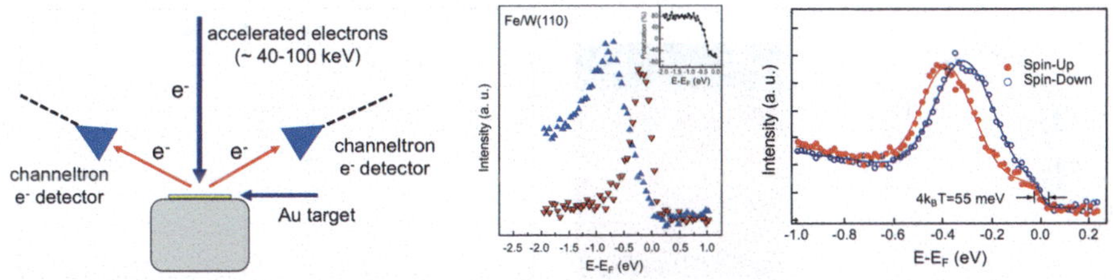

Fig. 11.8 Left: schematic of a Mott detector. Middle: spin-resolved PES for Fe on a W(1 1 0) surface. Exciting energy was 21.2 eV [535]. Right: spin-resolved PES for Au(1 1 1) on W(1 1 0) surface, using "LEX polarimeter" [536]

rate of backscattering into opposite directions (Fig. 11.8). The spin polarization of the electron beam is given by:

$$P = \frac{1}{S}\frac{I_A - I_B}{I_A + I_B} \tag{11.5}$$

where I_A and I_B are the measured intensities in opposing detectors and S is the so-called Sherman function, a measure of the ability of the device to distinguish opposite spins. The figure of merit for a device is given by:

$$\text{FOM} = S^2 \frac{I}{I_A + I_B} \tag{11.6}$$

Unfortunately, to achieve a useful asymmetry, electron kinetic energies on the order of 20–100 keV are required, and the cross sections for scattering events at these energies are extremely low. The Mott detector thus has several orders of magnitude decreased efficiency compared to conventional detectors.

Spin measurements are also being done based on the exchange interaction in low-energy scattering from ferromagnetic surfaces [536]. Since the cross sections at low (~20 eV) energies are orders of magnitude higher than at >20 keV, these devices offer much more efficient spin analysis.

11.6 Angle-Resolved Photoemission (ARPES)

The goal of an "*A*ngle-*R*esolved *P*hoto*E*mission *S*pectroscopy" or "ARPES" experiment is to understand the dispersion relation $E(\vec{k})$ for a solid based on measuring the energy E_{kin} and momentum $\vec{k}$ for emitted photoelectrons in vacuum. The geometry for such an experiment was shown in Fig. 11.1. A modern approach to an ARPES experiment is shown in Fig. 11.9, where a *2D* detector allows simultaneous measurement of a range of energies and momenta.

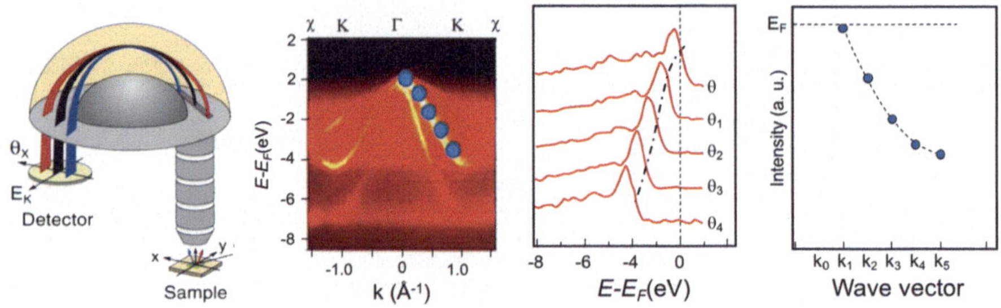

Fig. 11.9 Left: a modern ARPES system allowing simultaneous measurement of energy and angular distributions [537]. Right: alternate representations of ARPES data measured on annealed Ag(1 0 0) cleaned surface [538]

For a free electron in vacuum, the modulus of the wave vector $\vec{K} = \vec{p}/\hbar$ is given by:

$$K = \frac{\sqrt{2m_e E_{\text{kin}}}}{\hbar} \tag{11.7}$$

and referenced to the surface, the parallel ($\vec{K}_{\parallel} = \vec{K}_x + \vec{K}_y$) and perpendicular ($\vec{K}_{\perp} = \vec{K}_z$) components are obtained using the axes defined in Fig. 11.9 by [539]:

$$K_x = \frac{\sqrt{2m_e E_{\text{kin}}}}{\hbar} \sin\vartheta \, \cos\varphi \tag{11.8}$$

$$K_y = \frac{\sqrt{2m_e E_{\text{kin}}}}{\hbar} \sin\vartheta \, \sin\varphi \tag{11.9}$$

$$K_z = \frac{\sqrt{2m_e E_{\text{kin}}}}{\hbar} \cos\vartheta \tag{11.10}$$

It turns out that the parallel component of the electron momentum is conserved during the photoemission process, while the perpendicular momentum (assuming a nearly free electron [539]) requires adjustment by the inner potential:

$$k_{\parallel} = K_{\parallel} = \frac{\sqrt{2m_e E_{\text{kin}}}}{\hbar} \sin\vartheta \tag{11.11}$$

$$k_{\perp} = \frac{\sqrt{2m_e E_{\text{kin}} \cos^2\vartheta + V_0}}{\hbar} \tag{11.12}$$

ARPES has played an important role in understanding the electronic structure of high-T_c superconductors [537, 540–542] (Fig. 11.10).

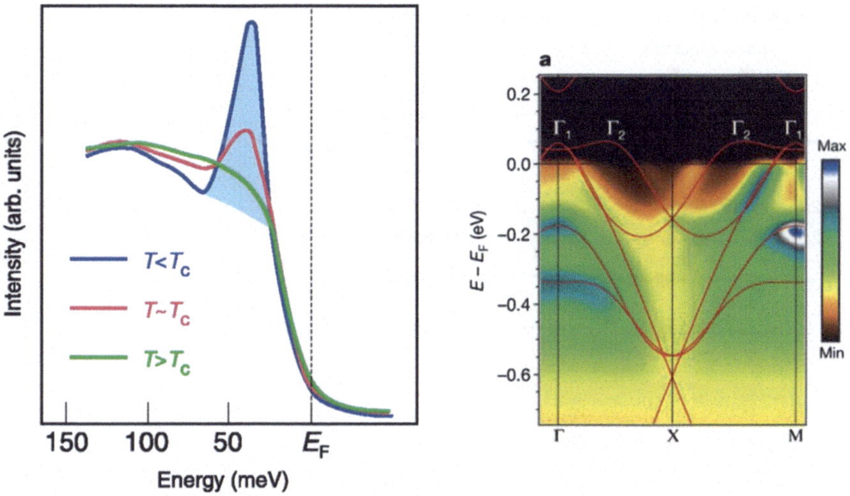

Fig. 11.10 Left: ARPES for an underdoped high-T_c superconductor at momentum near the $(0,\pi)$ or "antinodal" point of the Brillouin zone [540]. Right: ARPES for the iron oxypnictide superconductor LaOFeP ($T_c = 5.9$ K) [541]

11.7 Auger Electron Spectroscopy (AES)

When an atom is excited by an X-ray, it is promoted to an excited state with a core vacancy. Filling the core hole with a higher-level electron releases energy, and Auger electron emission is the process that competes with X-ray fluorescence for disposing of that energy. In the Auger process, instead of relaxing by emission of a photon, the atom ejects an electron (Fig. 11.11). "*A*uger *E*mission *S*pectroscopy" or "AES" records the energies of these electrons with the same instrumentation used for XPS. Because the energies of the levels are almost fixed, Auger electron energies are approximately fixed, and they can be extracted from published tables [543]. This is in contrast to absolute photoelectron energies, which track the energy of the exciting X-ray. Pierre Auger observed his eponymous effect in 1923, but in fact Lise Meitner had observed "Auger emission" a year earlier [523, 524].

There are three electron energy levels involved in an Auger transition, and the nomenclature for labeling these events reflects those levels. Supposing that the initial vacancy is in the A shell, the level filling that vacancy is the B shell, and the promoted electron is in the C shell, then the Auger electron and associated transition are labeled ABC. For example, a common case is to ionize a 1s electron, fill the vacancy with a 2p electron, and promote a second 2p electron into the continuum. This would be a KLL Auger transition, with additional subscripts for the type of 2p electron promoted. Valence electrons can also be promoted; they acquire the label V. For example, exciting Si at its $L_{2,3}$ edge leads to ejection of a core electron, filling the vacancy with a valence electron, and ejection of a valence electron to shed the extra energy—hence described as an $L_{2,3}VV$ process.

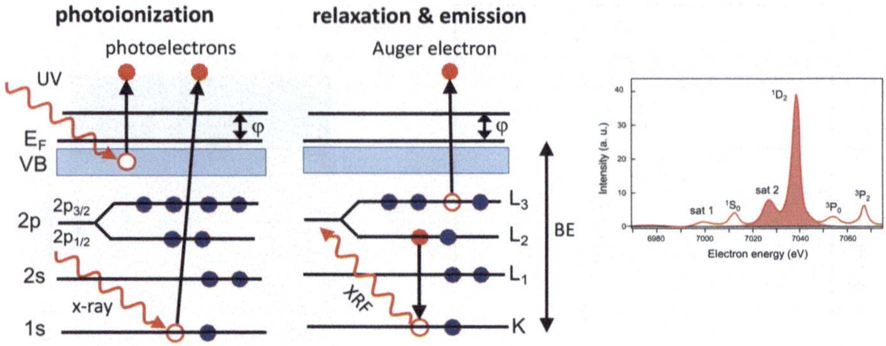

Fig. 11.11 Left: a comparison of photoemission and Auger emission. The competition between X-ray fluorescence and Auger emission yields was illustrated in Fig. 8.2. Right: high-energy KLL Auger spectrum for Cu metal, redrawn from [544]

A semi-rigorous calculation of the Auger electron energy would involve consideration of the initial state configuration with one core vacancy and the final state configuration with a second vacancy:

$$\mathrm{E}^{\mathrm{atom}}(a) \rightarrow \mathrm{E}^{\mathrm{atom}}(bc) \tag{11.13}$$

where (a) represents an atom with a vacancy in shell A and (bc) represents the same atom with vacancies in shells B and C. With two vacancies in the final state, there are invariably multiplet splittings to account for. Of course, if there is a partially filled valence shell, then there will be a variety of additional multiplet interactions for both initial and final states. The multiplet splittings for Cu KLL Auger emission are illustrated in Fig. 11.11, and some calculated values are reported in Table 11.1.

Modern theoretical calculations with the appropriate Slater integrals do a good job at predicting the relative energies of different Auger transitions, although they still need adjustment to get correct absolute values (Table 11.1). A less rigorous empirical formula (that ignores multiplet effects) for predicting the energy of Auger electrons has been offered by Chung and Jenkins [545]:

$$E_{\mathrm{ABC}} \cong E_{\mathrm{A}}(\mathrm{Z}) - \frac{1}{2}\{E_{\mathrm{B}}(\mathrm{Z}) + E_{\mathrm{B}}(\mathrm{Z}+1)\} - \frac{1}{2}\{E_{\mathrm{C}}(\mathrm{Z}) + E_{\mathrm{C}}(\mathrm{Z}+1)\} \tag{11.14}$$

Applying this to Cu as an example, and using binding energies from Bearden and Burr [546], yields the predictions summarized in Table 11.1.

We see that a semi-rigorous calculation using the appropriate Slater integrals is off by 7 eV in absolute energy, but it does an excellent job at predicting the relative energies of the different multiplets. The Chung and Jenkins formula is worse at absolute energies, but it too yields reasonable estimates for relative energies of the main features. Despite the success of such tabulated values, Auger electron energies

Table 11.1 Experimental and theoretical Cu Auger energies

Level	Energy (eV)	Level	Energy (eV)	Transition	Experiment (eV) Egri et al. [544]	Theory (eV) Larkin [543]	Simple (eV) Algorithm [545]
Cu K	8978.9	Zn K	9658.6	KL_2L_3 $^1D_2$	7038.6	7031.1	7006.6
Cu L_1	1096.6	Zn L_1	1193.6	KL_3L_3 $^3P_0$	+16.2	+16.3	+21.5
Cu L_2	951.0	Zn L_2	1042.8	KL_3L_3 $^3P_2$	+28.6	+27.6	
Cu L_3	931.1	Zn L_3	1019.7	KL_2L_2 $^1S_0$	−26.2	−24.2	−21.5

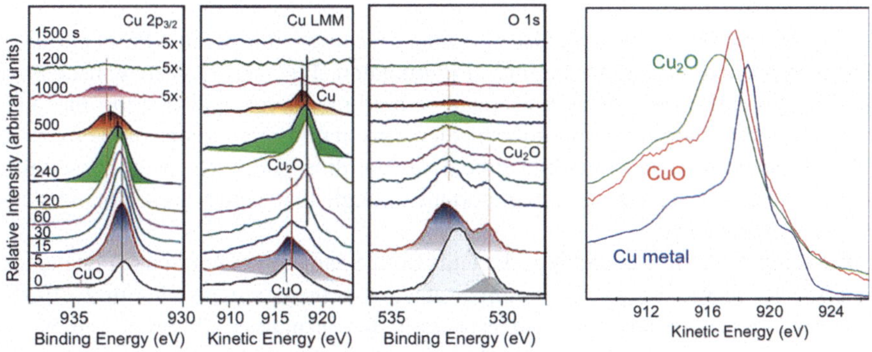

Fig. 11.12 Left: Cu $2p_{3/2}$ XPS, Cu LMM Auger, and O 1s XPS for Cu nanoparticles as-electrodeposited on Au [547]. Right: typical shifts for Cu LMM Auger spectra

exhibit shifts for different chemical states of an element, just as observed for absorption edges or XPS binding energies. The magnitude of these shifts allows characterization of surface species, as seen in Fig. 11.12.

11.7.1 Why Auger Spectroscopy? Why Synchrotron Radiation?

The most commonly employed Auger lines are in the 10–1000 eV range, and in this region, electron escape depths are in the nm range. Thus a benefit of Auger spectroscopy is its intrinsic and exquisite surface sensitivity. The extra benefit of synchrotron radiation is the ability to tune the excitation energy for maximum sensitivity to certain elements or even to tune to specific features on an absorption edge to conduct "resonant Auger spectroscopy." For more details, see [548].

11.8 More Is Not Always Better

We began this book describing the remarkable increases in brightness of synchrotron radiation over the past four decades. But, in the case of electron spectroscopy, more photons per unit time or per spatial area are not always better. Two issues are particular to methods that detect electrons: sample charging and space-charge effects. And of course, there is always the issue of sample damage, but this is common to all other X-ray methods.

11.8.1 Sample Charging

The loss of photoelectrons and also secondary electrons during an XPS experiment leads to a positive charge in the region illuminated (Fig. 11.13). This will create a retarding potential that raises the apparent binding energy and distorts the XPS spectrum. If the charging is nonuniform, electrons in different regions of the material will experience different retarding potentials, yielding additional peak broadening distortions. *Charging effects* are important enough that entire book chapters are devoted to them [551].

The magnitude of this problem depends on the sample conductivity. For metals, the positive charge is quickly compensated, and there are no charging effects. However, for semiconductors and especially for insulators, charging can be an issue. The key point is that the higher the photon flux density, the greater the possible distortion. At some point, extra synchrotron brightness will no longer be a benefit.

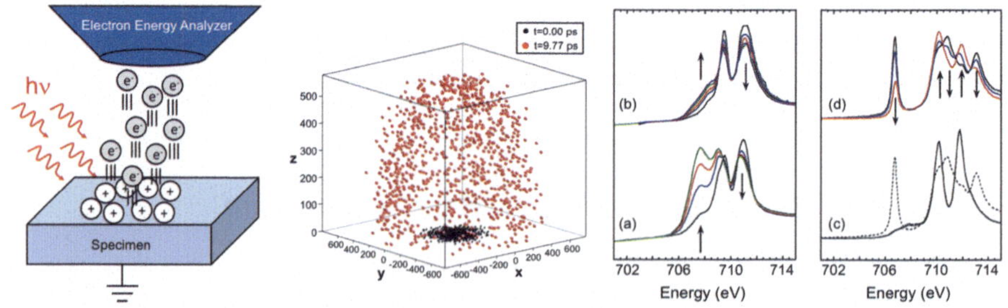

Fig. 11.13 Left to right: schematic of sample charging; development of space-charge effects [549]; typical radiation damage observed for $LFe(CO)_2$ complexes; and X-ray photoreduction of ferricyanide (top), compared with (bottom) spectra for ferricyanide (dashed line) and ferrocyanide (solid line) [550]

11.8.2 *Space-Charge Effects*

The pulse of photons from a synchrotron radiation source creates a cloud of electrons with a finite size on its way to the analyzer. The initial distribution of energies and momenta will be altered by the mutual Coulomb repulsions between the electrons—so-called space-charge effects (Fig. 11.13) [549].

Of course, the significance of space-charge effects depends on the energy or momentum resolution that is required. One estimate suggests that to keep the energy distortions below 50 meV, the linear photoelectron density per photon pulse must be limited to less than ~10^4 mm^{-1}. As with sample charging, these issues become especially important when using very short pulses such as from free-electron lasers described in Chap. 12.

11.8.2.1 Radiation Damage

By now, we should not need a reminder that X-rays are ionizing radiation. The degree to which different samples are susceptible to radiation damage varies enormously with chemical composition. Metals and many other materials such as Si and diamond are almost impervious to damage—witness their use in mirrors and monochromators. At the other extreme, organometallic complexes such as Fe carbonyl or cyanide complexes undergo rapid photochemistry (Fig. 11.13). In Chap. 12 we will learn about free-electron lasers that can literally vaporize samples. Any X-ray spectroscopy study should include an assessment of possible radiation damage effects, but this is particularly important for photon-in electron-out methods where the sample volume is small and the dose to sample ratio is large.

11.9 Suggested Exercises

1. Estimate the energy of a $2p_{3/2}$ photoelectron from Cu metal if the incident X-ray energy is 10 keV.
2. Use the TTM-2 algorithm to estimate the mean free path for a 1 keV electron in silicon.
3. Estimate by what factor the mean free path increases for a 5 keV electron in silicon, using the exponential approximation.
4. Use the Chung and Jenkins formula to estimate the KL_3L_2 Auger energy for Fe.
5. Estimate the peak photon intensity (photons s^{-1} mm^{-2}) at 10 keV on a Cu surface that would exceed the proposed space-charge limit of ~10^4 photoelectrons mm^{-1}.

11.10 Reference Books and Review Articles

1. *Photoelectron Spectroscopy—Principles and Applications*, third edition, S. Hüfner, Springer-Verlag, Berlin (2010) ISBN 3642075207—update of the 1995 text, softcover edition.
2. *Hard X-ray Photoelectron Spectroscopy,* J. Woicik, ed. Springer, Switzerland (2015) ISBN 3319240412.
3. *Spectroscopy of Complex Oxide Interfaces—Photoemission and Related Spectroscopies*, C. Cancellieri and V. N. Strocov, eds. Springer, Switzerland (2018) ISBN 978-3-319-74988-4—a book about applications of XPS to metal oxides.
4. *Photoelectron and Auger Spectroscopy*, Thomas A. Carlson, Plenum, New York (1975) ISBN: 978-1-4757-0120-3—original is inexpensive on Amazon and a good source of early literature, reprint available as a Springer softcover or e-book ISBN 978-1-4757-0118-0.
5. *Synchrotron Radiation News*, Vol. 27, No. 3, **2014**—an entire issue about the future of photoemission spectroscopy.
6. *Synchrotron Radiation News*, Vol. 31, No. 4, **2018**—an entire issue with news about HAXPES around the world.

** NIST provides databases useful for XPS applications (http://srdata.nist.gov/xps/), as described in [552].

Chapter 12
Free-Electron Lasers

12.1 Concepts

In our description of undulator radiation, we saw the enormous gains that could be achieved if radiation from separate magnetic poles were added coherently. As a reminder, when N equal sources are added coherently, the produced intensity I varies as N^2, compared to intensity gains of only N when the sources are added incoherently. In a conventional undulator, radiation from N_p individual poles is coherent; hence $I \propto N_p^2$. However, electrons within a given bunch still radiate incoherently with respect to each other, so the intensity varies linearly with the number of electrons N_e. (One exception is long wavelength synchrotron radiation—if the wavelength is long compared to the bunch length, in the far-IR or THz regimes, then coherent synchrotron radiation can be observed) [553].

The key feature of a "*f*ree-*e*lectron *l*aser" or "FEL" is the restructuring of the electron bunch into equally spaced "microbunches," so that the electrons also radiate coherently. Since the number of electrons N_e in a bunch is on the order of 10^8, the potential for additional intensity from N_e^2 vs. N_e is enormous. This is true for average brightness, but especially true for peak brightness, as illustrated in Fig. 12.1.

12.2 Peak Brightness vs. Average Brightness: A New Figure of Merit

Back in Chap. 1 when we defined the X-ray spectral brightness, we referred to the average over time, and we ignored the bunch structure of the storage ring source:

$$\mathcal{B}_S = \frac{\text{number of photons}}{(\text{s})(\text{mm}^2)(\text{mrad}^2)(0.1\%\Delta E/E)} \tag{12.1}$$

S. P. Cramer, *X-Ray Spectroscopy with Synchrotron Radiation*, Biological and Medical Physics, Biomedical Engineering, https://doi.org/10.1007/978-3-030-28551-7_12

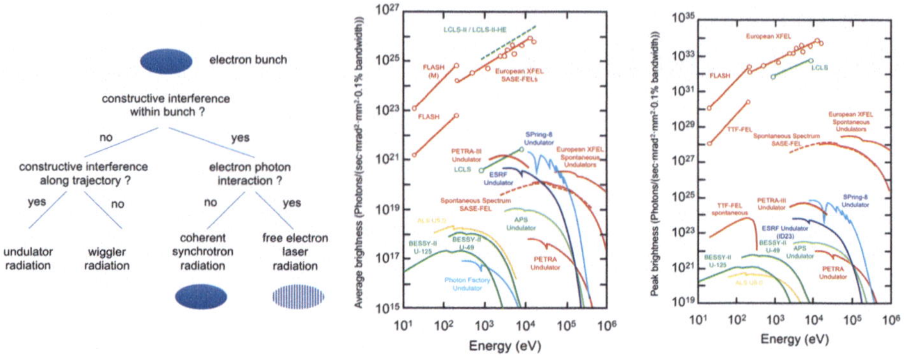

Fig. 12.1 Left: division of radiation regimes between wigglers, undulators, and free-electron lasers. Middle and right: average and peak spectral brightness of FELs compared to storage rings [554]

Consider a typical storage ring operating mode, the APS running with 24 bunches with a 33 ps bunch length and a separation of 153 ns between bunches. Assuming a flat distribution within the bunch, we see the duty cycle is approximately 33 ps out of 153 ns or ~2×10^{-4}. Thus, the peak brightness that occurs when the bunch is actually illuminating the beamline is higher than the average brightness by almost four orders of magnitude. The best storage rings have peak brightness on the order of 10^{25}. For comparison, FELs add another ten orders of magnitude!

12.3 Why Is Peak Brightness Important?

- For time-resolved work, it determines how brief a measurement can be achieved.
- For nonlinear spectroscopy, it allows observation of effects that depend on I^n.
- For high energy density work, it allows deposition of energy into small volumes.
- For diffraction and imaging experiments, high peak brightness of femtosecond pulses offers the possibility of capturing data before the sample is destroyed.

The development of free-electron lasers with high peak brightness has ushered in a second synchrotron radiation revolution, similar to the way the high average brightness of SR drew a new community of scientists into the X-ray field. Of course, peak brightness has enabled existing synchrotron scientists to do better, faster, and more sensitive experiments. The enormous change in peak brightness—ten orders of magnitude—has also attracted scientists from new fields such as nonlinear optics, high-density plasma research, and shock physics [555]. In short, FELs have quantitatively changed the time-scales of existing methods by orders of magnitude, but they have also qualitatively changed the type of science that can be done.

12.3.1 A Condensed FEL History

Hard X-ray FELs have only appeared in the last decade, but they represent the culmination of more than half a century of research. As far back as 1953, Motz used an electron bunch length shorter than the radiation wavelength to observe coherent radiation in the mm wavelength range from an undulator and a 5 MeV electron beam [556]. In the early 1970s, Madey's group demonstrated the FEL concept in the infrared using using a 24 MeV electron beam [557, 558]. Once "*S*elf-*A*mplified *S*pontaneous *E*mission" or "SASE" theory was developed, it became clear that mirror-free FELs were possible, and the next few decades saw a progression of SASE demonstrations from 1 cm to 1 Å. The world now has five hard X-ray FELs (Appendix Table M.1 and Fig. 12.8), doubtless with upgrades and additional sources to come. More on the history of FEL development can be found in reviews by Pellegrini [559, 560] and Madey [561].

A Selective History of FEL Highlights
1953—Motz observes coherent mm radiation from undulator [556]
1971–1973—Madey low-gain FEL theory [557, 558]
1976–1977—Madey demonstrates low-gain FEL at 10.6 μ [562, 563]
1980—Kondratenko and Saldin: High-gain SASE concept [564]
1984—Bonifacio, Pellegrini, and Narducci SASE theory [565]
1985—LBNL/Livermore SASE demonstration at 1 cm [566]
1998—UCLA/Los Alamos SASE demo at 12 μ [567]
2001—Argonne "LEUTL" SASE demonstration at 520 and 385 nm [568]
2002—DESY SASE at 98 nm [569]
2005—TESLA test facility (FLASH): Soft X-ray FEL at 32 and 4.2 nm [570]
2012–2016—FERMI HGHG seeded FELs: 65–20 nm to 4 nm [571]
2009—LCLS hard X-ray FEL at 1.5 Å [572]
2012—SACLA FEL at 0.6 Å [573]
2017—European XFEL: high repetition rate XFEL [574]

12.4 FEL Physics

Free-electron lasers offer enormous improvements in peak brightness. The fundamental mechanism behind the "FEL process" involves the interaction between the electrons oscillating through an undulator and the synchrotron radiation that these electrons produce. It is achieved by (1) using bunches that are three orders of magnitude shorter, (2) achieving an ultralow electron beam emittance that creates a diffraction limited source, and (3) having very long undulators that allow development of a microbunch structure on the overall bunch so that the electrons radiate coherently.

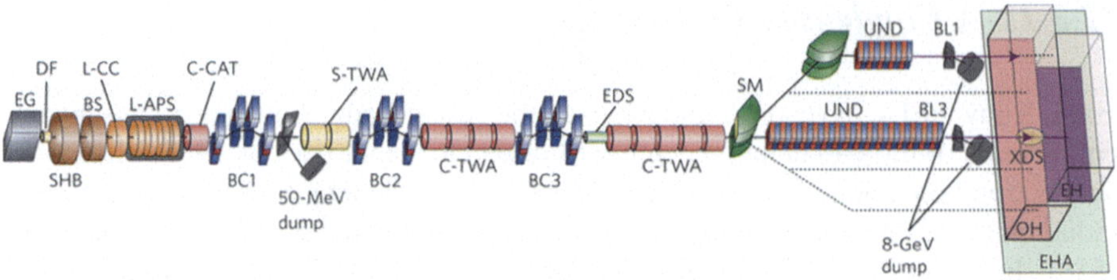

Fig. 12.2 A schematic for the SACLA FEL in Japan. EG is the electron gun, BCn are the bunch compressors, TWA are traveling-wave accelerating tubes, and UND are the undulators [575]

For the strength of this field-bunch interaction to be strong, the electron bunch needs to be as short as possible. This is accomplished by starting with a very short electron bunch and compressing the bunch further (by orders of magnitude) with electron optics. A typical configuration employs a series of "bunch compressors," as seen for SACLA in Fig. 12.2. The technology for producing ultrashort and low emittance electron beams was developed in part for high-energy linear collider physics.

Where does the microbunching come from? Assuming an undulator with a vertical magnetic field, electrons undergo horizontal motion in the undulator, and they experience a Lorentz force in the axial direction from the vertical magnetic field of the synchrotron radiation. Depending on the relative phase of the transverse motion with respect to the radiation, the electrons will be deflected forward or to the rear. Over the course of one undulator oscillation, the radiation field advances by one wavelength; hence the Lorentz forces act in the same manner over hundreds of undulator periods until the bunch structure is fully developed and the process saturates (Fig. 12.3).

Although many descriptions refer to electron acceleration or deceleration, it is more accurate to refer to energy gain or loss. Electrons that gain energy move forward in the bunch, because they are deflected less by the magnetic field of the undulator. Electrons that lose energy slip toward the rear of the bunch. Gradually, the initial bunch becomes modulated with a period corresponding to the undulator radiation wavelength (Fig. 12.3). As the bunching gets more profound, the coherence of the radiation from different regions improves, yielding a more intense radiation field. This in turn leads to ever-sharper microbunching and still stronger radiation. The net result is an exponential growth in radiation intensity until a saturation point is reached: Self-Amplified Spontaneous Emission (Fig. 12.3).

12.4.1 SASE Properties: The FEL or Pierce Parameter ρ

In Chap. 3 we were able to present (if not derive) analytic solutions to the power, intensity, and spectral distribution of bend magnet, wiggler, and undulator sources. Although there are analytical treatments [578, 579], in the high-gain (exponential

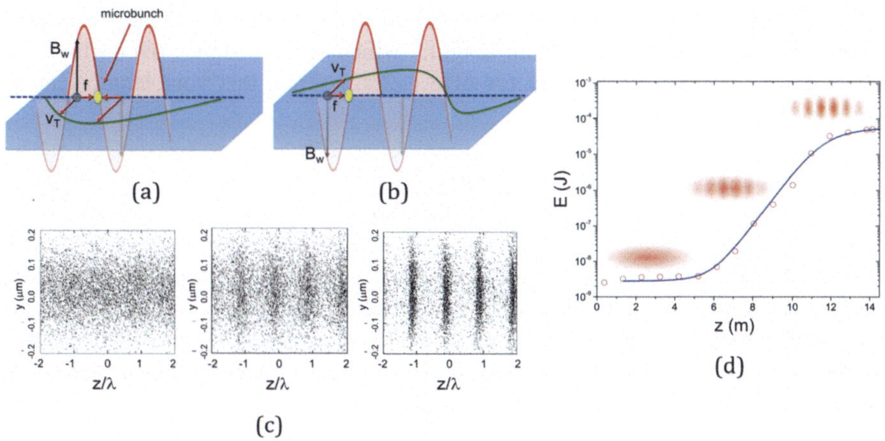

Fig. 12.3 Top left: the interaction of the electron beam with the synchrotron radiation field: in (**a**) the longitudinal Lorentz bunching forces result from the photon wave B-field B_{W} and the electron transverse velocity v_{T}; in (**b**) after the bunch travels one half undulator period, the reversed B-field interacts with the reversed electron transverse velocity to yield the same net Lorentz bunching forces, redrawn from [576]; (**c**) a numerical simulation of the gradual microbunch of the overall electron bunch and; (**d**) the exponential growth in radiation intensity as a function of distance through the undulators, redrawn from [577]

growth) regime, there are no complete analytic solutions, and it is necessary to employ numerical simulation codes. Still, there is one useful parameter that distills the essential physics of 1D FEL theory: the "Pierce parameter" or "FEL parameter" ρ.

The parameter ρ is a fundamental indicator of FEL performance—it represents the efficiency of the conversion of electron beam power to radiation power, and it has been called "the jack of all trades of 1D FEL theory" [580]. It is calculated from a collection of fundamental constants and system parameters:

$$\rho = \left[\frac{I_{\mathrm{e}}}{\gamma^3 I_{\mathrm{A}}} \frac{\lambda_{\mathrm{u}}^2}{2\pi\sigma_x\sigma_y} \frac{(K \times [\mathrm{JJ}])^2}{32\pi}\right]^{1/3} \tag{12.2}$$

where I_{e} is the peak electron current, $I_{\mathrm{A}} = \mathrm{e}c/r_{\mathrm{e}} \approx 17$ kA (the "Alfvén current"), γ is the ratio of the electron energy to its rest mass energy, σ_x and σ_y are the rms transverse sizes of the electron beam, λ_{u} is the undulator period, and K is the undulator parameter. JJ is a combination of Bessel function defined as $\mathrm{JJ} = [J_0(\xi) - J_1(\xi)]$, where $\xi = K^2/(4 + 2K^2)$.

Although ρ may be complicated to derive, it is easy to see how it depends on system parameters. Clearly, a larger ρ will result from (a) a smaller transverse beam size σ, (b) a longer undulator period λ_{u}, and (c) a lower beam energy $\sim\gamma$. On the other hand, ρ modestly improves with beam current I_{e}. Typical values for ρ are on the order of 10^{-3} .

Quite a few FEL properties can be estimated from the Pierce or FEL parameter. Suppose we define L_{G0} as the "gain length," which is the distance over which the radiation power increases by a factor of e (~2.72). Then this gain length can be expressed in terms of ρ as:

$$L_{G0} = \frac{\lambda_u}{4\pi\sqrt{3}\rho} \tag{12.3}$$

Thus, the larger the FEL parameter, the smaller the gain length L_{G0} and the more rapid the exponential increase in radiative power. Assuming a bunch energy E_b, other useful metrics that can be estimated from ρ include E_r, the total emitted radiation energy per pulse at saturation; L_s, the undulator length to reach saturation; and $\Delta\omega/\omega$, the SASE bandwidth at saturation:

$$E_r \sim 1.6 \cdot \rho \cdot E_b \tag{12.4}$$

$$L_s = \frac{\lambda_u}{\rho} \tag{12.5}$$

$$\frac{\Delta\omega}{\omega} \sim \rho \tag{12.6}$$

In more exact treatments, a degradation factor χ is introduced to account for additional beam properties and diffraction effects, so that the gain length is increased by a factor χ [568].

12.4.2 SASE Limitations

When electrons from the linac arrive at the first undulator of the FEL, they radiate independently and "stochastically." The position in the undulator where the amplification begins will vary, as will the precise wavelength of the radiation that is amplified. Thus, the output from the FEL will exhibit fluctuations in intensity, and the wavelength of that output can be anywhere within the natural bandwidth of the undulator radiation (Fig. 12.4). Although SASE FELs are an enormous improvement in peak brightness, the radiation would be even more useful if the pulses were always of the same intensity and wavelength.

12.4.3 Seeded FELs

The goal of seeded FELs is to remove the jitter in power and wavelength that results from the stochastic nature of the SASE process. At low energies (extreme ultraviolet), it is possible to start the FEL process by injecting radiation from a "high-gain

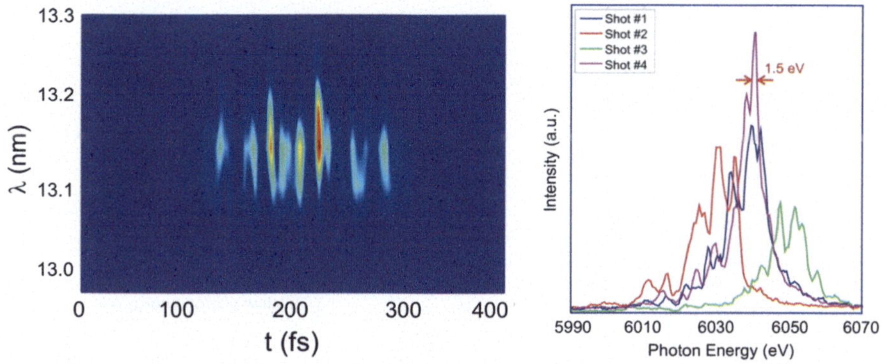

Fig. 12.4 Left: simulation of the time and wavelength structure of FEL output, redrawn from [555]. Right: measurement of the properties of individual FEL "shots" at the LCLS [581]

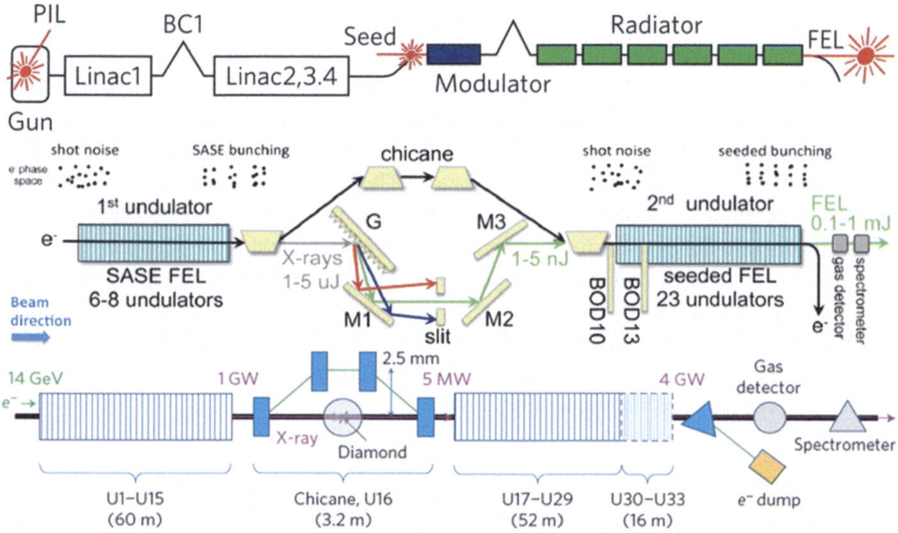

Fig. 12.5 Top: seeding approach for a UV FEL using an HGHG laser. Middle: seeding approach for a soft X-ray FEL using a grating monochromator (SXRSS system at LCLS) [582]. Bottom: seeding for a hard X-ray FEL using a diamond crystal monochromator [583]

harmonic generation" (HGHG) laser. The laser "seed" is upconverted within the first undulator called the "modulator," and it then is imposed on a delayed electron beam in a second undulator called the "radiator." At the low-energy source FERMI@Elettra, this yields a beam with much improved wavelength and amplitude stability in the 65–20 nm range (Fig. 12.5) [571].

At higher X-ray energies, out of the range of HGHG lasers, an alternative approach is to use the early FEL radiation, process it through a monochromator to

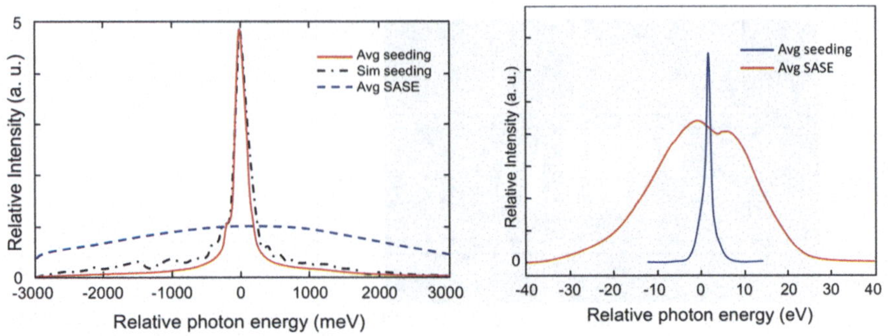

Fig. 12.6 Improvements in energy stability from seeding in (left) a soft X-ray FEL, redrawn from [582] and (right) a hard X-ray FEL, redrawn from [583]

select a single wavelength, and then recombine it with the delayed electron beam to stimulate the FEL gain process. This approach is called "self-seeding." In the soft X-ray regime, the monochromators are based on gratings, while seeding at higher energies employs diamond crystals to select the desired wavelength (Fig. 12.5).

Note the presence of a "chicane" in the self-seeding FELs. This is a set of magnets that will divert the electron beam around the monochromator and also delay it in time. The presence of a monochromator delays the radiation by ~5–10 ps in the soft X-ray range or ~10 fs in the hard X-ray region. The chicane delay matches the timing of the electron and photon beams so that they overlap following the seeding monochromator. Dispersion in the chicane also resets the electron beam to its initial shot noise distribution and washes out any of the original microbunching from the SASE process. The monochromatic X-rays and the electron beam then interact in a second undulator, restarting the FEL interaction at a controlled wavelength. The improvements in photon energy stability that result from self-seeding in the soft and hard X-ray regions are shown in Fig. 12.6.

12.5 The XFELO Option

There are proposals for even brighter XFEL sources, and one of these is an "*X*-ray *f*ree-*e*lectron *l*aser *o*scillator," or "XFELO" [504, 584, 585]. As a reminder, an oscillator FEL is a device where the emitted radiation is reflected multiple times back and forth through the undulator. For infrared radiation this is an old concept—recall it was demonstrated at 10.6 μ by Madey at Stanford around 1976 [562, 563]. Oscillator FELs are now common for the IR, visible, and UV regions, where low-loss normal incidence mirrors are readily available.

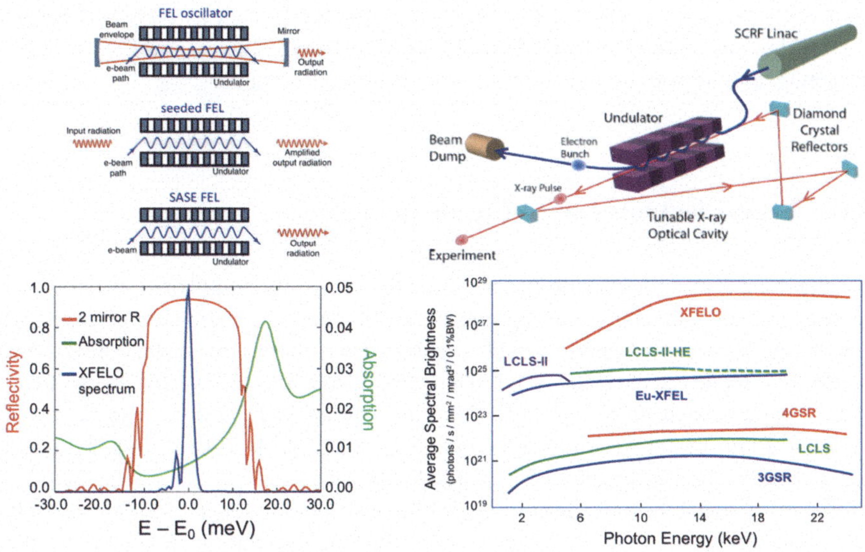

Fig. 12.7 Top left: a comparison of SASE FELs, seeded FELs, and XFELOs. Top right: simplified diagram for an XFELO. Lower left: the bandwidth of an XFELO. Lower right: the improvement in spectral brightness for an XFELO over other sources [504, 586]

Colella and Luccio first proposed to use crystals as reflectors for an XFELO in 1983, at the same workshop where SASE was proposed for X-ray FELs [587]. Following decades of progress in electron beam optics, the XFELO concept was revived by Kim and coworkers around 2007 [588]. The basic idea is to use a set of diamond crystals to recirculate the emitted X-rays through the FEL undulator many times (Fig. 12.7) [589]. The object is to achieve gain in the X-ray cavity as a fraction of the X-ray beam is reintroduced into the undulator over and over again.

An XFELO achieves extraordinary spectral brightness because the gain within the cavity is extremely frequency dependent. As described by Kim, another way to think about this is that the coherence length is increased during each pass through the undulator, so that the spectral bandwidth σ_ω decreases after n passes through the undulator with N_u poles by:

$$\left(\frac{\sigma_\omega}{\omega_1}\right)_n \sim \frac{1}{N_\mathrm{u}}\frac{1}{\sqrt{n}} \tag{12.7}$$

Recent calculations for an XFELO on LCLS-II HE suggest that a bandwidth of 3 meV could be achieved at the 14.4 keV Mössbauer resonance, while the bandwidth of SASE radiation is 2.5×10^3 broader. The average spectral brightness of an XFELO could be 1000× higher than SASE mode (Fig. 12.7). All of that improvement would translate to enormous improvements in high-resolution scattering spectroscopies such as RIXS, IXS, and nuclear methods such as NFS and NRVS. Now for the downside—an XFELO could be 10 or 15 years away, sad for old-timers but manageable if you are just starting your synchrotron career.

12.6 Free-Electron Laser Laboratories

Now that hard X-ray FEL feasibility and benefits are firmly established, new projects and upgrades are constantly coming on line. At the moment, there are only five truly hard X-ray facilities (Appendix M), compared to dozens of storage ring-based synchrotron facilities (Appendix B). There are also quite a few soft X-ray FELs, some of which are listed in Appendix M. The scale of FEL facilities can be seen in Fig. 12.8. Even the most compact X-ray FEL, SACLA, is 700 m long, while the largest facility, the European XFEL, extends for 3.4 km.

Of course, what is most important is the instrumentation inside. Most of the length is taken up by the electron accelerator, but the undulators are also more than an order of magnitude longer than in storage rings (Fig. 12.9). For example, the SASE1/2 undulators at the European XFEL stretch for over 175 m [573].

Fig. 12.8 Hard X-ray FEL sources around the world. Top left: SACLA at SPring-8. Top right: SwissFEL. Middle left: LCLS and plans for LCLS-II. Middle right: a map showing extent of European XFEL. Bottom: PAL-XFEL at Pohang Light Source

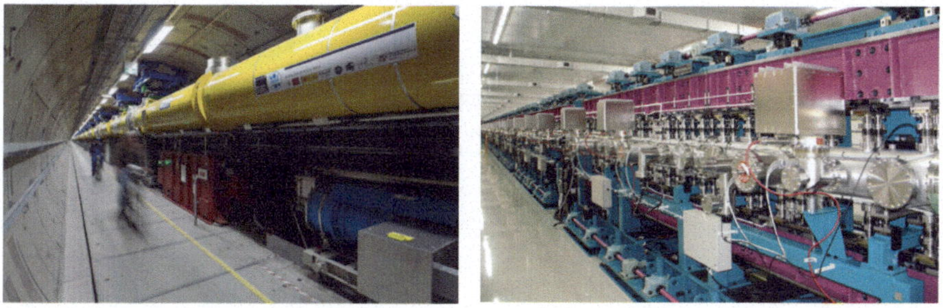

Fig. 12.9 Inside FEL sources around the world. Left: the linear accelerator hall at the European XFEL. Right: the undulator hall at SACLA

12.7 Applications of Peak Brightness

The diffraction-before-destruction applications of FELs are well-documented elsewhere [590]. There, the idea is to gather diffraction or imaging data before the sample is destroyed by the FEL pulse. Since our emphasis is on synchrotron spectroscopy, here we highlight a few of the methods that are simply not possible without the high instantaneous brightness of X-ray FELs. One clear example is nonlinear X-ray optics.

Nonlinear optical processes are those that depend on higher powers of the electric field—to be technical, the dielectric polarization P is no longer just a linear function of the electric field, but needs to be expressed in terms of higher-order polarizabilities. Research in this area has flourished since the advent of lasers made extremely strong electric fields routine, and much of the theory was already developed by 1965 [591]. These experiments are much harder for X-rays, since multi-photon interactions with n photons scale with the frequency ω as ω^{-2n+1} [592]. Nonlinear interactions in the X-ray region are therefore many many orders of magnitude weaker than their optical counterparts. Nevertheless, FELs now have the intensity required to observe nonlinear processes in the X-ray regime, including second-harmonic generation [593], two-photon absorption [592], stimulated X-ray emission [594], stimulated X-ray Raman [595, 596], sum-frequency generation [597], and nonlinear Compton scattering [598].

12.7.1 Second-Harmonic Generation

In a fundamental nonlinear process known as "second-harmonic generation" ("SHG"), two photons at one energy are converted into one photon at twice the energy. This is routine stuff with optical lasers; it is done in labs that convert 1064 nm Nd:YAG light into 532 nm radiation with greater than 50% efficiency. In

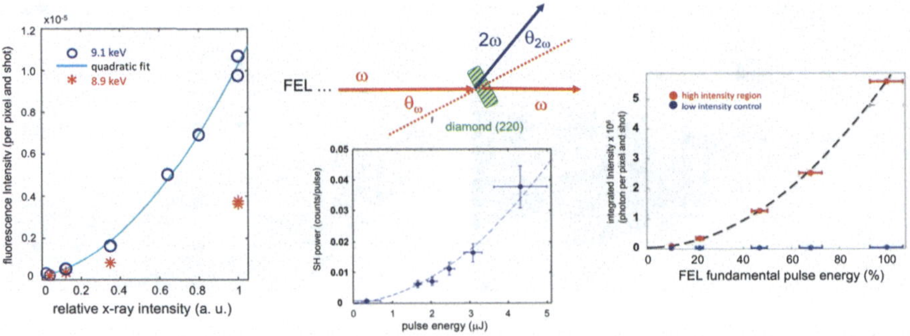

Fig. 12.10 Left: Zr K fluorescence intensity as a function of incident X-ray intensity when incident photon energy is above (9.1 keV) or below (8.9 keV)—the TPA threshold, redrawn from [592]. Middle: the configuration for second-harmonic generation and the quadratic dependence on pulse energy, redrawn from [593]. Right: the quadratic intensity dependence on pulse energy for nonlinear Compton scattering, redrawn from [598]

an X-ray SHG demonstration at SACLA, using 4 μJ FEL pulses with intensity $>10^{16}$ W/cm^2, 7.3 keV photons were converted to 14.6 keV photons with a maximum efficiency of $\sim 6 \times 10^{-11}$ [593] (Fig. 12.10). Although this was a beautiful proof of concept, with ten orders of magnitude lower yield, it will be a while before X-ray SHG has practical applications.

12.7.2 Two-Photon Absorption

In nonsequential two-photon absorption ("TPA"), two photons of energy *E* are absorbed simultaneously, allowing the system to reach an excited state of energy 2*E*. For biologists, it is a routine approach for fluorescence microscopy, allowing two near IR photons to excite visible fluorescence in a sample. As with SHG, TPA has become a feasible, albeit heroic, experiment with FELs, with demonstrations in the hard X-ray regime for Ge [599], Cu [600, 601], and Zr [592].

As shown in Fig. 12.10, at FEL intensities, the amount of Zr fluorescence (proportional to absorption) varies as the square of the X-ray intensity when the single photon energy is more than half the absorption edge energy. This indicates a two-photon process. X-ray TPA could prove useful in analysis of absorption edges, because it will have different selection rules from the one-photon process [600]. Features such as forbidden quadrupole transitions will be allowed in the TPA process. However, the TPA cross section is estimated to be only $\sim 10^{-34}$ of the one-photon value. Like SHG, X-ray TPA will probably require a few generations of FEL upgrades before it becomes "routine."

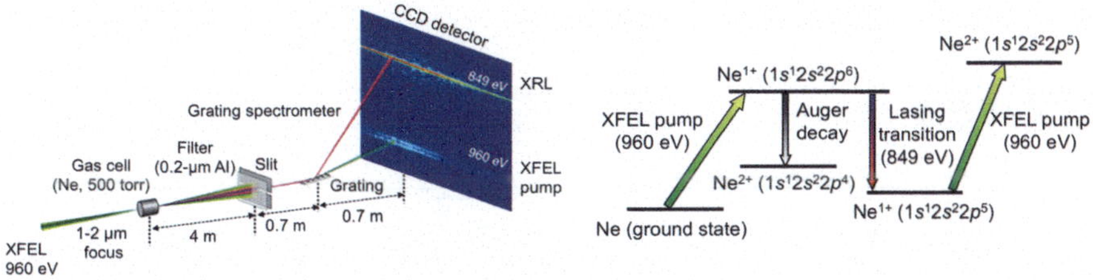

Fig. 12.11 Setup for lasing from Ne gas (left) and the atomic electronic transitions involved (right). Reproduced and redrawn from [594]

12.7.3 Stimulated Emission: X-ray Lasers

Free-electron lasers can be used to pump samples quickly enough to achieve population inversions for atomic core levels, which can then undergo stimulated emission like conventional optical lasers. With ~40–80 fs, ~0.02–0.27 mJ FEL pulses of 960 eV photons, a population inversion for the 1s vacancy was achieved in the resultant singly ionized neon (Fig. 12.11). Lasing at a photon energy of 849 eV was observed from the plasma column created by irradiation of the gas medium [594].

FEL-driven gas lasers have some distinct advantages over just using the SASE FEL radiation. Compared to the broad and jittery FEL pulses (10 eV FWHM), the Ne "X-ray laser" or "XRL" has a narrow (0.27 eV) Lorentzian shape with a fixed energy of 849 eV. In the future, lasing from more highly ionized Ne could yield even narrower lines ($\Delta E/E \sim 10^{-6}$). With foreseeable improvements in FEL sources, atomic XRLs show promise as sources for photoemission or inelastic scattering experiments.

12.7.4 Femtosecond XMCD

In Chap. 7 we saw how XMCD could provide an element selective probe of magnetic properties. The high peak brightness of FELs now allows those same measurements on a femtosecond time-scale [602]. In a representative case, a GdFeCo thin film was illuminated with 60 fs pulses from an 800 nm laser to cause transient demagnetization. From the time-dependent XMCD, the authors saw that Gd demagnetizes more slowly than Fe. In fact, Fe reverses magnetization at 870 fs, leading to a transient ferromagnetic-like state that exists until 3.7 ps, when Gd magnetization also reverses (Fig. 12.12).

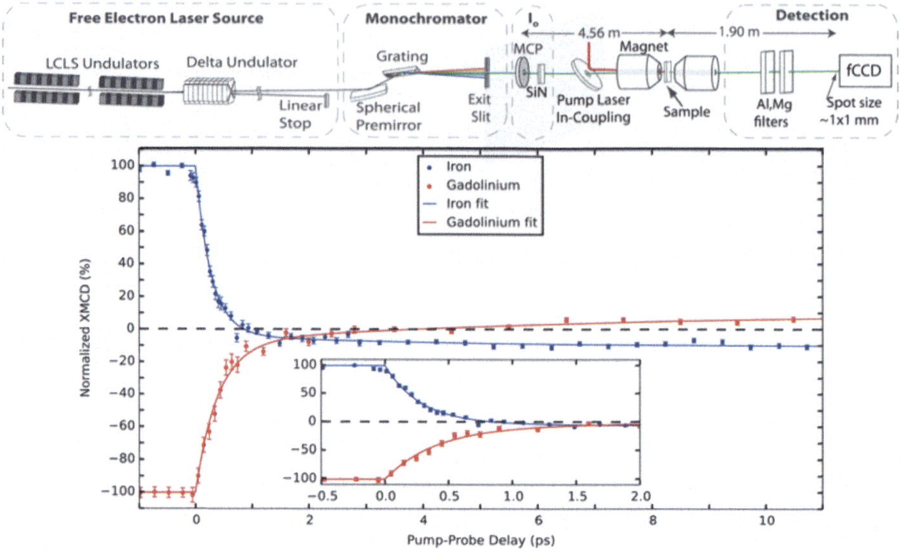

Fig. 12.12 Top: the setup for measuring XMCD on a femtosecond time-scale. Bottom: representative results on a GdFeCo thin-film system, reproduced from [603]

12.7.5 *Miscellaneous*

The rapidly growing field of FEL applications clearly merits a book of its own, and additional references are listed below. You can also consult a few review articles about recent applications, at LCLS [586], at SACLA [604], and at the European XFEL [605].

12.8 Are FELs Lasers?

For a late-night debate at the beamline, consider the question of whether FELs deserve to be called lasers. In Madey's original paper, an FEL was considered a source of stimulated emission (the SE in laser), "Stimulated emission of Bremsstrahlung in a Period Magnetic Field" [557], and a subsequent paper appeared with the title "A Free Electron Laser" [558]. However, according to Saldin, "the principle of its operation is similar to that of conventional vacuum tube devices" [606]. Ribic and Margaritondo agree, noting "the term 'laser' refers to optical amplification based on quantum stimulated emission . . . the X-FEL mechanism is different and based on classical effects" [607]. However, they admit "the seemingly unstoppable use of the name 'free-electron laser' can be, if not fully accepted, at least magnanimously tolerated". Have a drink and take sides . . .

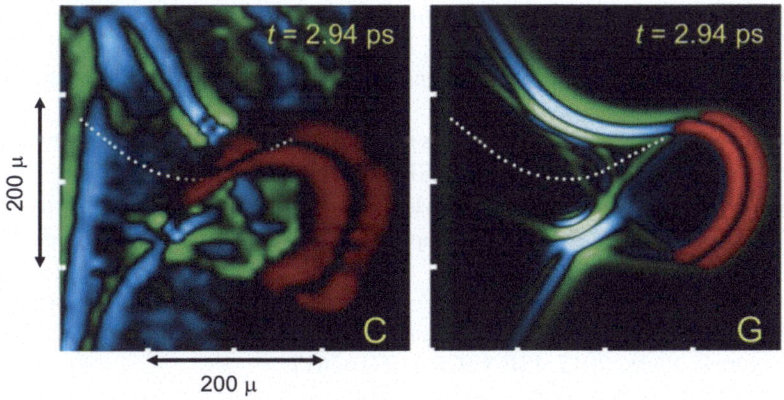

Fig. 12.13 Experimental (left) and theoretical simulations (right) for "synchrotron radiation" inside a $LiTiO_3$ crystal. See the original paper for details [608]

12.9 Is This Synchrotron Radiation?

Throughout this book we have grown accustomed to synchrotron radiation sources on a kilometer scale. A recent paper in *Science* reports a source on a sub-millimeter scale, using a metasurface to guide a pulse of 800 nm radiation along a 100-mμ-scale circular arc inside a $LiTiO_3$ crystal [608]. For a final late-night debate, is this beautiful picture from a true synchrotron radiation source? (Fig. 12.13).

12.10 Suggested Exercises

1. Calculate the Pierce parameter ρ for a SASE X-ray FEL assuming an electron current of 1 kA, a beam energy of 2.85 GeV, an undulator period of 29 mm, horizontal and vertical rms beam dimensions of 40 μ, and $K = 1.513$.
2. Calculate the gain length for the above Pierce parameter.
3. Estimate the length to saturation.
4. Estimate the emitted radiation per pulse at saturation.

12.11 Reference Books and Review Articles

1. *The Physics of Free Electron Lasers*; Saldin, E. L., Schneidmiller, E.V., Yurkov, M.V.; Springer, Berlin, 2000. ISBN 978-3-540-66266-2.
2. *Synchrotron Radiation and Free-Electron Lasers: Principles of Coherent X-Ray Generation*, Kim, K.-J.; Huang, Z.; Lindberg, R.; Cambridge University Press, Cambridge, 2017. ISBN 978-1-107-16261-1.

3. *Free-Electron Lasers in the Ultraviolet and X-Ray Regime: Physical Principles, Experimental Results, Technical Realization, 2nd Edition*; Schmüser, P., Dohlus, M., Rossbach, J., Behrens, C.; Springer, Heidelberg, 2014. ISBN 978-3319040806.
4. *Principles of Free Electron Lasers*, 3rd edition, Freund, H. P. and Antonsen, Jr. T.M. Springer Science, 2018. ISBN 978-3-319-75105-4.
5. *Elements of Synchrotron Light*, Margaritiondo, G. Oxford University Press, 2002. ISBN-978-0198509301. Chapter 4—Free Electron Lasers. A nice discussion of FELs at a very basic level.
6. *Synchrotron Radiation and Free-Electron Lasers: Principles of Coherent X-Ray Generation*, Kwang-Je Kim, Zhirong Huang, and Ryan Lindberg, Cambridge University Press, 2017. ISBN 978-1107162617.
7. *X-Ray Free Electron Lasers: Applications in Materials, Chemistry and Biology*, Uwe Bergmann, Vittal Yachandra, Junko Yano, eds., Royal Society of Chemistry, 2017. ISBN 978-1-84973-100-3.
8. "Short-wavelength free-electron laser sources and science: a review", Seddon, E. A.; Clarke, J. A.; Dunning, D. J.; Masciovecchio, C.; Milne, C. J.; Parmigiani, F.; Rugg, D.; Spence, J. C. H.; Thompson, N. R.; Ueda, K.; Vinko, S. M.; Wark, J. S.; Wurth, W. *Rep. Prog. Phys.* **80**, 73 (2017)
9. "A simplified description of X-ray free-electron lasers", G. Margaritondo and P. R, Ribic, *J. Syn. Rad.* **18**, 101–108 (2011).
10. "10 years of pioneering X-ray science at the Free-Electron Laser FLASH at DESY", J. Rossbach, J. Schneider, ands W. Wurth, *Physics Reports*, **808**, 1-74 (2019)

Afterward

Advice I should have heeded early on . . .

"Whatever it takes to finish things, finish. You will learn more from a glorious failure than you ever will from something you never finished."—Neil Gaiman

"If your book is taking longer than you'd like to write, stop researching the book and focus on writing it." Why You Must Finish Writing What You Started (And How to Do It).—How to Finish Writing What You Started—**Disconnect** from the internet while you write https://becomeawritertoday.com/finish-writing/

"Books aren't written—they're rewritten. Including your own. It is one of the hardest things to accept, especially after the seventh rewrite hasn't quite done it."—Michael Crichton

I remember hearing or reading once that you can either (a) attempt a totally up-to-date book, which will never be finished, or (b) resign yourself to a book that illustrates basic principles and has a chance for completion. Writing this book has convinced me of the truth of item (a). When I started, there were no hard X-ray FELS, no MBA lattices, no refractive X-ray optics, no TES detectors, and barely any synchrotron nuclear spectroscopy. I have finally chosen to go with option (b), with the solace that perhaps there can be a second edition. Stay tuned . . .

"The same iron rigor needed to begin the book is required to end it."—Gabriel Garcia Marquez

The End

October 2020, Los Altos, CA

S. P. Cramer, *X-Ray Spectroscopy with Synchrotron Radiation*, Biological and Medical Physics, Biomedical Engineering, https://doi.org/10.1007/978-3-030-28551-7

Appendix A: Fundamental Constants and Useful Conversion Factors

Table A.1 Fundamental constants

Symbol	Quantity	Magnitude	Units
e	Electron charge	1.6022×10^{-19}	C
h	Planck's constant	6.6260×10^{-34}	J s
$\hbar$	Reduced Planck's constant	1.0546×10^{-34}	J s
c	Speed of light	2.9979×10^{8}	m s^{-1}
m_e	Electron rest mass	9.1094×10^{-31}	kg
ε_0	Permittivity of vacuum	8.8542×10^{-12}	F m^{-1}
μ_0	Permeability of vacuum	$4\pi \times 10^{-7}$	H m^{-1}
$m_e c^2$	Electron rest mass energy	0.5110	MeV
N_A	Avogadro's number	6.0221×10^{23}	mol^{-1}
k_B	Boltzmann's constant	1.3806×10^{-23}	J K^{-1}
μ	Atomic mass unit	1.6605×10^{-27}	kg
R_H	Rydberg constant	13.606	eV
a_0	Bohr radius	0.5292	Å
μ_B	Bohr magneton	9.2740×10^{-24}	J T^{-1}
r_0	Classical electron radius	2.818×10^{-15}	m
λ_C	Compton scattering length	3.860×10^{-13}	m

S. P. Cramer, *X-Ray Spectroscopy with Synchrotron Radiation*, Biological and Medical Physics, Biomedical Engineering, https://doi.org/10.1007/978-3-030-28551-7

Table A.2 Useful conversion factors

Plane angle				
1 degree	17.45 mrad	1.745×10^{-2} rad	60 arcmin	3600 arcsec
57.296 degree	1000 mrad	1 radian	3437.7 arcmin	2.0626×10^{5} arcsec
0.057296 degree	1 mrad	10^{-3} rad	3.4377 arcmin	2.0626×10^{2} arcsec
2.778×10^{-4} degree	4.848×10^{-3} mrad	$4.848\ 10^{-6}$ rad	0.01667 arcmin	1 arcsec
Energy[a]				
1 eV	1.2398×10^{4} Å	8.0656×10^{3} cm^{-1}	1.1605×10^{4} K	96.486 kJ mol^{-1}
12.398 keV	1 Å	1×10^{8} cm^{-1}	1.4388×10^{8} K	1.1963×10^{6} kJ mol^{-1}
1.2398×10^{-4} eV	10^{8} Å	1 cm^{-1}	1.4388 K	0.011963 kJ mol^{-1}
8.6173×10^{-5} eV	1.4388×10^{8} Å	0.69503 cm^{-1}	1 K	8.3145×10^{-3} kJ mol^{-1}
0.010364 eV	1.1963×10^{6} Å	83.593 cm^{-1}	120.27 K	1 kJ mol^{-1}

[a]Units are obviously not dimensionally equivalent; "K" is kelvin, calculated by "temperature = (photon energy)/(Boltzmann constant)"

Appendix B: Storage Ring Synchrotron Radiation Facilities

Table B.1 SR sources active or planned in 2018 (for a more complete list, see Wikipedia ...)

#	Country/city	Name and/or acronym	*E* (GeV)	Circ. (m)
1[a]	Armenia/Yerevan	"CANDLE"	3	216
2	Australia/Melbourne	Australian Synchrotron	3	216
4	Brazil/Sao Paulo	Sirius	3	518
5	Canada/Saskatoon	CLS	2.9	147
6	China/Beijing	BSRF	2.5	240
7	China/Shanghai	SSRF	3.5	432
8[a]	China/Beijing	High Energy Photon Source	6	1360
9	France/Grenoble	ESRF	6	844
10	France/Paris	SOLEIL	3	354
12	Germany/Karlsruhe	ANKA	2.5	110
13	Germany/Berlin	BESSY II	1.7	240
14	Germany/Hamburg	PETRA III at DESY	6.0	2304
15	India/Indore	Indus 2	2.5	173
16	Iran/Qazvin	Iranian Light Source Facility	3	490
17	Italy/Trieste	ELLETRA	~2.2	260
18	Japan/Tsukuba	Photon Factory	2.5	187
19	Japan/Hyogo	SPring-8	8	1436
20	Japan/Okazaki	UVSOR	0.75	53
21	Jordan/Allaan	SESAME	2.5	133
22	Korea	Pohang Light Source II	3.0	282
23	Poland/Kraków	SOLARIS	1.5	96
24	Russia	SIBERIA-2	2.5	124
25	Spain	ALBA	3	270
26	Sweden	MAX IV Laboratory	3	528
27	Switzerland	Swiss Light Source	2.4	288
28	Taiwan	Taiwan Photon Source	3	518

(continued)

S. P. Cramer, *X-Ray Spectroscopy with Synchrotron Radiation*, Biological and Medical Physics, Biomedical Engineering, https://doi.org/10.1007/978-3-030-28551-7

Table B.1 (continued)

#	Country/city	Name and/or acronym	*E* (GeV)	Circ. (m)
29	United Kingdom	Diamond Light Source	3	562
30	USA/Berkeley	ALS	1.9	197
31	USA/Argonne	APS	7.0	1104
32	USA/Baton Rouge	CAMD	1.5	55
33	USA/Ithaca	CHESS	5.5	768
34	USA/Upton	NSLS-II	3	792
35	USA/Stanford	SSRL	3	234

[a]In planning stage

Appendix C: Properties of Magnets

Table C.1 Fields for different types of magnets[a]

Magnet / pole profile[b]	Upright		Skew[c]	
Dipole/$y = \pm G$	$B_x = 0$	$B_y = \kappa_x \frac{cp}{e}$	$B_x = -\frac{cp}{e}\kappa_y$	$B_y = 0$
Quadrupole / $xy = \pm\frac{1}{2}R^2$	$B_x = \frac{cp}{e}ky$	$B_y = \frac{cp}{e}kx$	$B_x = -\underline{k}x$	$B_y = \underline{k}y$
Sextupole / $3x^2y - y^3 = R^3$	$B_x = \frac{cp}{e}mxy$	$B_y = \frac{cp}{e}m(x^2 - y^2)$	$B_x = -\frac{cp}{2e}\underline{m}(x^2 - y^2)$	$B_y = \frac{cp}{e}\underline{m}xy$
Octupole / $4xy(x^2 - y^2) = R^4$	$B_x = \frac{cp}{6e}r(3x^2y - y^3)$	$B_y = \frac{cp}{6e}r(x^3 - 3xy^2)$	$B_x = -\frac{cp}{6e}\underline{r}(x^3 - 3xy^2)$	$B_y = \frac{cp}{6e}\underline{r}(3x^2y - y^3)$

[a]Wiedemann Tables 4.3 and 4.4
[b]Pole profiles are for upright fields
[c]Rotations of skew magnets are $\Phi_n = \pi/(2n)$, where n is the order of multipole

Table C.2 Magnetic multipole potentials[a]

Magnet	
Dipole	$-\frac{e}{cp}V_1 = -\kappa_y x + \kappa_x y$
Quadrupole	$-\frac{e}{cp}V_2 = -\frac{1}{2}\underline{k}\,(x^2 - y^2) + k\,x\,y$
Sextupole	$-\frac{e}{cp}V_3 = -\frac{1}{6}\underline{m}\,(x^3 - 3xy^2) + \frac{1}{6}m\,(3x^2y - y^3)$
Octupole	$-\frac{e}{cp}V_4 = -\frac{1}{24}\underline{r}\,(x^4 - 6x^2y^2 + y^4) + \frac{1}{6}r\,(x^3y - xy^3)$

[a]Wiedemann Table 4.1

S. P. Cramer, *X-Ray Spectroscopy with Synchrotron Radiation*, Biological and Medical Physics, Biomedical Engineering, https://doi.org/10.1007/978-3-030-28551-7

Appendix D: Special Functions for Synchrotron Radiation

Synchrotron radiation and spectroscopy employ a wide variety of special functions that are often unfamiliar to chemists and biologists. Here we limit ourselves to a few functions critical for description of synchrotron radiation. The more expansive descriptions in Wikipedia are so well done that it is worth making an exception to the "no-links" rule: https://en.wikipedia.org/wiki/Bessel_function. There are also excellent descriptions in Wolfram MathWorld: http://mathworld.wolfram.com/BesselFunctionoftheFirstKind.html and so on.

Appendix D.1: Bessel Functions

There are two regular Bessel functions, $J_n(z)$ and $Y_n(z)$, and they are independent solutions of the differential equation:

$$z^2y'' + zy' + \left(z^2 - n^2\right)y = 0 \tag{D.1}$$

Bessel functions of the first kind, $J_n(z)$, are needed to describe the properties of cylindrical waveguides and undulators, and they also appear in the description of nuclear forward scattering. In *Mathematica* they are defined by **BesselJ[n,z]**. For what it's worth, such Bessel functions are the radial part of the modes of vibration of a circular drum (Fig. D.1).

For integer n, they can be computed as sums over an infinite series:

$$J_n(x) = \left(\frac{x}{2}\right)^n \sum_{k=0}^{\infty} \frac{(-1)^k}{k!(n+k)!} \left(\frac{x}{2}\right)^{2k} \tag{D.2}$$

S. P. Cramer, *X-Ray Spectroscopy with Synchrotron Radiation*, Biological and Medical Physics, Biomedical Engineering, https://doi.org/10.1007/978-3-030-28551-7

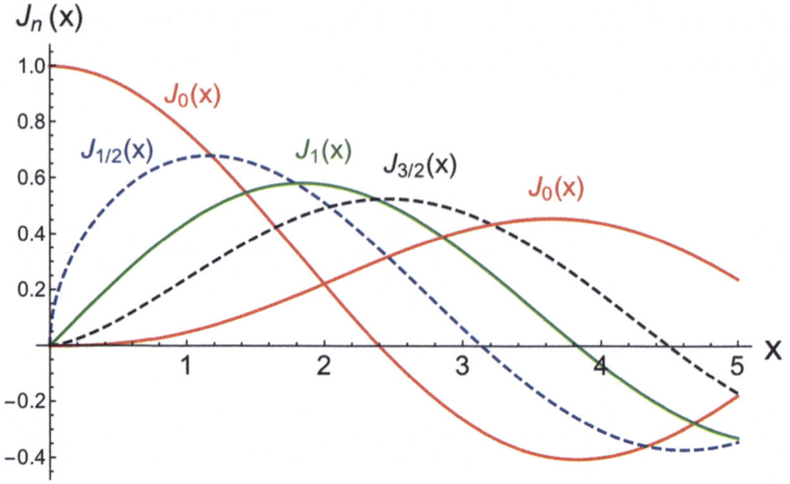

Fig. D.1 Bessel functions of the first kind

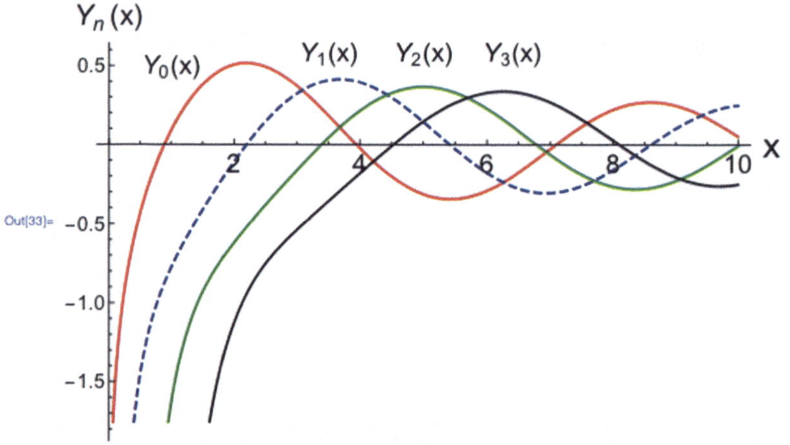

Fig. D.2 Bessel functions of the second kind

Bessel functions of the second kind, $Y_n(z)$, are needed to describe the properties of undulators, and in *Mathematica* they are defined by **BesselY[n,z]**. As seen in Fig. D.2, they have a singularity at the origin (Fig. D.2).

Appendix D.2:Modified Bessel Functions

As with regular Bessel functions, there are two modified Bessel functions, $I_n(z)$ and $K_n(z)$. They are independent solutions to the differential equation:

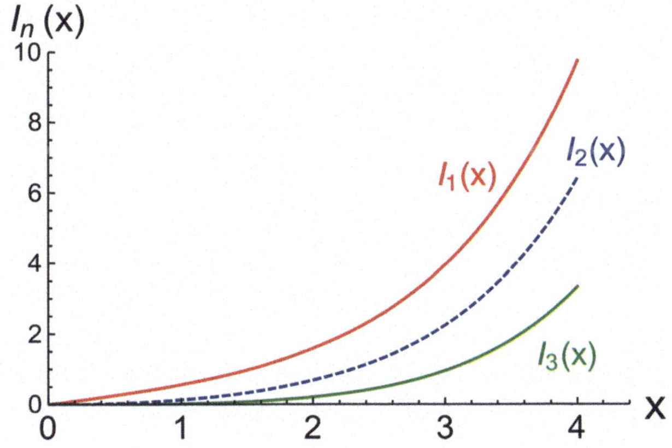

Fig. D.3 Modified Bessel functions of the first kind

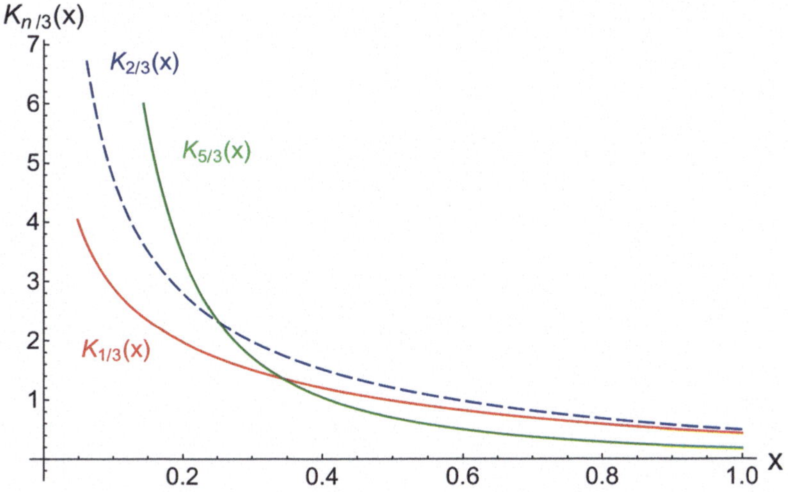

Fig. D.4 Modified Bessel function of the second kind

$$z^2y'' + zy' - \left(z^2 + n^2\right) y = 0 \tag{D.3}$$

Modified Bessel functions of the first kind, $I_n(z)$, are used in NRVS theory to calculate the fractional contribution of different transitions to the normalized excitation probability (see [481]). In *Mathematica* they are defined by **BesselI[n,z]** (Fig. D.3).

Modified Bessel functions of the second kind, $K_n(z)$, are needed to describe the spectral properties of bend magnets. In *Mathematica* they are defined by **BesselK[n, z]**. Like the regular Bessel functions of the second kind, they have a singularity as $z \to 0$ (Fig. D.4).

They too can be computed as sums over series:

$$I_\alpha(x) = \sum_{m=0}^{\infty} \frac{1}{m!\Gamma(m+\alpha+1)} \left(\frac{x}{2}\right)^{2m+\alpha} \tag{D.4}$$

and using:

$$K_\alpha(x) = \frac{\pi}{2} \frac{I_{-\alpha}(x) - I_\alpha(x)}{\sin \alpha\pi} \tag{D.5}$$

Since we only need two modified Bessel functions of the second kind, $K_{1/3}(x)$ and $K_{2/3}(x)$, we note with Kim that they can also be expressed as follows (Kim, 3.10):

$$K_{\frac{1}{3}}(x) = \sqrt{3} \int_0^{\infty} \tau \sin \left[\frac{3}{2} x \left(\tau + \frac{1}{3}\tau^3\right)\right] d\tau \tag{D.6}$$

and

$$K_{\frac{2}{3}}(x) = \sqrt{3} \int_0^{\infty} \tau \cos \left[\frac{3}{2} x \left(\tau + \frac{1}{3}\tau^3\right)\right] d\tau \tag{D.7}$$

Appendix E: X-ray Optics: Fresnel Equations and Stokes Parameters

Appendix E.1: Fresnel Equations

Source: Michette [102] (Fig. E.1)

The Fresnel equations are derived from four constraints at the interface between two media. At the boundary, the tangential components of the electric and magnetic vectors $\vec{E}$ and $\vec{H}$ are constrained to be equal. The normal components of $\epsilon\vec{E}$ and $\vec{B}$ must also be continuous. From these constraints one can derive four equations in four unknowns, which can be solved based on the optical constants of the materials involved (see Hecht or Michette for more complete derivations).

For X-rays, it is most convenient to use glancing angles for defining the various beams, and we can start with:

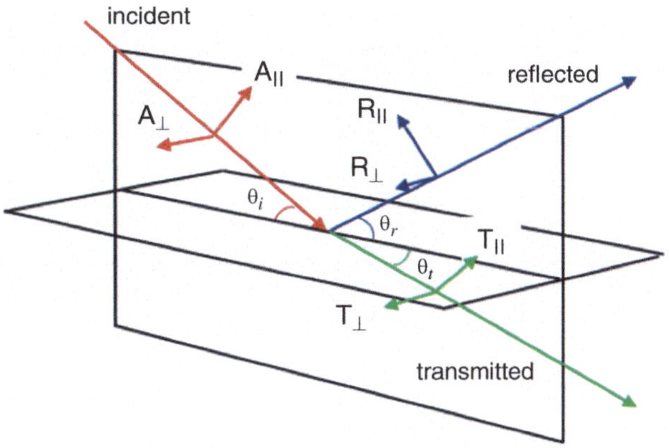

Fig. E.1 Definition of terms for the Fresnel equations. For clarity the directions of the ∥ field components have been reversed from those used in the following equations

S. P. Cramer, *X-Ray Spectroscopy with Synchrotron Radiation*, Biological and Medical Physics, Biomedical Engineering, https://doi.org/10.1007/978-3-030-28551-7

$$\left(A_{\parallel} - R_{\parallel}\right)\sin\theta_i = T_{\parallel}\sin\theta_t \tag{E.1}$$

$$A_{\perp} + R_{\perp} = T_{\perp} \tag{E.2}$$

$$(1-\delta_1)(A_{\perp} - R_{\perp})\sin\theta_i = (1-\delta_2)T_{\perp}\sin\theta_t \tag{E.3}$$

$$(1-\delta_1)\left(A_{\parallel} + R_{\parallel}\right) = (1-\delta_2)T_{\parallel} \tag{E.4}$$

These can be rewritten so that the reflected and transmitted components are expressed in terms of the incident components:

$$\frac{R_{\parallel}}{A_{\parallel}} = \frac{[(1-\delta_2)\sin\theta_i - (1-\delta_1)\sin\theta_t]}{[(1-\delta_2)\sin\theta_i + (1-\delta_1)\sin\theta_t]} \tag{E.5}$$

$$\frac{R_{\perp}}{A_{\perp}} = \frac{[(1-\delta_1)\sin\theta_i - (1-\delta_2)\sin\theta_t]}{[(1-\delta_1)\sin\theta_i + (1-\delta_2)\sin\theta_t]} \tag{E.6}$$

$$\frac{T_{\parallel}}{A_{\parallel}} = \frac{[2(1-\delta_1)\sin\theta_i]}{[(1-\delta_2)\sin\theta_i + (1-\delta_1)\sin\theta_t]} \tag{E.7}$$

$$\frac{T_{\perp}}{A_{\perp}} = \frac{[2(1-\delta_1)\sin\theta_i]}{[(1-\delta_1)\sin\theta_i + (1-\delta_2)\sin\theta_t]} \tag{E.8}$$

Making use of Snell's law in terms of glancing angles, we can write everything in terms of θ_i:

$$\frac{\cos\theta_i}{\cos\theta_t} = \frac{(1-\delta_2)}{(1-\delta_1)} \tag{E.9}$$

This allows us to transform the equations to:

$$\frac{R_{\parallel}}{A_{\parallel}} = \frac{(1-\delta_2)^2\sin\theta_i - (1-\delta_1)\left[(1-\delta_2)^2 - (1-\delta_1)^2\cos^2\theta_i\right]^{1/2}}{(1-\delta_2)^2\sin\theta_i + (1-\delta_1)\left[(1-\delta_2)^2 - (1-\delta_1)^2\cos^2\theta_i\right]^{1/2}} \tag{E.10}$$

$$\frac{R_{\perp}}{A_{\perp}} = \frac{(1-\delta_1)\sin\theta_i - \left[(1-\delta_2)^2 - (1-\delta_1)^2 - \cos^2\theta_i\right]^{1/2}}{(1-\delta_1)\sin\theta_i + \left[(1-\delta_2)^2 - (1-\delta_2)^2\cos^2\theta_i\right]^{1/2}} \tag{E.11}$$

$$\frac{T_{\parallel}}{A_{\parallel}} = \frac{2(1-\delta_1)(1-\delta_2)\sin\theta_i}{(1-\delta_2)^2\sin\theta_i + (1-\delta_1)\left[(1-\delta_2)^2 - (1-\delta_1)^2\cos^2\theta_i\right]^{1/2}} \tag{E.12}$$

$$\frac{T_{\perp}}{A_{\perp}} = \frac{2(1-\delta_1)\sin\theta_i}{(1-\delta_1)\sin\theta_i + \left[(1-\delta_2)^2 - (1-\delta_1)^2\cos^2\theta_i\right]^{1/2}} \tag{E.13}$$

The critical angle θ_c is the special case where:

$$\cos\theta_i = \frac{1-\delta_2}{1-\delta_1} \tag{E.14}$$

When this is the case:

$$\frac{R_{\parallel}}{A_{\parallel}} = \frac{R_{\perp}}{A_{\perp}} = \mathbf{1} \tag{E.15}$$

For synchrotron X-ray experiments, the initial region is under vacuum, so $\delta_1 = 0$, and using $\delta = \delta_2$, the equations simplify to:

$$\frac{R_{\parallel}}{A_{\parallel}} = \frac{(1-\delta)^2\sin\theta_i - \left[(1-\delta)^2 - \cos^2\theta_i\right]^{1/2}}{(1-\delta)^2\sin\theta_i + \left[(1-\delta)^2 - \cos^2\theta_i\right]^{1/2}} \tag{E.16}$$

$$\frac{R_{\perp}}{A_{\perp}} = \frac{\sin\theta_i - \left[(1-\delta)^2 - \cos^2\theta_i\right]^{1/2}}{\sin\theta_i + \left[(1-\delta)^2 - \cos^2\theta_i\right]^{1/2}} \tag{E.17}$$

$$\frac{T_{\parallel}}{A_{\parallel}} = \frac{2(1-\delta)\sin\theta_i}{(1-\delta)^2\sin\theta_i + \left[(1-\delta)^2 - \cos^2\theta_i\right]^{1/2}} \tag{E.18}$$

$$\frac{T_{\perp}}{A_{\perp}} = \frac{2\sin\theta_i}{\sin\theta_i + \left[(1-\delta)^2 - \cos^2\theta_i\right]^{1/2}} \tag{E.19}$$

In a real-world synchrotron X-ray experiment, there is always some absorption in the mirror. In this case, the most relevant form of the Fresnel equations is the case for radiation traveling from the vacuum with index of refraction unity to an absorbing medium with a complex index of refraction $\tilde{n} = 1 - \delta - i\beta$.

$$\frac{R_{\parallel}}{A_{\parallel}} = \frac{\tilde{n}^2\sin\theta_i - \left(\tilde{n}^2 - \cos^2\theta_i\right)^{1/2}}{\tilde{n}^2\sin\theta_i + \left(\tilde{n}^2 - \cos^2\theta_i\right)^{1/2}} \tag{E.20}$$

$$\frac{R_\perp}{A_\perp} = \frac{\sin\theta_i - \left(\tilde{n}^2 - \cos^2\theta_i\right)^{1/2}}{\sin\theta_i + \left(\tilde{n}^2 - \cos^2\theta_i\right)^{1/2}} \tag{E.21}$$

$$\frac{T_\parallel}{A_\parallel} = \frac{2\tilde{n}\sin\theta_i}{\tilde{n}^2\sin\theta_i + \left(\tilde{n} - \cos^2\theta_i\right)^{1/2}} \tag{E.22}$$

$$\frac{T_\perp}{A_\perp} = \frac{2\sin\theta_i}{\sin\theta_i + \left(\tilde{n}^2 - \cos^2\theta_i\right)^{1/2}} \tag{E.23}$$

Appendix E.2: Stokes Parameters and the Polarization Ellipse

The Stokes parameters are a way to describe the polarization of light based on measurable observables. They were first developed by Sir George Stokes in 1852, but only popularized about a century later. Measurement of the Stokes parameters assumes access to 4 filters, each of which transmits 50% of the incident light under natural illumination (Fig. E.2).

The first filter, let's call it "filter 0," has no polarization selectivity, while filters "1" and "2" are linear polarizers with transmission axes horizontal and at 45°, respectively. The final filter, "filter 3," is a circular polarizer that only passes right circularly polarized light. Imagine four successive measurements with each of these filters, yielding respective transmitted intensities I_0, I_1, I_2, and I_3. Then the operational definition of the Stokes parameters is:

$$S_0 = 2I_0 \tag{E.24}$$

$$S_1 = 2I_1 - 2I_0 \tag{E.25}$$

$$S_2 = 2I_2 - 2I_0 \tag{E.26}$$

$$S_3 = 2I_3 - 2I_0 \tag{E.27}$$

The first Stokes parameter, S_0, represents the total intensity of the light source, while S_1 describes the amount of linear horizontal or vertical polarization. S_2 describes the amount of linear polarization at ±45°, and S_3 quantifies the amount

Fig. E.2 The 4 measurements involved in determining the Stokes parameters

of circular polarization. Although the Stokes parameters apply equally to unpolarized, partially polarized, or completely polarized light, since we care about synchrotron radiation we will assume complete polarization, and in addition we will assume monochromatic radiation. In such a case we can describe the incoming X-rays at a given point chosen as $z = 0$ in terms of two components:

$$E_x(t) = E_{0x} \cos\left(\omega t + \delta_x\right) \tag{E.28}$$

$$E_y(t) = E_{0y} \cos\left(\omega t + \delta_y\right) \tag{E.29}$$

The Stokes measurements involve time averages of these fields, so we can use:

$$\left\langle E_x{}^2(t)\right\rangle = ½E_{0x}{}^2 \tag{E.29}$$

$$\left\langle E_y{}^2(t)\right\rangle = ½E_{0y}{}^2 \tag{E.30}$$

$$\left\langle E_x(t)E_y(t)\right\rangle = ½E_{0x}E_{0y} \cos \delta \tag{E.31}$$

We can now express the Stokes parameters in terms of electric fields that theory predicts in different cases for synchrotron radiation:

$$S_0 = E_{0x}{}^2 + E_{0y}{}^2 \tag{E.32}$$

$$S_1 = E_{0x}{}^2 - E_{0y}{}^2 \tag{E.33}$$

$$S_2 = 2E_{0x}E_{0y} \cos \delta \tag{E.34}$$

$$S_3 = 2\, E_{0x}E_{0y} \sin \delta \tag{E.35}$$

The *degree of polarization* P is defined as:

$$P = \frac{I_{\text{pol}}}{I_{\text{tot}}} = \frac{S_1^2 + S_2^2 + S_3^2}{S_0^2} \tag{E.36}$$

and for completely polarized light: $P = 1$.

The Stokes parameters can yield useful information about the polarization ellipse. For example, the orientation angle of the polarization ellipse, ψ, is given by:

$$\tan 2\psi = \frac{2\, E_{0x}E_{0y} \cos \delta}{E_{0x}^2 - E_{0y}^2} = \frac{S_2}{S_1} \tag{E.37}$$

and the ellipticity angle of the polarization ellipse, χ, is given by (Fig. E.3):

$$\sin 2\chi = \frac{2\, E_{0x}E_{0y} \sin \delta}{E_{0x}^2 + E_{0y}^2} = \frac{S_2}{S_1} \tag{E.39}$$

Fig. E.3 The polarization ellipse

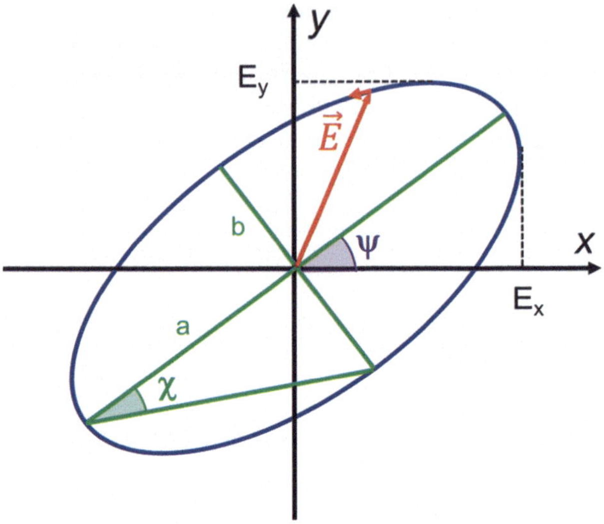

If the 4 Stokes parameters are assembled into a so-called Stokes vector (or Stokes column matrix), then the interaction of polarized light with an optical element or sample is sometimes described by the action of a so-called Mueller matrix:

$$\begin{pmatrix} S_0' \\ S_1' \\ S_2' \\ S_3' \end{pmatrix} = \begin{pmatrix} m_{00} & m_{01} & m_{02} & m_{03} \\ m_{10} & m_{11} & m_{12} & m_{13} \\ m_{20} & m_{21} & m_{22} & m_{23} \\ m_{30} & m_{31} & m_{32} & m_{33} \end{pmatrix} \begin{pmatrix} S_0 \\ S_1 \\ S_2 \\ S_3 \end{pmatrix} \tag{E.40}$$

But this is taking us too far afield. For more details about polarized light, Stokes parameters, Mueller matrices, and even Jones vectors and matrices, see:

[1.] *Optics, 5th Edition*, Eugene Hecht, 2017, Pearson, ISBN 978-0-13-397722-6.
[2.] *Polarized Light in Optics and Spectroscopy*, Kliger, Lewis, and Randall, 1990, Academic Press, ISBN 0-12-414975-8.
[3.] *Polarized Light, 3rd Edition*, Dennis Goldstein, CRC Press, Boca Raton, 2011, ISBN-13: 978–1439830406 – apparently plagiarized from:
[4.] *Polarized Light: Fundamentals and Applications*, Edward Collett, 1993, Marcel Dekker, ISBN 0824787293.

Appendix F: Detector Mathematics—Pileup

At most storage rings, there are so many buckets filled that the time between buckets is short compared to the processing time of the detection system. The photons can therefore be considered to arrive at random times. Any system that counts photons therefore has to consider what happens when two photons arrive at nearly the same time. There is a finite time called the "deadtime" during which a second photon cannot be properly processed. There are two kinds of deadtime (Fig. F.1):

[1] Nonextending or nonparalyzable
[2] Extending or paralyzable

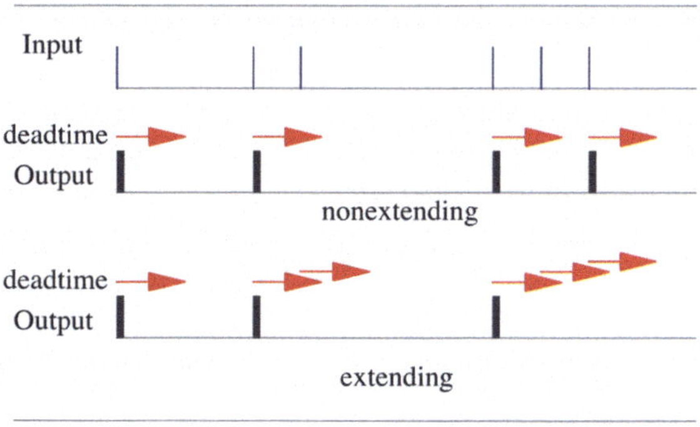

Fig. F.1 Comparison of nonextending and extending deadtime responses

S. P. Cramer, *X-Ray Spectroscopy with Synchrotron Radiation*, Biological and Medical Physics, Biomedical Engineering, https://doi.org/10.1007/978-3-030-28551-7

Appendix F.1: Non-paralyzable

A nonparalyzable system has a nonextending deadtime. The clock for the deadtime starts when the first event (photon) arrives. If another event (photon) arrives during this deadtime, it is not counted. However, the instrument recovers its processing ability once the original deadtime is over, regardless of how many events (photons) arrive during that period. A digital multiplexer with a capture and hold circuit behaves mostly like a nonparalyzable system. For a deadtime t_d and an incoming rate ρ, the output response r is:

$$r = \frac{\rho}{1 + \rho t_d} \tag{F.1}$$

Appendix F.2: Paralyzable

A "paralyzable" system has an "extending deadtime." The clock for the deadtime starts when the first event (photon) arrives. If another event (photon) arrives during this deadtime, it is not counted. Furthermore, the clock for the deadtime is restarted, and the instrument does not recover its processing ability until the new deadtime is over. If events keep coming at intervals shorter than the deadtime, the system essential shuts down. A detector-amplifier-SCA combination behaves mostly like a paralyzable system. For a deadtime t_d and an incoming rate ρ, the output response r is:

$$r = \rho e^{-\rho} t_d \tag{F.2}$$

Appendix F.3: Throughput and Loss Comparisons

We can compare the behavior of both types of systems in dimensionless terms by recording the input rate in units of the deadtime. The percent deadtime losses for a nonparalyzable system are:

$$\frac{\rho - r}{\rho} \times 100\% = 100\%(\rho t_d)/(1 + \rho t_d) \tag{F.3}$$

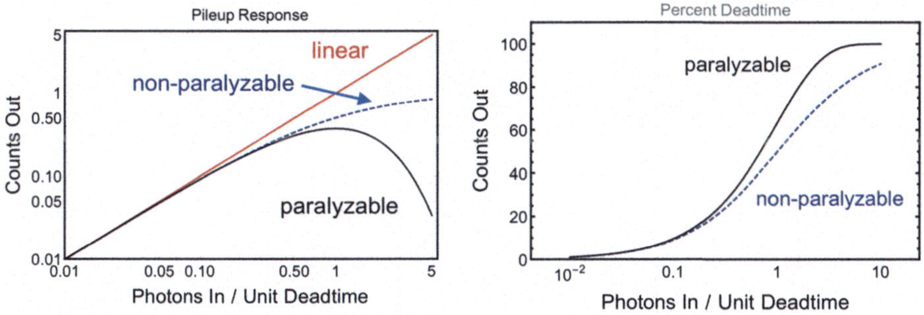

Fig. F.2 Left: comparison of the output from (**a**) a purely linear system, (**b**) a non-paralyzable system, and (**c**) a paralyzable system. Right: percentage deadtime losses for (**a**) a non-paralyzable system and (**b**) a paralyzable system

The percent deadtime losses for a paralyzable system are (Fig. F.2):

$$\frac{\rho - r}{\rho} \times 100\% = (1 - e^{-\rho t_d}) \times 100\% \tag{F.4}$$

Appendix G: Absorption Edge Energies and Linewidths

Source: CXRO X-ray Data Booklet: http://cxro.lbl.gov/X-ray-data-booklet

Linewidths: linewidths from: Krause & Oliver, *J. Phys. Chem. Ref. Data*, **8**, 329–338 (1979)

∗ Campbell and Papp, *Atomic Data and Nuclear Data Tables*, **77**, 1–56 (2001)

Table G.1 K-Edge and L-Edge energies/linewidths

Z	Element	*K* (eV)	L_1 (eV)	L_2 (eV)	L_3 (eV)
1	H	13.6			
2	He	24.6			
3	Li	54.7/0.03[a]			
4	Be	111.5			
5	B	188			
6	C	284.2/~0.08[a]			
7	N	409.9/0.13[a]	37.3		
8	O	543.1/0.14	41.6		
9	F	696.7			
10	Ne	870.2/0.21[a]	48.5	21.7	21.6
11	Na	1071/0.28[a]	63.5	30.4	30.5
12	Mg	1303/0.36	88.6/0.41	49.6/0.001	49.2/0.001
13	Al	1559/0.42	117.8/0.73	72.9/0.004	72.5/0.004
14	Si	1839/0.48	149.7/1.03	99.8/0.015	99.2/0.014
15	P	2149/0.53	189/1.26	136/0.032	135/0.033
16	S	2472/0.59	230.9/1.49	163.6/0.054	162.5/0.054
17	Cl	2833/0.64	270/1.58	202/0.083	200/0.087
18	Ar	3206/0.68	326.3/1.63	250.6/0.126	248.4/0.128
19	K	3608/0.74	378.6/1.92	297.3/0.152	294.6/0.156
20	Ca	4038/0.81	438.4/2.07	349.7/0.17	346.2/0.17
21	Sc	4492/0.86	498.0/2.21	403.6/0.19	398.7/ 0.19

(continued)

S. P. Cramer, *X-Ray Spectroscopy with Synchrotron Radiation*, Biological and Medical Physics, Biomedical Engineering, https://doi.org/10.1007/978-3-030-28551-7

Table G.1 (continued)

Z	Element	K (eV)	L_1 (eV)	L_2 (eV)	L_3 (eV)
22	Ti	4966/0.94	560.9/2.34	461.2/0.24	453.8/0.22
23	V	5465/1.01	626.7/2.41	519.8/0.26	512.1/0.24
24	Cr	5989/1.08	696/2.54	583.8/0.29	574.1/0.27
25	Mn	6539/1.16	769/2.62	649.9/0.34	638.7/0.32
26	Fe	7112/1.25	845/2.76	720/0.37	707/0.36
27	Co	7709/1.33	925/2.79	793/0.43	778/0.43
28	Ni	8333/1.44	1009/2.89	870/0.52	853/0.48
29	Cu	8979/1.55	1097/3.06	952/0.62	933/0.56
30	Zn	9659/1.67	1196/3.28	1045/0.72	1022/0.65
31	Ga	10,367/1.82	1299/3.38	1143/0.83	1116/0.76
32	Ge	11,103/1.96	1415/3.53	1248/0.95	1217/0.82
33	As	11,867/2.14	1527/3.79	1359/1.03	1324/0.94
34	Se	12,658/2.33	1652/3.94	1474/1.13	1434/1.10
35	Br	13,474/2.52	1782/4.11	1596/1.21	1550/1.08
36	Kr	14,326/2.75	1921/4.28	1731/1.31	1678/1.17
37	Rb	15,200/2.99	2065/4.44	1864/1.43	1804/1.27
38	Sr	16,105/3.25	2216/4.67	2007/1.54	1940/1.39
39	Y	17,038/3.52	2373/4.71	2156/1.65	2080/1.50
40	Zr	17,998/3.84	2532/4.78	2307/1.78	2223/1.57
41	Nb	18,986/4.14	2698/3.14	2465/1.87	2371/1.66
42	Mo	20,000/4.52	2866/4.25	2625/1.97	2520/1.78
43	Tc	21,044/4.91	3043/4.36	2793/2.08	2677/1.91
44	Ru	22,117/5.33	3224/4.58	2967/2.33	2838/2.00
45	Rh	23,220/5.77	3412/4.73	3146/2.35	3004/2.13
46	Pd	24,350/6.24	3604/4.93	3330/2.43	3173/2.25
47	Ag	25,514/6.75	3806/4.88	3524/2.62	3351/2.40
48	Cd	26,711/7.28	4018/4.87	3727/2.62	3538/2.50
49	In	27,940/7.91	4238/5.00	3938/2.72	3730/2.65
50	Sn	29,200/8.49	4465/2.97	4156/2.84	3929/2.75
51	Sb	30,491/9.16	4698/3.13	4380/3.00	4132/2.87
52	Te	31,814/9.89	4939/3.32	4612/3.12	4341/2.95
53	I	33,169/10.6	5188/3.46	4852/3.25	4557/3.08
54	Xe	34,561/11.4	5453/3.64	5104/3.40	4782/3.13
55	Cs	35,985/12.3	5714/3.78	5359/3.51	5012/3.25
56	Ba	37,441/13.2	5989/3.92	5624/3.57	5247/3.32
57	La	38,925/14.1	6266/4.06	5891/3.68	5483/3.41
58	Ce	40,443/15.1	6549/4.21	6164/3.80	5723/3.48
59	Pr	41,991/16.2	6835/4.34	6440/3.89	5964/3.60
60	Nd	43,569/17.3	7126/4.52	6722/3.97	6208/3.65
61	Pm	45,184/18.5	7428/4.67	7013/4.06	6459/3.75
62	Sm	46,834/19.7	7737/4.80	7312/4.15	6716/3.86
63	Eu	48,519/21.0	8052/4.91	7617/4.23	6977/3.91

(continued)

Table G.1 (continued)

Z	Element	K (eV)	L_1 (eV)	L_2 (eV)	L_3 (eV)
64	Gd	50,239/22.3	8376/5.05	7930/4.32	7243/4.01
65	Tb	51,996/23.8	8708/5.19	8252/4.43	7514/4.12
66	Dy	53,789/25.2	9046/5.25	8581/4.55	7790/4.17
67	Ho	55,618/26.8	9394/5.33	8918/4.66	8071/4.26
68	Er	57,486/28.4	9751/5.43	9264/4.73	8358/4.35
69	Tm	59,390/30.1	10,116/5.47	9617/4.79	8648/4.48
70	Yb	61,332/31.9	10,486/5.53	9978/4.82	8944/4.60
71	Lu	63,314/33.7	10,870/5.54	10,349/4.92	9244/4.68
72	Hf	65,351/35.7	11,271/5.63	10,739/5.02	9561/4.80
73	Ta	67,416/37.7	11,682/5.58	11,136/5.15	9881/4.88
74	W	69,525/39.9	12,100/5.61	11,544/5.33	10,207/4.98
75	Re	71,676/42.1	12,527/6.18	11,959/5.48	10,535/5.04
76	Os	73,871/44.4	12,968/7.26	12,385/5.59	10,871/5.16
77	Ir	76,111/46.8	13,419/8.30	12,824/5.69	11,215/5.25
78	Pt	78,395/49.3	13,880/9.39	13,273/5.86	11,564/5.31
79	Au	80,725/52.0	14,353/10.5	13,734/6.00	11,919/5.41
80	Hg	83,102/54.6	14,839/11.3	14,209/6.17	12,284/5.50
81	T1	85,530/57.4	15,347/12.0	14,698/6.32	12,658/5.65
82	Pb	88,005/60.4	15,861/12.2	15,200/6.48	13,055/5.81
83	Bi	90,524/63.4	16,388/12.4	15,711/6.67	13,419/5.98
84	Po	93,105/66.6	16,939/12.6	16,244/6.83	13,814/6.13
85	At	95,730/69.8	17,493/12.8	16,785/7.01	14,214/6.29
86	Rn	98,404/73.3	18,049/13.1	17,337/7.20	14,619/6.41
87	Fr	101,137/76.8	18,639/13.3	17,907/7.47	15,031/6.65
88	Ra	103,922/80.4	19,237/13.4	18,484/7.68	15,444/6.82
89	Ac	106,755/84.1	19,840/13.6	19,083/7.95	15,871/6.98
90	Th	109,651/88.0	20,472/13.7	19,693/8.18	16,300/7.13
91	Pa	112,601/91.9	21,105/14.3	20,314/8.75	16,733/7.33
92	U	115,606/96.1	21,757/14.0	20,948/9.32	17,166/7.43

Appendix H: XANES

Appendix H.1: Cross Sections

In the EXAFS Chap. 7 we blithely assumed that the interaction of X-rays with matter was primarily electric dipole in nature ...

Following Brouder, we write a more general interaction between a particle with charge q, mass m, gyromagnetic ratio g, and spin $\vec{s}$, in a potential $V(\vec{r})$ submitted to an external electromagnetic field $\left(\Phi\left(\vec{r}, t\right), \vec{A}\left(\vec{r}, t\right)\right)$ as:

$$\mathcal{H} = \left(\frac{1}{2m}\right)\left[-i\hbar\vec{\nabla} - q\vec{A}\right]^2 + V\left(\vec{r}\right) + q\Phi - \left(\frac{gq}{2m}\right)\vec{s}\cdot\vec{B}$$

which he expands and simplifies as:

$$\mathcal{H} = \mathcal{H}_0 + \left(\frac{i\hbar q}{m}\right)\vec{A}\cdot\vec{\nabla} + \left(\frac{i\hbar q}{m}\right)\vec{\nabla}\cdot\vec{A} + \left(\frac{q^2}{2m}\right)\vec{A}^2 + q\Phi - \left(\frac{gq}{2m}\right)\vec{s}\cdot\vec{B}$$

where:

$$\mathcal{H}_0 = -\left(\frac{\hbar^2}{2m}\right)\vec{\nabla}^2 + V\left(\vec{r}\right)$$

Assuming a plane wave electromagnetic field in the Coulomb gauge allows him $\Phi = 0$ and $\vec{\nabla}\cdot\vec{A} = 0$, leading to:

$$\mathcal{H} = \mathcal{H}_0 + \left(\frac{i\hbar q}{m}\right)\vec{A}\cdot\vec{\nabla} + \left(\frac{q^2}{2m}\right)\vec{A}^2 - \left(\frac{gq}{2m}\right)$$

S. P. Cramer, *X-Ray Spectroscopy with Synchrotron Radiation*, Biological and Medical Physics, Biomedical Engineering, https://doi.org/10.1007/978-3-030-28551-7

Brouder eventually goes on to argue that the $\vec{A}^2$ and $\vec{s} \cdot \vec{B}$ terms are negligible. From time-dependent perturbation theory, the transition probability per unit time for a perturbation $W(t) = W\exp(-i\omega t) + W^{*}\exp(i\omega t)$ is written as:

$$w = \left(\frac{2\pi}{\hbar}\right)\sum_{f} f \mid W \mid i^2 \delta(E_f - E_i - \hbar\omega)$$

and he writes the electromagnetic wave as:
$\vec{A}\left(\vec{r},t\right) = A_0\hat{\varepsilon}\exp\left[i\left(\vec{k}\cdot\vec{r} - \omega t\right] + A_0^{*}\hat{\varepsilon}^{*}\exp\left[-i\left(\vec{k}\cdot\vec{r} - \omega t\right]\right.\right.$

Following a Taylor series expansion of the electromagnetic plane wave, in which only dipole and quadrupole terms are retained, gets the absorption cross section σ as:

$$\sigma(\omega) = 4\pi^2\alpha\hbar\omega\sum_{if}\left\{\left|f|\hat{\varepsilon}\cdot\vec{r}|i\right|^2 + (1/4)\left|f|\hat{\varepsilon}\cdot\vec{r}\vec{k}\cdot\vec{r}|i\right|^2\right\}\delta(E_f - E_i - \hbar\omega)$$

This is a formula that is useful for describing the XANES cross section (Chap. 7), and only the first term needs to be kept for most EXAFS analyses (Chap. 6). The same matrix elements also enter the description of RIXS (Chap. 8). But, let's back up.

It turns out that the $\vec{A}^2$ and $\vec{s} \cdot \vec{B}$ terms are not always negligible. In fact, the $\vec{A}^2$ term gives rise to X-ray Raman scattering and inelastic X-ray scattering that make up the second half of Chap. 8. The $\vec{s} \cdot \vec{B}$ magnetic dipole term is often the major contributor to the nuclear cross sections that are the subject of Chaps. 9 and 10. So, in a broader perspective we can consider the interaction of X-rays with matter as governed by three terms that become significant under different conditions:

$$H_{\text{interaction}} = \underbrace{\left(\frac{i\hbar q}{m}\right)\vec{A}\cdot\vec{\nabla}}_{\substack{\text{XAS, XES,}\\ \text{\&RIXS}}} + \underbrace{\left(\frac{q^2}{2m}\right)\vec{A}^2}_{\text{XRS\&IXS}} - \underbrace{\left(\frac{q^2}{2m}\right)\vec{s}\cdot\vec{B}}_{\substack{\text{SMS, SRPAC,}\\ \text{\&IXS}}}$$

Appendix H.2: XANES Software

Links to Software for Calculation of XANES Spectra [609]

ADF	https://www.scm.com/product/adf/	[610]
CTM4XAS	http://www.anorg.chem.uu.nl/CTM4XAS/software.html	[330]
FEFF	http://monalisa.phys.washington.edu/index-feffproject.html	[228]
ORCA	https://orcaforum.cec.mpg.de	[272]
QuantumEspresso	http://www.quantum-espresso.org	[611]
StoBe	http://www.fhi-berlin.mpg.de/KHsoftware/StoBe/	[273]
Wien2K	http://www.wien2k.at http://susi.theochem.tuwien.ac.at	[282]

Appendix I: Fluorescence Energies, Yields, and Linewidths

Sources:

- Energies from: CXRO X-ray Data Booklet: http://cxro.lbl.gov/x-ray-data-booklet
- Originally: J. A. Bearden, "X-ray Wavelengths," *Rev. Mod. Phys.* 39, 78 (1967)
- Yields from: Krause & Oliver, *J. Phys. Chem. Ref. Data*, 8, 307-328 (1979)
- Linewidths from: Krause & Oliver, *J. Phys. Chem. Ref. Data*, 8, 329-338 (1979)
- Another online source: http://ftp.esrf.fr/pub/scisoft/xraylib/xraylib_tables_v2.3.pdf

Table I.1 Representative K & L fluorescence energies, yields, and linewidths

Z	Element	$K\alpha_1$ energy (keV)	Total K yield	$K\alpha_1$ Linewidth (eV)	$L\alpha_1$ energy (keV)	$L\alpha$ yield	L_3 linewidth (eV)
3	Li	0.0543	9.0E−05				
4	Be	0.1085	3.3E−04				
5	B	0.1833	7.0E−04				
6	C	0.277	1.4E−03				
7	N	0.3924	3.1E−03				
8	O	0.5249	5.8E−03				
9	F	0.6768	9.2E−03				
10	Ne	0.8486	0.018	0.24			
11	Na	1.0410	0.023	0.30			
12	Mg	1.2536	0.030	0.36			
13	Al	1.4867	0.039	0.43			
14	Si	1.7400	0.050	0.49			
15	P	2.0137	0.063	0.57			
16	S	2.3078	0.078	0.65			
17	Cl	2.6224	0.097	0.72			
18	Ar	2.9577	0.118	0.81			0.128

(continued)

S. P. Cramer, *X-Ray Spectroscopy with Synchrotron Radiation*, Biological and Medical Physics, Biomedical Engineering, https://doi.org/10.1007/978-3-030-28551-7

Table I.1 (continued)

Z	Element	$K\alpha_1$ energy (keV)	Total K yield	$K\alpha_1$ Linewidth (eV)	$L\alpha_1$ energy (keV)	$L\alpha$ yield	L_3 linewidth (eV)
19	K	3.3138	0.140	0.89			0.156
20	Ca	3.6917	0.163	0.98	0.3413	3.3E−04	0.17
21	Sc	4.0906	0.188	1.05	0.3954	8.4E−04	0.19
22	Ti	4.5108	0.214	1.16	0.4522	1.5E−03	0.22
23	V	4.9522	0.243	1.26	0.5113	2.6E−03	0.24
24	Cr	5.4147	0.275	1.35	0.5728	3.7E−03	0.27
25	Mn	5.8988	0.308	1.48	0.6374	5.0E−03	0.32
26	Fe	6.4038	0.340	1.61	0.7050	6.3E−03	0.36
27	Co	6.9303	0.373	1.76	0.7762	7.7E−03	0.43
28	Ni	7.4782	0.406	1.94	0.8515	9.3E−03	0.48
29	Cu	8.0478	0.440	2.11	0.9297	0.011	0.56
30	Zn	8.6389	0.474	2.32	1.01170	0.012	0.65
31	Ga	9.2517	0.507	2.59	1.09792	0.013	0.76
32	Ge	9.8864	0.535	2.78	1.18800	0.015	0.82
33	As	10.5437	0.562	3.08	1.2820	0.016	0.94
34	Se	11.2224	0.589	3.33	1.37910	0.018	1.00
35	Br	11.924	0.618	3.60	1.48043	0.020	1.08
36	Kr	12.649	0.643	3.92	1.5860	0.022	1.17
37	Rb	13.395	0.667	4.26	1.69413	0.024	1.27
38	Sr	14.165	0.690	4.63	1.80656	0.026	1.39
39	Y	14.958	0.710	5.02	1.92256	0.028	1.50
40	Zr	15.775	0.730	5.40	2.04236	0.031	1.57
41	Nb	16.615	0.747	5.80	2.16589	0.034	1.66
42	Mo	17.479	0.765	6.31	2.29316	0.037	1.78
43	Tc	18.367	0.780	6.82	2.424	0.040	1.91
44	Ru	19.279	0.794	7.33	2.55855	0.043	2.00
45	Rh	20.216	0.808	7.90	2.69674	0.046	2.13
46	Pd	21.177	0.820	8.49	2.83861	0.049	2.25
47	Ag	22.163	0.831	9.16	2.98431	0.052	2.40
48	Cd	23.174	0.843	9.79	3.13373	0.056	2.50
49	In	24.210	0.853	10.56	3.28694	0.060	2.65
50	Sn	25.271	0.862	11.2	3.44398	0.064	2.75
51	Sb	26.359	0.870	12.0	3.60472	0.069	2.87
52	Te	27.472	0.877	12.8	3.76933	0.074	2.95
53	I	28.612	0.884	13.7	3.93765	0.079	3.08
54	Xe	29.779	0.891	14.6	4.1099	0.085	3.13
55	Cs	30.973	0.897	15.6	4.2865	0.091	3.25
56	Ba	32.194	0.902	16.5	4.46626	0.097	3.32
57	La	33.442	0.907	17.6	4.65097	0.104	3.41
58	Ce	34.720	0.912	18.6	4.840.	0.111	3.48
59	Pr	36.026	0.917	19.8	5.037	0.118	3.60

(continued)

Table I.1 (continued)

Z	Element	$K\alpha_1$ energy (keV)	Total K yield	$K\alpha_1$ Linewidth (eV)	$L\alpha_1$ energy (keV)	$L\alpha$ yield	L_3 linewidth (eV)
60	Nd	37.361	0.921	20.9	5.2304	0.125	3.65
61	Pm	38.725	0.925	22.2	5.4325	0.132	3.75
62	Sm	40.118	0.929	23.6	5.6361	0.139	3.86
63	Eu	41.542	0.932	24.9	5.8457	0.147	3.91
64	Gd	42.996	0.935	26.4	6.0572	0.155	4.01
65	Tb	44.482	0.938	27.9	6.2728	0.164	4.12
66	Dy	45.998	0.941	29.4	6.4952	0.174	4.17
67	Ho	47.547	0.944	31.1	6.7198	0.182	4.26
68	Er	49.128	0.947	32.7	6.9487	0.192	4.35
69	Tm	50.742	0.949	34.6	7.1799	0.201	4.48
70	Yb	52.389	0.951	36.5	7.4156	0.210	4.60
71	Lu	54.070	0.953	38.4	7.6555	0.220	4.68
72	Hf	55.790	0.955	40.5	7.8990	0.231	4.80
73	Ta	57.532	0.957	42.6	8.1461	0.243	4.88
74	W	59.318	0.958	44.9	8.3976	0.255	4.98
75	Re	61.140	0.959	47.2	8.6525	0.268	5.04
76	Os	63.000	0.961	49.6	8.9117	0.281	5.16
77	Ir	64.896	0.962	52.1	9.1751	0.294	5.25
78	Pt	66.832	0.963	54.6	9.4423	0.306	5.31
79	Au	68.804	0.964	57.4	9.7133	0.320	5.41
80	Hg	70.819	0.965	60.1	9.9888	0.333	5.50
81	Tl	72.871	0.966	63.1	10.2685	0.347	5.65
82	Pb	74.969	0.967	66.2	10.5515	0.360	5.81
83	Bi	77.108	0.968	69.4	10.8388	0.373	5.98
84	Po	79.290	0.968	72.7	11.1308	0.386	6.13
85	At	81.520	0.969	76.1	11.4268	0.399	6.29
86	Rn	83.780	0.969	79.7	11.7270	0.411	6.41
87	Fr	86.100	0.970	83.4	12.0313	0.424	6.65
88	Ra	88.470	0.970	87.2	12.3397	0.437	6.82
89	Ac	90.884	0.971	91.1	12.6520	0.450	6.98
90	Th	93.350	0.971	95.2	12.9687	0.463	7.13
91	Pa	95.868	0.972	99.3	13.2907	0.476	7.33
92	U	98.439	0.972	103.5	13.6147	0.489	7.43

Appendix J: Nuclear Properties

Source: Rohlsberger: *Nuclear Condensed Matter Physics with Synchrotron Radiation*, List A

For online nuclear properties, see: http://ie.lbl.gov/toi.htm/ or http://atom.kaeri.re.kr/.

Z	Isotope	E_0 (keV)	τ_0	n.a. (%)	I_g	I_e
19	^{40}K	29.83	5.96 ns	0.012	4	3
21	^{45}Sc	12.40	458.78 ms	100.0	7/2	3/2
26	^{57}Fe	14.4124	141.11 ns	2.14	1/2	3/2
28	^{61}Ni	67.41	7.60 ns	1.19	3/2	5/2
30	^{67}Zn	93.31	13.22 μs	4.11	5/2	1/2
32	^{73}Ge	13.275	4.26 μs	7.76	9/2	5/2
32	^{73}Ge	68.752	2.51 ns	7.76	9/2	7/2
32	^{73}Ge	66.716	719.90 ms	7.76	9/2	1/2
36	^{83}Kr	9.396	212.08 ns	11.55	9/2	7/2
44	^{99}Ru	89.68	29.58 ns	12.72	5/2	3/2
50	^{119}Sn	23.871	25.61 μs	8.58	1/2	3/2
51	^{121}Sb	37.133	4.99 ns	57.25	5/2	7/2
52	^{125}Te	35.4919	2.14 ns	6.99	1/2	3/2
53	^{127}I	57.606	2.76 ns	100.0	5/2	7/2
54	^{129}Xe	39.578	1.40 ns	26.44	1/2	3/2
54	^{131}Xe	80.1853	692.4 ps	21.2	3/2	1/2
55	^{133}Cs	80.997	9.05 ns	100.0	7/2	5/2
57	^{138}La	72.57	167.35 ns	0.09	5	3
60	^{145}Nd	67.25	42.42 ns	8.30	7/2	3/2
60	^{145}Nd	72.5	1.04 ns	8.30	7/2	5/2
62	^{149}Sm	22.494	10.27 ns	13.83	7/2	5/2
62	^{154}Sm	81.976	4.36 ns	22.7	0	2
63	^{151}Eu	21.532	13.99 ns	47.82	5/2	7/2

(continued)

S. P. Cramer, *X-Ray Spectroscopy with Synchrotron Radiation*, Biological and Medical Physics, Biomedical Engineering, https://doi.org/10.1007/978-3-030-28551-7

Z	Isotope	E_0 (keV)	τ_0	n.a. (%)	I_g	I_e
63	^{153}Eu	83.3671	1.14 ns	52.18	5/2	7/2
63	^{153}Eu	97.4297	285.6 ps	52.18	5/2	5/2
64	^{158}Gd	79.5104	3.64 ns	24.84	0	2
64	^{155}Gd	60.0087	278.4 ps	14.73	3/2	5/2
64	^{155}Gd	86.546	9.13 ns	14.73	3/2	5/2
64	^{156}Gd	88.9666	3.19 ns	20.47	0	2
64	^{157}Gd	54.533	269.7 ps	15.68	3/2	5/2
64	^{157}Gd	63.917	663.64 ns	15.68	3/2	5/2
64	^{160}Gd	75.26	3.88 ns	21.86	0	2
66	^{162}Dy	80.66	3.27 ns	25.5	0	2
66	^{164}Dy	73.392	3.45 ns	28.2	0	2
66	^{161}Dy	25.65135	41.98 ns	18.9	5/2	5/2
66	^{160}Dy	86.7882	2.91 ns	2.34	0	2
66	^{161}Dy	74.56668	4.53 ns	18.9	5/2	3/2
66	^{161}Dy	43.8211	1.20 ns	18.9	5/2	7/2
68	^{166}Er	80.574	2.63 ns	33.6	0	2
68	^{168}Er	79.804	2.71 ns	26.8	0	2
68	^{167}Er	79.3219	165.9 ps	22.95	7/2	9/2
68	^{170}Er	78.68	2.73 ns	14.9	0	2
68	^{164}Er	91.40	2.12 ns	1.61	0	2
69	^{169}Tm	8.4103	5.89 ns	100.0	1/2	3/2
70	^{170}Yb	84.2551	2.32 ns	3.05	0	2
70	^{171}Yb	66.721	1.17 ns	14.31	1/2	3/2
70	^{171}Yb	75.878	2.37 ns	14.3	1/2	5/2
70	^{172}Yb	78.7435	2.38 ns	21.9	0	2
70	^{174}Yb	76.471	2.58 ns	31.84	0	2
70	^{176}Yb	82.13	2.54 ns	12.7	0	2
72	^{180}Hf	93.326	2.16 ns	35.1	0	2
72	^{178}Hf	93.180	2.14 ns	27.3	0	2
72	^{176}Hf	88.351	2.06 ns	5.21	0	2
73	^{181}Ta	6.238	8.73 μs	99.99	7/2	9/2
74	^{183}W	46.4839	271.2 ps	14.3	1/2	3/2
74	^{183}W	99.0793	1.11 ns	14.3	1/2	5/2
76	^{187}Os	75.04	3.12 ns	1.6	1/2	5/2
76	^{189}Os	36.202	764.6 ps	16.1	3/2	1/2
76	^{189}Os	69.537	2.34 ns	16.10	3/2	5/2
76	^{189}Os	95.254	331.8 ps	16.10	3/2	3/2
76	^{187}Os	9.746	3.43 ns	1.6	1/2	3/2
77	^{193}Ir	73.041	8.79 ns	62.70	3/2	1/2
77	^{191}Ir	82.407	5.89 ns	37.3	3/2	1/2
78	^{195}Pt	98.85	245.2 ps	33.80	1/2	3/2
79	^{197}Au	77.351	2.76 ns	100.0	3/2	1/2
80	^{201}Hg	1.556	144.27 ns	13.2	3/2	1/2

(continued)

Z	Isotope	E_0 (keV)	τ_0	n.a. (%)	I_g	I_e
80	^{201}Hg	32.138	288.5 ps	13.2	3/2	3/2
80	^{201}Hg	26.269	865.6 ps	13.2	3/2	1/2
92	^{238}U	44.915	324.6 ps	99.27	0	2

Appendix K: Nuclear Spectroscopy Beamlines

Table K.1 Nuclear spectroscopy beamlines active or planned in 2018

#	Facility	Beamline	Energy range (keV)	Isotopes
1	APS	3-ID	7–27	^{83}Kr, ^{57}Fe, ^{119}Sn, ^{151}Eu, ^{161}Dy
2		16-ID-D	14.4	^{57}Fe
3		30-ID-D	23–26	^{119}Sn
4	Beijing HEPS	ID-45	tbd	^{57}Fe, ^{119}Sn, . . .
5	ESRF	ID-18	7–90	^{181}Ta, ^{187}Os, ^{57}Fe, ^{151}Eu, ^{149}Sm, ^{119}Sn, ^{161}Dy, ^{129}I, ^{99}Ru
6	KEK accumulator ring	AR-NE1A	14.4	^{57}Fe
7	PETRA-III	P01	2.5–90	^{57}Fe, ^{119}Sn, ^{121}Sb, ^{125}Te, ^{193}Ir, ^{129}Xe
8	SPring-8	BL09XU	7–80	^{181}Ta, ^{73}Ge, ^{57}Fe, ^{151}Eu, ^{149}Sm, ^{174}Yb,
9		BL11XU	14.4	^{119}Sn, ^{40}K, ^{125}Te, ^{121}Sb, ^{61}Ni
10		BL19LXU	7–80	^{57}Fe
11	Taiwan Photon Source	BL23A	14.4	^{57}Fe

- Same monochromators as BL09XU
- https://www.esrf.eu/home/UsersAndScience/Experiments/MEx/ID18/beamline_layout/optics/HRM.html
- http://pfwww.kek.jp/users_info/users_guide_e/station_spec_e/ne1/ne1a.html
- http://www.spring8.or.jp/wkg/BL09XU/instrument/lang-en/INS-0000000302/instrument_summary_view
- http://www.spring8.or.jp/wkg/BL09XU/instrument/lang-en/INS-0000000302/instrument_summary_view

S. P. Cramer, *X-Ray Spectroscopy with Synchrotron Radiation*, Biological and Medical Physics, Biomedical Engineering, https://doi.org/10.1007/978-3-030-28551-7

Appendix L: Special XPS Beamlines

Appendix L.1: HAXPS Facilities

Table L.1 Representative HAXPS beamlines active or planned in 2018

#	Facility	Beamline	Properties
1	SPring-8	BL09XU	~6–10 keV Linear or circular polarization
2	SPring-8	BL47XU	~6, 8, 10 keV fixed 1×1 μ focus
3	SPring-8	BL19XLU	7.94 keV fixed Ps time-resolved HAXPES
4	SOLEIL	GALAXIES	≤12 keV Linear or circular polarization
5	DIAMOND	I09	~2.1–20 keV Soft X-rays also available
6	NSLS-II	SST-2	~0.1–7.5 keV Soft X-rays available
7	PETRA-III	P22	~2.4–15 keV Linear or circular polarization

S. P. Cramer, *X-Ray Spectroscopy with Synchrotron Radiation*, Biological and Medical Physics, Biomedical Engineering, https://doi.org/10.1007/978-3-030-28551-7

Appendix L.2: APXPS Facilities

Table L.2 Representative APXPS beamlines active or planned in 2018 [612]

#	Facility	Beamline	Source	Energy range (eV)	Spectrometer
1 2 3	ALS	9.3.2 9.3.1 11.0.2	Bend Magnet Bend Magnet EPU	200–800 2000–6000 90–1500	Scienta R4000 HiPP Scienta R4000 HiPP-2 Specs Phoibos NAP 150
4	ALBA	BL24	APPLE II	100–2000	Specs Phoibos NAP 150
5 6	BESSY-II	ISISS CAT@EMIL	Bend Magnet Undulator	80–2000 80–2000	Specs Phoibos NAP 150 HV –
7	Brazilian SLS	PGM	APPLE-II	100–1500	Scienta R4000 HiPP
8	DIAMOND	VERSOX	Bend Magnet	200–2800	–
9 10 11	MAX-IV	I511 SPECIES HIPPIE	UndulatorEPUEPU	50–1500 27–1500 263–2000	Specs Phoibos NAP 150 Specs Phoibos NAP 150 –
12 13	Photon Factory	7A 13A	Bend Magnet Undulator	100–1500 30–1600	Omicron EA12HP Roll-up
14	Shanghai SRF		Bend Magnet	50–2500	
15	SOLEIL	TEMPO	Undulator	50–1200	Specs Phoibos NAP 150
16 17	SPring-8	36XU 07LSU	Undulator Undulator	4500–35,000 250–2000	Scienta R4000 HiPP-2 Specs Phoibos NAP 150
18	SSRL	13.2	Undulator	200–1100	Scienta SES100
19 20	Swiss Light Source	SIM Phoenix	APPLE II APPLE II	150–1800 1500–8000	Scienta R4000 HiPP-2 Roll-up

Appendix M: X-ray Free Electron Laser Facilities

Table M.1 Current and planned hard X-ray free electron laser facilities

Country	Machine	Beam energy (GeV)	X-ray energy (keV)	Rep rate	First light
Switzerland	SwissFEL (Aramis)	2.1–5.8	1.8–12.4	100 Hz	2017
USA	LCLS	2.5–15	0.25–12.6	120 Hz	2009
USA	LCLS-II SXU	4	0.2–3	1 MHz	2020
USA	LCLS-II HXU	4	1–5	1 MHz	2020
USA	LCLS-II HXU/Cu	3.3–8	1–25	120	2020
USA	LCLS-II HE	3.3–8	5–12.8	1 MHz	2028
Korea	PAL-XFEL SX	2.6–3.2	0.28–1.24	60 Hz	2016
Korea	PAL-XFEL HX	4–10	2.1–21	60 Hz	2016
Japan	SACLA	8.5	5.0–20.7	60 Hz	2012
Germany	European XFEL HXR	8–17.5	3.1–25	27 KHz	2017

Table M.2 Current and planned soft X-ray free electron laser facilities

Country	Machine	Beam energy (GeV)	X-ray energy (keV)	Rep rate	First light
Switzerland	SwissFEL (Athos)	2.5–3.4	0.18–1.8	100 Hz	2019
USA	LCLS-II SXU	4	0.2–3	1 MHz	2020
Korea	PAL-XFEL SX	2.6–3.2	0.28–1.24	60 Hz	2016
Japan	SACLA	8.5	5.0–20.7	60 Hz	2012
Germany	FLASH1	<1.25	0.24–3.0	10 Hz	2012
Germany	FLASH2	<1.25	0.14–3.1	10 Hz	2012
Germany	European XFEL SXR	8–17.5	0.25–3.1	27 KHz	2017
China	SXFEL	0.8–1.5	0.18–0.6	50	2019

S. P. Cramer, *X-Ray Spectroscopy with Synchrotron Radiation*, Biological and Medical Physics, Biomedical Engineering, https://doi.org/10.1007/978-3-030-28551-7

References

1. "From the Psi to Charm—The Experiments of 1975 and 1976", Richter, B., in *Nobel Lectures in Physics*; Lundqvist, S., Ed.; World Scientific Publishing: Singapore, 1992
2. Born, M.; Wolf, E. *Principles of Optics*; Fifth ed.; Macmillan: New York**, 1974**
3. "Synchrotron Radiation—1873 to 1947", Blewett, J. P. *Nucl. Inst. Meth. A*, **1988**, *266*, 1–9
4. "Synchrotron Radiation Sources", Munro, I. H.; Marr, G. V., in *Handbook of Synchrotron Radiation*; Marr, G. V., Ed.; North-Holland: Amsterdam, 1987; Vol. 2, pp. 3–20
5. "On a new kind of rays", Röntgen, W. C. *Nature*, **1896**, *53*, 274–276
6. "On the maximal energy attainable in a betatron", Ivanenko, D.; Pomeranchuk, I. *Dokl. Akad. Nauk. SSSR*, **1944**, *44*, 315
7. "On the Classical Radiation of Accelerated Electrons", Schwinger, J. *Phys. Rev.*, **1949**, *75*, 1912–1925
8. "Spectral and Angular Distribution of Ultraviolet Radiation from the 300-Mev Cornell Synchrotron", Tomboulian, D. H.; Hartman, P. L. *Phys. Rev.*, **1956**, *102*, 1423–1447
9. Hallonsten, O., Ph. D. Thesis, Lund University, **2009**
10. "Operating Experience with the ALS Linac", Selph, F.; Massoletti, D., IEEE Part. Acc. Conf., San Francisco, CA, **1991**, pp. 2978–2980
11. "The Design of Linear Accelerators", Slater, J. C. *Rev. Mod. Phys.*, **1948**, *20*, 473–518
12. "Beam Dynamics of the 100 Mev Preinjector for the Spanish Synchrotron ALBA", Setty, A. *Proc. PAC07*, **2007**, 1472–1474
13. Wiedemann, H. *Particle Accelerator Physics—Basic Principles and Linear Beam Dynamics*; Third ed.; Springer: New York**, 2007**
14. "The MAX IV storage ring project", Tavares, P. F.; Leemann, S. C.; Sjostrom, M.; Andersson, A. *J. Syn. Rad.*, **2014**, *21*, 862–877
15. "MAX-IV lattice, dynamic properties and magnet system", Tarawneh, H.; Eriksson, M.; Lindgren, L. J.; Anderberg, B. *Nuc. Inst.Meth. A*, **2003**, *508*, 480–486
16. "MAX IV paves the way for ultimate X-ray microscope", Eriksson, M. *CERN Courier August*, **2016**
17. "NSLS-II Commissioning and Operation", Wang, G.; Shaftan, T.; Bassi, G.; Bengtsson, J.; Blednykh, A.; Blum, E.; Cheng, W.; Choi, J.; Davidsaver, M.; Doom, L.; Fliller, R.; Ganetis, G.; Guo, W.; Hidaka, Y.; Kramer, S.; Li, Y.; Podobedov, B.; Qian, K.; Rose, J.; Seletskiy, S.; Singh, O.; Smaluk, V.; Smith, R.; Summers, T.; Tanabe, T.; Willeke, F.; Yang, L.; Yang, X.; Yu, L. *Proc. 12th Int. Conf. Syn. Rad. Inst.*, **2016**, *1741*, 020007
18. "Preliminary Design of a Dedicated Synchrotron Radiation Facility", Chasman, R.; Green, K.; Rowe, E. *IEEE Trans.*, **1975**, *NS-22*, 1765–1767

S. P. Cramer, *X-Ray Spectroscopy with Synchrotron Radiation*, Biological and Medical Physics, Biomedical Engineering, https://doi.org/10.1007/978-3-030-28551-7

19. "The National Synchrotron Light Source, Part I: Bright Idea", Crease, R. P. *Phys. Perspect.*, **2008**, *10*, 438–467
20. "First multi-bend achromat lattice consideration", Einfeld, D.; Plesko, M.; Schaper, J. *J. Syn. Rad.*, **2014**, *21*, 856–861
21. "Design of the ALS-U Storage Ring Lattice", Sun, C.; Nishimura, H.; Robin, D.; Sannibale, F.; Steier, C.; Venturini, M.; Wan, W.; Jung, J. Y.; Swenson, C. A. *Proc. IPAC2017*, **2017**, 2827–2829
22. "The Multi-Bend Achromat Storage Rings", Eriksson, M. *AIP Conf. Proc.*, **2016**, *1741*
23. "DLSR design and plans: an international overview", Hettel, R. *J. Syn. Rad.*, **2014**, *21*, 843-855
24. "Touschek lifetime study based on the precise bunch-by-bunch BCM system at SSRF", Chen, F. Z.; Chen, Z. C.; Zhou, Y. M.; Zhang, N.; Gao, B.; Xu, X. Y.; Leng, Y. B. *Nucl. Sci. Tech.*, **2019**, *30*
25. "Sirius", Liu, L.; Sá, F.; Resende, X. *IPAC'16*, **2016**, 3413–3416
26. "Synchrotron Radiation Sources", Wille, K. *Rep. Prog. Phys.*, **1991**, *54*, 1005–1068
27. "On the Shaking of Ions in Electron Storage Rings", Bozoki, E.; Sagan, D. *Nucl. Inst. Meth. A*, **1994**, *340*, 259–271
28. "Vacuum System Design of the Third Generation Synchrotron Radiation Source PETRA III", Böspflug, R.; Boster, J.; Giesske, W.; Keese, D.; Köhler, R.; Mildne, N.; Nagorny, B.; Naujoks, U.; Remde, H.; Schulz, E.; Seidel, M.; Tiessen, J.; Wedekind, H. P.; Zapfe, K. *J. Phys.: Conf. Ser. 100 092012*, **2008**, *100*, 092012
29. "Successful Completion of the ALS Superbend Project", Robin, D.; Benjegerdes, R.; Biocca, A.; Bish, P.; Brown, W.; Byrne, W.; Calais, D.; Chin, M.; C. Corradi; Coulomb, D.; Vries, J. D.; DeMarco, R.; Fahmie, M.; Geyer, A.; Harkins, J.; Henderson, T.; J. Hinkson; Hoyer, E.; Hull, D.; Jacobson, S.; Krupnick, J.; Marks, S.; McDonald, J.; Molinari, P.; Mueller, R.; Nadolski, L.; Nishimura, K.; Ottens, F.; Paterson, J. A.; Pipersky, P.; Ritchie, A.; Rossi, S.; Salvant, B.; Schlueter, R.; Schwartz, A.; Spring, J.; Steier, C.; Taylor, C.; Thur, W.; Timossi, C.; Wandesforde, A.; Zbasnik, J. *Proc. EPAC*, **2002**, 215–217
30. "Periodic Magnets for Free Electron Lasers", Walker, R. P. *Nucl. Inst. Meth.*, **1985**, *A237* 366–380
31. "Design of Permanent Multipole Magnets with Oriented Rare Earth Cobalt Material", Halbach, K. *Nucl. Inst. Meth.*, **1980**, *169* 1–10
32. "A New Wiggler Beam Line for SSRL", Hoyer, E.; Bahr, C.; Chan, T.; Chin, J.; Elioff, T.; Halbach, K.; Harnett, G.; Humphries, D.; Hunt, D.; Kim, K.-J.; Lauritzen, T.; Lindel, D.; Shirley, D.; Tafelski, R.; Thompson, A.; Cramer, S.; Eisenberger, P.; Hewitt, R.; Stohr, J.; Boyce, R.; Brown, G.; Golde, A.; Gould, R.; Hower, N.; I. Lindau; Winick, H.; Yang, J.; Harris, J.; Scott, B. *Nucl. Inst. Meth.*, **1983**, *208*, 117–125
33. "The Beam Line VI REC-Steel Hybrid Wiggler For SSRL", Hoyer, E.; Chan, T.; Chin, J. W. G.; Halbach, K.; Kim, K. J.; Winick, H.; Yang, J. *IEEE Trans. Nuc. Sci.*, **1983**, *30*, 3118–3120
34. "Wiggler and Undulator Magnets - A Review", Brown, G.; Halbach, K.; Harris, J.; Winick, H. *Nucl. Inst. Meth.*, **1983**, *208*, 65–77
35. "Status of the PETRA III Insertion devices", Barthelmess, M.; Englisch, U.; Pflüger, J.; Schöps, A.; Skupin, J.; Tischer, M. *Proc. EPAC08*, **2008**, 2320–2322
36. "Cryogenic permanent magnet undulators", Hara, T.; Tanaka, T.; Kitamura, H.; Teruhiko Bizen; Maréchal, X.; Seike, T.; Kohda, T.; Matsuura, Y. *Phys. Rev. Special Topics*, **2004**, *7*, 050702
37. "Cryogenic In-Vacuum Undulator at Danfysik", Danfysik *Proc. IPAC '10*, **2010**, WEPD006
38. "Insertion Devices at the Advanced Photon Source (APS)", Moog, L., Proc. IDMAX2010, MAX-Lab, Lund, Sweden, **2010**, pp.
39. "Progress on the superconducting undulator for ANKA and on the instrumentation for R&D", Casalbuoni, S. *Proc. IDMAX2010*, **2010**

40. "Generation of various polarization states from insertion devices: A review", Elleaume, P. *Rev. Sci. Inst.*, **1989**, *60*, 1830–1833
41. "Production of Circularly Polarized Synchrotron Radiation", Kitamura, H. *Syn. Rad. News*, **1992**, *5*, 14–20
42. "Insertion devices—recent developments and future trends", Walker, R. P.; Diviacco, B. *Syn. Rad. News*, **2000**, *13*, 33–42
43. "Helical-undulator Beamline for Soft X-ray Spectromicroscopy at Saga LS", Yasumoto, M.; Takemoto, K.; Ishiguro, E.; Kihara, H.; Kamijo, N.; Chikaura, Y. *Proc. 8th Int. Conf. X-ray Micro.*, *7*, 73–75
44. "Special Multipole Wiggler Design Producing Circularly Polarized Synchrotron Radiation", Goulon, J.; Elleaume, P.; Raoux, D. *Nucl. Inst. Meth. A*, **1987**, *254*, 192–201
45. "Studies of insertion devices for producing circularly polarized radiation with variable helicity in ELETTRA", Walker, R.; Diviacco, B. *Rev. Sci. Inst.*, **1992**, *63*, 332–335
46. "A short-period helical wiggler as an improved source of synchrotron radiation", Kincaid, B. M. *J. App. Phys.*, **1977**, *48*, 2684–2691
47. "An experimental superconducting helical undulator", Caspi, S.; Taylor, C. *Nucl. Inst. Meth. A*, **1996**, *375*, 433–435
48. "Motion and Radiation of Relativistic Electrons in Special Undulator", Moiseev, M. B.; Nikitin, M. M.; Fedosov, N. I. *Izv.Vuz.Fiz.*, **1978**, *21*, 14–18
49. "A Synchrotron Radiation Source with Arbitrarily Adjustable Elliptical Polarization", Kim, K. J. *Nucl. Inst. Meth. A*, **1984**, *219*, 425–429
50. "Circularly Polarized Synchrotron Radiation from the Crossed Undulator at BESSY", Bahrdt, J.; Gaupp, A.; Gudat, W.; Mast, M.; Molter, K.; Peatman, W. B.; Scheer, M.; Schroeter, T.; Wang, C. *Rev. Sci. Inst.*, **1992**, *63*, 339–342
51. "Polarization properties of a crossed planar undulator", Li, Y.; Faatz, B.; Pflueger, J. *Nucl. Inst. Meth. A*, **2010**, *613*, 163–168
52. "Crossed undulator polarization control for X-ray FELs in the saturation regime", Geng, H.; Ding, Y.; Huang, Z. *Nucl.Inst. Meth. A*, **2010**, *622*, 276–280
53. "Design of a new type of planar undulator for generating variably polarized radiation", Sasaki, S.; Kakuno, K.; Takada, T.; Shimada, T.; Yanagida, K.-I.; Miyahara, Y. *Nucl. Inst. Meth. A*, **1993**, *331*, 763–767
54. "First commissioning results for the elliptically polarizing undulator beamline at the Advanced Light Source", Young, A. T.; Feng, J.; Arenholz, E.; Padmore, H. A.; Henderson, T.; Marks, S.; Hoyer, E.; Schlueter, R.; Kortright, J. B.; Martynov, V.; Steier, C.; Portmann, G. *Nucl. Inst. Meth.*, **2001**, *467*, 549–552
55. "A soft X-ray undulator beamline at the Advanced Light Source with circular and variable linear polarization for the spectroscopy and microscopy of magnetic materials", Young, A. T.; Arenholz, E.; Feng, J.; Padmore, H.; Marks, S.; Schlueter, R.; Hoyer, E.; Kelez, N.; Steier, C. *Surface Rev. Lett.*, **2002**, *9*, 549–554
56. "Installation and Commissioning of a 6-Tesla Superconducting Wavelength Shifter at Taiwan Light Source", Chang, C. H.; Chang, H. P.; Chen, J.; Chen, J. R.; Fan, T. C.; Hwang, C. S.; Hsiung, G. Y.; Hsu, K. T.; Kuo, C. C.; Luo, G. H.; Wang, D. J.; Wang, M. H. *Nucl. Inst. Meth.*, **2005**, *550*, 446–453
57. "Optimization of a Multipole Wiggler for TPS", J. C. Jan; Kuo, C. Y.; Chang, C. H.; Chung, T. Y.; Huang, J. C.; Hwang, C. S. *Proc. IPAC2016*, **2016**
58. "Cryogenic permanent magnet and superconducting undulators", Bahrdt, J.; Gluskin, E. *Nucl. Inst. Meth. A*, **2018**, *907*, 149–168
59. "Complete polarization analysis of an APPLE II undulator using a soft X-ray polarimeter", Wang, H.; Bencok, P.; Steadman, P.; Longhi, E.; Zhu, J.; Wang, Z. *J. Syn. Rad.*, **2012**, *19*, 944–948
60. "Magnetic Design of an Apple III Undulator for SwissFEL", Schmidt, T.; Anghel, A.; Böhler, P.; Brügger, M.; Calvi, M.; Danner, S.; Huber, P.; Keller, A.; Locher, M. *Proc. FEL2014*, **2014**, MOP043

61. "Delta undulator for Cornell energy recovery linac", Temnykh, A. B. *Phys. Rev. Spec. Top. - Acc.*, **2008**, *11*
62. "APPLE X Undulator for the SwissFEL Soft X-ray Beamline Athos", Schmidt, T.; Calvi, M. *Syn. Rad. News*, **2018**, *31*, 35–40
63. "Design and performance of the APPLE-Knot undulator", Ji, F. H.; Chang, R.; Zhou, Q. G.; Zhang, W.; Ye, M.; Sasaki, S.; Qiao, S. *J. Syn. Rad.*, **2015**, *22*, 901–907
64. "Performance upgrade in the JAEA actinide science beamline BL23SU at SPring-8 with a new twin-helical undulator", Saitoh, Y.; Fukuda, Y.; Takeda, Y.; Yamagami, H.; Takahashi, S.; Asano, Y.; Hara, T.; Shirasawa, K.; Takeuchi, M.; Tanaka, T.; Kitamura, H. *J. Syn. Rad.*, **2012**, *19*, 388–393
65. "Helios: a New Type of Linear/Helical Undulator", Elleaume, P. *J. Syn. Rad.*, **1994**, *1*, 19–26
66. "A Primer in Synchrotron Radiation: Everything You Wanted to Know About SEX (Synchrotron Emission of X-rays) but Were Afraid to Ask", Margaritondo, G. *J. Syn. Rad.*, **1995**, *2*, 148–154
67. "Characteristics of Synchrotron Radiation", Kim, K.-J., Physics of Particle Accelerators, **1989**, pp. 565–632
68. Green, G. K. "Brookhaven National Laboratory Report BNL-50522," **1976**
69. Jackson, J. D. *Classical Electrodynamics*; John Wiley & Sons: New York**, 1962**
70. Duke, P. J. *Synchrotron Radiation—Production and Properties*; Oxford University Press: Oxford**, 2000**
71. Feynman, R. P.; Leighton, R. B.; Sands, M. *The Feynman Lectures on Physics*; Addison-Wesley: Reading**, 1963**; Vol. 1
72. Kliger, D. S.; Lewis, J. W.; Randall, C. E. *Polarized Light in Optics and Spectroscopy*; Academic Press: San Diego**, 1990**
73. "Polarized Nature of Synchrotron Radiation", Kim, K.-J. *SPIE Proc.*, **1991**, *1548*, 73–79
74. "Introduction and Overview", Winick, H., in *Synchrotron Radiation Sources—A Primer*; World Scientific: Singapore, 1994; Vol. 1
75. "Optical and power characteristics of synchrotron radiation sources", Kim, K.-J. *Opt. Eng.*, **1995**, *34*, 342–352
76. "Undulators and wigglers for the production of radiation and other applications", Vinokurov, N. A.; Levichev, E. B. *Phys. Uspekhi*, **2015**, *58*, 850
77. Clarke, J. A. *The Science and Technology of Undulators and Wigglers*; Oxford University Press: New York**, 2004**
78. "1st Observation of Undulator Radiation from Apple-1", Sasaki, S.; Shimada, T.; Yanagida, K.; Kobayashi, H.; Miyahara, Y. *Nucl. Inst. Meth. A*, **1994**, *347*, 87–91
79. "Undulator Radiation", Elleaume, P., in *Undulators, Wigglers, and Their Applications*; Elleaume, H. O. A. P., Ed.; Taylor and Francis: New York, 2003, pp. 69–107
80. "Exotic Insertion Devices", Sasaki, S., in *Undulators, Wigglers, and Their Applications*; Elleaume, H. O. A. P., Ed.; Taylor and Francis: New York, 2003, pp. 237–254
81. "A New Undulator for Generating Variably Polarized Radiation", Sasaki, S.; Miyata, K.; Takada, T. *Jap. J. App. Phys. Part 2*, **1992**, *31*, L1794–L1796
82. "Polarizing Undulators and Wigglers", Onuki, H., in *Undulators, Wigglers, and Their Applications*; Elleaume, H. O. A. P., Ed.; Taylor and Francis: New York, 2003, pp. 214–236
83. "URGENT—A Computer-Program for Calculating Undulator Radiation Spectral, Angular, Polarization, and Power-Density Properties", Walker, R. P.; Diviacco, B. *Rev. Sci. Inst.*, **1992**, *63*, 392–395
84. "SPECTRA: a synchrotron radiation calculation code", Tanaka, T.; Kitamura, H. *J. Syn. Rad.*, **2001**, *8*, 1221–1228
85. "WAVE—A Computer Code for the Tracking of Electrons Through Magnetic Fields and the Calculation of Spontaneous Synchrotron Radiation", Scheer, M., ICAP2012, Rostock-Warnemünde, Germany, **2012**, pp. TUACC2
86. "XOP v2.4: Recent Developments of The X-ray Optics Software Toolkit", del Rio, M. S.; Dejus, R. J. *Proc. SPIE 8141*, **2011**, *8141*

87. "Wavefront propagation simulations for beamlines and experiments with 'Synchrotron Radiation Workshop'", Chubar, O.; Fluerasu, A.; Berman, L.; Kaznatcheev, K.; Wiegart, L. *J. Phys. Conf. Ser.*, **2013**, *425*
88. "Sirepo: an open-source cloud-based software interface for X-ray source and optics simulations", Rakitin, M. S.; Moeller, P.; Nagler, R.; Nash, B.; Bruhwiler, D. L.; Smalyuk, D.; Zhernenkov, M.; Chubar, O. *J. Syn. Rad.*, **2018**, *25*, 1877–1892
89. "X-ray interactions: photoabsorption, scattering, transmission, and reflection at E=50–30000 eV, Z=1–92," Henke, B. L.; Gullikson, E. M.; Davis, J. C. *Atomic Data Nuc. Data Tab.*, **1993**, *54*, 181–342
90. "Formation of optical images by X-rays", Kirkpatrick, P.; Baez, A. V. *J. Opt. Soc. Am.*, **1948**, *38*, 766–774
91. "Theory and practice of elliptically bent X-ray mirrors", Howells, M. R.; Cambie, D.; Duarte, R. M.; Irick, S.; MacDowell, A. A.; Padmore, H. A.; Renner, T. R.; Rah, S.; Sandler, R. *Opt. Eng.*, **2000**, *39*, 2748–2762
92. "Ultra-precise characterization of LCLS hard X-ray focusing mirrors by high resolution slope measuring deflectometry", Siewert, F.; Buchheim, J.; Boutet, S.; Williams, G. J.; Montanez, P. A.; Krzywinski, J.; Signorato, R. *Opt. Exp.*, **2012**, *20*, 4525–4536
93. "Focusing of X-ray free-electron laser pulses with reflective optics", Yumoto, H.; Mimura, H.; Koyama, T.; Matsuyama, S.; Tono, K.; Togashi, T.; Inubushi, Y.; Sato, T.; Tanaka, T.; Kimura, T.; Yokoyama, H.; Kim, J.; Sano, Y.; Hachisu, Y.; Yabashi, M.; Ohashi, H.; Ohmori, H.; Ishikawa, T.; Yamauchi, K. *Nat. Photonics*, **2013**, *7*, 43–47
94. "Simple Method for Focusing X-rays Using Tapered Capillaries", Stern, E. A.; Kalman, Z.; Lewis, A.; Lieberman, K. *Appl. Opt.*, **1988**, *27*, 5135–5139
95. "A compound refractive lens for focusing high-energy X-rays", Snigirev, A.; Kohn, V.; Snigireva, I.; Lengeler, B. *Nature*, **1996**, *384*, 49–51
96. "Highly efficient blazed grating with multilayer coating for tender X-ray energies", Senf, F.; Bijkerk, F.; Eggenstein, F.; Gwalt, G.; Huang, Q.; Kruijs, R.; Kutz, O.; Lemke, S.; Louis, E.; Mertin, M.; Packe, I.; Rudolph, I.; Schäfers, F.; Siewert, F.; Sokolov, A.; Sturm, J. M.; Waberski, C.; Wang, Z.; Wolf, J.; Zeschke, T.; Erko, A. *Opt. Express*, **2016**, *24*, 13220–13230
97. "Spectral tailoring of nanoscale EUV and soft X-ray multilayer optics", Huang, Q.; Medvedev, V.; Kruijs, R. V. D.; Yakshin, A.; Louis, E.; Bijkerk, F. *App. Phys. Rev.*, **2017**, *4*, 011104
98. "Optimized highly efficient multilayer-coated blazed gratings for the tender X-ray region", Sokolov, A.; Huang, Q. S.; Senf, F.; Feng, J. T.; Lemke, S.; Alimov, S.; Knedel, J.; Zeschke, T.; Kutz, O.; Seliger, T.; Gwalt, G.; Schafers, F.; Siewert, F.; Kozhevnikov, I. V.; Qi, R. Z.; Zhang, Z.; Li, W. B.; Wang, Z. S. *Opt. Express*, **2019**, *27*, 16833–16846
99. "Narrowband lamellar multilayer grating with low contrast $MoSi_2$/Si materials for the soft X-ray region", Huang, Q. S.; Feng, J. T.; Li, T. Z.; Wang, X. M.; Kozhevnikov, O. R.; Yang, Y.; Qi, R. Z.; Sokolov, A.; Sertsu, M. G.; Schafers, F.; Li, W. B.; Xie, C.; Zhang, Z.; Wang, Z. S. *J. Phys. D—App. Phys.*, **2019**, *52*
100. "On the manufacture and theory of diffraction gratings", Strutt, J. W.; Rayleigh, L. *Phil. Mag.*, **1874**, *47*, 193–205
101. Mills, D.; Padmore, H. "X-ray Optics for BES Light Source Facilities," **2013**
102. Michette, A. G. *Optical Systems for Soft X-rays*; Plenum: New York**, 1986**
103. "Performance of the Dragon Soft X-ray Beamline", Chen, C. T.; Sette, F. *Rev. Sci. Inst.*, **1989**, *60*, 1616–1621
104. "Performance and capabilities of the Canadian Dragon: The SGM beamline at the Canadian Light Source", Regier, T.; Krochak, J.; Sham, T. K.; Hu, Y. F.; Thompson, J.; Blyth, R. I. R. *Nucl. Inst. Meth. A*, **2007**, *582*, 93–95
105. "Upgrade of beamline BL08B at Taiwan Light Source from a photon-BPM to a double-grating SGM beamline", Yuh, J.-Y.; Lin, S.-W.; Huang, L.-J.; Fung, H.-S.; Lee, L.-L.; Chen, Y.-J.; Cheng, C.-P.; Chin, Y.-Y.; Lin, H.-J. *J. Syn. Rad.*, **2015**, *22*, 1312–1318
106. "The BESSY constant length Rowland circle monochromator", Schwarzkopf, O.; Borchert, M.; Eggenstein, F.; Flechsig, U.; Kalus, C.; Lammert, H.; Menthel, U.; Pietsch, M.; Reichardt,

G.; Rotter, P.; Senf, F.; Zeschke, T.; Peatman, W. B. *J. Elect. Spect. Rel. Phen.*, **1999**, *101*, 997–1001
107. "Application of a Varied-Line-Spacing Grating in a High-Performance Soft X-ray Monochromator", Yan, Y.; Shigemasa, E.; Yagishita, A. *J. Syn. Rad.*, **1998**, *5*, 246–251
108. "First results from the online variable line spacing grating spectrometer at FLASH", Brenner, G.; Kapitzki, S.; Kuhlmann, M.; Ploenjes, E.; Noll, T.; Siewert, F.; Treusch, R.; Tiedtke, K.; Reininger, R.; Roper, M. D.; Bowler, M. A.; Quinn, F. M.; Feldhaus, J. *Nucl. Inst. Meth. A*, **2011**, *635*, S99-S103
109. "A new extra-focus monochromator designed for high-performance VUV beamlines", Xue, C. F.; Wu, Y. Q.; Zou, Y.; Xue, L.; Wang, Y.; Xu, Z. J.; Tai, R. Z. *J. Syn. Rad.*, **2015**, *22*, 328–335
110. "Tools of the Trade: Diamonds Are Forever . . . for Focusing Intense X-rays", Ross, M. *SLAC Today*, **2011**, September 22, 2011
111. "High-efficiency multilevel zone plates for keV X-rays", Di Fabrizio, E.; Romanato, F.; Gentili, M.; Cabrini, S.; Kaulich, B.; Susini, J.; Barrett, R. *Nature*, **1999**, *401*, 895–898
112. "High-efficiency zone-plate optics for multi-keV X-ray focusing", Mohacsi, I.; Karvinen, P.; Vartiainen, I.; Guzenko, V. A.; Somogyi, A.; Kewish, C. M.; Mercere, P.; David, C. *J. Syn. Rad.*, **2014**, *21*, 497–501
113. "Nanofocusing of hard X-ray free electron laser pulses using diamond based Fresnel zone plates", David, C.; Gorelick, S.; Rutishauser, S.; Krzywinski, J.; Vila-Comamala, J.; Guzenko, V. A.; Bunk, O.; Farm, E.; Ritala, M.; Cammarata, M.; Fritz, D. M.; Barrett, R.; Samoylova, L.; Grunert, J.; Sinn, H. *Scientific Reports*, **2011**, *1*, 57–61
114. "Phase Zone Plates for X-Rays and the Extreme UV", Kirz, J. *J. Opt. Soc. Amer.*, **1974**, *64*, 301
115. "X-Ray Optics for Third-Generation Synchrotron Radiation Sources", Mills, D. M., in *Third Generation Hard X-Ray Synchrotron Radiation Sources: Source Properties, Optics, and Experimental Techniques*; Mills, D. M., Ed.; John Wiley & Sons: New York, 2002, pp. 41–99
116. "Spherical analyzers and monochromators for resonant inelastic hard X-ray scattering: a compilation of crystals and reflections", Gog, T.; Casa, D. M.; Said, A. H.; Upton, M. H.; Kim, J.; Kuzmenko, I.; Huang, X.; Khachatryan, R. *J. Syn. Rad.*, **2013**, *20*, 74–79
117. "New materials for high-energy-resolution X-ray optics", Yavas, H.; Sutter, J. P.; Gog, T.; Wille, H. C.; Baron, A. Q. R. *MRS Bull.*, **2017**, *42*, 424–429
118. "The sapphire backscattering monochromator at the Dynamics beamline P01 of PETRA III", Alexeev, P.; Asadchikov, V.; Bessas, D.; Butashin, A.; Deryabin, A.; Dill, F. U.; Ehnes, A.; Herlitschke, M.; Hermann, R. P.; Jafari, A.; Prokhorov, I.; Roshchin, B.; Rohlsberger, R.; Schlage, K.; Sergueev, I.; Siemens, A.; Wille, H. C. *Hyp. Int.*, **2016**
119. "Resonant inelastic X-ray scattering spectrometer with 25 meV resolution at the Cu K-edge", Ketenoglu, D.; Harder, M.; Klementiev, K.; Upton, M.; Taherkhani, M.; Spiwek, M.; Dill, F.-U.; Wille, H.-C.; Yavas, H. *J. Syn. Rad.*, **2015**, *22*, 961–967
120. "The high-resolution diffraction beamline P08 at PETRA III", Seeck, O. H.; Deiter, C.; Pflaum, K.; Bertam, F.; Beerlink, A.; Franz, H.; Horbach, J.; Schulte-Schrepping, H.; Murphy, B. M.; Greve, M.; Magnussen, O. *J. Syn. Rad.*, **2012**, *19*, 30–38
121. "X-Ray Monochromator System for Use with Synchrotron Radiation Sources", Golovchenko, J. A.; Levesque, R. A.; Cowan, P. L. *Rev. Sci. Inst.*, **1981**, *52*, 509–516
122. "High-resolution X-ray monochromators", Ishikawa, T.; Tamasaku, K.; Yabashi, M. *Nucl. Inst. Meth. A*, **2005**, *547*, 42–49
123. "An X-ray scattering beamline for studying dynamics", Baron, A. Q. R.; Tanaka, Y.; Goto, S.; Takeshita, K.; Matsushita, T.; Ishikawa, T. *J. Phys. Chem. Sol.*, **2000**, *61*, 461–465
124. "Milli-electronvolt monochromatization of hard X-rays with a sapphire backscattering monochromator", Sergueev, I.; Wille, H. C.; Hermann, R. P.; Bessas, D.; Shvyd'ko, Y. V.; Zajac, M.; Ruffer, R. *J. Syn. Rad.*, **2011**, *18*, 802–810

125. "High-quality quartz single crystals for high-energy-resolution inelastic X-ray scattering analyzers", Honnicke, M. G.; Huang, X. R.; Cusatis, C.; Koditwuakku, C. N.; Cai, Y. Q. *J. App. Cryst.*, **2013**, *46*, 939–944
126. "Ultra-stable sub-meV monochromator for hard X-rays", Toellner, T. S.; Collins, J.; Goetze, K.; Hu, M. Y.; Preissner, C.; Trakhtenberg, E.; Yan, L. *J. Syn. Rad.*, **2015**, *22*, 1155–1162
127. "X-ray monochromator with an energy resolution of 8×10^{-9} at 14.41 keV", Yabashi, M.; Tamasaku, K.; Kikuta, S.; Ishikawa, T. *Rev. Sci. Instrum.*, **2001**, *72*, 4080–4083
128. "Novel diamond X-ray crystal optics for synchrotrons and X-ray free-electron lasers", Stoupin, S. *Diamond Rel. Mat.*, **2014**, *49*, 39–47
129. "Performance of a silicon monochromator under high heat load", Chumakov, A. I.; Sergeev, I.; Celse, J. P.; Ruffer, R.; Lesourd, M.; Zhang, L.; del Rio, M. S. *J. Syn. Rad.*, **2014**, *21*, 315–324
130. "High-Heat-Load Monochromator Options for the RIXS Beamline at the APS with the MBA Lattice", Liu, Z.; Gog, T.; Stoupin, S. A.; Upton, M. H.; Ding, Y.; Kim, J.-H.; Casa, D. M.; Said, A. H.; Carter, J. A.; Navrotski, G. *AIP Conf. Proc.*, **2015**, *1741*, 040046-040041–040046-040045
131. "Improving the performance of high-resolution X-ray spectrometers with position-sensitive pixel detectors", Huotari, S.; Vanko, G.; Albergamo, F.; Ponchut, C.; Graafsma, H.; Henriquet, C.; Verbeni, R.; Monaco, G. *J. Syn. Rad.*, **2005**, *12*, 467–472
132. "MERIX—Next generation medium energy resolution inelastic X-ray scattering instrument at the APS", Shvyd'ko, Y. V.; Hill, J. P.; Burns, C. A.; Coburn, D. S.; Brajuskovic, B.; Casa, D.; Goetze, K.; Gog, T.; Khachatryan, R.; Kim, J. H.; Kodituwakku, C. N.; Ramanathan, M.; Roberts, T.; Said, A.; Sinn, H.; Shu, D.; Stoupin, S.; Upton, M.; Wieczorek, M.; Yavas, H. *J. Elect. Spect. Rel. Phen.*, **2013**, *188*, 140–149
133. "Quarter-Wave Plates and X-ray Magnetic Circular Dichroism on ID24 at the ESRF", Pizzini, S.; Bonfim, M.; Baudelet, F.; Tolentino, H.; San Miguel, A.; Mackay, K.; Malgrange, C.; Hagelstein, M.; Fontaine, A. *J. Syn. Rad.*, **1998**, *5*, 1298–1303
134. "Perfect Crystal X-Ray Phase Retarders", Hirano, K.; Ishikawa, T.; Kikuta, S. *Nucl. Inst. Meth. A*, **1993**, *336*, 343–353
135. "Performance of phase plates on the XMaS beamline at the ESRF", Bouchenoire, L.; Brown, S. D.; Thompson, P.; Duffy, J. A.; Taylor, J. W.; Cooper, M. J. *J. Syn. Rad.*, **2003**, *10*, 172–176
136. "Energy and Polarization-Tunable X-Ray Quarter-Wave Plates for Energy-Dispersive Absorption Spectrometer", Giles, C.; Malgrange, C.; Goulon, J.; Debergevin, F.; Vettier, C.; Fontaine, A.; Dartyge, E.; Pizzini, S. *Nucl. Inst. Meth.*, **1994**, *349*, 622–625
137. "ID20: a beamline for magnetic and resonant X-ray scattering investigations under extreme conditions", Paolasini, L.; Detlefs, C.; Mazzoli, C.; Wilkins, S.; Deen, P. P.; Bombardi, A.; Kernavanois, N.; de Bergevin, F.; Yakhou, F.; Valade, J. P.; Breslavetz, I.; Fondacaro, A.; Pepellin, G.; Bernard, P. *J, Syn. Rad.*, **2007**, *14*, 301–312
138. "Bragg Transmission Phase Plates for the Production of Circularly-Polarized X-rays", Lang, J. C.; Srajer, G. *Rev. Sci. Inst.*, **1995**, *66*, 1540–1542
139. "An X-Ray Phase Plate Using Bragg-Case Diffraction", Hirano, K.; Izumi, K.; Ishikawa, T.; Annaka, S.; Kikuta, S. *Jap. J. App. Phys. Lett.*, **1991**, *30*, L407–L410
140. "Double phase-retarder set-up at beamline P09 at PETRA III", Francoual, S.; Strempfer, J.; Reuther, D.; Shukla, D. K.; Skaugen, A. *J. Phys. Conf. Ser.*, **2013**, *425*, 132010
141. "Cr/B4C multilayer mirrors: Study of interfaces and X-ray reflectance", Burcklen, C.; Soufli, R.; Dennetiere, D.; Polack, F.; Capitanio, B.; Gullikson, E.; Meltchakov, E.; Thomasset, M.; Jerome, A.; de Rossi, S.; Delmotte, F. *J. App. Phys.*, **2016**, *119*
142. Thompson et al. *X-Ray Data Booklet*; 2nd ed.; Lawrence Berkeley National Laboratory: Berkeley, **2001**
143. "Analysis of periodic Mo/Si multilayers: Influence of the Mo thickness", Maury, H.; Andre, J. M.; Le Guen, K.; Mahne, N.; Giglia, A.; Nannarone, S.; Bridou, F.; Delmotte, F.; Jonnard, P. *Surf. Sci.*, **2009**, *603*, 407–411
144. "The BALDER Beamline at the MAX IV Laboratory", Klementiev, K.; Norén, K.; Carlson, S.; Clauss, K. G. V. S.; Persson, I. *J. Phys. Conf. Ser.*, **2016**, *712*, 012023

145. "Nuclear resonant scattering beamline at SPring-8", Yoda, Y.; Yabashi, M.; Izumi, K.; Zhang, X. W.; Kishimoto, S.; Kitao, S.; Seto, M.; Mitsui, T.; Harami, T.; Imai, Y.; Kikuta, S. *Nucl. Inst. Meth. A*, **2001**, *467*, 715–718
146. "High-resolution soft X-ray beamline ADRESS at the Swiss Light Source for resonant inelastic X-ray scattering and angle-resolved photoelectron spectroscopies", Strocov, V. N.; Schmitt, T.; Flechsig, U.; Schmidt, T.; Imhof, A.; Chen, Q.; Raabe, J.; Betemps, R.; Zimoch, D.; Krempasky, J.; Wang, X.; Grioni, M.; Piazzalunga, A.; Patthey, L. *J. Syn. Rad.*, **2010**, *17*, 631–643
147. "A Practical Guide for Nuclear Resonance Vibrational Spectroscopy (NRVS) of Biochemical Samples and Model Compounds", Wang, H.; Alp, E. E.; Yoda, Y.; Cramer, S. P., in *Metalloproteins: Methods and Protocols*; Fontecilla-Camps, J. C. and Nicolet, Y., Ed.; Springer - Humana Press, 2014; Vol. 1122, pp. 125–138
148. "Construction of Focusing Soft-X-Ray Beamline Bl1A at the UVSOR", Hiraya, A.; Horigome, T.; Okada, N.; Mizutani, N.; Sakai, K.; Matsudo, O.; Hasumoto, M.; Fukui, K.; Watanabe, M. *Rev. Sci. Inst.*, **1992**, *63*, 1264–1268
149. "Study of Different SX-700 Monochromator Designs for the Undulator Beamline (Bl51) at Max-Lab", Aksela, S.; Kivimaki, A.; Nyholm, R.; Svensson, S. *Rev. Sci. Inst.*, **1992**, *63*, 1252–1255
150. "Discussion and Improvement of the SX-700 Beamline", Lu, L. J.; Chen, J. Y. *Nucl. Inst. Meth.*, **1991**, *309*, 581–584
151. "Bessy SX-700 Uhv Monochromator—Design Features and Kinematic Concept", Riemer, F.; Torge, R. *Nucl. Inst. Meth.*, **1983**, *208*, 313–314
152. "Bessy SX-700—A Monochromator System Covering the Spectral Range 3 eV to 700 eV ", Petersen, H.; Baumgartel, H. *Nucl. Inst. Meth.*, **1980**, *172*, 191–193
153. "High-resolution RIXS with the SAXES Spectrometer at the ADRESS Beamline of the Swiss Light Source", Ghiringhelli, G.; Braicovich, L.; Schmitt, T.; Strocov, V. *Syn. Rad. News*, **2012**, *25*, 16–22
154. "The in-focus variable line spacing plane grating monochromator", Reininger, R. *Nucl. Inst. Meth.*, **2011**, *649*, 139–143
155. "The intermediate energy X-ray beamline at the APS", McChesney, J. L.; Reininger, R.; Ramanathan, M.; Benson, C.; Srajer, G.; Abbamonte, P.; Campuzano, J. C. *Nucl. Inst. Meth. A*, **2014**, *746*, 98–105
156. "Concept and Design Procedure for Cylindrical Element Monochromators for Synchrotron Radiation", Chen, C. T. *Nucl. Inst. Meth. A*, **1987**, *256*, 595–604
157. "Soft-X-Ray Spectroscopy Beamline 6-M High-Energy Spherical Grating Monochromator At SRCC - Optical Design and First Performance Tests", Chung, S. C.; Chen, C. I.; Tseng, P. C.; Lin, H. F.; Dann, T. E.; Song, Y. F.; Huang, L. R.; Chen, C. C.; Chuang, J. M.; Tsang, K. L.; Chang, C. N. *Rev. Sci. Inst.*, **1995**, *66*, 1655–1657
158. "Commissioning of the Spherical Grating Monochromator Soft X-ray Spectroscopy Beamline at the Canadian Light Source", Regier, T.; Paulsen, J.; Wright, G.; Coulthard, I.; Tan, K.; Sham, T. K.; Blyth, R. I. R. *AIP Conf. Proc.*, **2007**, *879*, 473–476
159. "OASYS (OrAnge SYnchrotron Suite): an open source graphical environment for X-ray virtual experiments", Rebuffi, L.; del Rio, M. S. *Proc. SPIE*, **2017**, *10388*
160. "Ray Tracing of Recent VUV Monochromator Designs", Cerrina, F. *Proc. Soc. Photo-Optical Inst. Eng.*, **1984**, *503*, 68–77
161. "SHADOW3: a new version of the synchrotron X-ray optics modelling package", del Rio, M. S.; Canestrari, N.; Jiang, F.; Cerrina, F. *J. Syn. Rad.*, **2011**, *18*, 708–716
162. "A new SHADOW update: integrating diffraction effects into ray-tracing", Shi, X. B.; del Rio, M. S.; Reininger, R. *Proc. SPIE*, **2014**, *9209*
163. "ShadowOui: a new visual environment for X-ray optics and synchrotron beamline simulations", Rebuffi, L.; Sanchez del Rio, M. *J, Syn. Rad.*, **2016**, *23*, 1357–1367
164. "Über eine neue Art von Sthrahlen", Röntgen, W. C. *Sitzungsber. der Würzburger Physik.-Medic. Gesellschaft*, **1895**

165. "X-Ray Burn Paper and Imaging Screen for Beam Finding", Keister, J. NSLS II Technical Note: NSLSII-ASD-TN-099, 2018
166. "Röntgenspektroskopische Messungen der L-Absorption der Elemente 74 Wolfram bis 92 Uran", Sandstrom, A. *Z. Phys.*, **1930**, *65*, 632–655
167. "Mechanism of photostimulated luminescence in BaFX:Eu^{2+} (X = Cl,Br)", Takahashi, K.; Khoda, K.; Miyahara, J.; Kanemitsu, Y.; Amitani, K.; Shionoya, S. *J. Luminescence*, **1984**, *31–32*, 266–268
168. "Fano factor in xenon", Ishida, N.; Kikuchi, J.; Doke, T.; Kase, M. *Phys. Rev. A*, **1992**, *46*, 1676–1679
169. "The Fano Factor for Electrons in Gas Mixtures", Bronic, I. K.; Kimura, M.; Inokuti, M.; Dillon, M. A. *Nucl. Inst. Meth. B*, **1992**, *71*, 366–370
170. "Ionization Yield of Radiations. II. The Fluctuations of the Number of Ions", Fano, U. *Phys. Rev.*, **1947**, *72*, 26–29
171. "New measurement of the Fano factor of mercuric iodide", Ricker, G. R.; Vallerga, J. V.; Dabrowski, A. J.; Iwanczyk, J. S.; Entine, G. *Rev. Sci. Inst.*, **1982**, *53*, 700–701
172. "An electrical method of counting the number of α particles from radioactive substances", Rutherford, E. G., H. *Proc. Royal Soc. A*, **1908**, *81*, 141–161
173. "Elektronenzählrohr zur Messung schwächster Aktivitäten" [Electron counting tube for measurement of weakest radioactivities]", Geiger, H. M., W. *Naturwissen.*, **1928**, *16*, 617–618
174. "Spin-Polarized and Site-Selective X-Ray Absorption—Demonstration with Fe Porphyrins and K-Beta Detection", Wang, X.; Randall, C. R.; Peng, G.; Cramer, S. P. *Chem. Phys. Lett.*, **1995**, *243*, 469–473
175. "Performance of a single photon counting microstrip detector for strip pitches down to 10 μm", Bergamaschi, A.; Broennimann, C.; Dinapoli, R.; Eikenberry, E.; Gozzo, F.; Henrich, B.; Kobas, M.; Kraft, P.; Patterson, B.; Schmitt, B. *Nucl. Inst. Meth.*, **2008**, *591*, 163–166
176. "Development of a silicon microstrip detector with single photon sensitivity for fast dynamic diffraction experiments at a synchrotron radiation beam", Arakcheev, A.; Aulchenko, V.; Kudashkin, D.; Shekhtman, L.; Tolochko, B.; Zhulanov, V. *J, Instrum.*, **2017**, *12*, C06002
177. "Development of the microstrip silicon detector for imaging of fast processes at a synchrotron radiation beam", Aulchenko, V.; Pruuel, E.; Shekhtman, L.; Ten, K.; Tolochko, B.; Zhulanov, V. *Nuc. Inst. Meth. A*, **2017**, *845*, 169–172
178. "Scintillation detectors for X-rays", Nikl, M. *Meas. Sci. Tech.*, **2006**, *17*, R37-R54
179. "Recent R&D Trends in Inorganic Single-Crystal Scintillator Materials for Radiation Detection", Nikl, M.; Yoshikawa, A. *Adv. Opt. Mat.*, **2015**, *3*, 463–481
180. "New Fluorescence Detection System for X-ray Absorption Spectroscopy", Cramer, S. P.; Scott, R. A. *Rev. Sci. Inst.*, **1981**, *52*, 395–399
181. "A 13-Element Ge Detector for Fluorescence EXAFS", Cramer, S. P.; Tench, O.; Yocum, M.; George, G. N. *Nuc. Inst. Meth. A*, **1988**, *266*, 586–591
182. "Maia X-ray fluorescence imaging: Capturing detail in complex natural samples", Ryan, C. G.; Siddons, D. P.; Kirkham, R.; Li, Z. Y.; de Jonge, M. D.; Paterson, D. J.; Kuczewski, A.; Howard, D. L.; Dunn, P. A.; Falkenberg, G.; Boesenberg, U.; De Geronimo, G.; Fisher, L. A.; Halfpenny, A.; Lintern, M. J.; Lombi, E.; Dyl, K. A.; Jensen, M.; Moorhead, G. F.; Cleverley, J. S.; Hough, R. M.; Godel, B.; Barnes, S. J.; James, S. A.; Spiers, K. M.; Alfeld, M.; Wellenreuther, G.; Vukmanovic, Z.; Borg, S. *J. Phys. Conf. Ser.*, **2014**, *499*
183. "Fluorescence XAFS Array Detectors for 0.3-20 keV", Cramer, S. P.; George, S. J.; Kraner, H.; Rogers, L.; Rescia, S.; Radeka, V.; Yocum, M.; Colaresi, J.; Tench, O.; Mullins, O. C., in *X-Ray Absorption Fine Structure*; Hasnain, S. S., Ed.; Ellis Horwood: West Sussex, 1991, pp. 640–645
184. "A New Prototype X-ray Fluorescence Detector System with Silicon Drift Detector Array", Chen, W.; DeGeronimo, G.; Elliott, D.; Giacomini, G.; Kuczewski, A. J.; Mcad, J.; Pinelli, D.; Rumaiz, A. K.; Siddons, D. P.; Smith, G.; Vernon, E. O., IEEE Nuc. Sci. Symp. (NSS/MIC/RTSD), Strasbourg, FRANCE, **2016**, pp.

185. "Next generation of pnCCDs for X-ray spectroscopy and imaging", Meidinger, N.; Andritschke, R.; Haelker, O.; Hartmann, R.; Herrmann, S.; Holl, P.; Lutz, G.; Kimmel, N.; Schaller, G.; Schnecke, M.; Schopper, F.; Soltau, H.; Struder, L. *Nucl. Inst. Meth. A*, **2006**, *568*, 141–148
186. "Dual crystal X-ray spectrometer at 1.8 keV for high repetition-rate single-photon counting spectroscopy experiments", Gamboa, E. J.; Bachmann, B.; Kraus, D.; MacDonald, M. J.; Bucher, M.; Carron, S.; Coffee, R. N.; Drake, R. P.; Emig, J.; Ferguson, K. R.; Fletcher, L. B.; Glenzer, S. H.; Gorkhover, T.; Hau-Riege, S. P.; Krzywinski, J.; Levitan, A. L.; Meiwes-Broer, K. H.; Osipov, T.; Pardini, T.; Peltz, C.; Skruszewicz, S.; Bostedt, C.; Fennel, T.; Doppner, T. *J. Instrum.*, **2016**, *11*, 14
187. "Silicon avalanche photodiodes for direct detection of X-rays", Baron, A. Q. R.; Kishimoto, S.; Morse, J.; Rigal, J.-M. *J. Syn. Rad.*, **2006**, *13*, 131–142
188. "Recent progress of avalanche photodiodes in high-resoution X-rays and gamma-rays detection", Kataoka, J.; Saito, T.; Kuramoto, Y.; Ikagawa, T.; Yatsu, Y.; Kotoku, J.; Arimoto, M.; Kawai, N.; Ishikawa, Y.; Kawabata, N. *Nucl. Inst. Meth. A*, **2005**, *541*, 398–404
189. "Fast X-ray detector system with simultaneous measurement of timing and energy for a single photon", Masuda, T.; Okubo, S.; Hara, H.; Hiraki, T.; Kitao, S.; Miyamoto, Y.; Okai, K.; Ozaki, R.; Sasao, N.; Seto, M.; Uetake, S.; Yamaguchi, A.; Yoda, Y.; Yoshimi, A.; Yoshimura, K. *Rev. Sci. Inst.*, **2017**, *88*
190. "A fast detector using stacked avalanche photodiodes for X-ray diffraction experiments with synchrotron radiation", Kishimoto, S.; Ishizawa, N.; Vaalsta, T. P. *Rev. Sci. Inst.*, **1998**, *69*, 384–391
191. "Measurement of nuclear resonant scattering on Ni-61 with fast scintillation detector using proportional-mode silicon avalanche photodiode", Inoue, K.; Kobayashi, Y.; Yoda, Y.; Koshimizu, M.; Nishikido, F.; Haruki, R.; Kishimoto, S. *Nuc. Inst. Meth.*, **2018**, *880*, 87–91
192. "An ultrafast front-end ASIC for APD array detectors in X-ray time-resolved experiments", Zhou, Y. F.; Li, Q. J.; Liu, P.; Fan, L.; Xu, W.; Tao, Y.; Li, Z. J. *Chi. Phys. C*, **2017**, *41*
193. "Characterization of a subpicosecond X-ray streak camera for ultrashort laser-produced plasmas experiments", Gallant, P.; Forget, P.; Dorchies, F.; Jiang, Z.; Kieffer, J. C.; Jaanimagi, P. A.; Rebuffie, J. C.; Goulmy, C.; Pelletier, J. F.; Sutton, M. *Rev. Sci. Inst.*, **2000**, *71*, 3627–3633
194. "High dynamic range streak camera for subpicosecond time-resolved X-ray spectroscopy", Bonté, C.; Harmand, M.; Dorchies, F.; Magnan, S.; Pitre, V.; Kieffer, J. C.; Audebert, P.; Geindre, J. P. *Rev. Sci. Inst.*, **2007**, *78*, 043503
195. "Future optical detectors based on Al superconducting tunnel junctions", Brammertz, G.; Verhoeve, P.; Martin, D.; Peacock, A.; Venn, R., in *Optical and Infrared Detectors for Astronomy*; Garnett, J. D. and Beletic, J. W., Ed., 2004; Vol. 5499, pp. 269–280
196. "Superconducting tunnel junctions", Martin, D. D. E.; Verhoeve, P. *ISSI Sci. Rep.*, **2010**, *9*, 441–457
197. "A Multichannel Superconducting X-ray Spectrometer for High-Resolution Spectroscopy of Dilute Samples", Friedrich, S.; Funk, T.; Drury, O.; S. E. Labov; Cramer, S. P. *Rev. Sci. Inst.*, **2002**, *73*, 1629–1631
198. "A multichannel superconducting tunnel junction detector for high-resolution X-ray spectroscopy of magnesium diboride films", Friedrich, S.; Vailionis, A.; Drury, O.; Niedermayr, T.; Funk, T.; Kang, W. N.; Choi, E. M.; Kim, H. J.; Lee, S. I.; Cramer, S. P.; Kim, C.; Labov, S. E. *IEEE Trans. App. Supercon.*, **2003**, *13*, 1114–1119
199. "A 36-pixel superconducting tunnel junction soft X-ray detector for environmental science applications", Friedrich, S.; Drury, O. B.; Cramer, S. P.; Green, P. G. *Nucl. Inst. Meth. A*, **2006**, *559*, 776–778
200. "112-Pixel Arrays of High-Efficiency STJ X-Ray Detectors", Friedrich, S.; Harris, J.; Warburton, W. K.; Carpenter, M. H.; Hall, J. A.; Cantor, R. *J. Low Temp. Phys.*, **2014**, *176*, 553–559

201. "Toward 1000-Pixel Superconducting-Tunnel-Junction Detectors for X-Ray Absorption Fine Structure Spectroscopy", Fujii, G.; Ukibe, M.; Shiki, S.; Ohkubo, M. *J. Low Temp. Phys.*, **2014**, *176*, 255–260
202. "A practical superconducting-microcalorimeter X-ray spectrometer for beamline and laboratory science", Doriese, W. B.; Abbamonte, P.; Alpert, B. K.; Bennett, D. A.; Denison, E. V.; Fang, Y.; Fischer, D. A.; Fitzgerald, C. P.; Fowler, J. W.; Gard, J. D.; Hays-Wehle, J. P.; Hilton, G. C.; Jaye, C.; McChesney, J. L.; Miaja-Avila, L.; Morgan, K. M.; Joe, Y. I.; O'Neil, G. C.; Reintsema, C. D.; Rodolakis, F.; Schmidt, D. R.; Tatsuno, H.; Uhlig, J.; Vale, L. R.; Ullom, J. N.; Swetz, D. S. *Rev. Sci. Instrum.*, **2017**, *88*, 23
203. "Review of superconducting transition-edge sensors for X-ray and gamma-ray spectroscopy", Ullom, J. N.; Bennett, D. A. *Supercon. Sci. Tech.*, **2015**, *28*, 084003
204. "Development of Low-Temperature Detectors For Generation-X and Other Missions Requiring High-Resolution, Large-Format, X-ray Detector Array", Bandler, S. *EOS White Paper*, **2009**
205. "Transition-Edge Sensors", Irwin, K. D.; Hilton, G. C., in *Cryogenic Particle Detection*; Enss, C., Ed.; Springer Berlin Heidelberg: Berlin, Heidelberg, 2005, pp. 63–150
206. Knoll, G. F. *Radiation Detection and Measurement*; John Wiley & Sons, Inc.: New York**, 1999**
207. "Digital pulse processing for NaI(T1) detectors", Di Fulvio, A.; Shin, T. H.; Hamel, M. C.; Pozzi, S. A. *Nucl. Inst. Meth. A*, **2016**, *806*, 169–174
208. "Fluorescence Detection of EXAFS: Sensitivity Enhancement for Dilute Species and Thin Films", Jaklevic, J.; Kirby, J. A.; Klein, M. P.; Robertson, A. S.; Brown, G. S.; Eisenberger, P. *Sol. State Comm.*, **1977**, *23*, 679–682
209. "Fluorescence Yield Detection - Why It Does Not Measure the X-Ray Absorption Cross Section", de Groot, F. M. F.; Arrio, M.-A.; Sainctavit, P.; Cartier, C.; Chen, C. T. *Sol. State Comm.*, **1994**, *92*, 991–995
210. "Total 3 s Emission Yield as Bulk-Sensitive Probe for a True Soft X-ray Absorption Spectrum?", Miedema, P. S.; Beye, M. *J. Phys. Chem. Lett.*, **2018**, *9*, 2579–2583
211. Carpenter, M., Ph. D., University of California—Davis, **2015**
212. Stöhr, J. *NEXAFS Spectroscopy*; Springer-Verlag: New York**, 2003**
213. "Extended X-ray absorption fine structure—its strengths and limitations as a structural tool", Lee, P. A.; Citrin, P. H.; Eisenberger, P.; Kincaid, B. M. *Rev. Mod. Phys.*, **1981**, *53*, 769–806
214. "Thickness effect on the extended-X-ray-absorption-fine-structure amplitude", Stern, E. A.; Kim, K. *Phys. Rev. B*, **1981**, *23*, 3781–3787
215. "Design of an EXAFS Experiment", Heald, S. M., in *X-Ray Absorption*; Koningsberger, D. C. and Prins, R., Ed.; John Wiley: New York, 1988, pp. 87–118
216. "On Experimental Attentuation Factors of the Amplitude of EXAFS Oscillations in Absorption, Reflectivity, and Luminescence Measurements", Goulon, J.; Goulon-Ginet, C.; Cortes, R.; Dubois, J. M. *J. Physique*, **1982**, *43*, 539–548
217. "Minimizing Glitches in XAFS Data - A Model for Glitch Formation", Bridges, F.; Wang, X.; Boyce, J. B. *Nucl. Inst. Meth. A*, **1991**, *307*, 316–324
218. "Brief Introduction of X-Ray Multiple Diffraction", Chuan-Zheng, Y.; Jian-Min, H.; Guang-Wen, P. *Rigaku J.*, **2000**, *17*
219. "Rotation of X-ray polarization in the glitches of a silicon crystal monochromator", Sutter, J. P.; Boada, R.; Bowron, D. T.; Stepanov, S. A.; Diaz-Moreno, S. *J. App. Cryst.*, **2016**, *49*, 1209–1222
220. "Umweg—A Computer-Program for Calculation and Graphical Representation of Umweganregung-Patterns", Rossmanith, E. *Zeit. Kristall.*, **1985**, *171*, 253–254
221. "Synchrotron Radiation Studies of K-Edge Photoabsorption Spectra of Kr, Br_2, And $GeCl_4$—Comparison of Theory and Experiment", Kincaid, B. M.; Eisenberger, P. *Phys. Rev. Lett.*, **1975**, *34*, 1361–1364

222. "Theory of EXAFS", Stern, E. A., in *X-Ray Absorption—Principles, Applications, Techniques of EXAFS, SEXAFS, and XANES*; Koningsberger, D. C. and Prins, R., Ed.; Wiley-Interscience: New York, 1988, pp. 3–51
223. "Theoretical approaches to X-ray absorption fine structure", Rehr, J. J.; Albers, R. C. *Rev. Mod. Phys.*, **2000**, *72*, 621–654
224. Als-Nielsen, J.; McMorrow, D. *Elements of Modern X-Ray Physics*; John Wiley & Sons: New York**, 2001**
225. "Ab Initio Calculations of Amplitude and Phase Functions for Extended X-ray Absorption Fine Structure Spectroscopy", Teo, B.-K.; Lee, P. A. *J. Am. Chem. Soc.*, **1979**, *101*, 2815–2832
226. "Improved ab Initio Calculations of Amplitude and Phase Functions for Extended X-ray Absorption Fine Structure Spectroscopy", McKale, A. G.; Veal, B. W.; Paulikas, A. P.; Chan, S.-K.; Knapp, G. S. *J. Am. Chem. Soc.*, **1988**, *110*, 3763–3768
227. "Magnetic Dichroism in the X-Ray-Absorption Branching Ratio", van der Laan, G.; Thole, B. T. *Phys. Rev. B*, **1990**, *42*, 6670–6674
228. "Parameter-free calculations of X-ray spectra with FEFF9", Rehr, J. J.; Kas, J. J.; Vila, F. D.; Prange, M. P.; Jorissen, K. *Phys. Chem. Chem. Phys.*, **2010**, *12*, 5503–5513
229. Atkins, P. W.; Friedman, R. S. *Molecular Quantum Mechanics*; Third ed.; Oxford University Press: Oxford**, 1997**
230. Cyvin, S. J. *Molecular Vibrations and Mean Square Amplitudes*; Elsevier: Amsterdam**, 1968**
231. "Mean Amplitudes of Vibration of the Halogen Molecules", Baran, E. J. *Z. Naturforsch.*, **2003**, *58a*, 36–38
232. "Lattice dynamics and correlated atomic motion from the atomic pair distribution function", Jeong, I.-K.; Heffner, R. H.; Graf, M. J. *Phys. Rev. B*, **2003**, *67*, 104301:104301–104309
233. Atkins, P. W. *Quanta—A Handbook of Concepts*; Oxford University Press: Oxford**, 1974**
234. "Manganese Oxyl Radical Intermediates and O-O Bond Formation in Photosynthetic Oxygen Evolution and a Proposed Role for the Calcium Cofactor in Photosystem II", Siegbahn, P. E. M.; Crabtree, R. H. *J. Am. Chem. Soc.*, **1999**, *121*, 117–127
235. "Extended X-Ray Absorption Fine Structure Debye-Waller Factors. I. Monatomic Crystals", Sevillano, E.; Meuth, H.; Rehr, J. J. *Phys. Rev. B*, **1979**, *20*, 4908–4911
236. "Multiple-electron excitation in X-ray absorption: a simple generic model", Roy, M.; Lindsay, J. D.; Louch, S.; Gurman, S. J. *J. Synchrotron Rad.*, **2001**, *8*, 1103–1108
237. "Single Crystal EXAFS of Nitrogenase", Cramer, S. P.; Flank, A. M.; Weininger, M.; Mortenson, L. E. *J. Am. Chem. Soc.*, **1986**, *108*, 1049–1055
238. "Anisotropic X-Ray Absorption in Layered Compounds", Heald, S. M.; Stern, E. A. *Phys. Rev. B*, **1977**, *16*, 5549–5559
239. "Angular Dependence of X-Ray Absorption Spectra", Brouder, C. *J. Phys. Cond. Matt.*, **1990**, *2*, 701–738
240. "The Study of Disordered Systems by EXAFS: Limitations", Eisenberger, P.; G. S, B. *Sol. State Comm.*, **1979**, *29*, 481–484
241. "Novel Method for Angle Determinations by EXAFS via a New Multiple-Scattering Formalism", Teo, B. K. *J. Am. Chem. Soc.*, **1981**, *103*, 3990–4001
242. "Multiple-Scattering Calculations of X-Ray-Absorption Spectra", Zabinsky, S. I.; Rehr, J. J.; Ankudinov, A.; Albers, R. C.; Eller, M. J. *Phys. Rev. B*, **1995**, *52*, 2995–3009
243. "The Calculation of X-Ray Mass Absorption Coefficients", Victoreen, J. A. *J. App. Phys.*, **1949**, *20*, 1141–1147
244. "Data Analysis", Sayers, D. E.; Bunker, B. A., in *X-Ray Absorption*; Koningsberger, D. C. and Prins, R., Ed.; John Wiley: New York, 1988, pp. 211–253
245. "A Systematic X-ray Absorption Study of Molybdenum Complexes. The Accuracy of Structural Information from Extended X-ray Absorption Fine Structure", Cramer, S. P.; Hodgson, K. O.; Stiefel, E. I.; Newton, W. E. *J. Am. Chem. Soc.*, **1978**, *100*, 2748–2761
246. "Crystal-Structure of Sodium Molybdate Dihydrate, $Na_2MoO_4.2H_2O$", Matsumoto, K.; Kobayashi, A.; Sasaki, Y. *Bull. Chem. Soc. Jap.*, **1975**, *48*, 1009–1013

247. "Nitrogen Fixation: The Mechanism of the Mo-Dependent Nitrogenase", Igarashi, R. Y.; Seefeldt, L. C. *Crit. Rev. Biochem. Mol. Biol.*, **2003**, *38*, 351–384
248. "Structural basis of biological nitrogen fixation", Rees, D. C.; Tezcan, F. A.; Haynes, C. A.; Walton, M. Y.; Andrade, S.; Einsle, O.; Howard, J. B. *Phil. Trans. R. Soc. A*, **2005**, *363*, 971–984
249. "Substrate interactions with the nitrogenase active site", Dos Santos, P. C.; Igarashi, R. Y.; Lee, H.-I.; Hoffman, B. M.; Seefeldt, L. C.; Dean, D. R. *Acc. Chem. Res.*, **2005**, *38*, 208–214
250. "The Molybdenum Site of Nitrogenase—Preliminary Structural Evidence from X-ray Absorption Spectroscopy", Cramer, S. P.; Hodgson, K. O.; Gillum, W. O.; Mortenson, L. E. *J. Am. Chem. Soc.*, **1978**, *100*, 3398–3407
251. "GNXAS. Extended Suite of Programs for Advanced X-Ray Absorption Data-Analysis: Methodology and Practice ", Di Cicco, A. **2009**, ISBN 978-983-908112-908118-908114
252. "ATHENA, ARTEMIS, HEPHAESTUS: data analysis for X-ray absorption spectroscopy using IFEFFIT", Ravel, B.; Newville, M. *J. Syn. Rad.*, **2005**, *12*, 537–541
253. EXAFSPAK: a Suite of Computer Programs for Analysis of X-ray Absorption Spectra, http://ssrl.slac.stanford.edu/exafspak.html
254. Bunker, G. *Introduction to XAFS - A Practical Guide to X-Ray Absorption Fine Structure Spectroscopy*; Cambridge University Press: Cambridge**, 2010**
255. "X-ray Absorption Spectroscopy at the Sulfur K-Edge: A New Tool to Investigate the Biochemical Mechanisms of Neurodegeneration", Hackett, M. J.; Smith, S. E.; Paterson, P. G.; Nichol, H.; Pickering, I. J.; George, G. N. *ACS Chem. Neuro.*, **2012**, *3*, 178–185
256. "Characterization of Arsenic in Oil Shale and Oil Shale Derivatives by X-Ray Absorptionn Spectroscopy", Cramer, S. P.; Siskin, M.; Brown, L. D.; George, G. N. *Energy & Fuels*, **1988**, *2*, 175–180
257. "X-Ray Resonance Absorption Lines in the Argon K Spectrum", Parratt, L. G. *Phys. Rev.*, **1939**, *56*, 295–297
258. "New Autoionizing Atomic Energy Levels in He, Ne, and Ar", Madden, R. P.; Codling, K. *Phys. Rev. Lett.*, **1963**, *10*, 516–518
259. "Extensive Double-Excitation States in Atomic Helium", Domke, M.; Xue, C.; Puschmann, A.; Mandel, T.; Hudson, E.; Shirley, D. A.; Kaindl, G.; Greene, C. H.; Sadeghpour, H. R.; Petersen, H. *Phys. Rev. Lett.*, **1991**, *66*, 1306–1309
260. "Differential redox and sorption of Cr(III/VI) on natural silicate and oxide minerals: EXAFS and XANES results", Peterson, M. L.; Brown, G. E.; Parks, G. A.; Stein, C. L. *Geochim. Cosmochim. Acta*, **1997**, *61*, 3399–3412
261. "Sulfur K-edge X-ray Absorption Spectroscopy: From Molecules to Whole Cells", Pickering, I. J
262. "K-shell photoabsorption of the N_2 molecule", Chen, C. T.; Ma, Y.; Sette, F. *Phys. Rev. A*, **1989**, *40*, 6737–6740
263. "Fe-N_2/CO complexes that model a possible role for the interstitial C atom of FeMo-cofactor (FeMoco)", Rittle, J.; Peters, J. C. *Proc. Nat. Acad. Sci.*, **2013**, *110*, 15898–15903
264. "Shape Resonances in K Shell Photoionization of Diatomic Molecules", Dehmer, J. L.; Dill, D. *Phys. Rev. Lett.*, **1975**, *35*, 213–215
265. "X-ray Absorption Near-Edge Structure (XANES) Spectroscopy", Henderson, G. S.; de Groot, F. M. F.; Moulton, B. J. A., in *Spectroscopic Methods in Mineralology and Materials Sciences*; Henderson, G. S., Neuville, D. R. and Downs, R. T., Ed., 2014; Vol. 78, pp. 75–138
266. "Reappraisal of the existence of shape resonances in the series C_2H_2, C_2H_4, and C_2H_6", Kempgens, B.; Koppe, H. M.; Kivimaki, A.; Neeb, M.; Maier, K.; Hergenhahn, U.; Bradshaw, A. M. *Phys. Rev. Lett.*, **1997**, *79*, 35–38
267. "On the correct identification of shape resonances in NEXAFS", Kempgens, B.; Koppe, H. M.; Kivimaki, A.; Neeb, M.; Maier, K.; Hergenhahn, U.; Bradshaw, A. M. *Surf. Sci.*, **1999**, *425*, L376–L380
268. "The neverending story of shape resonances", Piancastelli, M. N. *J. Elect. Spect. Rel. Phen.*, **1999**, *100*, 167–190

269. "Recent advances in X-ray absorption spectroscopy", Wende, H. *Rep. Prog. Phys.*, **2004**, *67*, 2105–2181
270. "Correlation between intramolecular bond lengths and K-shell σ-shape resonances in gas-phase molecules", Sette, F.; Stöhr, J.; Hitchcook, A. P. *Chem. Phys. Lett.*, **1984**, *110*, 517–520
271. "Shape resonances of oriented molecules: ab initio theory and experiment on hydrocarbon molecules", Haack, N.; Ceballos, G.; Wende, H.; Baberschke, K.; Arvanitis, D.; Ankudinov, A. L.; Rehr, J. J. *Phys. Rev. Lett.*, **2000**, *84*, 614–617
272. "The ORCA program system", Neese, F. *WIREs Comput. Mol. Sci.*, **2012**, *2*, 73–78
273. StoBe, http://www.fhi-berlin.mpg.de/KHsoftware/StoBe/StoBeMAN.html
274. "X-ray magnetic circular dichroism—A versatile tool to study magnetism", van der Laan, G.; Figueroa, A. I. *Coord. Chem. Rev.*, **2014**, *277*, 95–129
275. "Comparison of X-ray absorption spectra between water and ice: New ice data with low pre-edge absorption cross-section", Sellberg, J. A.; Kaya, S.; Segtnan, V. H.; Chen, C.; Tyliszczak, T.; Ogasawara, H.; Nordlund, D.; Pettersson, L. G. M.; Nilsson, A. *J. Chem. Phys.*, **2014**, *141*, 034507
276. "A proposal for the structure of high- and low-density fluctuations in liquid water", Camisasca, G.; Schlesinger, D.; Zhovtobriukh, I.; Pitsevich, G.; Pettersson, L. G. M. *J. Chem. Phys.*, **2019**, *151*, 034508
277. "X-ray emission spectroscopy of hydrogen bonding and electronic structure of liquid water", Guo, J. H.; Luo, Y.; Augustsson, A.; Rubensson, J. E.; Sathe, C.; Agren, H.; Siegbahn, H.; Nordgren, J. *Phys. Rev. Lett.*, **2002**, *89*
278. "X-ray Raman Spectroscopy at the Oxygen K-Edge of Water and Ice: Implications on Local Structural Models", Bergmann, U.; Wernet, P.; Glatzel, P.; Cavalleri, M.; Pettersson, L. G. M.; Nilsson, A.; Cramer, S. P. *Phys. Rev. B*, **2002**, *6*, 2104–2107
279. "Compatibility of quantitative X-ray spectroscopy with continuous distribution models of water at ambient conditions", Niskanen, J.; Fondell, M.; Sahle, C. J.; Eckert, S.; Jay, R. M.; Gilmore, K.; Pietzsch, A.; Dantz, M.; Lu, X.; McNally, D. E.; Schmitt, T.; Vaz da Cruz, V.; Kimberg, V.; Gel'mukhanov, F.; Föhlisch, A. *Proc. Nat. Acad. Sci.*, **2019**, *116*, 4058–4063
280. "Electronic structure of silicon nitride", Gritsenko, V. A. *Phys. Uspek.*, **2012**, *55*, 498–507
281. "First-principles study of gamma-Si_3N_4 with generalized gradient approximation and local density approximation", Ding, Y. C.; Zhou, H. P.; Xu, M.; Shen, Y. B.; Chen, Q. Y.; Zhu, W. J.; He, H. L. *Int. J. Mod. Phys. B*, **2008**, *22*, 2155–2165
282. "Solid state calculations using WIEN2k", Schwarz, K.; Blaha, P. *Comp. Mat. Sci.*, **2003**, *28*, 259–273
283. "Multiplet effects in X-ray spectroscopy", de Groot, F. *Coord. Chem. Rev.*, **2005**, *249*, 31–63
284. "Ligand Field Strengths and Oxidation States from Manganese L-Edge Spectroscopy", Cramer, S. P.; Degroot, F. M. F.; Ma, Y.; Chen, C. T.; Sette, F.; Kipke, C. A.; Eichhorn, D. M.; Chan, M. K.; Armstrong, W. H.; Libby, E.; Christou, G.; Brooker, S.; Mckee, V.; Mullins, O. C.; Fuggle, J. C. *J. Am. Chem. Soc.*, **1991**, *113*, 7937–7940
285. de Groot, F. M. F.; Kotani, A. *Core Level Spectroscopy of Solids*; CRC Press: Boca Raton, **2008**
286. "Calculation of Multiplet Structure of Core p-vacancy Levels. II", Gupta, R. P.; Sen, S. K. *Phys. Rev. B*, **1975**, *12*, 15–19
287. "Multiplet Structure in X-Ray P-Shell Photoelectron and K-Emission Spectra of Nickel Compounds", Asada, S.; Satako, C.; Sugano, S. *J. Phys. Soc. Japan*, **1975**, *37*, 855
288. "Multiplet Structure in the $L_{2,3}$ X-Ray Absorption Spectra: A Fingerprint for High- and Low-spin Ni^{2+} Compounds", van der Laan, G.; Thole, B. T.; Sawatzky, G. A.; Verdaguer, M. *Phys. Rev. B*, **1988**, *37*, 6587–6589
289. Butler, P. H. *Point Group Symmetry. Applications, Methods and Tables*; Plenum: New York, **1981**
290. Cowan, R. D. *The Theory of Atomic Structure and Spectra*; University of California Press: Berkeley, **1981**

291. "X-Ray Absorption and Dichroism of Transition Metals and their Compounds", de Groot, F. M. F. *J. Elect. Spec. Rel. Phen.*, **1994**, *67*, 529–622
292. "2p absorption spectra of the 3d elements", Fink, J.; Muller-Heinzerling, T.; Scheerer, B.; Speier, W.; Hillebrecht, F. U.; Fuggle, J. C.; Zaanen, J.; Sawatzky, G. A. *Phys. Rev. B*, **1985**, *32*, 4899–4904
293. "$L_{2,3}$ X-Ray Absorption Edges of d^0 Compounds - K^+, Ca^{2+}, Sc^{3+} and Ti^{4+} in O_h (Octahedral) Symmetry", de Groot, F. M. F.; Fuggle, J. C.; Thole, B. T.; Sawatzky, G. A. *Phys. Rev. B*, **1990**, *41*, 928–937
294. "Spin-mixed ground state of Fe phthalocyanine and the temperature-dependent branching ratio in X-ray absorption spectroscopy", Thole, B. T.; van der Laan, G.; Butler, P. H. *Chem. Phys. Lett.*, **1988**, *149*, 295–299
295. "Bioinorganic Applications of X-ray Multiplets - The Impact of Theo Thole's Work", Cramer, S. P.; Ralston, C. Y.; Wang, H. X.; Bryant, C. *J. Elect. Spec. Rel. Phenom.*, **1997**, *86*, 175–183
296. "Iron L-edge X-ray absorption Spectroscopy of Myoglobin Complexes and Photolysis Products", Wang, H. X.; Peng, G.; Miller, L. M.; Scheuring, E. M.; George, S. J.; Chance, M. R.; Cramer, S. P. *J. Am. Chem. Soc.*, **1997**, *119*, 4921–4928
297. "Characterization of Heterogeneous Nickel Sites in CO Dehydrogenases from *Clostridium thermoaceticum* and *Rhodospirillum rubrum* by Nickel L-Edge X-Ray Spectroscopy", Ralston, C. Y.; Wang, H.; Ragsdale, S. W.; Kumar, M.; Spangler, N. J.; Ludden, P. W.; Gu, W.; Jones, R. M.; Patil, D. S.; Cramer, S. P. *J. Am. Chem. Soc.*, **2000**, *122*, 10553–10560
298. "Copper L-Edge Spectral Studies - A Direct Experimental Probe of the Ground-State Covalency in the Blue Copper Site in Plastocyanin", George, S. J.; Lowery, M. D.; Solomon, E. I.; Cramer, S. P. *J. Am. Chem. Soc.*, **1993**, *115*, 2968–2969
299. "Potential-Barrier Effects in Photoabsorption. I. General Theory", Starace, A. F. *Phys. Rev. B*, **1972**, *5*, 1773–1784
300. "Sum rules for X-ray absorption and dichroism", Carra, P. J. *Elect. Spec. Rel. Phen.*, **1997**, *86*, 139–142
301. "A Quantitative Description of the Ground State Wavefunction of Cu_A by X-ray Absorption Spectroscopy: Comparison to Plastocyanin and Relevance to Electron Transfer Function", George, S. D.; Metz, M.; Szilagyi, R. K.; Wang, H.; Cramer, S. P.; Lu, Y.; Tolman, W. B.; Hedman, B.; Hodgson, K. O.; Solomon, E. I. *J. Am. Chem. Soc.*, **2001**, *123*, 5757–5767
302. "Counting Holes in Ni Complexes Using Integrated X-Ray L Absorption Spectra", Wang, H.; Ge, P.; Riordan, C. G.; Brooker, S.; Woomer, C. G.; Collins, T.; Melendres, C.; Graudejus, O.; Bartlett, N.; Cramer, S. P. *J. Phys. Chem. B*, **1998**, *102*, 8343–8346
303. "Magnetic properties of transition metal multilayers studied with X-ray magnetic circular dichroism spectroscopy", Stöhr, J.; Nakajima, R. *IBM J. Res. Develop.*, **1998**, *42*, 73–88
304. "Absorption of Circularly Polarized X Rays in Iron", Schütz, G.; Wagner, W.; Wilhelm, W.; Kienle, P.; Zeller, R.; Frahm, R.; Materlik, G. *Phys. Rev. Lett.*, **1987**, *58*, 737
305. "Soft X-Ray Magnetic Circular Dichroism of the $L_{2,3}$ Edges of Nickel", Chen, C. T.; Sette, F.; Ma, Y.-J.; Modesti, S. *Phys. Rev. B*, **1990**, *42*, 7262–7265
306. "X-ray Magnetic Circular Dichroism Spectroscopy of Transition Metal Thin Films", Stöhr, J. *J. Electron Spec. Rel. Phenom.*, **1995**, *75*, 253–272
307. "Light Radiation in a Magnetic Field", Zeeman, P. *Nobel Lecture*, **1902**
308. "Principles of X-Ray Magnetic Dichroism Spectromicroscopy", Stöhr, J.; Pademore, H. A.; Anders, S.; Stammler, T.; Scheinfein, M. R. *Surf. Rev. Lett.*, **1998**, *5*, 1297–1308
309. "Detecting magnetic flux distributions in superconductors with polarized x rays", Stahl, C.; Audehm, P.; Grafe, J.; Ruoss, S.; Weigand, M.; Schmidt, M.; Treiber, S.; Bechtel, M.; Goering, E.; Schutz, G.; Albrecht, J. *Phys. Rev. B*, **2014**, *90*, 7
310. "Photoinduced Magnetization on Mo Ion in Copper Octacyanomolybdate: An X-ray Magnetic Circular Dichroism Investigation", Arrio, M. A.; Long, J.; Moulin, C. C. D.; Bachschmidt, A.; Marvaud, V.; Rogalev, A.; Mathoniere, C.; Wilhelm, F.; Sainctavit, P. *J. Phys. Chem. C*, **2010**, *114*, 593–600
311. Cullity, B. D. *Introduction to Magnetic Materials*; Addison-Wesley: Reading, Mass.**, 1972**

312. "On the exact crystal structure of exchange-biased Fe_3O_4–CoO nanoaggregates produced by seed-mediated growth in polyol", Gaudisson, T.; Sayed-Hassan, R.; Yaacoub, N.; Franceschin, G.; Nowak, S.; Grenèche, J. M.; Menguy, N.; Sainctavit, P.; Ammar, S. *Cryst. Eng. Comm.*, **2016**, *18*, 3799–3807
313. "X-Ray Magnetic Circular Dichroism - A Primer for Chemists", Cramer, S. P., in *Paramagnetic Resonance of Metallobiomolecules*; Telser, J., Ed.; American Chemical Society: Washington, D. C., 2003; Vol. 858, pp. 358–402
314. "X-Ray Magnetic Circular Dichroism: Basic Concepts and Theory for 3d Transition Metal Atoms", Stöhr, J.; Wu, Y., in *New Directions in Research with Third-Generation Soft X-Ray Synchrotron Radiation Sources*; Schlacter, A. S. and Wuilleumier, F. J., Ed.; Kluwer: Amsterdam, 1994, pp. 221–250
315. "Inner-Core Excitation-Spectra of Transition-Metal Compounds: II. p-d Absorption Spectra", Yamaguchi, T.; Shibuya, S.; Suga, S.; Shin, S. *J. Phys. C*, **1982**, *15*, 2641
316. "2p X-Ray Absorption of 3d Transition Metal Compounds—An Atomic Multiplet Description Including the Crystal Field", de Groot, F. M. F.; Fuggle, J. C.; Thole, B. T.; Sawatzky, G. A. *Phys. Rev. B*, **1990**, *42*, 5459–5468
317. "Strong Magnetic X-Ray Dichroism in 2p Absorption Spectra of 3d Transition-Metal Ions", van der Laan, G.; Thole, B. T. *Phys. Rev. B*, **1991**, *43*, 13401–13411
318. "Determination of Spin-Moment and Orbital-Moment Anisotropies in Transition Metals by Angle-Dependent X-Ray Magnetic Circular Dichroism", Stöhr, J.; Konig, H. *Phys. Rev. Lett.*, **1995**, *75*, 3748–3751
319. "X-ray circular dichroism as a Probe of Orbital and Spin Magnetizations", Carra, P. *Syn. Rad. News*, **1992**, *5*, 21–24
320. "Experimental Confirmation of the X-Ray Magnetic Circular Dichroism Sum Rules for Iron and Cobalt", Chen, C. T.; Idzerda, Y. U.; Lin, H. J.; Smith, N. V.; Meigs, G.; Chaban, E.; Ho, G.; Pellegrin, E.; Sette, F. *Phys. Rev. Lett.*, **1995**, *75*, 152–155
321. "First principles investigation of the validity and range of applicability of the X-ray magnetic circular dichroism sum rule", Wu, R. Q.; Wang, D.; Freeman, A. J. *Phys. Rev. Lett.*, **1993**, *71*, 3581–3584
322. "Limitation of the Magnetic-Circular-Dichroism Spin Sum Rule for Transition Metals and Importance of the Magnetic Dipole Term", Wu, R. Q.; Freeman, A. J. *Phys. Rev. Lett.*, **1994**, *73*, 1994–1997
323. "Analytic Calculation of the Spin Sum Rule at the $L_{2,3}$ Edges of Cu^{2+}", Sainctavit, P.; Arrio, M. A.; Brouder, C. *Phys. Rev. B*, **1995**, *52*, 12766–12769
324. "The importance of the magnetic dipole term in magneto-circular X-ray absorption dichroism for 3d transition metal compounds", Crocombette, J. T.; Thole, B. T.; Jollett, F. *J. Phys.: Condens. Matter*, **1996**, *8*, 4095–4105
325. "Accuracy of the spin sum rule in XMCD for the transition-metal L edges from manganese to copper", Piamonteze, C.; Miedema, P.; de Groot, F. M. F. *Phys. Rev. B*, **2009**, *80*
326. "The accuracy of the spin sum rule in XMCD", Piamonteze, C.; Miedema, P.; de Groot, F. M. F., in *14th International Conference on X-Ray Absorption Fine Structure*; DiCicco, A. and Filipponi, A., Ed., 2009; Vol. 190
327. "Sum rule distortions in fluorescence-yield X-ray magnetic circular dichroism", Liu, B. Y.; Piamonteze, C.; Delgado-Jaime, M. U.; Wang, R. P.; Heidler, J.; Dreiser, J.; Chopdekar, R.; Nolting, F.; de Groot, F. M. F. *Phys. Rev. B*, **2017**, *96*
328. "Magnetic Proximity Effect as a Pathway to Spintronic Applications of Topological Insulators", Vobornik, I.; Manju, U.; Fujii, J.; Borgatti, F.; Torelli, P.; Krizmancic, D.; Hor, Y. S.; Cava, R. J.; Panaccione, G. *Nano Lett.*, **2011**, *11*, 4079–4082
329. "Phase Coexistence and Kinetic Arrest in the Magnetostructural Transition of the Ordered Alloy FeRh", Keavney, D. J.; Choi, Y.; Holt, M. V.; Uhlir, V.; Arena, D.; Fullerton, E. E.; Ryan, P. J.; Kim, J. W. *Sci. Rep.*, **2018**, *8*
330. "The CTM4XAS program for EELS and XAS spectral shape analysis of transition metal L edges", Stavitski, E.; Groot, F. M. F. d. *Micron*, **2010**, *41*, 687–694

331. Siegbahn, M. *Spektroskopik der Röntgenstrahlen*; Springer: Berlin, **1931**
332. Jenkins, R. *X-Ray Fluorescence Spectrometry*; 2nd ed.; Wiley-Interscience: New York, **1999**; Vol. 152
333. "Theoretical Analysis of the Magnetic Circular Dichroism in the 2p3d and 2p4d X-ray emission of Gd", de Groot, F. M. F.; Nakazawa, M.; Kotani, A.; Krisch, M. H.; Sette, F. *Phys. Rev. B*, **1997**, *56*, 7285–7292
334. "High resolution 1 *s* core hole X-ray spectroscopy in 3*d* transition metal complexes - electronic and structural information", Glatzel, P.; Bergmann, U. *Coord. Chem. Rev.*, **2005**, *249*, 65–95
335. "Charge transfer multiplet calculations of the K beta X-ray emission spectra of divalent nickel compounds", de Groot, F. M. F.; Fontaine, A.; Kao, C. C.; Krisch, M. *J. Phys. Cond. Matt.*, **1994**, *6*, 6875–6884
336. "Multiplet structure in Ni K beta X-ray fluorescence spectra of nickel compounds", Kawai, J.; Takami, M.; Satoko, C. *Phys. Rev. Lett.*, **1990**, *65*, 2193–2196
337. "Anisotropic valence → core X-ray fluorescence from a $[Rh(en)_3][Mn(N)(CN)_5]$. H_2O single crystal: Experimental results and density functional calculations", Bergmann, U.; Bendix, J.; Glatzel, P.; Gray, H. B.; Cramer, S. P. *J. Chem. Phys.*, **2002**, *116*, 2011–2015
338. "Multiple Excitations in the K Fluorescence Emission of Mn, Fe, and Ni", Glatzel, P.; Bergmann, U.; de Groot, F. M. F.; Cramer, S. P., in *X-Ray and Inner Shell Processes*; Bianconi, A., Marcelli, A. and Saini, N. L., Ed.; American Institute of Physics: Melville, N. Y., 2003; Vol. 652, pp. 250–255
339. "Magnetic circular dichroism of lanthanoid $L\beta_{2,15}$ fluorescence spectra of magnetic lanthanoid compounds", Iwazumi, T.; Kobayashi, K.; Kishimoto, S.; Nakamura, T.; Nanao, S.; Ohsawa, D.; Katano, R.; Isozumi, Y.; Maruyama, H. *J. Electr. Spect. Rel. Phen.*, **1998**, *92*, 257–260
340. "Hard X-ray emission spectroscopy: a powerful tool for the characterization of magnetic semiconductors", Rovezzi, M.; Glatzel, P. *Semi. Sci. Technol.*, **2014**, *29*, 023002
341. "Magnetic circular dichroism of resonant X-ray emission spectroscopy for Sm $L_3M_{4,5}$ and L_2M_4 in a $SmM_{21}Co_{79}$ amorphous alloy", Nakamura, T.; Shoji, H.; Hirai, E.; Nanao, S.; Fukui, K.; Ogasawara, H.; Kotani, A.; Iwazumi, T.; Harada, I.; Katano, R.; Isozumi, Y. *Phys. Rev. B*, **2003**, *67*, 094439
342. "Magnetic circular dichroism of the X-ray-emission spectra for the 2p → 1s decay in Co metal", Nakamura, T.; Shoji, H.; Nanao, S.; Iwazumi, T.; Kishimoto, S.; Katano, R.; Isozumi, Y. *Phys. Rev. B*, **2000**, *62*, 5301–5304
343. "Magnetic circular dichroism in the 2p4d X-ray emission of EuO", Wittkop, C.; Schulke, W.; de Groot, F. M. F. *Phys, Rev. B*, **2000**, *61*, 7176–7179
344. "Range-extended EXAFS at the L edge of rare earths using high-energy-resolution fluorescence detection: A study of La in LaOCl", Glatzel, P.; de Groot, F. M. F.; Manoilova, O.; Grandjean, D.; Weckhuysen, B. M.; Bergmann, U.; Barrea, R. *Phys. Rev. B*, **2005**, *72*, 014117
345. "Pressure-induced high-spin to low-spin transition in FeS evidenced by X-ray emission spectroscopy", Rueff, J. P.; Kao, C. C.; Struzhkin, V. V.; Badro, J.; Shu, J.; Hemley, R. J.; Hao, H. K. *Phys. Rev. Lett.*, **1999**, *82*, 3284–3287
346. "Magnetism in FeO at megabar pressures from X-ray emission spectroscopy", Badro, J.; Struzhkin, V. V.; Shu, J. F.; Hemley, R. J.; Mao, H. K.; Kao, C. C.; Rueff, J. P.; Shen, G. Y. *Phys. Rev. Lett.*, **1999**, *83*, 4101–4104
347. "Magnetic and structural alpha-epsilon phase transition in Fe monitored by X-ray emission spectroscopy", Rueff, J. P.; Krisch, M.; Cai, Y. Q.; Kaprolat, A.; Hanfland, M.; Lorenzen, M.; Masciovecchio, C.; Verbeni, R.; Sette, F. *Phys. Rev. B*, **1999**, *60*, 14510–14512
348. "Reflections on hard X-ray photon-in/photon-out spectroscopy for electronic structure studies", Glatzel, P.; Weng, T. C.; Kvashnina, K.; Swarbrick, J.; Sikora, M.; Gallo, E.; Smolentsev, N.; Mori, R. A. *J. Elect. Spect. Rel. Phen.*, **2013**, *188*, 17–25
349. "Absence of Mn-Centered Oxidation in the $S_2 \rightarrow S_3$ Transition: Implications for the Mechanism of Photosynthetic Water Oxidation", Messinger, J.; Robblee, J. H.; Bergmann, U.; Fernandez, C.; Glatzel, P.; Visser, H.; Cinco, R. M.; Holman, K., M.; Bellacchio, E.; Pizarro,

S. A.; Cramer, S. P.; Sauer, K.; Klein, M. P.; Yachandra, V. K. *J. Am. Chem. Soc.*, **2001**, *123*, 7804–7820
350. "Photon-in photon-out hard X-ray spectroscopy at the Linac Coherent Light Source", Alonso-Mori, R.; Sokaras, D.; Zhu, D. L.; Kroll, T.; Chollet, M.; Feng, Y. P.; Glownia, J. M.; Kern, J.; Lemke, H. T.; Nordlund, D.; Robert, A.; Sikorski, M.; Song, S.; Weng, T. C.; Bergmann, U. **2015**, *22*, 612–620
351. "Direct Detection of Oxygen Ligation to the Mn4Ca Cluster of Photosystem II by X-ray Emission Spectroscopy", Pushkar, Y.; Long, X.; Glatzel, P.; Brudvig, G. W.; Dismukes, G. C.; Collins, T. J.; Yachandra, V. K.; Yano, J.; Bergmann, U. **2010**, *49*, 800–803
352. "X-ray Emission Spectroscopy Evidences a Central Carbon in the Nitrogenase Iron-Molybdenum Cofactor", Lancaster, K. M.; Roemelt, M.; Ettenhuber, P.; Hu, Y. L.; Ribbe, M. W.; Neese, F.; Bergmann, U.; DeBeer, S. *Science*, **2011**, *334*, 974–977
353. "Elimination of the Inner-Shell Lifetime Broadening in X-Ray-Absorption Spectroscopy", Hamaläinen, K.; Siddons, D. P.; Hastings, J. B.; Berman, L. E. *Phys. Rev. Lett.*, **1991**, *67*, 2850–2853
354. "Primary structure of hydrogenase from *Clostridium pasteurianum*", Meyer, J.; Gagnon, J. *Biochemistry*, **1991**, *30*, 9697–9704
355. "Spectral sharpening of the Pt L edges by high-resolution X-ray emission", de Groot, F. M. F.; Krisch, M. H.; Vogel, J. *Phys. Rev. B*, **2002**, *66*
356. "Site-Selective EXAFS in Mixed-Valent Compounds Using High Resolution Fluorescence Spectroscopy: A Study of Fe in Prussian Blue", Glatzel, P.; Jacquemet, L.; Bergmann, U.; deGroot, F. M. F.; Cramer, S. P. *Inorg. Chem.*, **2002**, *41*, 3121–3127
357. "Spin-dependent X-ray absorption of MnO and MnF_2", Hamalainen, K.; Kao, C. C.; Hastings, J. B.; Siddons, D. P.; Berman, L. E.; Stojanoff, V.; Cramer, S. P. *Phys. Rev. B*, **1992**, *46*, 14274–14277
358. "Resonant inelastic X-ray scattering studies of elementary excitations", Ament, L. J. P.; van Veenendaal, M.; Devereaux, T. P.; Hill, J. P.; van den Brink, J. *Rev. Mod. Phys.*, **2011**, *83*, 705–767
359. "Über die Streuung von Strahlung durch Atome (On the dispersal of radiation by atoms)", Kramers, H. H.; Heisenberg, W. *Z. Physik*, **1925**, *31*, 681–708
360. "Resonant inelastic X-ray scattering in d and f electron systems", Kotani, A. *Eur. Phys. J. B*, **2005**, *47*, 3–27
361. "Electronic Structure of Ni Complexes by X-Ray Resonance Raman Spectroscopy (Resonant Inelastic X-Ray Scattering)", Glatzel, P.; Bergmann, U.; Gu, W.; Wang, H.; Stepanov, S.; Mandimutsira, B. S.; Riordan, C. G.; Horwitz, C. P.; Collins, T.; Cramer, S. P. *J. Am. Chem. Soc.*, **2002**, *124*, 9668 -9669
362. "Charge transfer at very high pressure in NiO", Shukla, A.; Rueff, J. P.; Badro, J.; Vanko, G.; Mattila, A.; de Groot, F. M. F.; Sette, F. *Phys. Rev. B*, **2003**, *67*, 4
363. "NiO as a test case for high resolution resonant inelastic soft X-ray scattering", Ghiringhelli, G.; Matsubara, M.; Dallera, C.; Fracassi, F.; Gusmeroli, R.; Piazzalunga, A.; Tagliaferri, A.; Brookes, N. B.; Kotani, A.; Braicovich, L. *J. Phys. Condens. Matter*, **2005**, *17*, 5397–5412
364. "Crystal field and low energy excitations measured by high resolution RIXS at the L_3 edge of Cu, Ni and Mn", Ghiringhelli, G.; Piazzalunga, A.; Wang, X.; Bendounan, A.; Berger, H.; Bottegoni, F.; Christensen, N.; Dallera, C.; Grioni, M.; Grivel, J. C.; Sala, M. M.; Patthey, L.; Schlappa, J.; Schmitt, T.; Strocov, V.; Braicovich, L. *Eur. Phys. J. Spec. Top.*, **2009**, *169*, 199–205
365. "Local spin-flip spectral distribution obtained by resonant X-ray Raman scattering", de Groot, F. M. F.; Kuiper, P.; Sawatzky, G. A. *Phys. Rev. B*, **1998**, *57*, 14584–14587
366. Figgis, B. N.; Hitchman, M. A. *Ligand Field Theory and Its Applications*; Wiley-VCH: New York, **2000**
367. "Magnetic Studies with Copper(II) Salts. 1. Anomalous Paramagnetism and Delta-Bonding in Anhydrous and Hydrated Copper(II) Acetates", Figgis, B. N.; Martin, R. L. *J. Chem. Soc.*, **1956**, 3837–3846

368. "Variable-Temperature Raman Infrared Spectra of the Copper-Acetate Dimer $Cu_2(O_2CCH_3)_4(H_2O)_2$ and Its Derivatives", Mathey, Y.; Greig, D. R.; Shriver, D. F. *Inorg. Chem.*, **1982**, *21*, 3409–3413
369. "Direct Observation of Singlet-Triplet Separation in Dimeric Copper(II) Acetate by Neutron Inelastic Scattering Spectroscopy", Güdel, H. U.; Stebler, A.; Furrer, A. *Inorg. Chem.*, **1979**, *18*, 1021–1023
370. "Observation of Two Nondispersive Magnetic Excitations in NiO by Resonant Inelastic Soft-X-Ray Scattering", Ghiringhelli, G.; Piazzalunga, A.; Dallera, C.; Schmitt, T.; Strocov, V. N.; Schlappa, J.; Patthey, L.; Wang, X.; Berger, H.; Grioni, M. *Phys. Rev. Lett.*, **2009**, *102*
371. "Magnetic, transport, and optical properties of monolayer copper oxides", Kastner, M. A.; Birgeneau, R. J.; Shirane, G.; Endoh, Y. *Rev. Mod. Phys.*, **1998**, *70*, 897–928
372. "Dispersion of Magnetic Excitations in the Cuprate La_2CuO_4 and $CaCuO_2$ Compounds Measured Using Resonant X-Ray Scattering", Braicovich, L.; Ament, L. J. P.; Bisogni, V.; Forte, F.; Aruta, C.; Balestrino, G.; Brookes, N. B.; Luca, G. M.; Medaglia, P. G.; Granozio, F. M.; Radovic, M.; Salluzzo, M.; van den Brink, J.; Ghiringhelli, G. *Phys. Rev. Lett.*, **2009**, *102*, 167401
373. "Magnetic nature of the 500 meV peak in $La_{2-x}Sr_xCuO_4$ observed with resonant inelastic X-ray scattering at the Cu K-edge", Ellis, D. S.; Kim, J.; Hill, J. P.; Wakimoto, S.; Birgeneau, R. J.; Shvyd'ko, Y.; Casa, D.; Gog, T.; Ishii, K.; Ikeuchi, K.; Paramekanti, A.; Kim, Y.-J. *Phys. Rev. B*, **2010**, *81*, 085124
374. "Momentum and polarization dependence of single-magnon spectral weight for Cu L_3-edge resonant inelastic X-ray scattering from layered cuprates", Braicovich, L.; Moretti Sala, M.; Ament, L. J. P.; Bisogni, V.; Minola, M.; Balestrino, G.; Di Castro, D.; De Luca, G. M.; Salluzzo, M.; Ghiringhelli, G.; van den Brink, J. *Phys. Rev. B*, **2010**, *81*
375. "Magnetic Excitations and Phase Separation in the Underdoped $La_{2-x}Sr_xCuO_4$ Superconductor Measured by Resonant Inelastic X-Ray Scattering", Braicovich, L.; van den Brink, J.; Bisogni, V.; Moretti Sala, M.; Ament, L. J. P.; Brookes, N. B.; De Luca, G. M.; Salluzzo, M.; Schmitt, T.; Strocov, V. N.; Ghiringhelli, G. *Phys. Rev. Lett.*, **2010**, *104*, 077002
376. "Spin–orbital separation in the quasi-one-dimensional Mott insulator Sr_2CuO_3", Schlappa, J.; Wohlfeld, K.; Zhou, K. J.; Mourigal, M.; Haverkort, M. W.; Strocov, V. N.; Hozoi, L.; Monney, C.; Nishimoto, S.; Singh, S.; Revcolevschi, A.; Caux, J. S.; Patthey, L.; Rønnow, H. M.; van den Brink, J.; Schmitt, T. *Nature*, **2012**, *485*, 82
377. "Demonstration of resonant inelastic X-ray scattering as a probe of exciton-phonon coupling", Geondzhian, A.; Gilmore, K. *Phys. Rev. B*, **2018**, *98*
378. "Resonant Inelastic Scattering Spectra of Free Molecules with Vibrational Resolution", Hennies, F.; Pietzsch, A.; Berglund, M.; Fohlisch, A.; Schmitt, T.; Strocov, V.; Karlsson, H. O.; Andersson, J.; Rubensson, J. E. *Phys. Rev. Lett.*, **2010**, *104*, 193002–193005
379. "Observation of phonons with resonant inelastic X-ray scattering", Yavas, H.; van Veenendaal, M.; van den Brink, J.; Ament, L. J. P.; Alatas, A.; Leu, B. M.; Apostu, M. O.; Wizent, N.; Behr, G.; Sturhahn, W.; Sinn, H.; Alp, E. E. *J. Phys. Cond. Matt.*, **2010**, *22*
380. "A change of wave-length in light scattering", Raman, C. V. *Nature*, **1928**, *121*, 619–619
381. "The Molecular Scattering of Light—Nobel Lecture, December 11, 1930", Raman, C. V., in *Nobel Lectures, Physics 1922–1941*; Elsevier Publishing Company: Amsterdam, 1965
382. "X-Ray Raman Scattering. Experiment. I", Suzuki *J. Phys. Soc. Japan*, **1967**, *22*, 1139
383. "A large-solid-angle X-ray Raman scattering spectrometer at ID20 of the European Synchrotron Radiation Facility", Huotari, S.; Sahle, C. J.; Henriquet, C.; Al-Zein, A.; Martel, K.; Simonelli, L.; Verbeni, R.; Gonzalez, H.; Lagier, M. C.; Ponchut, C.; Sala, M. M.; Krisch, M.; Monaco, G. *J. Syn. Rad.*, **2017**, *24*, 521–530
384. "The LERIX user facility", Seidler, G.; T. Fister, T.; Cross, J.; P. Nagle, K. *AIP Conf. Proc.*, **2007**, *882*, 911
385. "X-ray-Raman-scattering-based EXAFS beyond the dipole limit", Huotari, S.; Pylkkanen, T.; Soininen, J. A.; Kas, J. J.; Hamalainen, K.; Monaco, G. *J. Synchrotron Rad.*, **2012**, *19*, 106–113

386. “Probing electronic correlations in actinide materials using multipolar transitions”, Bradley, J. A.; Sen Gupta, S.; Seidler, G. T.; Moore, K. T.; Haverkort, M. W.; Sawatzky, G. A.; Conradson, S. D.; Clark, D. L.; Kozimor, S. A.; Boland, K. S. *Phys. Rev. B*, **2010**, *81*, 193104
387. “X-ray Raman Scattering as a Substitute for Soft X-Ray Extended X-Ray Absorption Fine Structure”, Tohji, K.; Udagawa, Y. *Phys. Rev. B*, **1989**, *39*, 7590–7594
388. “Theory of X-Ray Raman Scattering”, Mizuno, Y.; Ohmura, Y. *J. Phys. Soc. Jap.*, **1967**, *22*, 445
389. “Planning, performing and analyzing X-ray Raman scattering experiments”, Sahle, C. J.; Mirone, A.; Niskanen, J.; Inkinen, J.; Krisch, M.; Huotari, S. *J. Syn. Rad.*, **2015**, *22*, 400–409
390. Schülke, W. *Electron Dynamics by Inelastic X-Ray Scattering*; Oxford University Press: New York, **2007**
391. “Correlations in Space and Time and Born Approximation Scattering in Systems of Interacting Particles”, Van Hove, L. *Phys. Rev.*, **1954**, *95*, 249–262
392. “Nonresonant Inelastic X-ray Scattering and X-ray Raman Scattering”, Fister, T. T., in *XAFS Techniques for Catalysts, Nanomaterials, and Surfaces*; Iwasawa, Y., Asakura, K. and Tada, M., Ed.; Springer, 2016, pp. 237–250
393. “Phonon spectroscopy by inelastic X-ray scattering”, Burkel, E. *Rep. Prog. Phys.*, **2000**, *63*, 171–232
394. “High-Resolution Inelastic X-Ray Scattering I & II”, Baron, A. Q. R., in *Synchrotron Light Sources and Free-Electron Lasers: Accelerator Physics, Instrumentation and Science*; Jaeschke, E., Khan, S., Schneider, J. R. and Hastings, J. B., Ed.; Springer International Publishing, 2018, pp. 1643–1757
395. “X-Ray Raman Spectroscopy—A New Tool to Study Local Structure of Aromatic Hydrocarbons and Asphaltenes”, Bergmann, U.; Groenzin, H.; Mullins, O. C.; Glatzel, P.; Fetzer, J.; Cramer, S. P. *Pet. Sci. Tech.*, **2004**, *22*, 863–875
396. “The Structure of the First Coordination Shell in Liquid Water”, Wernet, P.; Nordlund, D.; Bergmann, U.; Cavalleri, M.; Odelius, M.; Ogasawara, H.; Näslund, L. Å.; Hirsch, T. K.; Ojamäe, L.; Glatzel, P.; Pettersson, L. G. M.; Nilsson, A. *Science*, **2004**, *304*, 995–999
397. “Pressure induced spin transition revealed by iron $M_{2,3}$-edge spectroscopy”, Nyrow, A.; Tse, J. S.; Hiraoka, N.; Desgreniers, S.; Büning, T.; Mende, K.; Tolan, M.; Max Wilke; Sternemann, C. *Appl. Phys. Lett.*, **2014**, *104*, 262408
398. “In situ X-ray Raman scattering spectroscopy of a graphite electrode for lithium-ion batteries”, Nonaka, T.; Kawaura, H.; Makimura, Y.; Nishimura, Y. F.; Dohmae, K. *J. Power Sources*, **2019**, *419*, 203–207
399. “One hundred years of diffuse scattering”, Welberry, T. R.; Weber, T. *Crystall. Rev.*, **2016**, *22*, 2–78
400. “Determination of phonon dispersion relations by X-ray thermal diffuse scattering”, Xu, R. Q.; Chiang, T. C. *Z. Kristall.*, **2005**, *220*, 1009–1016
401. “Phonons in Crystals using Inelastic X-Ray Scattering”, Baron, A. Q. R. *J. Spect. Soc. Japan*, **2009**, *58*, 205–214 (Japanese) arXiv 0910.5764 (English)
402. “Determination of phonon dispersion curves by means of inelastic X-ray scattering”, Burkel, E. *J. Phys. Cond. Matt.*, **2001**, *13*, 7627–7644
403. “Inelastic X-ray scattering: a new spectroscopy tool to investigate the atomic dynamics in condensed matter”, Ruocco, G.; Sette, F., Synchrotron Radiation: Fundamentals, Methodologies and Applications, Bologna, **2003**, pp. 1–24
404. “Theory of inelastic X-ray scattering from condensed matter”, Sinha, S. K. *J. Phys. Cond. Matt.*, **2001**, *13*, 7511–7523
405. “Inelastic X-ray scattering from phonons”, Krisch, M.; Sette, F., in *Light Scattering in Solids IX*; Cardona, M. and Merlin, R., Ed.; Springer-Verlag: Berlin, 2007; Vol. 108, pp. 317–369
406. “Molecular vibrational spectroscopy by inelastic X-ray scattering: Experimental determination of the absolute vibrational cross section in liquid nitrogen”, Monaco, G.; Nardone, M.; Sette, F.; Verbeni, R. *Phys. Rev. B*, **2001**, *64*

407. "Collective Dynamics in Water by High Energy Resolution Inelastic X-Ray Scattering", Sette, F.; Ruocco, G.; Krisch, M.; Bergmann, U. *Phys. Rev. Lett.*, **1995**, *75*, 850–853
408. "Collective dynamics in water by inelastic X-rays scattering", Sette, F.; Ruocco, G.; Krisch, M.; Masciovecchio, C.; Verbeni, R. *Phys. Scripta*, **1996**, *1996*, 48
409. "The high-frequency dynamics of liquid water", Ruocco, G.; Sette, F. *J. Phys. Cond. Matt.*, **1999**, *11*, R259
410. "High-frequency dynamics of liquid and supercritical water", Bencivenga, F.; Cunsolo, A.; Krisch, M.; Monaco, G.; Ruocco, G.; Sette, F. *Phys. Rev. E*, **2007**, *75*, 051202
411. "Seeing real-space dynamics of liquid water through inelastic X-ray scattering", Iwashita, T.; Wu, B.; Chen, W.-R.; Tsutsui, S.; Baron, A. Q. R.; Egami, T. *Sci. Adv.*, **2017**, *3*
412. "Soft and isotropic phonons in $PrFeAsO_{1-y}$", Fukuda, T.; Baron, A. Q. R.; Nakamura, H.; Shamoto, S.; Ishikado, M.; Machida, M.; Uchiyama, H.; Iyo, A.; Kito, H.; Mizuki, J.; Arai, M.; Eisaki, H. *Phys. Rev. B*, **2011**, *84*, 064504
413. "Superconductivity at 39 K in magnesium diboride", Nagamatsu, J.; Nakagawa, N.; Muranaka, T.; Zenitani, Y.; Akimitsu, J. *Nature*, **2001**, *410*, 63–64
414. "Correlation between crystal structures of CaAlSi with and without superlattice and superconducting properties", Kuroiwa, S.; Sagayama, H.; Kakiuchi, T.; Sawa, H.; Noda, Y.; Akimitsu, J. *Phys. Rev. B*, **2006**, *74*
415. "Soft-phonon-driven superconductivity in CaAlSi as seen by inelastic X-ray scattering", Kuroiwa, S.; Baron, A. Q. R.; Muranaka, T.; Heid, R.; Bohnen, K. P.; Akimitsu, J. *Phys. Rev. B*, **2008**, *77*, 140503
416. "Phonon anomalies and lattice dynamics in the superconducting oxychlorides $Ca_{2-x}CuO_2Cl_2$", d'Astuto, M.; Yamada, I.; Giura, P.; Paulatto, L.; Gauzzi, A.; Hoesch, M.; Krisch, M.; Azuma, M.; Takano, M. *Phys. Rev. B*, **2013**, *88*, 014522
417. "Strongly Anisotropic Magnesiowustite in Earth's Lower Mantle", Finkelstein, G. J.; Jackson, J. M.; Said, A.; Alatas, A.; Leu, B. M.; Sturhahn, W.; Toellner, T. S. *J. Geophys. Res. - Solid Earth*, **2018**, *123*, 4740–4750
418. "Identification of Carbonate-Rich Outcrops on Mars by the Spirit Rove", Morris, R. V.; Ruff, S. W.; Gellert, R.; Ming, D. W.; Arvidson, R. E.; Clark, B. C.; Golden, D. C.; Siebach, K.; Klingelhöfer, G.; Schröder, C.; Fleischer, I.; Yen, A. S.; Squyres, S. W. *Science*, **2010**, *329*, 421–424
419. Nuclear Web Site, such as: http://ie.lbl.gov/toi.htm/ and http://atom.kaeri.re.kr/
420. Röhlsberger, R. *Nuclear Condensed Matter Physics with Synchrotron Radiation - Basic Principles, Methodology and Applications*; Springer-Verlag: Berlin Heidelberg**, 2004**
421. "Appendix", Röhlsberger, R., in *Nuclear Condensed Matter Physics with Synchrotron Radiation - Basic Principles, Methodology and Applications*; Springer-Verlag: Berlin/Heidelberg, 2004, pp. 285–312
422. "On the Application of Interference Methods to Spectroscopic Measurements, II", Michelson, A. A. *Phil. Mag.*, **1892**, *34*, 280–299
423. "Recent advances in spectroscopy", Michelson, A. A. *Nobel Lecture*, **1907**
424. "Chap. 13 - Hyperfine Interactions", Yang, F.; Hamilton, J. H., in *Modern Atomic and Nuclear Physics*; McGraw-Hill: New York, 1996, pp. 622–648
425. "Calibration of the isomer shift for the 14.4-keV transition in ^{57}Fe using the full-potential linearized augmented plane-wave method", Wdowik, U. D.; Ruebenbauer, K. *Phys. Rev. B*, **2007**, *76*
426. "Nuclear and Nucleon Scattering of High Energy Electrons", Hofstadter, R. *Ann. Rev. Nuc. Sci.*, **1957**, *7*, 231–316
427. "Bis(tetramethylammonium) tetra-μ_3-sulfido-tetrakis[(2,2-dimethylpropanethiolato-S)iron]: a heterocubane complex anion with crystallographically imposed 4(overbar)2 m symmetry", Hauptmann, R.; Schneider, J.; Kökerling, M.; Henkel, G. *Acta Cryst.*, **1999**, *C55*, 190–192
428. "Overlap Contribution to Isomer Shift of Iron Compounds", Simánek, E.; Sroubek, Z. *Phys. Rev.*, **1967**, *163*, 275–279

429. "Mössbauer spectroscopy - An indispensable tool in solid state research", Gütlich, P.; Schröder, C.; Schünemann, V. *Spect. Eur.*, **2012**, *24*, 21–32
430. "Nitrogen Fixation via a Terminal Fe(IV) Nitride", Thompson, N. B.; Green, M. T.; Peters, J. C. *J. Am. Chem. Soc.*, **2017**, *139*, 15312–15315
431. "Mössbauer Spectrometry", Fultz, B., in *Characterization of Materials*; Kaufmann, E., Ed.; John Wiley, 2011
432. "Nuclear Diffraction of Resonant Gamma Radiation by an Antiferromagnetic Crystal", Smirnov, G. V.; Sklyarev.Vv; Voskanya.Ra; Artemev, A. N. *JETP Lett. USSR*, **1969**, *9*, 70
433. "Synchrotron radiation ^{57}Fe Mössbauer spectroscopy using nuclear monochromator", Mitsui, T.; Masuda, R.; Hirao, N.; Mibu, K.; Seto, M. *Hyp. Int.*, **2012**, *204*, 97–100
434. "Studies on spintronics-related thin films using synchrotron-radiation-based Mossbauer spectroscopy", Mibu, K.; Seto, M.; Mitsui, T.; Yoda, Y.; Masuda, R.; Kitao, S.; Kobayashi, Y.; Suharyadi, E.; Tanaka, M.; Tsunoda, M.; Yanagihara, H.; Kita, E. *Hyp. Int.*, **2013**, *217*, 127–135
435. "Synchrotron Radiation Based Mössbauer Spectroscopy", Seto, M.; Masuda, R.; Higashitaniguchi, S.; Kitao, S.; Kobayashi, Y.; Inaba, C.; Mitsui, T.; Yoda, Y. *Phys. Rev. Lett.*, **2009**, *102*, 217602
436. "Evolution of synchrotron-radiation-based Mossbauer absorption spectroscopy for various isotopes", Seto, M.; Masuda, R.; Kobayashi, Y.; Kitao, S.; Kurokuzu, M.; Saito, M.; Hosokawa, S.; Ishibashi, H.; Mitsui, T.; Yoda, Y.; Mibu, K. *Hyp. Int.*, **2017**, *238*, 78 ...
437. "Synchrotron-based Nickel Mössbauer Spectroscopy", Gee, L. B.; Lin, C.-Y.; Jr., F. E. J.; Yoda, Y.; Masuda, R.; Saito, M.; Kobayashi, Y.; Tamasaku, K.; Seto, M.; Charles G. Riordan; Ploskonka, A.; Power, P. P.; Cramer, S. P.; Lauterbach, L. *Inorg. Chem.*, **2016**, *551*, 6866–6872
438. "Nuclear Forward Scattering of Synchrotron Radiation by ^{99}Ru", Bessas, D.; Merkel, D. G.; Chumakov, A. I.; Ruffer, R.; Hermann, R. P.; Sergueev, I.; Mahmoud, A.; Klobes, B.; McGuire, M. A.; Sougrati, M. T.; Stievano, L. *Phys. Rev. Lett.*, **2014**, *113*, 147601
439. "Detectors for nuclear resonant scattering experiments", Baron, A. Q. R. *Hyperfine. Interact.*, **2000**, *125*, 29–42
440. "Array of avalanche photodiodes as a position-sensitive X-ray detector", Kishimoto, S.; Yoda, Y.; Seto, M.; Kitao, S.; Kobayashi, Y.; Haruki, R.; Harami, T. *Nucl. Inst. Meth. A*, **2003**, *513*, 193–196
441. "Coherent γ-ray optics", Hannon, J. P.; Trammell, G. T. *Hyperfine Interact.*, **1999**, *123–124*, 127–274
442. "Nuclear resonant scattering of synchrotron radiation", Smirnov, G. V. *Hyp. Int.*, **1996**, *97/98*, 551–588
443. Baron, A. Q. R., Ph. D. Thesis, Stanford University, **1995**
444. Lübbers, R., Ph. D. Thesis, Universität - GH - Paderborn, **2000**
445. "Coherent Elastic Nuclear Resonant Scattering", Röhlsberger, R., in *Nuclear Condensed Matter Physics with Synchrotron Radiation - Basic Principles, Methodology and Applications*; Springer-Verlag: Berlin/Heidelberg, 2004, pp. 67–180
446. "Nuclear resonance forward scattering of X rays: Time and space picture", Shvyd'ko, Y. V. *Phys. Rev. B*, **1999**, *59*, 9132–9143
447. "Electronic and Magnetic Transitions in Europium Compounds Studied by Nuclear Forward Scattering of Synchrotron Radiation", Leupold, O.; Rupprecht, K.; Wortmann, G. *Struct. Chem.*, **2003**, *14*, 97–107
448. "Nuclear forward scattering vs. conventional Mössbauer studies of atomically tailored Eu-based materials", Konjhodzic, A.; Adamczyk, A.; Vagizov, F.; Hasan, Z.; Alp, E. E.; Sturhahn, W.; Zhao, J. Y.; Carroll, J. J. *Hyp. Int.*, **2006**, *170*, 83–89
449. "Nuclear resonant spectroscopy", Sturhahn, W. *J. Phys. Cond. Matt.*, **2004**, *16*, S497–S530
450. "Very low sound velocities in iron-rich (Mg,Fe)O: Implications for the core-mantle boundary region", Wicks, J. K.; Jackson, J. M.; Sturhahn, W. *Geophys. Res. Lett.*, **2010**, *37*, L15304

451. Steffen, R. M.; Frauenfelder, H., in *Alpha-, Beta- and Gamma-Ray Spectroscopy*; Siegbahn, K., Ed.; North-Holland: Amsterdam, 1965
452. "Nuclear Quadrupole Interactions Studied by Time Differential Perturbed Angular Correlations of γ-Rays", Butz, T. *Z. Naturforsch.*, **1996**, *51a*, 396–410
453. "Biological Applications of Perturbed Angular Correlations of γ-Ray Spectroscopy ", Hemmingsen, L.; Sas, K. N.; Danielsen, E. *Chem. Rev.*, **2004**, *104*, 4027 – 4061
454. "Synchrotron-radiation-based perturbed angular correlations used in the investigation of rotational dynamics in soft matter", Sergueev, I.; van Bürck, U.; Chumakov, A. I.; Asthalter, T.; Smirnov, G. V.; Franz, H.; Rüffer, R.; Petry, W. *Phys. Rev. B*, **2006**, *73*, 024203
455. "Synchrotron-radiation-based perturbed angular correlations from ^{119}Sn", Strohm, C.; Sergueev, I.; van Bürck, U. *Eur. Phys. Lett.*, **2008**, *81*, 52001
456. "Hyperfine interactions in ^{61}Ni with synchrotron-radiation-based perturbed angular correlations", Sergueev, I.; Leupold, O.; Wille, H.-C.; Roth, T.; Chumakov, A. I.; R., R. *Phys. Rev. B*, **2008**, *78*
457. "The Scattering of α and β Particles by Matter and the Structure of the Atom", Rutherford, E. *Phil. Mag.*, **1911**, *21*, 669–688
458. Cathcart, B. *The Fly in the Cathedral*; Farrar, Straus & Giroux, **2004**
459. "On a Diffuse Reflection of the α-Particles", Geiger, H.; Marsen, E. *Proc. Royal Soc.*, **1909**, *82*, 495–500
460. "Recoilless Nuclear Resonance Absorption of Gamma Radiation", Mössbauer, R. L., in *Nobel Lectures, Physics 1942–1962*; Elsevier Publishing Company: Amsterdam, 1964
461. Marmier, P.; Sheldon, E. *Physics of Nuclei and Particles*; Academic Press: New York, **1969**; Vol. I
462. "The Hard X-Ray Components of Scattered Gamma-Rays", Moon, P. B. *Proc. Royal Soc.*, **1950**, *63*, 1189-1196
463. "Kernresonanzfluoreszenz von Gammastrahlung in ^{191}Ir", Mössbauer, R. L. *Z. Physik*, **1958**, *151*, 124–143
464. "Kernresonanzabsorption von gamma-Strahlung in ^{191}Ir ", Mössbauer, R. L. *Naturwissen.*, **1958**, *45*, 538
465. "Kernresonanzabsorption von γ-Strahlung in Ir^{191}", Mössbauer, R. L. *Z. Naturforschg.*, **1959**, *14a*, 211–216
466. "Resonant Absorption of the 14.4-keV gamma Ray from 0.10-μsec ^{57}Fe", Pound, R. V.; G. A. Rebka, J. *Phys. Rev. Lett.*, **1959**, *3*, 554–556
467. "Recoilless Resonance Absorption of Gamma Rays in ^{57}Fe", Schiffer, J. P.; Marshall, W. *Phys. Rev. Lett.*, **1959**, *3*, 556–557
468. "On Directional Correlation of Successive Quanta", Hamilton, D. R. *Phys. Rev.*, **1940**, *58*, 122–131
469. "Angular Correlation of Successive Gamma-Ray Quanta", Brady, E. L.; Deutsch, M. *Phys. Rev.*, **1947**, *72*, 870–871
470. "Interaction Quadrupolaire dans ^{111}Cd", Lehmann, P.; Miller, J. *Comptes Rendus*, **1955**, *240*, 298–299
471. "Mössbauer Experiments Without Conventional Sources", Ruby, S. L. *J. Phy. Colloq.*, **1974**, *35*, 209–211
472. "Scientific legacy of Stanley Ruby ", Shenoy, G. K. *Hyperfine Interact.*, **2006**, *170*, 5–13
473. "Historical overview and introduction", Gerdau, E.; van Bürck, U.; Rüffer, R. *Hyp. Interact.*, **1999**, *123/124*, 3–10
474. "Nuclear Bragg diffraction of synchrotron radiation in yttrium iron garnet", Gerdau, E.; Rüffer, R.; Winkler, H.; Tolksdorf, W.; Klages, C. P.; Hannon, J. P. *Phys. Rev. Lett.*, **1985**, *54*, 835
475. "Mössbauer Spectroscopy Using Synchrotron Radiation", Hastings, J. B.; Siddons, D. P.; van Bürck, U.; Hollatz, R.; Bergmann, U. *Phys. Rev. Lett.*, **1991**, *66*, 770–773

476. "Observation of Nuclear Resonant Scattering Accompanied by Phonon Excitation Using Synchrotron Radiation", Seto, M.; Yoda, Y.; Kikuta, S.; Zhang, X. W.; Ando, M. *Phys. Rev. Lett.*, **1995**, *74*, 3828–3831
477. "Single-nucleus quantum beats excited by synchrotron radiation", Baron, A. Q. R.; Chumakov, A. I.; Rüffer, R.; Grünsteudel, H. F.; Leupold, O. *Europhys. Lett.*, **1996**, *34*, 331–336
478. "Mössbauer sum rules for use with synchrotron sources", Lipkin, H. J. *Phys. Rev. B*, **1995**, *52*, 10073–10079
479. "CONUSS and PHOENIX: evaluation of nuclear resonant scattering data", Sturhahn, W. *Hyp. Int.*, **2000**, *125*, 149–172
480. "A combined NRVS and DFT study of Fe(IV) = O model complexes: a diagnostic method for the elucidation of non-heme iron enzyme intermediates", Bell III, C. B.; Wong, S. D.; Xiao, Y.; Klinker, E. J.; Tenderholt, A. L.; Smith, M. C.; Rohde, J. U.; Que. Jr., L.; Cramer, S. P.; Solomon, E. I. *Angew. Chem. Int. Ed.*, **2008**, *47*, 9071–9074
481. "Nuclear resonance vibrational spectroscopy of a protein active-site mimic", Sage, J. T.; Paxson, C.; Wyllie, G. R. A.; Sturhahn, W.; Durbin, S. M.; Champion, P. M.; Alp, E. E.; Scheidt, W. R. *J. Phys.: Condens. Matter*, **2001**, *13*, 7707–7722
482. "Quantitative Vibrational Dynamics of Iron in Carbonyl Porphyrins", Leu, B. M.; Silvernail, N. J.; Zgierski, M. Z.; Wyllie, G. R. A.; Ellison, M. K.; Scheidt, W. R.; Zhao, J.; Sturhahn, W.; Alp, E. E.; Sage, J. T. *Biophys. J.*, **2007**, *92*, 3764–3783
483. "Nuclear resonant scattering from the subnanosecond lifetime excited state of ^{201}Hg", Ishikawa, D.; Baron, A. Q. R.; Ishikawa, T. *Phys. Rev. B*, **2005**, *72*, 140301
484. "Direct determination of the phonon density of states in beta-Sn", Barla, A.; Rüffer, R.; Chumakov, A. I.; Metge, J.; Plessel, J.; Abd-Elmeguid, M. M. *Phys. Rev. B*, **2000**, *61*, R14881-R14884
485. "Phonon Density of States of Metallic Sn at High Pressure", Giefers, H.; Tanis, E. A.; Rudin, S. P.; Greeff, C.; Ke, X.; Chen, C.; Nicol, M. F.; Pravica, M.; Pravica, W.; Zhao, J.; Alatas, A.; Lerche, M.; Sturhahn, W.; Alp, E. *Phys. Rev. Lett.*, **2007**, *98*, 245502
486. "Partial phonon density of states of dysprosium and its compounds measured using inelastic nuclear resonance scattering", Brown, D. E.; Toellner, T. S.; Sturhahn, W.; Alp, E. E.; Hu, M.; Kruk, R.; Rogacki, K.; Canfield, P. C. *Hyp. Int.*, **2004**, *153*, 17–24
487. "Vibrational dynamics of biological molecules: Multi-quantum contributions", Leu, B. M.; Zgierski, M. Z.; Wyllie, G. R. A.; Mary K. Ellison; Scheidt, W. R.; Sturhahn, W.; Alp, E. E.; Durbin, S. M.; Sage, J. T. *J. Phys. Chem. Solids*, **2005**, *66* 2250–2256
488. "Probing Vibrational Anisotropy with Nuclear Resonance Vibrational Spectroscopy", Pavlik, J. W.; Barabanschikov, A.; Oliver, A. G.; Alp, E. E.; Sturhahn, W.; Zhao, J.; Sage, J. T.; Scheidt, W. R. *Angew. Chem. Int. Ed.*, **2010**, *49*, 4400–4404
489. "Measuring velocity of sound with nuclear resonant inelastic X-ray scattering", Hu, M. Y.; Sturhahn, W.; Toellner, T. S.; Mannheim, P. D.; Brown, D. E.; Zhao, J. Y.; Alp, E. E. *Phys. Rev. B*, **2003**, *67*, 094304
490. "Moments in nuclear resonant inelastic X-ray scattering and their applications", Hu, M. Y.; Toellner, T. S.; Dauphas, N.; Alp, E. E.; Zhao, J. Y. *Phys. Rev. B*, **2013**, *87*
491. "Nuclear resonant inelastic absorption of synchrotron radiation in an anisotropic single crystal", Kohn, V. G.; Chumakov, A. I.; Ruffer, R. *Phys. Rev. B*, **1998**, *58*, 8437–8444
492. "Bulk Modulus of a Protein Active-Site Mimic", Leu, B. M.; Sage, J. T.; Silvernail, N. J.; Scheidt, W. R.; Alatas, A.; Alp, E. E.; Sturhahn, W. *J. Phys. Chem. B*, **2011**, *115*, 4469–4473
493. "Spin crossover equation of state and sound velocities of $(Mg_{0.65}Fe_{0.35})O$ ferropericlase to 140 GPa", Chen, B.; Jackson, J. M.; Sturhahn, W.; Zhang, D.; Zhao, J.; Wicks, J. K.; Murphy, C. A. *J. Geophys. Res.*, **2012**, *117*, B08208
494. "Lattice Dynamics in Bi_2Te_3 and Sb_2Te_3: Te and Sb density of phonon states", Bessas, D.; Sergueev, I.; Wille, H. C.; Persson, J.; Ebling, D.; Hermann, R. P. *Phys. Rev. B*, **2012**, *86*, 224301

495. "Lattice dynamics in the thermoelectric Zintl compound $Yb_{14}MnSb_{11}$", Möchel, A.; Sergueev, I.; Wille, H. C.; Juranyi, F.; Schober, H.; Schweika, W.; Brown, S. R.; Kauzlerich, S. M.; Hermann, R. P. *Phys. Rev. B*, **2011**, *84*, 184303
496. "DOS: Evaluation of phonon density of states from nuclear resonant inelastic absorption", Kohn, V. G.; Chumakov, A. I. *Hyp. Interact.*, **2000**, *125*, 205–221
497. "Normal Mode Analysis of $[FeCl_4]^-$ and $[Fe_2S_2Cl_4]^{2-}$ via Vibrational Mössbauer, Resonance Raman, and FT-IR Spectroscopy", Smith, M. C.; Xiao, Y.; Wang, H.; George, S. J.; Coucovanis, D.; Koutmos, M.; Sturhahn, W.; Alp, E. E.; Zhao, J.; Cramer, S. P. *Inorg. Chem.*, **2005**, *44*, 5562–5570
498. "Direct observation of an iron bound terminal hydride 0intermediate in [FeFe] hydrogenase", Reijerse, E. J.; Pham, C. C.; Pelmenschikov, V.; Gilbert-Wilson, R.; Adamska-Venkatesh, A.; Siebel, J. F.; Gee, L. B.; Yoda, Y.; Tamasaku, K.; Lubitz, W.; Rauchfuss, T. B.; Cramer, S. P. *J. Am. Chem. Soc.*, **2017**, *139*, 4306–4309
499. "Elasticity and Constitution of the Earth Interior", Birch, F. *J. Geophys. Res.*, **1952**, *57*, 227–286
500. "Atomic dynamics of low-lying rare-earth guest modes in heavy fermion filled skutterudites ROs_4Sb_{12} (R = light rare-earth)", Tsutsui, S.; Uchiyama, H.; Sutter, J. P.; Baron, A. Q. R.; Mizumaki, M.; Kawamura, N.; Uruga, T.; Sugawara, H.; Yamaura, J.; Ochiai, A.; Hasegawa, T.; Ogita, N.; Udagawa, M.; Sato, H. *Phys. Rev. B*, **2012**, *86*, 195115
501. "Anharmonic motions of Kr in the clathrate hydrate", Tse, J. S.; Klug, D. D.; Zhao, J. Y.; Sturhahn, W.; Alp, E. E.; Baumert, J.; Gutt, C.; Johnson, M. R.; Press, W. *Nature Mat.*, **2005**, *4*, 917–921
502. "^{129}Xe nuclear resonance scattering on solid Xe and ^{129}Xe clathrate hydrate", Klobes, B.; Desmedt, A.; Sergueev, I.; Schmalzl, K.; Hermann, R. P. *EPL*, **2013**, *103*, 36001
503. "Scientific opportunities in nuclear resonance spectroscopy from source-driven revolution", Shenoy, G. K.; Rohlsberger, R. *Hyp. Int.*, **2008**, *182*, 157–172
504. "Scientific Opportunities with an X-ray Free-Electron Laser Oscillator", Adams, B.; Aeppli, G.; Baron, A. Q. R.; Bucksbaum, P.; Chumakov, A.; Corder, C.; Cramer, S. P.; Ding, Y.; Evers, J.; Frisch, J.; Fuchs, M.; Grübel, G.; Harris, S.; Hastings, J.; Heyl, C.; Holberg, L.; Huang, Z.; Ishikawa, T.; Jones, R. J.; Kaldun, A.; Kim, K.-J.; Kolodziej, T.; Krzywinski, J.; Li, Z.; Liao, W.-T.; Lindberg, R.; Madsen, A.; Maxwell, T.; Monaco, G.; Nelson, K.; Palffy, A.; Porat, G.; Qin, W.; Raubenheimer, T.; Reis, D. A.; Röhlsberger, R.; Santra, R.; Schoenlein, R.; Schünemann, V.; Shpyrko, O.; Shvydko, Y.; Shwartz, S.; Singer, A.; Sinha, S.; Sutton, M.; Tamasaku, K.; Wille, H.-C.; Yabashi, M.; Ye, J.; Zhu, D. *J. Syn. Rad.*, **2019**, *https://arxiv.org/abs/1903.09317, accepted - in press*
505. "Photoelectron Diffraction", Hüfner, S., in *Photoelectron Spectroscopy: Principles and Applications*; Springer Berlin Heidelberg: Berlin, Heidelberg, 2003, pp. 597–634
506. "X-ray photoelectron spectroscopy and diffraction in the hard X-ray regime: Fundamental considerations and future possibilities", Fadley, C. S. *Nucl. Inst. Meth. A*, **2005**, *547*, 24–41
507. "Photoelectron Diffraction and Photoelectron Holography", Suga, S.; Sekiyama, A., in *Photoelectron Spectroscopy: Bulk and Surface Electronic Structures*, 2014; Vol. 176, pp. 331–338
508. "Photoelectron Diffraction", Woodruff, D. P., in *Reference Module in Materials Science and Materials Engineering*; Elsevier, 2016
509. "X-ray photoelectron spectroscopy: From origins to future directions", Fadley, C. S. *Nucl. Inst. Meth. A*, **2009**, *601*, 8–31
510. "Ueber einen Einfluss des ultravioletten Lichtes auf die electrische Entladung", Hertz, H. *Ann. Phys. (Leipzig)*, **1887**, *17*, 983–1000
511. "Generation and conversion of light with regard to a heuristic point of view", Einstein, A. *Ann. Phys.*, **1905**, *17*, 132–148
512. "Electron Spectroscopy for Chemical Analysis", Hagstrom, S.; Nordling, C.; Siegbahn, K. *Phys. Lett.*, **1964**, *9*, 235–236
513. "Electron Spectroscopic Determination of Chemical Valence State", Hagstrom, S.; Nordling, C.; Siegbahn, K. *Z. Phys.*, **1964**, *178*, 439–444

514. "Electron Spectroscopy for Atoms, Molecules and Condensed Matter - Nobel Prize in Physics Lecture - 1981", Siegbahn, K. M., in *Nobel Lectures, Physics 1981–1990*; World Scientific Publishing Co.: Singapore, 1993
515. "Vibrational and Lifetime Line Broadenings in ESCA", Gelius, U.; Svensson, S.; Siegbahn, H.; Basilier, E.; Faxalv, A.; Siegbahn, K. *Chem. Phys. Lett.*, **1974**, *28*, 1–7
516. "X-ray photoelectron spectroscopy: Progress and perspectives", Fadley, C. S. *J. Elec. Spec. Rel. Phen.*, **2010**, *178*, 2–32
517. "X-Ray Photoemission Spectroscopy", Lindau, I.; Pianetta, P.; Doniach, S.; Spicer, W. E. *Nature*, **1974**, *250*, 214–215
518. "Extracting Chemical Information from XPS Spectra: A Perspective", Bagus, P. S.; Ilton, E.; Nelin, C. J. *Cat. Lett.*, **2018**, *148*, 1785–1802
519. "Iron contamination in silicon technology", Istratov, A. A.; Hieslmair, H.; Weber, E. R. *App. Phys. A*, **2000**, *70*, 489–534
520. "High-efficiency Silicon Heterojunction Solar Cells: A Review", De Wolf, S.; Descoedres, A.; Holman, Z.; Ballif, C. *Green*, **2012**, *2*, 7–24
521. "Analysis of silicon wafer surface preparation for heterojunction solar cells using X-ray photoelectron spectroscopy and effective minority carrier lifetime", Nsofor, U. J.; Zhang, L.; Somali, A.; Goodwin, C. M.; Liu, H.; Dobson, K. D.; Das, U. K.; Beebe, T. P.; Hegedus, S. *Solar Energy Mat.Solar Cells*, **2018**, *183*, 205–210
522. "Carbon 1 s photoelectron spectrum of methane: Vibrational excitation and core-hole lifetime", Carroll, T. X.; Berrah, N.; Bozek, J.; Hahne, J.; Kukk, E.; Saethre, L. J.; Thomas, T. D. *Phys. Rev. A*, **1999**, *59*, 3386–3393
523. "Über die Entstehung der β-Strahl-Spektren radioaktiver Substanzen", Meitner, L. *Z. Phys.*, **1922**, *9*, 131–144
524. "Sur les rayons β secondaires produits dans un gaz par des rayons X", Auger, P. *Comptes Rendus*, **1923**, *177*, 169–171
525. "Precision Method for Obtaining Absolute Values of Atomic Binding Energies", Nordling, C.; Sokolowski, E.; Siegbahn, K. *Phys. Rev.*, **1957**, *105*, 1676–1677
526. "Observation of Electron Polarization in Photoemission", Busch, G.; Campagna, M.; Cotti, P.; Siegmann, H. C. *Phys. Rev. Lett.*, **1969**, *22*, 597–600
527. "X-ray photoelectron spectroscopy under real ambient pressure conditions", Takagi, Y.; Nakamura, T.; Yu, L. W.; Chaveanghong, S.; Sekizawa, O.; Sakata, T.; Uruga, T.; Tada, M.; Iwasawa, Y.; Yokoyama, T. *App. Phys. Exp.*, **2017**, *10*
528. "Hard X-ray photoemission spectroscopy", Kobayashi, K. *Nucl. Instr. Meth. A*, **2009**, *601*, 32–47
529. "Quantifying the effective attenuation length in high-energy photoemission experiments", Sacchi, M.; Offi, F.; Torelli, P.; Fondacaro, A.; Spezzani, C.; Cautero, M.; Cautero, G.; Huotari, S.; Grioni, M.; Delaunay, R.; Fabrizioli, M.; Vanko, G.; Monaco, G.; Paolicelli, G.; Stefani, G.; Panaccione, G. *Phys. Rev. B*, **2005**, *71*
530. "Inelastic Mean Free Paths, Mean Escape Depths, Information Depths, and Effective Attenuation Lengths for Hard X-ray Photoelectron Spectroscopy", Powell, C. J.; Tanuma, S., in *Hard X-Ray Photoelectron Spectroscopy*; Woicik, J. C., Ed., 2016; Vol. 59, pp. 111–140
531. "Calculations of electron inelastic mean free paths. IX. Data for 41 elemental solids over the 50 eV to 30 keV range", Tanuma, S.; Powell, C. J.; Penn, D. R. *Surf. Int. Anal.*, **2011**, *43*, 689–713
532. "Calculation of electron inelastic mean free paths (IMFPs) VII. Reliability of the TPP-2 M IMFP predictive equation", Tanuma, S.; Powell, C. J.; Penn, D. R. *Surf. Int. Anal.*, **2003**, *35*, 268–275
533. "Synchrotron-based Ambient Pressure X-ray Photoelectron Spectroscopy", Shavorskiy, A.; Karslioglu, O.; Zegkinoglou, I.; Bluhm, H. *Syn. Rad. News*, **2014**, *27*, 14–23
534. "Spin-polarized photoemission", Johnson, P. D. *Rep. Prog. Phys.*, **1997**, *60*, 1217–1304

535. "Direct observation of spin-resolved full and empty electron states in ferromagnetic surfaces", Berti, G.; Calloni, A.; Brambilla, A.; Bussetti, G.; Duo, L.; Ciccacci, F. *Rev. Sci. Instrum.*, **2014**, *85*, 07390
536. "A high-efficiency spin-resolved photoemission spectrometer combining time-of-flight spectroscopy with exchange-scattering polarimetry", Jozwiak, C.; Graf, J.; Lebedev, G.; Andresen, N.; Schmid, A. K.; Fedorov, A. V.; El Gabaly, F.; Wan, W.; Lanzara, A.; Hussain, Z. *Rev. Sci. Inst.*, **2010**, *81*
537. "Angle-resolved photoemission studies of the cuprate superconductors", Damascelli, A.; Hussain, Z.; Shen, Z.-X. *Rev. Mod. Phys.*, **2003**, *75*, 473–541
538. "Introduction: Interfaces as an Object of Photoemission Spectroscopy", Cancellieri, C.; Strocov, V. N., in *Spectroscopy of Complex Oxide Interfaces — Photoemission and Related Spectroscopies*; Cancellieri, C. and Strocov, V. N., Ed.; Springer: Switzerland, 2018
539. "Probing the electronic structure of complex systems by ARPES", Damascelli, A. *Phys. Scripta*, **2004**, *T109*, 61–74
540. "Advances in the physics of high-temperature superconductivity", Orenstein, J.; Millis, A. J. *Science*, **2000**, *288*, 468–474
541. "Electronic structure of the iron-based superconductor LaOFeP", Lu, D. H.; Yi, M.; Mo, S. K.; Erickson, A. S.; Analytis, J.; Chu, J. H.; Singh, D. J.; Hussain, Z.; Geballe, T. H.; Fisher, I. R.; Shen, Z. X. *Nature*, **2008**, *455*, 81–84
542. "Energy gaps in high-transition-temperature cuprate superconductors", Hashimoto, M.; Vishik, I. M.; He, R. H.; Devereaux, T. P.; Shen, Z. X. *Nat. Phys.*, **2014**, *10*, 483–495
543. "Semiempirical Auger-Electron Energies for Elements $10 \leq Z \leq 100$", Larkins, F. P. *Atomic Data Nucl. Data Tables*, **1977**, *20*, 311–387
544. "Analysis of KLL Auger spectra excited by X-rays from Ni and Cu metal surfaces", Egri, S.; Kövér, L.; Cserny, I.; Novák, M.; Drube, W. *Nucl. Inst. Meth. B*, **2016**, *369*, 15–20
545. "Auger Electron Energies of Outer Shell Electrons", Chung, M. F.; Jenkins, L. H. *Surf. Sci.*, **1970**, *22*, 479–485
546. "Reevaluation of X-Ray Atomic Energy Levels", Bearden, J. A.; Burr, A. F. *Rev. Mod. Phys.*, **1967**, *39*, 125–141
547. "Anomalous Oxidation Resistance of "Core-Only" Copper Nanoparticles Electrochemically Grown on Gold Nanoislands Prefunctionalized by 1,4-phenylene Diisocyanide", Sohn, Y.; Pradhan, D.; Zhao, L. Y.; Leung, K. T. *Electrochem. Sol. State Lett.*, **2012**, *15*, K35-K39
548. "Resonant Auger spectroscopy at the L_2,L_3 shake-up thresholds as a probe of electron correlation effects in nickel", Magnuson, M.; Wassdahl, N.; Nilsson, A.; Fohlisch, A.; Nordgren, J.; Martensson, N. *Phys. Rev. B*, **1998**, *58*, 3677–3681
549. "Space-charge effects in high-energy photoemission", Verna, A.; Greco, G.; Lollobrigida, V.; Offi, F.; Stefani, G. *J. Elect. Spect. Rel. Phen.*, **2016**, *209*, 14–25
550. "X-Ray Photochemistry in Iron Complexes from Fe(0) to Fe(IV) – Can a Bug Become a Feature?", George, S. J.; Fu, J.; Guo, Y.; Drury, O.; Friedrich, S.; Rauchfuss, T.; Volkers, P. I.; Peters, J. C.; Scott, V.; Brown, S. D.; Thomas, C. M.; Cramer, S. P. *Inorg. Chim. Acta*, **2008**, *361*, 1157–1165
551. "Charging effects in X-ray photoelectron spectroscopy", Colón Santana, J. A., in *Quantitative Core Level Photoelectron Spectroscopy*; Morgan & Claypool Publishers, 2015, pp. 4-1–4-11
552. "Development of the Web-based NIST X-ray Photoelectron Spectroscopy (XPS) Database", Lee, A. Y.; Blakeslee, D. M.; Powell, C. J.; Rumble, J. R. *Data Sci. J.*, **2002**, *1*, 1
553. "Steady-State Far-Infrared Coherent Synchrotron Radiation detected at BESSY II", Abo-Bakr, M.; Feikes, J.; Holldack, K.; Wüstefeld, G.; Hübers, H. W. *Phys. Rev. Lett.*, **2002**, *88*, 254801
554. Schmüser, P.; Dohlus, M.; Rossbach, J.; Behrens, C. *Free-Electron Lasers in the Ultraviolet and X-Ray Regime: Physical Principles, Experimental Results, Technical Realization*; Springer: Heidelberg, **2014**
555. "Short-wavelength free-electron laser sources and science: a review", Seddon, E. A.; Clarke, J. A.; Dunning, D. J.; Masciovecchio, C.; Milne, C. J.; Parmigiani, F.; Rugg, D.; Spence,

J. C. H.; Thompson, N. R.; Ueda, K.; Vinko, S. M.; Wark, J. S.; Wurth, W. *Rep. Prog. Phys.*, **2017**, *80*, 115901

556. "Experiments on Radiation by Fast Electron Beams", Motz, H.; Thon, W.; Whitehurst, R. N. *J. App. Phys.*, **1953**, *24*, 826–833
557. "Stimulated Emission of Bremsstrahlung in a Periodic Magnetic Field", Madey, J. M. J. *J. App. Phys.*, **1971**, *42*, 1906–1913
558. "Free-Electron Laser", Madey, J. M. J.; Schwettman, H. A.; Fairbank, W. M. *IEEE Trans. Nuc. Sci.*, **1973**, *NS20*, 980–983
559. "The history of X-ray free-electron lasers", Pellegrini, C. *Eur. Phys. J.*, **2012**, *37*, 659–708
560. "X-ray free-electron lasers: from dreams to reality", Pellegrini, C. *Phys. Scrip.*, **2016**, *2016*, 014004
561. "The free electron laser: conceptual history", Madey, J.; Scully, M. O.; Sprangle, P. *Phys. Scripta*, **2016**, *91*
562. "Observation of Stimulated Emission of Radiation by Relativistic Electrons in a Spatially Periodic Transverse Magnetic-Field", Elias, L. R.; Fairbank, W. M.; Madey, J. M. J.; Schwettman, H. A.; Smith, T. I. *Phys. Rev. Lett.*, **1976**, *36*, 717–720
563. "First Operation of a Free-Electron Laser", Deacon, D. A. G.; Elias, L. R.; Madey, J. M. J.; Ramian, G. J.; Schwettman, H. A.; Smith, T. I. *Phys. Rev. Lett.*, **1977**, *38*, 892–894
564. "Generation of coherent radiation by a relativistic electron beam in an ondulator", Kondratenko, A. M.; Saldin, E. L. *Part. Accel.*, **1980**, *10*, 207–216
565. "Collective Instabilities and High-Gain Regime in a Free-Electron Laser", Bonifacio, R.; Pellegrini, C.; Narducci, L. M. *Opt. Comm.*, **1984**, *50*, 373–378
566. "Microwave radiation from a high-gain free-electron laser amplifier", Orzechowski, T. J.; Anderson, B.; Fawley, W. M.; Prosnitz, D.; Scharlemann, E. T.; Yarema, S.; Hopkins, D.; Paul, A. C.; Sessler, A. M.; Wurtele, J. *Phys. Rev. Lett.*, **1985**, *54*, 889–892
567. "Measurements of Gain Larger than 10^5 at 12 μ in a Self-Amplified Spontaneous-Emission Free-Electron Laser", Hogan, M. J.; Pellegrini, C.; Rosenzweig, J.; Anderson, S.; Frigola, P.; Tremaine, A.; Fortgang, C.; Nguyen, D. C.; Sheffield, R. L.; Kinross-Wright, J.; Varfolomeev, A.; Varfolomeev, A. A.; Tolmachev, S.; Carr, R. *Phys. Rev. Lett.*, **1998**, *81*, 4867–4870
568. "Exponential gain and saturation of a self-amplified spontaneous emission free-electron laser", Milton, S. V.; Gluskin, E.; Arnold, N. D.; Benson, C.; Berg, W.; Biedron, S. G.; Borland, M.; Chae, Y. C.; Dejus, R. J.; Den Hartog, P. K.; Deriy, B.; Erdmann, M.; Eidelmann, Y. I.; Hahne, M. W.; Huang, Z.; Kim, K. J.; Lewellen, J. W.; Li, Y.; Lumpkin, A. H.; Makarov, O.; Moog, E. R.; Nassiri, A.; Sajaev, V.; Soliday, R.; Tieman, B. J.; Trakhtenberg, E. M.; Travish, G.; Vasserman, I. B.; Vinokurov, N. A.; Wang, X. J.; Wiemerslage, G.; Yang, B. X. *Science*, **2001**, *292*, 2037–2041
569. "Generation of GW Radiation Pulses from a VUV Free-Electron Laser Operating in the Femtosecond Regime", Ayvazyan, V.; Baboi, N.; Bohnet, I.; Brinkmann, R.; Castellano, M.; Castro, P.; Catani, L.; Choroba, S.; Cianchi, A.; Dohlus, M.; Edwards, H. T.; Faatz, B.; Fateev, A. A.; Feldhaus, J.; Flöttmann, K.; Gamp, A.; Garvey, T.; Genz, H.; Gerth, C.; Gretchko, V.; Grigoryan, B.; Hahn, U.; Hessler, C.; Honkavaara, K.; Hüning, M.; Ischebeck, R.; Jablonka, M.; Kamps, T.; Körfer, M.; Krassilnikov, M.; Krzywinski, J.; Liepe, M.; Liero, A.; Limberg, T.; Loos, H.; Luong, M.; Magne, C.; Menzel, J.; Michelato, P.; Minty, M.; Müller, U. C.; Nölle, D.; Novokhatski, A.; Pagani, C.; Peters, F.; Pflüger, J.; Piot, P.; Plucinski, L.; Rehlich, K.; Reyzl, I.; Richter, A.; Rossbach, J.; Saldin, E. L.; Sandner, W.; Schlarb, H.; Schmidt, G.; Schmüser, P.; Schneider, J. R.; Schneidmiller, E. A.; Schreiber, H. J.; Schreiber, S.; Sertore, D.; Setzer, S.; Simrock, S.; Sobierajski, R.; Sonntag, B.; Steeg, B.; Stephan, F.; Sytchev, K. P.; Tiedtke, K.; Tonutti, M.; Treusch, R.; Trines, D.; Türke, D.; Verzilov, V.; Wanzenberg, R.; Weiland, T.; Weise, H.; Wendt, M.; Will, I.; Wolff, S.; Wittenburg, K.; Yurkov, M. V.; Zapfe, K. *Phys. Rev. Lett.*, **2002**, *88*, 104802
570. "First Lasing at 32 nm of the VUV-FEL at DESY", Schreiber, S. *Proc 27th Int. Free Electron Laser Conf.*, **2005**, 12–18

571. "Highly coherent and stable pulses from the FERMI seeded free-electron laser in the extreme ultraviolet", Allaria, E.; Appio, R.; Badano, L.; Barletta, W. A.; Bassanese, S.; Biedron, S. G.; Borga, A.; Busetto, E.; CastronovoD.; Cinquegrana, P.; Cleva, S.; Cocco, D.; Cornacchia, M.; Craievich, P.; Cudin, I.; D'Auria, G.; Dal Forno, M.; Danailov, M. B.; De Monte, R.; De Ninno, G.; Delgiusto, P.; Demidovich, A.; Di Mitri, S.; Diviacco, B.; Fabris, A.; Fabris, R.; Fawley, W. M.; Ferianis, M.; Ferrari, E.; S., F.; Froehlich, L.; Furlan, P.; Gaio, G.; Gelmetti, F.; Giannessi, L.; Giannini, M.; Gobessi, R.; Ivanov, R.; Karantzoulis, E.; Lonza, M.; Lutman, A.; Mahieu, B.; Milloch, M.; Milton, S. V.; Musardo, M.; Nikolov, I.; Noe, S.; Parmigiani, F.; Penco, G.; Petronio, M.; Pivetta, L.; Predonzani, M.; Rossi, F.; Rumiz, L.; Salom, A.; Scafuri, C.; Serpico, C.; Sigalotti, P.; Spampinati, S.; Spezzani, C.; Svandrlik, M.; Svetina, C.; Tazzari, S.; Trovo, M.; Umer, R.; Vascotto, A.; Veronese, M.; R., V.; Zaccaria, M.; Zangrando, D.; Zangrando, M. *Nat. Photon.*, **2012**, *6*, 699–704
572. "First lasing and operation of an angstrom-wavelength free-electron laser", Emma, P.; Akre, R.; Arthur, J.; Bionta, R.; Bostedt, C.; Bozek, J.; Brachmann, A.; Bucksbaum, P.; Coffee, R.; Decker, F. J.; Ding, Y.; Dowell, D.; Edstrom, S.; Fisher, A.; Frisch, J.; Gilevich, S.; Hastings, J.; Hays, G.; Hering, P.; Huang, Z.; Iverson, R.; Loos, H.; Messerschmidt, M.; Miahnahri, A.; Moeller, S.; Nuhn, H. D.; Pile, G.; Ratner, D.; Rzepiela, J.; Schultz, D.; Smith, T.; Stefan, P.; Tompkins, H.; Turner, J.; Welch, J.; White, W.; Wu, J.; Yocky, G.; Galayda, J. *Nat. Photon.*, **2010**, *4*, 641–647
573. "Baseline Parameters of the European XFEL", Schneidmiller, E. A.; Yurkov, M. V. *Proc. FEL 2917*, **2017**, MOP033
574. "The European XFEL is go!", Wilson, R. *FELs Europe News*, **2017**, *5*, 8
575. "A compact X-ray free-electron laser emitting in the sub-angstrom region", Tanaka, H.; Yabashi, M.; al., e. *Nat. Photon.*, **2012**, *6*, 540–544
576. "A simplified description of X-ray free-electron lasers", Margaritondo, G.; Rebernik Ribic, P. *J. Syn. Rad.*, **2011**, *18*, 101–108
577. **The XFEL Principle, http://photon-science.desy.de/sites/site_photonscience/content/e62/e189219/e187240/e208351/e187350/felbasics_eng.pdf
578. "3-Dimensional Analysis of Coherent Amplification and Self-Amplified Spontaneous Emission in Free-Electron Lasers", Kim, K. J. *Phys. Rev. Lett.*, **1986**, *57*, 1871–1874
579. "Review of X-ray free-electron laser theory", Huang, Z. R.; Kim, K. J. *Phys. Rev. Spec. Top.*, **2007**, *10*, 034801
580. "Introduction to single pass FELs for UV and X-ray production", Di Mitri, S. *US Part. Accel. School*, **2015**
581. "Single-shot analysis of hard X-ray laser radiation using a noninvasive grating spectrometer", Karvinen, P.; Rutishauser, S.; Mozzanica, A.; Greiffenberg, D.; Juranic, P. N.; Menzel, A.; Lutman, A.; Krzywinski, J.; Fritz, D. M.; Lemke, H. T.; Cammarata, M.; David, C. *Opt. Lett.*, **2012**, *37*, 5073–5075
582. "Experimental Demonstration of a Soft X-Ray Self-Seeded Free-Electron Laser", Ratner, D.; Abela, R.; Amann, J.; Behrens, C.; Bohler, D.; Bouchard, G.; Bostedt, C.; Boyes, M.; Chow, K.; Cocco, D.; Decker, F. J.; Ding, Y.; Eckman, C.; Emma, P.; Fairley, D.; Feng, Y.; Field, C.; Flechsig, U.; Gassner, G.; Hastings, J.; Heimann, P.; Huang, Z.; Kelez, N.; Krzywinski, J.; Loos, H.; Lutman, A.; Marinelli, A.; Marcus, G.; Maxwell, T.; Montanez, P.; Moeller, S.; Morton, D.; Nuhn, H. D.; Rodes, N.; Schlotter, W.; Serkez, S.; Stevens, T.; Turner, J.; Walz, D.; Welch, J.; Wu, J. *Phys. Rev. Lett.*, **2015**, *114*, 054801
583. "Demonstration of self-seeding in a hard-X-ray free-electron laser", Amann, J.; Berg, W.; Blank, V.; Decker, F. J.; Ding, Y.; Emma, P.; Feng, Y.; Frisch, J.; Fritz, D.; Hastings, J.; Huang, Z.; Krzywinski, J.; Lindberg, R.; Loos, H.; Lutman, A.; Nuhn, H. D.; Ratner, D.; Rzepiela, J.; Shu, D.; Shvyd'ko, Y.; Spampinati, S.; Stoupin, S.; Terentyev, S.; Trakhtenberg, E.; Walz, D.; Welch, J.; Wu, J.; Zholents, A.; Zhu, D. *Nat. Phot.*, **2012**, *6*, 693–698
584. "Feasibility Study for an X-Ray FEL Oscillator at the LCLS-II", Maxwell, T. J.; Arthur, J.; Ding, Y.; Fawley, W. M.; Frisch, J.; Hastings, J.; Huang, Z.; Krzywinski, J.; Kim, K.-J.; Lindberg, R. R.; Shu, D.; Shvyd'ko, Y.; Stoupin, S. *Proc. IPAC2015*, **2015**, 1897–1900

585. "FEL Oscillators and Coherent Hard X-Rays", Kim, K.-J., in *Synchrotron Radiation and Free-Electron Lasers*; Kim, K.-J., Huang, Z. and Lindberg, R., Ed.; Cambridge University Press: Cambridge, 2017, pp. 197–214
586. "The Linac Coherent Light Source: Recent Developments and Future Plans", Schoenlein, R. W.; Boutet, S.; Minitti, M. P.; Dunne, A. M. *App. Sci. Basel*, **2017**, *7*
587. "Proposal for a Free-Electron Laser in the X-Ray Region", Colella, R.; Luccio, A. *Opt. Comm.*, **1984**, *50*, 41–44
588. "A proposal for an X-ray free-electron laser oscillator with an energy-recovery linac", Kim, K. J.; Shvyd'ko, Y.; Reiche, S. *Phys. Rev. Lett.*, **2008**, *100*
589. "Tunable optical cavity for an X-ray free-electron-laser oscillator", Kim, K. J.; Shvyd'ko, Y. V. *Phys. Rev. Spec. Top. - Acc. Beams*, **2009**, *12*
590. "Diffraction before destruction", Chapman, H. N.; Caleman, C.; Timneanu, N. *Phil. Trans. Royal Soc.*, **2014**, *369*, 20130313
591. Bloembergen, N. *Nonlinear Optics*; World Scientific: Singapore**, 1996 (1st edition 1965)**
592. "Nonsequential two-photon absorption from the K shell in solid zirconium", Ghimire, S.; Fuchs, M.; Hastings, J.; Herrmann, S. C.; Inubushi, Y.; Pines, J.; Shwartz, S.; Yabashi, M.; Reis, D. A. *Phys. Rev. A*, **2016**, *94*, 6
593. "X-Ray Second Harmonic Generation", Shwartz, S.; Fuchs, M.; Hastings, J. B.; Inubushi, Y.; Ishikawa, T.; Katayama, T.; Reis, D. A.; Sato, T.; Tono, K.; Yabashi, M.; Yudovich, S.; Harris, S. E. *Phys. Rev. Lett.*, **2014**, *112*, 163901
594. "Atomic inner-shell X-ray laser at 1.46 nanometres pumped by an X-ray free-electron laser", Rohringer, N.; Ryan, D.; London, R. A.; Purvis, M.; Albert, F.; Dunn, J.; Bozek, J. D.; Bostedt, C.; Graf, A.; Hill, R.; Hau-Riege, S. P.; Rocca, J. J. *Nature*, **2012**, *481*, 488
595. "Stimulated Electronic X-Ray Raman Scattering", Weninger, C.; Purvis, M.; Ryan, D.; London, R. A.; Bozek, J. D.; Bostedt, C.; Graf, A.; Brown, G.; Rocca, J. J.; Rohringer, N. *Phys. Rev. Lett.*, **2013**, *111*, 233902
596. "Stimulated X-ray Raman scattering - a critical assessment of the building block of nonlinear X-ray spectroscopy", Kimberg, V.; Sanchez-Gonzalez, A.; Mercadier, L.; Weninger, C.; Lutman, A.; Ratner, D.; Coffee, R.; Bucher, M.; Mucke, M.; Agaker, M.; Sathe, C.; Bostedt, C.; Nordgren, J.; Rubensson, J. E.; Rohringer, N. *Far. Disc.*, **2016**, *194*, 305–324
597. "X-ray and optical wave mixing", Glover, T. E.; Fritz, D. M.; Cammarata, M.; Allison, T. K.; Coh, S.; Feldkamp, J. M.; Lemke, H.; Zhu, D.; Feng, Y.; Coffee, R. N.; Fuchs, M.; Ghimire, S.; Chen, J.; Shwartz, S.; Reis, D. A.; Harris, S. E.; Hastings, J. B. *Nature*, **2012**, *488*, 603
598. "Anomalous nonlinear X-ray Compton scattering", Fuchs, M.; Trigo, M.; Chen, J.; Ghimire, S.; Shwartz, S.; Kozina, M.; Jiang, M.; Henighan, T.; Bray, C.; Ndabashimiye, G.; Bucksbaum, P. H.; Feng, Y.; Herrmann, S.; Carini, G. A.; Pines, J.; Hart, P.; Kenney, C.; Guillet, S.; Boutet, S.; Williams, G. J.; Messerschmidt, M.; Seibert, M. M.; Moeller, S.; Hastings, J. B.; Reis, D. A. *Nat. Phys.*, **2015**, *11*, 964
599. "X-ray two-photon absorption competing against single and sequential multiphoton processes", Tamasaku, K.; Shigemasa, E.; Inubushi, Y.; Katayama, T.; Sawada, K.; Yumoto, H.; Ohashi, H.; Mimura, H.; Yabashi, M.; Yamauchi, K.; Ishikawa, T. *Nat. Photonics*, **2014**, *8*, 313–316
600. "Establishing nonlinearity thresholds with ultraintense X-ray pulses", Szlachetko, J.; Hoszowska, J.; Dousse, J. C.; Nachtegaal, M.; Blachucki, W.; Kayser, Y.; Sa, J.; Messerschmidt, M.; Boutet, S.; Williams, G. J.; David, C.; Smolentsev, G.; van Bokhoven, J. A.; Patterson, B. D.; Penfold, T. J.; Knopp, G.; Pajek, M.; Abela, R.; Milne, C. J. *Sci. Rep.*, **2016**, *6*
601. "Nonlinear Spectroscopy with X-Ray Two-Photon Absorption in Metallic Copper", Tamasaku, K.; Shigemasa, E.; Inubushi, Y.; Inoue, I.; Osaka, T.; Katayama, T.; Yabashi, M.; Koide, A.; Yokoyama, T.; Ishikawa, T. *Phys. Rev. Lett.*, **2018**, *121*, 083901
602. "Ultrafast magnetodynamics with free-electron lasers", Malvestuto, M.; Ciprian, R.; Caretta, A.; Casarin, B.; Parmigiani, F. *J. Phys. Cond. Matt.*, **2018**, *30*

603. "Femtosecond X-ray magnetic circular dichroism absorption spectroscopy at an X-ray free electron laser", Higley, D. J.; Hirsch, K.; Dakovski, G. L.; Jal, E.; Yuan, E.; Liu, T.; Lutman, A. A.; MacArthur, J. P.; Arenholz, E.; Chen, Z.; Coslovich, G.; Denes, P.; Granitzka, P. W.; Hart, P.; Hoffmann, M. C.; Joseph, J.; Guyader, L. L.; Mitra, A.; Moeller, S.; Ohldag, H.; Seaberg, M.; Shafer, P.; Stöhr, J.; Tsukamoto, A.; Nuhn, H.-D.; Reid, A. H.; Dürr, H. A.; Schlotter, W. F. *Rev. Sci. Inst.*, **2016**, *87*, 033110
604. "Overview of the SACLA facility", Yabashi, M.; Tanaka, H.; Ishikawa, T. *J. Syn. Rad.*, **2015**, *22*, 477–484
605. "Structural biology: doors open at the European XFEL", Marx, V. *Nature Meth.*, **2017**, *14*, 843–846
606. Saldin, E. L.; Schneidmiller, E. V.; Yurkov, M. V. *The Physics of Free Electron Lasers*; Springer: Berlin**, 2000**
607. "Status and prospects of X-ray free-electron lasers (X-FELs): a simple presentation", Ribic, P. R.; Margaritondo, G. *J. Phys. D*, **2012**, *45*
608. "Synchrotron radiation from an accelerating light pulse", Henstridge, M.; Pfeiffer, C.; Wang, D.; Boltasseva, A.; Shalaev, V. M.; Grbic, A.; Merlin, R. *Science*, **2018**, *362*, 439-+
609. XAFS Software, http://www.ixasportal.net/wiki/doku.php?id=software:software#theory_and_analysis_programs
610. ADF, http://www.scm.com/Products/Capabilities/X-ray_absorption_spectroscopy.html
611. "Advanced capabilities for materials modelling with QUANTUM ESPRESSO", Giannozzi, P.; Andreussi, O.; Brumme, T.; Bunau, O.; Nardelli, M. B.; Calandra, M.; Car, R.; Cavazzoni, C.; Ceresoli, D.; Cococcioni, M.; Colonna, N.; Carnimeo, I.; Dal Corso, A.; de Gironcoli, S.; Delugas, P.; DiStasio, R. A.; Ferretti, A.; Floris, A.; Fratesi, G.; Fugallo, G.; Gebauer, R.; Gerstmann, U.; Giustino, F.; Gorni, T.; Jia, J.; Kawamura, M.; Ko, H. Y.; Kokalj, A.; Kucukbenli, E.; Lazzeri, M.; Marsili, M.; Marzari, N.; Mauri, F.; Nguyen, N. L.; Nguyen, H. V.; Otero-de-la-Roza, A.; Paulatto, L.; Ponce, S.; Rocca, D.; Sabatini, R.; Santra, B.; Schlipf, M.; Seitsonen, A. P.; Smogunov, A.; Timrov, I.; Thonhauser, T.; Umari, P.; Vast, N.; Wu, X.; Baroni, S. *J. Phys. - Cond. Matt.*, **2017**, *29*
612. "Synchrotron-based Ambient Pressure Photoelectron Spectroscopy", Shavorskiy, A.; Karslioglu, O.; Zegkinoglou, I.; Bluhm, H. *Syn. Rad. News*, **2014**, *27*, 14–16

Index

S. P. Cramer, *X-Ray Spectroscopy with Synchrotron Radiation*, Biological and Medical Physics, Biomedical Engineering, https://doi.org/10.1007/978-3-030-28551-7

L

M

N

O

P

Q

R

S

The manufacturer's authorised representative in the EU is Springer Nature Customer Service Centre GmbH, Europaplatz 3, 69115 Heidelberg, Germany. If you have any concerns regarding our products, please contact ProductSafety@springernature.com

Printed and bound by CPI Group (UK) Ltd, Croydon, CR0 4YY
30/09/2026
02225692-0012